Pergamon Series in Analytical Chemistry
Volume 8

General Editors: R. Belcher[†] (Chairman), D. Betteridge & L. Meites

The Analysis of Plastics

Related Pergamon Titles of Interest

Full details of all Pergamon publications/free specimen copy of any Pergamon journal available on request from your nearest Pergamon office.

The Analysis of Plastics

by

T. R. CROMPTON
M.Sc., F.R.C.S., C.Chem., Dipl. Appl. Chem.

Formerly Head of the Analytical Research Department, Plastics Laboratory, Shell Research Ltd., Carrington, Cheshire, U.K.

PERGAMON PRESS
OXFORD · NEW YORK · TORONTO · SYDNEY · PARIS · FRANKFURT

7226 -2187

CHEMISTRY

U.K.	Pergamon Press Ltd., Headington Hill Hall, Oxford OX3 0BW, England
U.S.A.	Pergamon Press Inc., Maxwell House, Fairview Park, Elmsford, New York 10523, U.S.A.
CANADA	Pergamon Press Canada Ltd., Suite 104, 150 Consumers Road, Willowdale, Ontario M2J 1P9, Canada
AUSTRALIA	Pergamon Press (Aust.) Pty. Ltd., P.O. Box 544, Potts Point, N.S.W. 2011, Australia
FRANCE	Pergamon Press SARL, 24 rue des Ecoles, 75240 Paris, Cedex 05, France
FEDERAL REPUBLIC OF GERMANY	Pergamon Press GmbH, Hammerweg 6, D-6242 Kronberg-Taunus, Federal Republic of Germany

First edition 1984

Library of Congress Cataloging in Publication Data
Crompton, T. R. (Thomas Roy)
The analysis of plastics.
(Pergamon series in analytical chemistry ; v. 8)
1. Plastics — Analysis. I. Title II. Series.
TP1140.C75 1984 668.4'1 84–1060

British Library Cataloguing in Publication Data
Crompton, T. R.
The analysis of plastics. — Pergamon series in analytical chemistry; v.8
1. Plastics
I. Title
668.4 TP1120
ISBN 0-08-026251-1

Printed in Great Britain by A. Wheaton & Co. Ltd., Exeter

PREFACE

TP1140
C75
1984
CHEM

This book aims to cover all aspects of the analysis of plastics using chemical and physical methods of analysis. Its contents are based on the author's experience as previous head of the Analytical Research Department concerned with plastics at a major plastic producer and also a complete review of world literature on the subject.

In many instances, and certainly in the case of many major techniques the book gives sufficient detail of methods to enable the readers to work direct from the book without recourse to the original literature.

The subject matter is broken down into eight chapters each dealing with a particular polymer or group of polymers. Copolymers and terpolymers are also fully discussed. In addition to dealing with the analysis of the polymers the book also deals at the end of chapter or section, with the analysis of non-polymer components of the polymer, whether these be adventitious (water, solvents, monomers, catalyst residues etc.) or deliberately added, (processing additives, antioxidants, plasticizers etc.)

Within each chapter the subject matter is broken down in as logical a manner as possible. Firstly, there is a discussion of the purely chemical methods of analysis available, dealing principally with the determination of elements and functional groups and including, where relevant, a discussion on end-group analysis. The remainder of the sections within each chapter deal with the wide variety of physical analytical techniques now available to the chemist for the elucidation of the composition, structure and microstructure of polymers. Firstly, techniques such as infrared and Raman, NMR, PMR, ESR and EPR spectroscopy are discussed showing how these techniques have already provided much information on polymer and copolymer structure and indicating the way to their use in solving future problems. Pyrolysis followed by gas chromatography or a combination of gas chromatography and mass spectrometry is also revealing itself as an extremely useful tool in the elucidation of polymer structure. Combinations of the techniques mentioned so far are being used, increasingly, in the elucidation of problems concerned with aspects such as side-group analysis and oxidative stability studies on polymers and copolymers. Where relevant, other techniques concerned with structure elucidation such as neutron scattering, radiochemical methods, mass spectrometry, ultraviolet spectroscopy, x-ray diffraction and electron probe micro-

v

analysis are discussed in detail.

Means of separating polymers into fractions for possible further analysis
including, column chromatography, gel permeation chromatography and gas and
thin-layer chromatography are also discussed and illustrated by examples of
the successful application of these techniques to polymer problems. Polymer
fractionation and reviews of methods of molecular weight determination are
discussed in separate sections.

Other miscellaneous techniques concerned include autoradiography of polymers
for inorganic inclusions and various thermal methods for examining polymers
such as differential thermal analysis, thermogravimetric analysis and differential scanning calorimetry.

The book is aimed at giving an up to date and thorough exposition of the
present state of the art concerning polymer analysis, and, as such, should be
of great interest to all those engaged in this subject in industry, university,
research establishments and general education. The book is also intended for
undergraduate and graduate chemistry students and those engaged on plastics
courses in plastics technology, engineering chemistry, materials science and
industrial chemistry.

CONTENTS

vii

CHAPTER 1

POLYOLEFINS

POLYETHYLENE

Chemical and Physical Methods for Determination of Trace Non-metallic Elements

These methods were particularly developed for the analysis of polyolefins. Most of the methods would also be applicable to other types of polymers.

Determination of sulphur. A procedure is described below for the determination of sulphur in amounts from 500 ppm upwards in polyolefins and in other organic materials. The repeatability of the method is ± 40% of the determined sulphur content at the 500 ppm sulphur level, improving to ± 2% at the 1% level. Chlorine and nitrogen concentrations in the sample may exceed the sulphur concentration several times over without causing interferences. Fluorine does not interfere unless present in concentrations exceeding 30% of the sulphur content. Phosphorus interferes even when present in moderate amounts. Metallic constituents also interfere when present in moderate amounts.

Summary of method. The sample is wrapped in a piece of filter paper and burnt in a closed conical flask filled with oxygen at atmospheric pressure. The sulphur dioxide produced in the reaction reacts with dilute hydrogen peroxide solution contained in the reaction flask to produce an equivalent amount of sulphuric acid. The sulphuric acid is estimated by photometric or visual titration with N/1000 or N/100 barium perchlorate using Thorin indicator.

Apparatus. Combustion flask Pyrex 500 ml capacity conical flask with B24 conical ground joint.
Stopper B24 with a fixed-in platinum wire (30 mm long, 0.8 mm diameter) carrying a 15 x 20 mm piece of 40 mesh platinum gauze or carrying a ⅝ in. long x ¼ in. diameter bucket fabricated in 40 mesh platinum gauze.
Safety jacket for combustion flask to serve as a protection during the combustion. Conical shaped, detachable metal wire gauze fitting round the conical flask (Fig. 1)

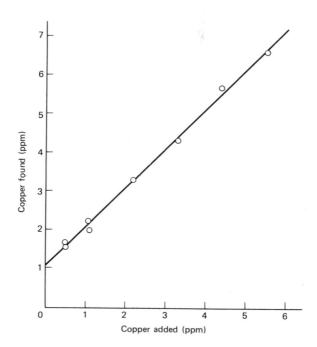

Fig. 1   Recovery of added copper to polyethylene.

<u>Reagents</u>. <u>Barium perchlorate solution standard 0.001 normal</u> Dissolve 0.17 g
perchlorate (anhydrous) laboratory reagent grade in 200 ml of deionised water
and make up to 1 litre with redistilled isopropanol. Using a pH meter adjust
to pH 3.5 by additions of 10% perchloric acid solution.
<u>Barium perchlorate solution standard 0.01 normal</u>. Dissolve 1.7 g barium per-
chlorate in 200 ml of deionised water and make up to 1 litre with redistilled
isopropanol. Adjust to pH 3.5 by additions of 10% perchloric acid.
<u>Determination of purity of solid barium perchlorate</u>. Pipette 25 ml of 0.01 N
sulphuric acid into a conical flask and add 2 ml of 0.02% Thorin indicator.
Titrate with 0.01 N barium perchlorate to the pink coloured end-point. Calculate the purity (usually 90-95%) of the solid barium perchlorate and allow for
this in the calculation of the normalities of the 0.001 and 0.01 barium perchlorate solutions referred to above.
<u>Filter paper for sample wrapping</u>. Any grade with sulphur content less than
100 ppm e.g. Whatman No. 41 or 42. Cut out paper, sized and shaped according
to Fig. 1. Fold them along the dotted line to a U shape and store in a closed
bottle to protect from any sulphur present in the atmosphere. As the paper
contains a small amount of sulphur it is necessary to ensure that the same
weight (30 to 40 mg) is used in the sample and the blank combustions.
<u>Hydrogen peroxide</u>. 100 volume Microanalytical Reagent Grade.
<u>Isopropyl alcohol</u>. Redistilled.
<u>Oxygen</u> cylinder, free from sulphur compounds.
<u>Perchloric acid</u>. 0.01 N Microanalytical Reagent Grade in glacial acetic acid.
<u>Sulphuric acid</u>. 0.01 N prepared by accurate dilution of 0.1 normal acid.
<u>Sulphuric acid</u>. 0.0005 N, prepared by accurate dilution of 0.01 N acid.

Prepare the 0.0005 N reagent at least once daily as required.
Thorin indicator solution. 0.02% aqueous 'Thorin' (sodium salt of 2(2 hydroxy-3,6-disulpho 1-napthylazo) benzene arsonic acid).
Water. Deionised, sulphate content less than 0.05 ppm.

Sampling. Solid organic compounds or polymers The weight of sample taken for analysis is governed by its sulphur content and should be sufficient to produce a titration of about 4 ml of 0.01 or 0.001 N barium perchlorate. The maximum permissible sample weight for complete combustion is 30 mg. Place a filter strip on the balance pan and weigh, transfer on to the middle of the paper the appropriate weight of sample and re-weigh. Weigh a blank piece of filter paper of similar size and check that its weight is within $\pm$ 5% of the weight of the paper used in the sample analysis. Wrap the sample up in the following way. First cover the sample with the raised edges of the paper strip, and then roll up the body of the strip towards the narrow strip at the end which will serve as a fuse. Now clamp the packet in the platinum gauze or bucket on the flask stopper, keeping the fuse free from, and in line with the platinum suspension wire. Repeat this operation with the blank piece of filter paper.

Procedure. Pipette 4.0 $\pm$ 0.5 ml deionised water and 0.15 ml hydrogen peroxide solution into a clean combustion flask. Place the flask in the safety jacket. Replace the air in the conical flask with a rapid stream of oxygen near the bottom for 30 seconds and stopper immediately. Use a long glass tube to introduce oxygen near the bottom of the flask. Place the flame of the lighter close to the top of the conical flask, ignite the end of the fuse of the sample packet and immediately insert the sample into the combustion flask. Keep the flask firmly closed and keep it upside down for a minute.
Remove the safety jacket, shake the flask for one minute and allow it to stand for 15 minutes. When all the mist has disappeared, wet the rim of the flask with isopropyl alcohol, carefully open the flask and transfer the flask contents quantitatively to a 100 ml beaker by means of 65 ml isopropyl alcohol and 12 ml water. To the beaker add 2 ml 0.02% Thorin indicator, 3 drops of 0.1 N perchloric acid and 1 ml N 0.0005 N sulphuric acid (by pipette). Carry out a blank combustion including the paper and all the reagents but omitting the sample. Carry out the titration photometrically or visually. Titrate the sample and blank solutions with 0.01 N or 0.001 N barium perchlorate solution (Table 1) using a 5 ml syringe.

Calculations. Calculate the sulphur content of the sample as follows:

$$\text{Sulphur (ppm)} \quad \frac{(V-B) \times N \times 16 \times 10^3}{W}$$

Where V = volume of barium perchlorate solution consumed in actual determination, millilitres.
  B = volume of barium perchlorate solution consumed in blank determination, millilitres.
  W = weight of sample in g.
  N = normality of barium perchlorate.

Determination of Chlorine

The following method is applicable to the determination of chlorine down to 30 ppm in polyethylene.

In the method polyethylene is mixed with pure sodium carbonate and ashed in a muffle furnace at 500°C. The residual ash is dissolved in aqueous nitric acid

TABLE 1   Determination of Chlorine by X-Ray Fluorescence Procedure

Number of samples taken from whole batch of polymer	Polymer not treated with alcoholic potassium hydroxide before analysis ppm chlorine				Polymer treated with alcoholic potassium hydroxide before analysis ppm chlorine				Difference between average chlorine contents obtained on potassium hydroxide treated and untreated samples
	X-ray fluorescence on polymer discs.				X-ray fluorescence on polymer discs.				
	Side 1.	Side 2.	Average of Sides 1 & 2	Average of two discs(A)	Side 1.	Side 2.	Average of Sides 1 & 2	Average of two discs(B)	(B) - (A)
S1 ptn a	510	450	480	510	710	1020	865	840	330
S1 ptn b	570	510	540		750	880	815		
S2 ptn a	350	380	365	422	400	670	535	552	130
S2 ptn b	460	500	480		520	620	570		
S3 ptn a	440	430	435	440	720	850	785	730	290
S3 ptn b	450	440	445		680	670	675		
S4 ptn a	390	460	425	497	580	670	625	650	153
S4 ptn b	560	580	570		570	780	675		
S5 ptn a	390	390	390	460	820	970	895	882	422
S5 ptn b	490	570	530		770	970	870		

ptn = portion

and then diluted with acetone. This solution is titrated potentiometrically
with standard silver nitrate.

Reagents. Silver nitrate. N/10
Silver nitrate. N/100
Nitric acid. 30% aqueous, dilute 30 ml. concentrated nitric acid (M.A.R.) to
100 ml with distilled water. Microanalytical reagent (M.A.R.) available from
British Drug Houses Limited, Poole, Dorset..
Sodium carbonate. solid microanalytical reagent grade.
Xylene cyanol/methyl orange mixed indicator. alcoholic solution.
Acetone. A.R. grade.
Nitric acid. 30%
Acid Buffer solution. to approximately 200 ml of distilled water in a 500 ml
volumetric flask add 100 ml glacial acetic acid and 6.5 ml concentrated
nitric acid (sp gr 1.42) dilute to the mark with distilled water.
Gelatine solution. add to 250 ml of deionised water, 2.5 g gelatine and 0.5 g
thymol blue. Heat slowly to the boil and stir until solution is complete.
Add 0.5 g thymol as preservative and dilute to 500 ml with distilled water.
The solution is stable for up to 3 months at room temperature.

Procedure 1 - Determination of chlorine in polyolefins. Accurately weigh 5 g
of polymer into a metal crucible and cover the polymer with a layer of 2 g of
sodium carbonate (Note 1). Place the crucible in a cold muffle furnace and
allow the temperature to increase gradually to 500°C to 550°C, maintaining
this temperature for four hours. Allow the crucibles to cool then dissolve
the residue in a minimum volume of distilled water, (not more than 20 ml).
Transfer this solution with crucible washings to a 100 ml beaker. Adjust the
final volume of water to 30 ml. Add 5 drops of screened methyl orange indicator solution and neutralise the mixture by dropwise addition of 30% nitric
acid to the purple red coloured end point. Add a further 10 drops of 30%
nitric acid solution to the beaker and then add 30 ml of acetone. Titrate the
resultant solution potentiometrically with silver nitrate solution (N/100)
using an automatic titrator equipped with a silver measuring electrode and a
glass reference electrode. Operate the instrument on the millivolt scale and
set the instrument to obtain full scale deflection for 250 millivolts. Carry
out a blank determination exactly as described above, omitting only the sample
addition.

Calculation. ppm (w/w) chlorine in polymer

$$= \frac{(T_S - T_B) \times N \times 35.46 \times 10^3}{W}$$

where, T_B = Titration of silver nitrate (mls) in blank
            determination..
       T_S = Titration of silver nitrate (mls) in sample
            determination
       N = Normality of silver nitrate solution.
       W = Weight of polymer sample in grams.

If a low pressure polyolefin sample does not contain any free residual alkali
left in from the manufacturing process then there exists a danger that some
chlorine will be lost during the ignition process and low chlorine analyses
will result. If this is suspected to be the case the polymer sample (50 g)
should first be contacted with twice its volume of 2% W/V alcoholic potassium
hydroxide solution and left in an oven at 70°C until the polymer is dry. The
above method is then applied. The effect of pretreatment with alcoholic
potassium hydroxide on the chlorine content obtained is shown in Table 1

in this case the polymers were analysed by X-ray fluorescence spectography of hot pressed discs of the polymer. Chlorine contents up to 50% higher are obtained in alkali treated samples.

Oxygen bomb and Schöniger oxygen flask combustion methods have been used to determine traces of chlorine in polyolefins at levels between 0 and 500 ppm. The oxygen bomb method involves the combustion of 0.5 to 1 gram of polymer in a conventional oxygen bomb at 25 to 30 atm oxygen pressure. Water (1 ml) and 2 drops of 30% hydrogen peroxide serve as the absorption medium for the chlorides formed during the combustion. After cooling the chloride is potentiometrically titrated in the bomb in acetic acid/acetone medium using 0.01 N silver nitrate solution. The Schöniger oxygen flask combustion technique requires 0.1 gram sample and the use of a one-litre conical flask. Chlorine-free polyethylene foil is employed to wrap the sample and a small amount of water is used as the absorbent. Combustion takes place at atmospheric pressure in oxygen. The chloride formed is potentionmetrically titrated in nitric acid/acetone medium in a small beaker using 0.01 N silver nitrate solution added from a syringe microlitre burette.

Gouveneur et al (1) have studied a neutron activation method, an X-ray fluorescence method and a chemical method for the determination of trace amounts of chlorine in polyolefin samples. A reasonable good agreement was obtained between the values obtained for samples tested by each of these methods.

Crompton (2) has compared chlorine contents obtained by the X-ray method with those obtained by a chemical method of analysis involving fusion at 500°C with sodium carbonate. Analyses were carried out on samples which had been previously treated with alcoholic potassium hydroxide and untreated samples. It is seen in Table 2 that, when carried out on the same sample, this chemical method gives considerably more reproducible duplicate results than those obtained by the X-ray method, (compare columns A, C, E and G). In the case of alcoholic potassium hydroxide treated samples the average of duplicate analyses carried out by the X-ray and the chemical methods agree satisfactorily within ± 15% of each other. In the case of untreated samples the chemical method tends to give lower chlorine contents than the X-ray method.

The superior reproducibility of the chemical method of analysis is believed to be due to the fact that a considerably higher weight of sample is taken for analysis than is used in the X-ray method of analysis. Thus, whereas in the chemical method one obtains the average chlorine content of 5 g of polymer, in the X-ray method the amount analysed is only a few microns thickness of the disc of polymer taken for analysis. Any local variations in the chlorine content of the polymer will be picked up much more readily, therefore, by the X-ray method of analysis.

It is seen that although the chemical method gives good duplicate analysis on the same sample a considerably greater variation of chlorine contents is obtained when the results are compared for the five alkali treated and for the five untreated samples, i.e. the sampling error was considerably greater than the analytical error. It is concluded that in each of the two sets of five samples there is a true variation of chlorine content from sample to sample. To obtain a more representative sample for chlorine analyses of a large batch of polymer it is obviously desirable to take several samples from different parts of the batch and mix these well prior to analysis.

The chlorine content of low pressure polyethylene may be present in several different forms. The polymer may contain free hydrochloric acid and/or aluminium and titanium chlorides which are loosely held in the polymer and can be

TABLE 2   Comparison of Chlorine Contents by X-Ray Method and Chemical Method

Sample	Polymer not treated with alcoholic potassium hydroxide before analysis, ppm chlorine				Polymer treated with alcoholic potassium hydroxide before analysis, ppm chlorine		
	X-ray on discs (av. in brackets) "A"	Chemical method on same disc as used for X-ray anal. "B"	Chemical method on powder (av. in brackets) "C"	Diff. (%) between av. X-ray & chem. anal. "D" = $\frac{("A"-"C")}{"C"} \times 100$	X-ray on discs (av. in brackets) "E"	Chemical method on same disc as used for X-ray anal. "F"	Chemical method on powder (av. in brackets) "G"
(1)	480,540 (510)	456	502,499 (500)	+2	865,815 (840)	700	786,761 (773)
(2)	365,480 (422)	334	380,355 (367)	+15	535,570 (552)	606	636,651 (643)
(3)	435,445 (440)	338	349,353 (351)	+25	785,675 (730)	598	650,654 (652)
(4)	425,570 (497)	392	395,404 (400)	+24	625,675 (650)	600	637,684 (660)
(5)	390,530 (460)	371	416,399 (408)	+13	895,870 (882)	733	828,816 (822)

easily removed from the polymer by water washing. The polymer may also contain these chlorine compounds bound within the polymer particles; i.e. much
more difficult to remove by washing procedure. In addition, non-extractable
chlorinated polyethylene may also be present. The X-ray and chemical methods
of analysis both determine the total amount of chlorine present in the polymer,
regardless of the form in which it might exist.

An alternate method involving shaking the polyethylene powder with isopropyl
alcohol, followed by acidification with nitric acid and silver nitrate titration determines only the loosely held ionic chlorine in the polymer. It is
seen in Table 3 that chlorine contents determined by this procedure are always
considerably lower than those obtained by the sodium carbonate fusion method.
The fact that the isopropyl alcohol extraction method reproducibly gives a
lower chlorine content than is obtained by the sodium carbonate fusion method
for total chlorine suggests that the isopropyl extraction method determines
only the more "loosely bound chlorine" content of polyethylene powder samples.
On this assumption it is possible to obtain by difference, from the "total
chlorine content" (by sodium carbonate fusion), and the "loosely held chlorine"
content, (by isopropyl alcohol extraction), an estimate of the "tightly held
chlorine" content of the polyethylene samples, (see Table 3, columns C and F).

Various other workers (3-9) have described methods for the determination of
traces of chlorine in polypropylene.

Determination of Organic Nitrogen

Scope. A procedure is described below for determining nitrogen in amounts
down to 50 ppm in polyethylene and polypropylene. The method is useful for
determining the concentration of nitrogen containing additives in polymers.

Method summary. A sample of polymer (up to 5 g) is digested with a mixture
of sulphuric acid, mercuric oxide, copper sulphate and selenium catalysts to
convert nitrogen containing organic compounds to ammonium sulphate. Digestion
is completed by adding sufficient potassium sulphate to the digestion mixture
to raise its boiling point to 380°C followed by heating to completely remove
any carbonaceous matter present.

The solution is then transferred to a distillation apparatus and excess sodium
hydroxide added to convert ammonium sulphate to free ammonia. The ammonia is
distilled off and collected in boric acid solution and the ammonia content of
the steam distillate is determined by titration with standard hydrochloric
acid solution.

Apparatus..Kjeldahl digestion flasks, 500 ml. Kjeldahl flask heater, 120 watts.
Mount flasks in heater at an angle of 30 degrees to the horizontal. Protect
the flask from draughts during digestion. Distillation apparatus
Distillation apparatus, Micro burette, capacity 10 ml.

Reagents. Boric acid. 4% w/v.
Mercuric oxide/cupric sulphate catalyst. grind together 10 g mercuric oxide
and 5 g cupric sulphate (anhydrous).
Selenium powder.
Ferric chloride. 40% w/v aqueous, to 100 ml. of this solution add 5 drops
concentrated sulphuric acid.
Sulphuric acid. 98%, use nitrogen-free grade. Keep well stoppered to ensure
minimum contamination of this reagent by nitrogen containing impurities in the
laboratory atmosphere.
Potassium sulphate. 'ANALAR' grade.

TABLE 3   Estimates of "Loosely Bound" and "Tightly Bound" Chlorine in Polyethylene

Sample	Polymer not treated with alcoholic potassium hydroxide before analysis			Polymer treated with alcoholic potassium hydroxide before analysis			Additional Cl retained in polymer due to preliminary potassium hydroxide treatment	
	Total Cl by Na$_2$CO$_3$ fusion at 500-550°C "A" ppm	Loosely bound Cl by iso-propanol extr. method (av. in brackets) "B" ppm	Tightly bound Cl by diff. "C" = ("A"-"B") ppm	Total Cl by Na$_2$CO$_3$ fusion at 500-550°C "D" ppm	Loosely bound Cl by iso-propanol extr. method (av. in brackets) "E" ppm	Tightly bound Cl by diff. "F" = ("D"-"E") ppm	Loosely held Chlorine "G" = ("E"-"B") ppm	Tightly held Chlorine "H" = ("F"-"C") ppm
(1)	500	217	283	773	376	397	159	114
(2)	367	223	144	643	276	367	53	223
(3)	351	197	154	652	327	325	130	171
(4)	400	269	131	660	325	335	56	204
(5)	408	267	141	822	396	426	129	285

Sodium hydroxide. 40% w/v aqueous, made from 'ANALAR' reagent.
Hydrochloric acid. 0.02 normal.
Tashiro indicator. dissolve 0.125 g methyl red and 0.083 g methylene blue in
100 ml. ethyl alcohol.

Method. Weigh out a known amount of the polymer (maximum 5 g) and carefully
transfer into a 500 ml. Kjeldahl digestion flask. Into a second (blank)
flask transfer the same weight of nitrogen-free polymer, (Note 1). To each
flask add 0.2 g mercuric oxide/cupric sulphate mixture, 10 mg selenium, some
glass beads and 15 ml. concentrated sulphuric acid. In addition, to each
flask add a further 10 ml. concentrated sulphuric acid per g of polymer present in the sample flask. Place the digestion flasks on the electric heaters
and raise to the boiling point as rapidly as possible with occasional swirling.
Continue heating until the sample solution becomes yellow/brown in colour
(this usually takes 2 to 4 hours).. If necessary add further similar quantities of concentrated sulphuric acid to the sample and blank flasks to maintain
a minimum volume of 20 ml. of acid.. Leave the flasks to cool and note the
weight of the flask contents (i.e. difference between initial weight of empty
digestion flask and weight of cold flask after digestion). To the sample and
blank flask add 0.72 g potassium sulphate per g of flask contents. Again
place the flasks in the heating mantle and boil for a further 90 minutes to
complete the digestion process. Allow the flasks to cool and to each add
approximately five volumes of deionised water per volume of acid present
(care). Add 2 drops ferric chloride reagent.

Steam distillation. Fill the steam generator with water and add to it a few
pellets of sodium hydroxide. Connect the generator to the steam distillation
apparatus and steam out the apparatus for 20 minutes, (run steam distillate
to waste). Transfer the sulphuric acid digestion mixture into the 1 litre
sample flask and pass steam through the sample for a further twenty minutes,
(removes any volatile acids present in the sulphuric acid digest). Into a
250 ml. conical flask (B24 socket) pipette 25 ml. 4% boric acid solution and
add 100 ml. deionised water and 5 drops of Tashiro indicator. If necessary
add a few drops of N/10 sodium hydroxide solution to produce a green colour
and then back titrate each flask with N/50 hydrochloric acid to a turquoise
blue colour. Transfer 30 ml. of this solution into each of two 250 ml. clean
stoppered conical flasks. Retain one of the flasks (Flask A) as a comparison
standard for the final ammonia titration. At the end of the steam purging,
connect the second 250 ml. titration flask (Flask B) to the freshly purged
steam distillation apparatus. Continue passing steam through the apparatus
and fill the separatory funnel on the steam distillation apparatus with 40%
sodium hydroxide. Run this solution into the hot acid mixture until a slight
excess .of alkali is present, (indicated by formation of brown precipitate of
cupric hydroxide). Run in a further 5 to 10 ml. excess of sodium hydroxide.
Continue the steam distillation until about 130 ml. of liquid are present in
conical flask 'B'. Remove this flask from the distillation apparatus and
leave until about 130 ml. of liquid are present in the flask. Clean out the
distillation apparatus as described in Note 2. Titrate the contents of the
sample conical flask B with 0.02 N hydrochloric acid until it has the same
shade (turquoise or bluish grey) as the contents of the comparison flask A.
Note the volume of hydrochloric acid used in the titration.

Calculation. Calculate the nitrogen content of the polymer as follows:

$$\text{Nitrogen (ppm)} = \frac{(V1 - V2) \times F \times 14 \times 10^6}{1000 \times W}$$

where V1   =   titration (mls.) of standard hydrochloric acid in

		sample determination
V2	=	titration (mls.) of standard hydrochloric acid in reagent blank determination.
F	=	normality of hydrochloric acid
W	=	weight (g) of sample taken for analysis.

Note 1. Blank digestion. It is desirable to include nitrogen-free polymer in the blank digestion in order to ensure that identical conditions are used in the digestion of the sample and the blank. Use nitrogen-free (i.e. 10 ppm nitrogen) polyethylene or polypropylene in the blank run as appropriate.

Note 2. Cleaning out distillation apparatus. When an analysis is completed turn the three-way stop-cock on the steam generator to atmosphere. The contents of the sample flask are then drawn into the steam condensation trap from which it can be run to waste.

Hernandez (10) has described an alternate procedure based on pyrochemilumin-escence which he applied to the determination of 250 to 1500 ppm nitrogen in polyethylene. In this technique the nitrogen in the sample is subjected to oxidative pyrolysis to produce nitric oxide. This when contacted with ozone produces a metastable nitrogen dioxide molecule which, as it relaxes to a stable state emits a photon of light. This emission is measured quantita-tively at 700-900 nm.

Determination of Oxygen

Neutron activation analysis is a very useful technique for determining total oxygen in polyethylene at all concentration levels.

Determination of Trace Metals

Many methods are available for the determination of traces of metals in poly-olefins involving both chemical and instrumental techniques. Some examples of metals analysis are quoted below.

Sodium in polyolefins. During the manufacture of high-density polyethylene and polypropylene, sodium containing alkalies are added to the polymer to neutralise residual acidity originating from the catalyst system. It was found that replicate sodium contents determined on the same sample by a flame photometric procedure were frequently widely divergent. Neutron activation analysis offers an independent method of checking the sodium contents which does not involve ashing. Both methods are discussed below.

In the flame photometric procedure the sample is dry-ashed in a nickel cruc-ible and the residue dissolved in hot water before determining sodium by evaluating the intensity of the line emission occurring at 589 nm.

Neutron activation analysis (flux 10^{12} neutrons cm^{-1} S^{-1}) for sodium was carried out on polyethylene and polypropylene moulded discs containing up to about 550 ppm sodium which had been previously analysed by the flame photo-metric method. The results obtained in these experiments (Table 4) show that significantly higher sodium contents are usually obtained by neutron activa-tion analysis and this suggested that sodium is being lost during the ashing stage of the latter method. Sodium can also be determined by a further inde-pendent method, namely emission spectrographic analysis which involves ashing the sample at 500°C in the presence of an ashing aid consisting of sulphur and the magnesium salt of a long-chain fatty acid compared to dry-ashing at 650-800°C without an aid as used in the flame photometric procedure. In

TABLE 4   Interlaboratory Variation of Flame Photometric
          Sodium Determinations in Polyolefins

Sodium content, ppm

Neutron Activation analysis		Flame photometry			
Powder	Discs	Analysis 1	Analysis 2	Analysis 3	Analysis 4
		Polyethylene			
207	211,204	35	165	140	-
177	175,172	100	140	148	-
266	267,263	85	210	221	-
203	197,191	70	160	150	-
		Polypropylene			
165	151,161	50	130	133	-
198	186,191	95	-	173	-
322	333,350	95	-	318	-

TABLE 5   Comparison of Sodium Determination in Polyolefins
          by Neutron Activation Analysis, Emission Spectro-
          graphy and Flame Photometry

Sodium, ppm

Sample description	By neutron activation analysis	By emission spectrography	By flame photometry
Polyethylene	99,96,99	95	60,76,55
Polyethylene	256,247,256	258,259	160,178,271
Polyethylene	343,321,339	339,287	250,312
Polyethylene	213,210,212	218,212	140,196
Polypropylene	194,189,192	209,198	80,158,229
Polypropylene	186,191,198	191,191	95,173

Table 5 results by neutron activation and emission spectrography are shown to agree well with each other, thereby confirming that low results can be obtained by flame photometry when a simple ashing process is employed. Unlike the neutron activation technique, emission spectrography involved a preliminary ashing of the sample at 500°C, thereby involving the risk of sodium losses. Despite this, the technique gives correct results.

In the flame photometric method the sample was ashed in nickel without an ashing aid at a maximum temperature of between 650° and 800°C. In the ashing procedure preliminary to the emission spectrographic procedure the sample was mixed with magnesium AC dope and sulphur and ashed in platinum at 500°C. Unlike the flame photometric ashing procedure, the ash obtained in the emission spectrographic method is supported on a magnesium sulphate matrix and this presumably prevents losses of sodium by volatilisation and/or retention in the crucible in a water-insoluble form. Alternatively the losses in the flame photometric ashing procedure may have been caused by the maximum ashing temperature used exceeding that used in the emission spectrographic method by some 150°C to 300°C. This suggested that it might be possible to obtain more reliable sodium contents by flame photometry if the magnesium AC dope ashing procedure could be used in conjunction with flame photometry.

The results in Table 6 show clearly that flame photometry following dope ashing at 500°C gives a quantitative recovery of sodium. Direct ashing without an ashing aid at 500°C causes losses of 10% or more of the sodium whilst direct ashing at 800°C causes even greater losses.

TABLE 6   The Effects of Modification of Ashing Procedure
on the Flame Photometric Determination of Sodium

Sodium ppm

By neutron activation	By emission Spectrography	By flame photometry		
		Original (ashed between 650°C and 800°C	Dope Ash at 500°C	Direct Ash at 500°C
99,96,99	95	60,75,55	100	75
256,247,259	258,259	160,178,271	225	208
343,321,339	339,287	250,312	282	265
213,210,212	218,212	140,196	210	191
194,189,192	209,198	80,158,229	196	169
186,191,198	191,191	95,173	193	173

Lithium in Polyolefins

The flame photometric procedure described below determines lithium in amounts down to 0.2 ppm in polyolefins. Sodium in amounts up to 10 ppm do not interfere.

Reagents. <u>Lithium stock solution, 100 micrograms per ml</u>. Dissolve 0.9219 g
lithium sulphate monohydrate in 1 N nitric acid and make up to 1 litre with
1 N nitric acid.
<u>Nitric acid, 1 N solution</u>. Dilute 63 ml of analytical reagent grade nitric
acid to 1 litre with distilled water.
<u>Water</u>. Distilled or deionised.
<u>Calibration solutions</u>. 0 to 10 ppm Li. Prepare standard solutions contain-
ing 3,7 and 10 ppm Li by diluting to 100 ml, 3, 7 and 10 ml aliquots of the
stock solution using 1 N nitric acid.
<u>0 to 1 ppm Li</u>. Prepare standard solutions containing 0.3, 0.7 and 1.0 ppm Li
by diluting to one-tenth strength, the standard solutions prepared above
using 1 N nitric acid.

Sample preparation. Clean the platinum dish by boiling in hydrochloric acid
(1 part concentrated acid plus 1 part water), washing with distilled water
and drying. Weigh 10 g of the polymer sample into the dish. Place the dish
in the cold electric muffle which is then programmed as follows:-

> Time from start: 0 to 1 hour. Heat to $200^{\circ}C$
>   1 to 3 hr.   Hold at $200^{\circ}C$
>   3 to 5 hr.   Heat to $450^{\circ}C$
>   5 to 8 hr.   Hold at $450^{\circ}C$

Remove the dish from the muffle and allow to cool in desiccator. When cool
add 5 ml 1 N nitric acid and warm the dish on a hot plate to ensure complete
dissolution of soluble salts. Transfer the solution to a 10 ml volumetric
flask. Rinse the dish with further 1 ml portions of nitric acid and add each
to the volumetric flask. When the flask and the contents are cool, make up
the volume to 10 ml and mix thoroughly.

Measurement. Set the flame photometer wavelength to 670.9 nm and adjust the
slit, zero and gain controls of the flame photometer until 0% and 100% scale
readings are obtained for 1 N nitric acid solution and the 1 ppm Li standard
solution (or the 10 ppm Li solution as required) respectively. Obtain the
scale readings for each standard solution in turn, followed by the sample,
at 670.9, 685.9 and 655.9 nm.

Calculation. Subtract the mean of the 685.9 and 655.9 nm scale readings from
the value obtained at 670.9 nm to give a background corrected scale reading
for each solution. Plot a calibration graph of the background corrected scale
readings for the standard solutions against lithium concentrations under the
conditions given above, the standard solutions given above are equivalent to
0.3 to 10 ppm lithium in polymer. Interpolate the sample background corrected
scale reading into the calibration to obtain the concentration of lithium in
the polymer.

Iron, Aluminium, Titanium in Polyolefins

Methods are described below for the determination of traces of iron, aluminium
and titanium in polyolefins.

Scope. The method describes procedures for the determination of trace amounts
of titanium, aluminium and iron in polyolefins. Analysis of a 5 g sample
enables aluminium, iron and titanium to be determined in amounts down to 1 ppm
with an accuracy of $\pm$ 1 ppm.

Summary. Polymer samples are ashed to remove organic material. Aqueous and
solvent samples are evaporated to dryness on a hot plate. The ash residues

are fused with potassium hydrogen sulphate to effect solution of the metals.
The resulting ash is dissolved in hot water containing sulphuric acid and
diluted to a standard volume. Suitable aliquots of the aqueous solutions are
reacted with organic reagents under controlled conditions i.e. titanium is
determined spectrophotometrically with 'Tiron', aluminium is determined spectrophotometrically by the 'Aluminon' procedure and iron is determined spectrophotometrically by a procedure using thioglycollic acid.

Apparatus. Muffle furnace. maximum operating temperature 900°C.
Silica crucibles. 50 ml capacity.
Visible spectrophotometer.
Standard volumetric glassware.

Reagents. Potassium hydrogen sulphate, AR grade.
50% v/v and 0.5% v/v sulphuric acid solution.
'Tiron solution'. Dissolve 4 g of 1.2 dihydroxybenzene, 3:5 sulphonic acid
sodium salt in 100 ml distilled water.
Buffer solution. pH 4.7, acetic acid/sodium acetate. Dissolve 136 g of
$CH_3COOHa3H_2O$ in 1 litre of water. Dilute 57 ml of glacial acetic acid to
1 litre with water. The pH 4.7 buffer solution is prepared by mixing equal
volumes of these two solutions.
Sodium hydrosulphite. solid AR grade
Ammonia solution. Mix equal volumes of 0.880 ammonia and water.
Potassium titanyl oxalate. pure grade.
'Aluminon' solution. separately dissolve each of the following reagents in
minimum volume of distilled water: 77 g Ammonium acetate AR grade, 2.5 ml
conc. hydrochloric acid, AR grade, 0.2 g 'Aluminon' reagent, B.D.H. Poole,
Dorset, 0.5 g gum acacia. Transfer the solutions in the above order to a
500 ml standard volumetric flask and dilute to the mark with distilled water.
Aluminium powder. AR grade.
Thioglycollic acid. 98% AR grade.
Ammonia. specific gravity 0.880, AR grade.
Iron wire. AR grade.

Procedure. Weigh accurately about 5 g of polymer into a silica crucible (see
Note 1) and ignite the sample in a muffle furnace. Allow the temperature to
increase gradually to 800°C over a period of 4 hours, maintain this temperature for 1 hour.

Remove the crucibles from the muffle furnace when cold and add 5 g of fused
potassium hydrogen sulphate. Heat the mixture over a bunsen burner until a
clear melt is obtained, cool and dissolve in 25 mls of hot water containing
1 ml of 50% w/v sulphuric acid. Transfer the resultant solution quantitatively
into a 100 ml standard volumetric flask, allow to cool, and dilute to the mark
with distilled water. Carry out a blank under identical conditions omitting
the sample. To determine titanium, aluminium and iron, continue as described
below.

Titanium. Pipette an aliquot of the test and blank solutions (maximum 35 ml)
into two separate 50 ml standard volumetric flasks. If the volume added is
less than 40 ml then adjust the volume to 40 ml with distilled water delivered
from a burette. Add 5 ml of 'Tiron' to each flask and neutralize the solutions to Congo red paper with 50% ammonia solution. Add 5 ml of sodium ace-
tate-acetic acid buffer solution (pH 4.7) to each flask and dilute the solution to 50 ml with distilled water and mix thoroughly.

Transfer the coloured solutions to 1 cm cells and add 5 mgs of sodium hydrosulphite (see Note 3), mix with a glass rod. Immediately measure the optical

density of the sample solution against the blank solution in the comparison
cell at a wavelength of 410 nm.

Read off the titanium content of the solution from the calibration graph, and
calculate the titanium content of the original sample.

Aluminium. Pipette an aliquot of the test and blank solutions (maximum 20 ml)
into two separate 50 ml standard volumetric flasks. If the volume added is
less than 20 ml adjust the volume to 20 ml with distilled water delivered from
a burette. Add 20 ml of 'Aluminon' reagent and immerse the flasks in a boiling water bath for 10 mins. to allow the colour to develop. Cool the flasks
under running water and dilute to 50 ml with distilled water.

Measure the optical density of the sample solution against the blank solution
in the comparison cell, using the spectrophotometer at a wavelength of 535 mu
in 1 cm cells. Read off the concentration of aluminium from the appropriate
calibration graph. Calculate the aluminium content of the original sample,
(see Note 4). The 'Aluminon' reagent solution, which gradually darkens on
storage must be discarded when the optical density of the blank solution,
when prepared by the above procedure, exceeds 0.09 compared with distilled
water.

Iron. Transfer an aliquot of the sample and blank solutions (maximum 40 ml)
into two separate 50 ml standard volumetric flasks and add 0.2 ml of thioglycollic acid by means of a graduated 0.5 ml pipette. Adjust the volume of
the solutions to 40 ml with distilled water and add 5 ml of 0.880 ammonia.
Check that the solution is alkaline to litmus and dilute to 50 ml with distilled water. Mix thoroughly and measure the optical density of the sample
solution versus the blank solution in the comparison cell at a wavelength of
535 mu using 4 cm glass cells.

Read off the iron content of the sample solution from the calibration graph
and calculate the iron content of the original sample.

Calculations.   Ti or Al or Fe ppm   = $\dfrac{M \times 100}{W \times V}$

Where M = weight (microgram) of titanium, or aluminium
          or iron found from calibration graph.
      W = weight (g) of sample taken for analysis.
      V = volume (ml) of 100 ml of ash solution taken
          from colour development.

Calibration procedure for determination of titanium. Weigh 0.7360 g of pure
potassium titanyl-oxalate into a 500 ml Kjeldahl flask, and add 8 g of ammonium sulphate and 100 ml of concentrated sulphuric acid. Gradually heat to
boiling and boil for 5-10 minutes. Cool and pour the solution into 700-800
ml of distilled water. After cooling, dilute to 1 litre, this solution contains 100 μg, Ti/ml. Transfer 50 mls of 100 g/ml stock solution to a 1 litre
standard volumetric flask and dilute to the mark with distilled water. This
solution contains 5 μg Ti/ml.

Into six 50 ml standard volumetric flasks pipette the following volumes of the
5 g/ml titanium solution.

0 ml	2 ml	5 ml	10 ml	20 ml	25 ml
0	10	25	50	100	125 μg Ti

Add the required volume of distilled water to make the volume of each solution 35 ml. Add 5 ml of 'Tiron' reagent and neutralize to Congo red paper
with 50% ammonia solution, add 5 ml of pH 4.7 acetic acid/sodium acetate
buffer solution and dilute to the mark with distilled water. Evaluate the
optical density of the solutions. Transfer the solutions to 1 cm cells and
read off the optical density at a wavelength of 410 nm using the solution
containing 0 µg titanium as the reference solution. Plot a graph of optical
density against µg of titanium contained in the 50 ml standard volumetric
flasks.

Calibration procedure for determination of aluminium. Weigh out 0.1500 g of
aluminium powder into a 100 ml beaker, add 30 ml of distilled water and 10 ml
of concentrated hydrochloric acid. Allow the aluminium to dissolve and transfer quantitatively to a 1 litre standard volumetric flask with distilled water.
Make up to the mark and mix thoroughly, this solution contains 150 g Al/ml.
Transfer 50 ml of the 150 µg/ml aluminium solution into a 1 litre standard
volumetric flask and dilute to the mark with distilled water and mix thoroughly. This solution contains 7.5 µg Al/ml.

Into seven sets of 50 ml volumetric flasks pipette the volumes of 7.5 µg/ml
aluminium solution and of 100 µg/ml titanium solution shown in Table 1 (Note
4). Take the seven flasks comprising Set 1. Add to each flask sufficient
distilled water to make the volume up to 20 ml. Add 20 ml 'Aluminon' reagent
and immerse the seven flasks in a boiling water bath for 10 minutes to allow
the colour to develop. Cool the flasks under running water and dilute to
50 ml with distilled water and mix well.

Measure the optical density at 535 mu of the solutions in 1 cm cells against
the aluminium and titanium free reagent blank solution (flask 1) in the comparison cell. Plot a graph of µg aluminium present in the 50 ml solution
versus determined optical density.

Repeat the operations described in the previous paragraph on each of the seven
flasks comprising Sets 2 to 7. Plot the seven aluminium calibration graphs
and indicate on each graph the constant amount of titanium (micrograms) in
each calibration solution (i.e.between 0 and 300 µg of titanium). Recalibrate
the procedure each time a new batch of 'Aluminon' reagent is used in the
analysis.

Calibration procedure for determination of iron. Weight out 0.1000 g of iron
wire and add 50 ml of a 1:3 nitric acid:water solution. Boil to expel oxides
of nitrogen and to dissolve the wire. Transfer the solution quantitatively
to a 1 litre standard volumetric flask and dilute to the mark with distilled
water, mix thoroughly. This solution contains 100 µg Fe/ml.

Pipette 10 ml of the 100 µg/ml iron solution into a 100 ml standard volumetric
flask and dilute to the mark with 0.2 nitric acid solution, mix thoroughly.
This solution contains 10 µg/ Fe/ml.

Into six 50 ml standard volumetric flasks pipette the following volumes of
10 µg/ml iron solution.

    0   1   2   5   10   15 ml

    0   10   20   50   100   150 µg Fe.

Add the required volume of 0.2 N nitric acid solution to make the volume of
each solution 40 ml. Add 5 ml of 0.880 ammonia solution and check that the

solution is alkaline to litmus paper. Dilute to 50 ml with distilled water
and mix thoroughly. Measure the optical density of the solutions. Transfer
the solutions to 4 cm glass cells and read off the optical density at a wavelength of 535 nm using the solution containing 0 µg iron as the reference
solution. Plot a graph of optical density against µg of iron present in the
50 ml standard volumetric flask.

Trace amounts of aluminium in polyethylene have been determined by gas chromo-
tography of volatile aluminium complexes with trifluoroacetylacetone or
pivaloyltrifluoroacetone (11).

Copper in Polyolefins

Gorsuch (12, 13) has shown in studies using radioactive copper isotope that
when organic materials containing copper are ashed losses of up to some 10%
of the copper will occur due to retention in the crucible and this could not
be removed by acid washing. Virtually no retention of copper in the silica
crucible occurred, however, when copper was ashed under the same conditions
in the absence of added organic matter. This was attributed to reduction of
copper to the metal by organic matter present, followed by partial diffusion
of the copper metal into the crucible wall.

The technique used in the method described below for reducing losses of copper
by diffusion into the crucible during ashing involves the use of an ashing
aid. This procedure consists of intimately mixing with the polymer before
ignition a solution of magnesium nitrate which leaves an ash in the crucible
at the end of the ignition step. This ash tends to retain the copper within
itself, thereby decreasing the losses of copper occurring by diffusion into
the crucible.

Apparatus. Silica crucibles. ('Vitrosil' is suitable), preferably new and
used only for the determination of copper. Cleaned periodically by soaking
in concentrated nitric acid. Diameter mouth 2½", base 1¼", height 1¾".

Miscellaneous volumetric glassware. Pipettes miscellaneous pipette 5 ml
Volumetric flask 50 ml, 25 ml. Graduated cylinder 10 ml.
Separatory funnels. 50 ml Ungreased stockcock. Cleaned by soaking in
concentrated nitric acid followed by rinsing with distilled water before use.
Electric hotplate. with heat control.
Absorption spectrophotometer. The Unicam SP 600 instrument is suitable or
an equivalent instrument.

Reagents. Distilled water. Water freshly distilled from 1 g of potassium
permanganate and 20 ml 2N sulphuric acid in an all glass apparatus fitted with
a steam trap to be used throughout for the preparation of reagents and test
solutions.
Sodium diethyldithiocarbomate solution 1% Dissolve 1 g of sodium diethyl-
dithiocarbomate 'ANALAR' in 100 ml water and filter.
EDTA - citrate solution. Dissolve 18.2 g of citric acid in distilled water
and add the theoretical quantity of 0.880 S.G. ammonia (both B.D.H. lead-free
reagents for foodstuff analysis grade) to produce the triammonium salt. The
final solution should be slightly alkaline to litmus, add 5 g of disodium
EDTA 'ANALAR' and dilute to 100 ml with distilled water.
Ammonium hydroxide (1:3). Mix one volume of S.G. 0.880 ammonia (B.D.H. lead-
free reagents for foodstuff analysis grade) with 3 volumes of distilled water.
Cresol red indicator 0.04%. Dissolve 0.04 g cresol red in 100 ml redistilled
ethanol.
Carbon tetrachloride. redistilled ex British Drug Houses-passes dithizone
test

Standard copper solution. Stock solution (1 mg copper per ml). Dissolve 0.1000 g of pure copper wire (Johnson Matthey spectrographically standardized copper wire 1/16" in diameter) in 1 ml of (1+4) nitric acid (use B.D.H. lead-free reagent for foodstuff analysis grade nitric acid). Boil gently to expel nitrous fumes, cool and dilute to 100 with distilled water. Dilute solution for calibration. Dilute 5 ml of the stock copper solution to 500 ml as required. This dilute solution to be used only on the day of preparation as its strength decreases appreciably on standing due to absorption of copper on the surface of the flask. 1 ml dilute solution 1=10 microgram Cu.

Magnesium nitrate solution. 7%. Dissolve 70 g hydrated magnesium nitrate $(Mg(NO_3)_26H_2O)$ 'ANALAR' in 700 ml distilled water contained in a 1000 ml volumetric flask. Make up to the 1000 ml mark with redistilled methanol.

Sulphuric-nitric acid mixture 1:3. Mix one volume of concentrated sulphuric acid with three volumes of concentration nitric acid (both B.D.H. lead-free reagents for foodstuffs analysis grade).

Potassium cyanide. 5% aqueous solution (B.D.H. lead-free reagent for foodstuff analysis grade).

Procedure. Ashing procedure. Weigh out 10 g sample into a silica crucible. Pipette in 10 ml magnesium nitrate solution. Stir the mixture with a thin glass rod until a thorough contact between the polymer and the solution is obtained. In order to obtain thorough mixing of polymer and magnesium nitrate solution the addition of a further 5 ml of distilled water may be necessary at this stage.

Support the crucible on a pipe-clay triangle and heat strongly (about $600^{\circ}C$) on a bunsen until most of the carbon is removed. Variable copper contents are obtained when ashing is performed in an electric muffle. This may be due to contamination of the ash by specks of copper containing firebrick falling from the oven roof onto the crucible. Allow the crucible to cool. Run 2 ml of 1:3 sulphuric nitric acid mixture down the wall of the crucible using a thin glass rod to transfer any remaining carbon to the bottom of the crucible. Heat again on the bunsen until sulphuric acid fumes are produced. Allow to cool and repeat the treatment with a further 2 ml of sulphuric-nitric acid mixture. Complete oxidation of carbon is usually obtained at this stage. Ensure that some sulphuric acid remains in the crucible at the end of each acid treatment i.e. do not heat to dryness.

Dilute the contents of the crucible with 10 ml of cold distilled water and boil gently on an electric hot plate. Quantitatively transfer this solution, together with two or three subsequent hot distilled water crucible washings, to a 50 ml volumetric flask. Allow the contents of the flask to reach room temperature and make up to the 50 ml mark with distilled water. Filter this solution if necessary.

Perform a reagent blank determination in exactly the same manner as described above omitting only the addition of polymer. Make the blank solution also up to 50 ml.

Determination of copper. Transfer a suitable volume of the sample extract containing not more than 30 microgram of copper and the same volume of the reagent blank solution to two clean 50 ml separatory funnels. Suitable volumes from the 50 ml of dissolved ash solution obtained from polymers containing various amounts of copper are shown below. Add sufficient distilled water to the sample and the blank solution to make the volumes up to 40 ml.

Volume of sample solution required for colour development.

Copper content of polymer sample, ppm	Volume of the 50 ml of sample solution required for colour development, ml
less than 2	40
5	15
10	10
50	2
100	1

Pipette 10 ml of EDTA-citrate solution and add 2 drops of cresol-red indicator into the sample and blank solutions and then add 1:3 ammonia from a dropping pipette until the solution becomes purple-red in colour, i.e. pass through the rose-pink colour stage.

Allow the mixture to cool and add 1 ml of sodium diethyldithiocarbomate solution. Pipette in 10 ml carbon tetrachloride and shake for 2 minutes. Allow the layers to separate: then filter the lower layer through dry filter paper into a 25 ml volumetric flask. Extract the aqueous layer with further 5 ml portions of carbon tetrachloride until no further colour is extracted. Use these further 5 ml carbon tetrachloride extracts to wash any colour remaining in the original filter paper into the 25 ml flask. Make the combined filtrates up to the mark with carbon tetrachloride. Filter this solution through dry filter paper again if cloudy.

Determine the optical density of the sample solution approximately 30 minutes after the addition of sodium diethyl-dithiocarbamate reagent with a suitable spectrophotometer.

Spectrophotometer conditions.

Instrument: Unicam SP 600 (using blue photocell or equivalent instrument.
Wavelength: 432 nm
Cells: 4 cm glass
Blank solution: this is the reagent which has been through the whole test procedure from the ashing stage.

Preparation of calibration graph. Pipette 40 ml distilled water into several clean 50 ml separatory funnels and accurately pipette in 0.5, 1.0, 2.0, 2.5, 3.0 and 3.5 ml of freshly prepared dilute (10 microgram per ml) standard copper solution, i.e. between 5 and 35 microgram copper. Include a further separatory funnel containing 40 ml distilled water only i.e. no copper addition, to serve as a blank to allow for copper impurity in the reagents used. Continue with the addition of 10 ml EDTA citrate solution and 2 drops of cresol-red as described under determination of copper. Using the spectrophotometer conditions described above, measure the optical densities of the carbon tetrachloride extracts obtained from the various standard copper solutions. Employ the extract obtained in the reagent blank determination in the comparison cell. The effect of copper impurity present in the reagents on the calibration graph is, thereby, overcome. Construct a graph relating these optical densities to the number of micrograms of copper present in the copper calibration solutions. The copper diethyl-dithiocarbomate complex obeys Beers Law in the region 0 to 50 microgram of copper.

Calculations. Refer the optical density given by the carbon tetrachloride

extract obtained from the sample to the copper calibration graph and read off
the number of micrograms of copper present in 25 ml of copper tetrachloride
solution.

Calculate the copper content of the polymer sample from the following
equation:

$$\text{Copper, ppm} \quad = \quad \frac{N \times 10^5 \times 50}{10^6 \times V} \quad = \quad \frac{5\ N}{V}$$

assuming 10 g of polymer is ashed, where

N = number of micrograms of copper present in 25 ml of
    carbon tetrachloride solution.

V = volume (ml) of aqueous solution taken for extraction
    with carbon tetrachloride from the 50 ml dissolved
    polymer ash solution.

Distinctly higher copper determinations were obtained on polyolefins by the
procedure involving the use of an ashing aid than are obtained without an
ashing aid or by the use of a molten potassium bisulphate fusion technique to
take up the polymer ash. This is confirmed in Fig. 1 which shows copper
recoveries obtained by the above technique when known additions of copper are
made to a polyethylene whose original copper content was 1·1 ppm.

Cadmium, Selenium Pigments

X-ray fluorescence spectrometry has been used extensively for the determination of traces of metals and non-metals in polyolefins. The technique has
also been used to the determination of major metallic constituents in polymers
such as cadmium selenide pigment determinations.

The X-ray fluorescence spectrometry method has several advantages over other
methods. The analysis is non-destructive, specimen preparation is simple,
measurement time is usually less than for other methods and X-rays interact
with elements as such i.e. the intensity measurement of a constituent element
is independent of its state of chemical combination. However, the technique
does have some drawbacks and these are evident in the measurement of cadmium
and selenium e.g..absorption effects of other elements present, for instance
the carbon and hydrogen of the polyethylene matrix, excitation of one element
by X-rays from another e.g. cadmium and selenium mutually affect one another.

Carbon Number Distribution of Polyethylene Waxes

Scope. The method is applicable to the determination of the weight and/or
molar carbon number distribution of the straight chain alkanes in samples of
high density polyethylene wax. The determination of the concentration of each
n-alkane is also possible.

Method summary. The sample is dissolved in a suitable hot paraffinic solvent
and an aliquot of the solution chromatographed under suitable programmed
temperature conditions. Component identities are established by re-analysing
the sample in the presence of added known n-alkanes. Carbon number distribution values are obtained by peak area normalisation. For quantitative
evaluation, a similar solution is examined in the presence of a suitable odd
carbon number n-alkane as the internal standard.

Operating conditions.

Instrument	F, and M. Model 810, operated as a dual column instrument, or equivalent instrument.
Column	Stainless steel (¼ in. O.D. x 3/16 in. I.D.) packed with 20% w/w Silicone S.E. 30 on 60-80 Chromosorb (pre-heated at 400°C for 12 hr. prior to use). Length: 4 ft.
Gases	Air: 20 psig, Hydrogen: 20 psig (fully open), Helium: 50 psig (each rotameter at 3.5)
Temperatures	Injection: 300°C Detector: 350°C Column: Automatically programmed under suitable conditions for the sample, e.g. between 100 and 400°C at 20°C per min.
Range/ Attenuation	Set to suitable positions for the sample.
Chart speed	Set to a suitable position to give a chromatogram with easily measurable peaks.

Determination of carbon number distribution. Dissolve about 0.01 g sample in the minimum volume of a suitable paraffinic solvent. This latter will usually be a C6-C12 n-alkane, its carbon number being at least two lower than that of the first member in the wax sample. The volume of solvent used to dissolve the wax will be about 0.1 µl and solution will only be affected when hot. Using a heated 10 µl syringe, inject 1 µl of the solution into the chromatograph and programme the temperature under suitable conditions for the sample. Usually the lower and upper temperatures will lie between 100°C and 400°C and the programming rate will be between 10°C/min. and 30°C/min. Develop the chromatogram until all measurable components have emerged, the tallest peak being preferably just "on scale".

Measure each relative peak area by constructing the escribed triangle and obtaining the product of peak height and peak width at the peak base. In making the measurements, allow for components other than the n-alkanes by drawing in the "true" chromatogram base-line, depicted in Fig. 2 by the dotted line.

Determination of component concentrations. Accurately weigh about 0.01 g of wax into a specimen tube and add to this exactly 0.1 ml of a solution of an exactly known amount of a suitable odd carbon number distribution analysis. A convenient odd carbon number n-alkane to use is usually n-nonadecane and its concentration in the solvent is between 0.25% w/v and 0.50% w/v, depending on the sample being examined. Inject 1.0 µl of the solution into the chromatograph and programme the temperature under the conditions used for the carbon number distribution analysis. Develop the chromatogram until all measurable components have emerged, the tallest peak being preferably just "on scale". Measure as already described the peak area of the internal standard (e.g. n-nonadecane) and that of the largest of the sample components.

Identification of components. Repeat the carbon number distribution analysis as described except that the selected solvent should now contain about 0.25% w/v each of two even carbon number n-alkanes, n-eicosane and n-docosane being usually suitable. Once these two components have been identified the identities of the remaining components are readily obtained since the distribution is in increments of two carbon atoms.

Kalinina and Doroshina (14) reviewed chemical methods for qualitative and quantitative analyses of polyethylenes.

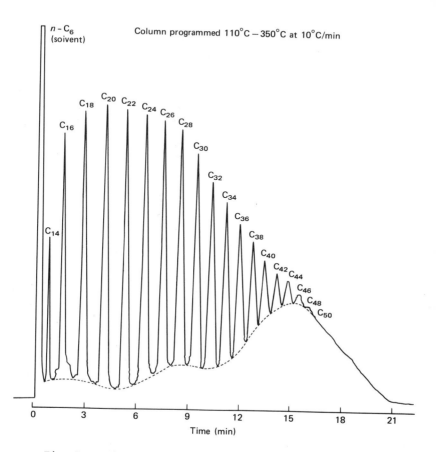

Fig. 2 Typical gas chromatogram of polyethylene wax

Infrared Spectroscopy

Branching. Stanescu (18) has investigated in detail the infrared spectroscopy of polyethylene and polypropylene and included numerous absorption frequencies for these polymers and the measurement of crystallinity.

Painter et al (42, 43) assigned an infrared absorption band at 716 cm^{-1} in the spectrum of polyethylene to a monoclinic arrangement of chains and an absorption band at 1346 cm^{-1} to a regular tight fold chain structure.

Infrared absorption spectra of the surface layers and of the bulk of low density polyethylene show a smaller content of crystallites in the surface layers compared to the polymer bulk (45).

The crystal structure of polyethylene has been determined by Krimm (46) and by Fraser et al (47). Frank and Wulf (48) heated polyethylene above 75°C and observed that the strong infrared absorption peak at 73 cm^{-1} was displaced to lower frequencies and its intensity decreased.

AOP–B

Assignments in the vibrational spectra of linear crystalline polyethylene
have been reported by Hendra et al (49).

Willis and Cudby (50) have described the use of low frequency spectroscopy in
the far infrared region for studying the crystallinity of polyethylene and
polypropylene.

A regression analysis of infrared, differential thermal analysis and X-ray
diffraction data by Laiber et al (27) for low pressure polyethylene showed
that, as synthesis conditions varies, the number of methyl groups varied from
0 to 15 per 1000 carbon atoms and the degree of crystallinity varied from 84
to 61%.

Mcrae et al (41) followed changes in the relative concentrations of methylene
groups in crystalline regions, in gauche conformations, and in tie chains in
amorphous regions of cold drawn high density polyethylene by unpolarized and
polarized infrared spectroscopy.

Luongo (29) studied crystalline orientation effects in the 720/730 cm^{-1} doublet of transcrystalline polyethylene. The effects of defects in the infrared
spectrum of polyethylene crystals was established by comparison with polyethylene in the extended chain form (30).

Saturated hydrocarbons evolved during electron irradiation of polyethylene are
characteristic of short side chains in the polymer. Salovey and Pascale (21)
showed that a convenient analysis is effected by programmed temperature gas
chromatography. In order to minimize the relative concentrations of extraneous hydrocarbons, i.e. those not arising from selective scission of complete
side chains, it is necessary to irradiate at low temperatures and doses.
Such analyses of a high pressure polyethylene indicates that the two to three
methyls per 100 carbon atoms detected in infrared absorption (low pressure
polyethylenes at least on order of magnitude lower) are probably equal amounts
of ethyl and butyl branches. These arise by intramolecular chain transfer
during polymerization. At a dose of 10 Mrad about 1 to 4% of the alkyl group
are removed. Methane is the only hydrocarbon detected on irradiation of polypropylene, indicating little combination of methyl radicals to form ethane
during irradiation.

The measurement of the methyl absorption at 1378 cm^{-1} in polyethylene can
serve as a good estimation of branching. However, interference from the
methylene absorption at 1368 cm^{-1} makes it difficult to measure the 1368 cm^{-1}
band especially in the case of relatively low methyl contents.

A method (16) has been developed at du Pont which utilizes the suggestion by
Neilson and Holland (17). They associated the amorphous phase absorption at
1368 cm^{-1} and 1304 cm^{-1} with the trans-trans conformation of the polymer chain
about the methylene group. Therefore the intensity of these two absorptions
are proportional to one another. By placing an annealed film (ca 10-15 mil)
of high density polyethylene in the reference beam of a double beam spectrometer and a thin, quenched film of the sample in the sample beam, most of the
interference at 1368cm^{-1} can be removed. The method has the advantage that it
is not necessary to have complete compensation for the 1368 cm^{-1} band since a
correction for uncompensation at 1378 cm^{-1} can be applied based on the intensity of the 1368 cm^{-1} absorption.

A calibration for the methyl absorption based on mass spectrometric
studies of such gaseous products produced during electron bombardment of
polyethylene have demonstrated irradiation induced detachment of complete

alkyl units (19). In addition to saturated alkanes characteristic of the branches, small quantities of methane, other paraffins, and olefins were simultaneously evolved. It was suggested that "extraneous" paraffins result from cleavage of the main chain (19, 20).

The branch length and number of side chains in polyethylene has been determined by Seeger and Barrall (39).

The degree of branching of polyethylene has been determined by Bezdadea et al (36).

Tasumi et al (25) carried out an analysis of infrared spectra and structures of polymethylene chains consisting of CH_2, CHD and CD_2 groups.

Nerheim (26) has described a circular calibrated polymethylene wedge for the compensation of CH_2 interferences in the determination of methyl groups in polyethylene by infrared spectroscopy.

Methyl group content of low density polyethylene has been determined with a standard deviation of 0.8% provided methylene group absorptions were compensated by polyethylene of similar structure (28).

Unsaturation. Luongo and Salovey (22) have examined the infrared spectrum of 1 mev electron irradiated linear and branched polyethylene.

The irradiation of polyethylene causes a number of molecular rearrangements in the chemical structure of the polymer. In addition to the significant changes in the type and distribution of unsaturated groups, an infrared comparison of the radiation induced chemical changes that occur in air and in a vacuum showed the presence of oxygen has a marked influence on the structural rearrangements that occur on irradiation. The infrared spectra of irradiated single crystals prepared by two different means indicated the location of vinyl groups in the crystals and their protection from radiation by aromatic structure. From their data they postulated a number of reaction mechanisms.

In Fig. 3 are shown the infrared spectra of the branched polyethylene before and after irradiation in vacuum and air. Strongly absorbing trans type unsaturation (CH=CH) bands at 964 cm^{-1} appear in both the vacuum and air-irradiated sample spectra. Vinylidene decay on irradiation is shown by the decrease in the $R_1R_2C=CH_2$ band at 888 cm^{-1}.

Irradiation in vacuum produces a significant decrease in the methyl (-CH$_3$) content (1373 cm^{-1}), whereas in the bombardment in air, there appears to be only a negligible decrease in -CH$_3$, if any. In addition a comparison of the 720-730 cm^{-1} doublet shows that only the 720 cm^{-1} component remains in the spectra of the vacuum sample, whereas there is only a slight decrease of the 730 cm^{-1} component in the air-irradiated sample. Additional evidence of structural changes are shown in the spectra of the air-irradiated sample. Here both -OH and C=O bands appear, and there is a general depression of the spectrum background from 1300 to 900 cm^{-1}.

In Fig. 4 is shown the unsaturation region of the spectra. The top two traces show this region for branched polyethylene before and after 6 MR irradiation. The lower traces are those of linear polyethylene before and after similar irradiation. In branched polyethylene before irradiation, most of the unsaturation is of the external vinylidene type. After a dose of 6 MR, trans unsaturation (at 964 cm^{-1}) increases and the vinylidene (at 888 cm^{-1}) decreases.

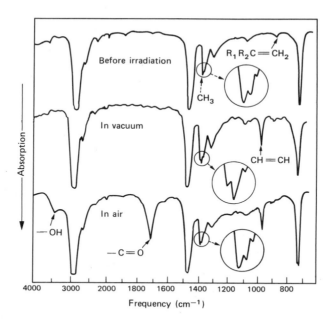

Fig. 3   The effect of irradiation (500M) Rad) in air and
         vacuum on branches polyethylene.

In linear polyethylene almost all the saturation is of the terminal vinyl
type $(CH=CH_2)$ as shown by the bands at 990 and 910 cm^{-1} Here again, after
only a 6 MR dose, the trans groups form rapidly, and the vinyl groups at
990 and 910 cm^{-1} decrease. Because of the rapid increase of trans groups
during irradiation and the simultaneous decrease of the other unsaturated
groups, it might appear that the trans groups are being formed from a reaction
involving the sacrifice of the other unsaturated groups in the polymer.   In
order to determine the validity of this observation, Luongo and Salovey (22)
exposed to similar doses of irradiation a sample of polymethylene which has
no infrared-detectable unsaturation or branching. In Fig. 4 (lower curves)
it is seen that in polymethylene, the trans unsaturation band still forms
strongly after irradiation in either air or vacuum. This means that the trans
groups come from a reaction that is independent of either unsaturation or
branching. As for the vinyl and vinylidene decay, although there is no con-
clusive mechanism to explain their disappearance, they probably become saturated by atomic hydrogen in the system or become crosslinking sites.

Unsaturation in low density polyethylene has been estimated to $\overset{+}{-}$ 0.003 C=C/
10^3C atoms by compensating with brominated polymer of the same thickness (28).

Reuda (44) has used infrared spectroscopy to measure vinyl, vinylidene and
internal cis or trans olefinic end-groups in polyethylene.

Dankovics (37) determined the degree of branching, the length of the side
chains (methyl, ethyl and butyl branches) and the degree of unsaturation of
low density polyethylene. The total degree of unsaturation in polyethylene
was determined by summing the vinyl, vinylene and vinylidene unsaturation

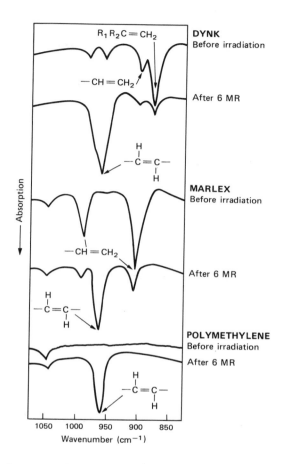

Fig. 4   Unsaturation region of three polyethylenes before
         and after irradiation at 6 M Rad.

derived from the differential infrared spectra using an unbrominated polyethylene film as the sample and a brominated film as the reference (38).

Oxygenated functional groups.   Extensive infrared studies have been carried
out on oxidized polyethylene (115-121).

Infrared absorption spectrometry has been widely used to determine the oxidation products and the rate of formation of these products during the thermal
or photo oxidation of polyethylene (122-124).   Acids, ketones and aldehydes,
the end-products reported from these oxidations have similar spectra in the
$1819-1666$ cm^{-1} region.   It is only in this carbonyl stretching region that
the products have suitable absorptivity to give quantitative data.   The
absorption band of the acid (1712 cm^{-1}, ketone (1721 cm^{-1}) and aldehyde
(1733 cm^{-1}) groups present in oxidized polyethylene are so overlapped as
to give only a broad band on standard laboratory spectrometers.   Interpretation

of these data, based on the increase in total carbonyl rather than on a
single chemical moiety could lead to incorrect conclusions because of the
large differences in the absorptivity of the various oxidation products.
Acid absorptivity has been reported (125) to be 2.4 times greater than that
of ketones and 3.1 times greater than that of aldehydes.

Rugg et al (125) using a grating spectrometer for increased resolution have
demonstrated that the carbonyl groups formed by heat oxidation are mainly
ketonic, while in highly photooxidized polyethylene the amounts of aldehyde,
ketone and acid are approximately equal. This procedure is adequate for
qualitative but suitable for accurate quantitative data.
An infrared study of oxidative crystallization of polyethylene was made from
examination of the 1894 cm^{-1} "crystallinity" band and 1303 cm^{-1} "amorphous"
band and of carbonyl absorption at 1715 cm^{-1} (31). The distribution of oxidized groups on the surface of polyethylene and polypropylene have been determined by attenuated total reflectance (32). Miller and coworkers (33) used
polarized infrared spectra obtained by Fourier-transform spectroscopy to
study several absorptions of polyethylene crystallized by orientation and
pressure in a capillary viscometer.
Tabb and coworkers (40) used Fourier transform infrared to study the effect
of irradiation on polyethylene. Aldehydic carbonyl and vinyl groups decreased
and the ketonic carbonyl and trans-vinylene double bonds increased on irradiation.

Infrared reflection was used in studies of oxidation of polyethylene at a
copper surface in the presence and absence of an inhibitor, N,N-diphenyl-
oxamide (35).

Cooper and Prober (126) have used alcoholic sodium hydroxide to convert the
acid groups to sodium carboxylate (1563 cm^{-1}) to analyse polyethylene oxidized with corona discharge in the presence of oxygen and oxone. This procedure requires five days and has been found to extract the low molecular
weight acids from the film.

Heacock (127) has described a method for the determination of carboxyl groupς
in oxidized polyolefins without interference by carbonyl groups. This procedure is based upon the relative reactivities of the various carbonyl groups
present, in oxidized polyethylene film, to sulphur tetrafluoride gas. The
quantity of the carboxyl groups in the film is then measured as a function
of the absorption at 1835 cm^{-1}

$$RC\overset{\displaystyle O}{\diagup}\diagdown_{CH} + SF_4 \longrightarrow RC\overset{\displaystyle O}{\diagup}\diagdown_{F} + HF + SOF_2$$

Raman Spectroscopy

Laser-Raman spectroscopy has been reviewed by Brandmueller and Schroetter
(51), including studies on polyethylene.

Strobl and Hagedorn (52) developed a Raman method for determining the crystallinity of polyethylene. Maxfield et al (53) used polarized Raman spectroscopy to study the crystallinity and amorphous orientation in polyethylene.

Kobayashi (34) reviewed the Raman spectroscopy of polyethylene crystals and
polymer solutions.

Lindberg and coworkers (23) obtained infrared and laser-Raman spectra of chlorinated polyethylene and compared them with spectra of polyethylene, PVC and chlorinated PVC.

Nuclear Magnetic Resonance Spectroscopy

Woodward (54) has utilized proton NMR spectroscopy in his study of relaxation phenomena in polyethylene.

Broad line proton NMR spectra of drawn high density polyethylene shows three component lines: a broad component assigned to crystalline regions; an inter- mediate component assigned to high molecular weight molecules interconnec- ting the crystalline regions; and an isotropic narrow component assigned to mobile low molecular weight material and the ends of molecules rejected from the crystalline regions (55).

High field ^{13}C Fourier transform proton NMR spectra have indicated -40°C as the upper limit to the Tg of linear polyethylene (56). Pulsed NMR measure- ments of ultra high molecular weight linear polyethylene indicated a second transition of longer relaxation time appearing at the temperature character- istic of the gamma relaxation. The gamma relaxation meets most of the cri- teria for assignment of Tg (57).

Sobottka and coworkers (58) have described a method for determining activa- tion temperatures of various kinds of molecular motion and degrees of order in crystalline polymers such as polyethylene using broad line proton NMR.

Multiple-pulse proton NMR has been used by Pembleton et al (59) to determine the amorphous fraction of polyethylene.

Proton NMR investigations of solution crystallized polyethylene showed that the lammellar crystallites are composed of approximately 80% crystalline material with a noncrystalline overlayer of about 15%. The phase structure exists independent of molecular weight (60).

High resolution proton NMR. High resolution proton NMR has been used in studies on radiolytic reactions in polyethylene (61). Proton NMR studies of low molecular weight polyethylene confirmed that it is predominantly composed of lamellar crystalline regions with a minor amount of interfacial regions and no liquid-like interzonal regions (62).

NMR measurements of branching. Average conformational energy of polyethylene has been calculated from proton NMR data as represented by ratio of trans and gauche units (64). NMR measurements on molten polyethylene gave values for vinyl, vinylidene and transvinylene content with a sensitivity of ca. 10^{-3}/C at 60 MHz in a 5-mm tube and 10^{-4}/C at 100 MHz in a 10-mm tube, using diphenyl- methane as reference (65). NMR results compared favourably with infrared data. Dorman and coworkers (66) used ^{13}C NMR spectra for low density poly- ethylene and model compounds to confirm n-butyl groups as primary short branches. Randall (67) examined several ethylene-1-olefin copolymers as model compounds for isolated short-chain branches in low density polyethylene; a series of ^{13}C resonances were examined and assignments made for methyl through amyl branches. Zackmann (68) obtained information on crystalline order and molecular mobility from high resolution NMR spectra which could be compared with results by pulse techniques.

Ahmad and Charlesby (74) used broad line proton NMR to study the effect of branching and structural changes on the glass transition temperature and

activation energy of molecular motion of high and low density polyethylene
and ethylene-acrylic acid block copolymers.

Cudby and Bunn (75) used eicosane as the standard for determining the optimum
conditions for relative intensity measurements of ethyl, butyl and longer
chain branches in low density polyethylene. The average number of branch
points per 1000 carbon atoms for ethylene-propylene copolymers, 1-butane-
ethylene copolymers, ethylene-1-hexene copolymers and low density polyethylene was calculated from the ^{13}C NMR spectra. The branches of low density
polyethylene were shown to consist of butyl groups (76).

Nishioka et al (77) determined the degree of chain branching in low density
polyethylene using proton NMR Fourier transform NMR at 100 MHz and ^{13}C
Fourier transform NMR at 25 MHz with concentrated solutions at approximately
100°C. The methyl concentrations agreed well with those of infrared based
on the absorbance at 1378 cm^{-1}.

Cutler et al (78) determined the distribution of side chains in branched polyethylene by ^{13}C NMR of the nitric acid degradation product. Ethyl and butyl
side chains were found to be excluded from the crystalline zones. Amyl and
hexyl branches were identified from the ^{13}C NMR spectra as short chain branches
in low density polyethylene (79).

Bennett et al (80) investigated the direct detection of cross links in polyethylene.

Various workers have reported that butyl groups generally predominate in the
branch distribution in low density polyethylene (81-84). Cudby and Bunn (81)
claimed that they could not recognize separately by ^{13}C NMR any amyl branches
from ethyl, butyl and longer chain branches in low density polyethylene.
Randall (82, 85) has claimed to have separately identified by ^{13}C NMR amyl
branches from ethyl, butyl and longer chain branches in ethylene-1-olefin
copolymers.

It has been well established from infrared measurements that low density
polyethylenes possess appreciable quantities of ethyl and butyl branches
(86-88) but it was not until C-13 NMR became available that an absolute
identification, both qualitatively and quantitatively, of the short branches
became possible (89-93). Long chain branching is also present in low density
polyethylenes and carbon-13 NMR was useful here also in establishing the
identity and relative amounts of long versus short chain branches (94-96).

High density (low pressure) polyethylenes are usually linear although the
physical and rheological properties of some high density polyethylenes have
suggested the presence of long chain branching (97) at a level one to two
orders of magnitude below that found for low density polyethylenes prepared
by a high pressure process. A measurement of long chain branching in high
density polyethylenes has been elusive because of the concentrations involved
(98) and can only be directly provided by high field, high sensitivity NMR
spectrometers. Randall (99) has reviewed the history of structural studies
of polyethylene and shows where these recent advances in C-13 NMR instrumentation have greatly enhanced our knowledge about polyethylene structure.

High density polyethylenes prepared with a Ziegler type, titanium based catalyst have predominantly n-alkyl or saturated end-groups. Those prepared with
chromium based catalysts have a propensity toward more olefinic end-groups.
As will be seen later, the ratio of olefinic to saturated end-groups for polyethylenes prepared with chromium based catalysts is approximately unity. The

end-group distribution is therefore, another structural feature of interest in polyethylenes because it can be related to the catalyst employed and possibly the extent of long chain branching. Prior to the availability of C-13 NMR, there was no technique for measuring directly the saturated end-group concentration. Now it is possible not only to measure concentrations of saturated end-groups, but also the olefinic end-groups and, subsequently, an end-group distribution.

Short chain branches can be introduced deliberately in a controlled manner into polyethylenes by copolymerizing ethylene with a 1-olefin. The introduction of 1-olefins allows the density to be controlled and butene-1 and hexene-1 are commonly used for this purpose. Once again, as in the case of high pressure process low density polyethylenes, C-13 NMR can be used to measure ethyl and butyl branch concentrations independently of the saturated end-groups. This result gives C-13 NMR a distinct advantage over corresponding infrared measurements because the latter technique can only detect methyl groups irrespective of whether the methyl group belongs to a butyl branch or a chain end. C-13 NMR also has a disadvantage in branching measurements because only branches five carbons in length and shorter can be discriminated independently of longer chain branches (90 & 94). Branches six carbons in length and longer give rise to the same C-13 NMR spectral pattern independently of the chain length. This lack of discrimination among the longer side-chain branches is not a deterring factor, however, in the usefulness of C-13 NMR in a determination of long chain branching.

By far the most difficult structural measurement and, as stated previously, the most elusive, has been long chain branching. In low density polyethylenes, the concentration of long chain branches is such (> 0.5 per 1000 carbons) that characterization through size exclusion chromatography in conjunction with either low angle laser light scattering or intrinsic viscosity measurements becomes feasible (94-96, 98, 101-102). When carbon-13 NMR measurements have been compared to results from polymer solution property measurements, good agreement has been obtained between long chain branching from solution properties with the concentration of branches six carbons long and longer (94 & 95). Unfortunately, these techniques utilizing solution properties do not possess sufficient sensitivity to detect long chain branching in a range of one in ten thousand carbons, the level suspected in high density polyethylenes. The availability of superconducting magnet systems has made measurements of long chain branching by C-13 NMR a reality because of a greatly improved sensitivity. An enhancement by factors between 20 to 30 over conventional NMR spectrometers has been achieved through a combination of higher field strengths, 20 mm probes, and the ability to examine polymer samples in essentially a melt state. The data discussed by Randall (82, 85) have been obtained from both conventional iron magnet spectrometers with field strengths of 23.5 kG and superconducting magnet systems operating at 47 kG.

C-13 NMR structure sensitivity of polyethylene long chain branching. The C-13 NMR spectra from a homologous series of six linear ethylene 1-olefin copolymers beginning with 1-propene and ending with 1-octene are reproduced in Figs. 5 and 6. The side-chain branches are, therefore, linear and progress from one to six carbons in length. Also, the respective 1-olefin concentrations are less than 3%; thus, only isolated branches are produced. Unique spectral fingerprints are observed for each branch length. The chemical shifts, which can be predicted with the Grant and Paul parameters (90 & 103) are given in Table 7 for this series of model ethylene-1-olefin copolymers. The nomenclature, used to designate those polymer backbone and side-chain carbons discriminated by C-13 NMR, is as follows:

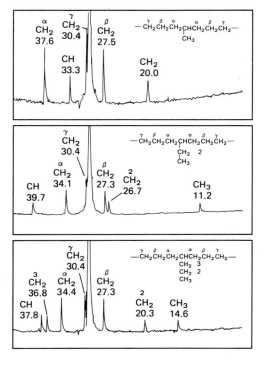

Fig. 5 C$_{13}$ NMR at 25.2 M Hz of (top) an ethylene-1-propene
copolymer, (middle) and ethylene-butene copolymer,
and (bottom) an ethylene-1-pentene copolymer.

$$
\begin{array}{cccccccc}
\gamma & \beta & \alpha & 1 & \alpha & \beta & \gamma \\
-CH_2 & -CH_2 & -CH_2 & -CH_2 & -CH-CH_2 & -CH_2 & -CH_2 & -CH_2 & -CH_2
\end{array}
$$

$$
\begin{array}{cc}
2 & CH_2 \\
3 & CH_2 \\
4 & CH_2 \\
5 & CH_2 \\
6 & CH_2
\end{array}
$$

The distinguishable backbone carbons are designated by Greek symbols while
the side-chain carbons are numbered consecutively starting with the methyl
group and ending with the methylene carbon bonded to the polymer backbone (90).
The identity of each resonance is indicated in Figs. 5 and 6. It should be
noticed in Fig. 6 that the "6" carbon resonance for the hexyl branch is the
same as α , the "5" carbon resonance is the same as β and the "4" carbon
resonance is the same as γ Resonances 1, 2 and 3, likewise, are the same
as the end group resonances observed for a .linear polyethylene. Thus a six
carbon branch produces the same C-13 spectral pattern as any subsequent branch
of greater length. Carbon-13 NMR, alone, therefore cannot be used to distin-

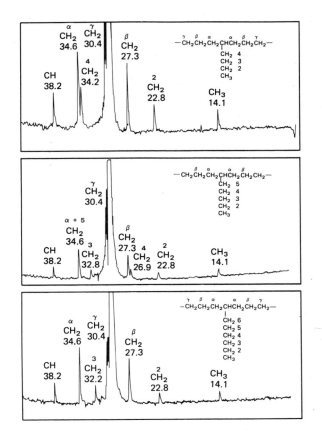

Fig. 6   C₁₃ NMR spectra at 25.2 M Hz of (top) an ethylene-
1-hexene copolymer, (middle) an ethylene-1-heptene
copolymer and (bottom) an ethylene-1-octene
copolymer.

guish a linear six carbon from a branch of some intermediate length or a true
long chain branch.

The capability for discerning the length of short chain branches has made C-13
NMR a powerful tool for characterizing low density polyethylenes produced from
free radical, high pressure processes. The C-13 NMR spectrum of such a polyethylene is shown in Fig. 7. It is evident that the major short chain
branches are butyl, amyl and ethyl. Others are also present, and Axelson,
Mandelkern and Levy, in a comprehensive study (91) have concluded that no
unique structure can be used to characterize low density polyethylenes. They
have found nonlinear short chain branches as well as 1, 3 paired ethyl
branches. Bovey, Schilling, McCracken and Wagner (94) compared the content
of branches six and longer in low density polyethylenes with the long chain
branching results obtained through a combination of gel permeation chromato-

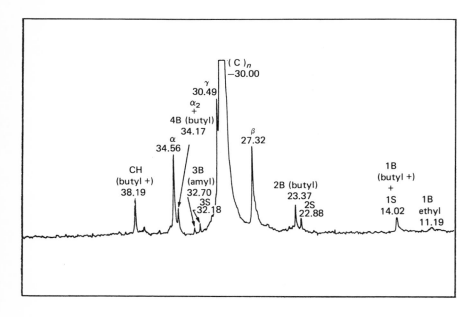

Fig. 7   C$_{13}$ spectrum at 25.2 M Hz of a low density
polyethylene from a high pressure process.

TABLE 7   Polyethylene Backbone and Side-Chain C-13 Chemical
Shifts in ppm from TMS ($\pm$0.1) as a Function of
Branch Length (Carbon Chemical Shifts, which
occur near 30.4 ppm, are not given because they
are often obscured by the major 30 ppm resonance
for the "n" equivalent, recurring methylene
carbons).   Solvent: 1,2,4-trichlorobenzene.
Temperature: 125°C

Branch Length	Methine		6	5	4	3	2	1
1	33.3	37.6	27.5	20.0				
2	39.7	34.1	27.3	11.2	26.7			
3	37.8	34.4	27.3	14.6	20.3	36.8		
4	38.2	34.6	27.3	14.1	23.4	-	34.2	
5	38.2	34.6	27.3	14.1	22.8	32.8	26.9	34.6
6	38.2	34.6	27.3	14.1	22.8	32.2	30.4	27.3	34.6

graphy and intrinsic viscosity. An observed good agreement led to the conclusion that the principal short chain branches contained fewer than six carbons
and the six and longer branching content could be related entirely to long
chain branching. Others have now reported similar observations in studies
where solution methods are combined with C-13 NMR (95). However, as a result

of the possible uncertainty of the branch lengths, associated with the resonances for branches six carbons and longer, C-13 NMR should be used in conjunction with independent methods to establish true long chain branching.

From the results we have seen thus far, it is easy to predict the C-13 NMR
spectrum anticipated for essentially linear polyethylenes containing a small
degree of long chain branching. An examination of a C-13 NMR spectrum from
a completely linear polyethylene, containing both terminal olefinic and saturated end groups, shows that only five resonances are produced. A major
resonance at 30 ppm arises from equivalent, recurring methylene carbons,
designated as "n", which are four or more removed from an end-group or a
branch. Resonances at 14.1, 22.9 and 32.3 ppm are from carbons 1, 2 and 3,
respectively, from the saturated, linear end-group. A final resonance which
is observed at 33.9 ppm, arises from an allylic carbon, designated as "a",
from a terminal olefinic end-group. These resonances, depicted structurally
below, are fundamental to the spectra of all polyethylenes.

$$CH_3-CH_2-CH_2-CH_2-CH_2- \qquad -(CH_2)_n- \qquad -CH_2-CH=CH_2$$

$$1 \quad 2 \quad 3 \qquad\qquad\qquad "n" \qquad\qquad "a"$$

An introduction of branching, either long or short, will create additional
resonances to those described above.

From the observed C-13 NMR spectrum of the ethylene-1-octene copolymer (Fig.
6) Randall found that the α and β and methine resonances associated with
branches six carbons and longer occur at 34.56, 27.32 and 38.17 ppm respectively. Thus in high density polyethylenes, where long chain branching is
essentially the only type present, carbon-13 NMR can be used to establish
unequivocally the presence of branches six carbons long and longer. If no
comonomer has been used during polymerization, it is very likely that the
presence of such resonances will be indicative of true long chain branching.
In any event, C-13 NMR can be used to pinpoint the absence of long chain
branching and place an upper limit upon the long chain branch concentration
whenever branches six carbons and longer are detected.

The need for a complementary measurement to C-13 NMR in studies of long chain
branching should be apparent. A promising possibility appears to be flow
activation energies obtained from dynamic shear moduli as a function of temperature. Flow activation energies range from approximately 6.0 kcal/mol for
linear systems to around 13.5 kcal/mol for systems containing extensive long
chain branching. Four polyethylenes, labelled "A" through "D" and selected
for C-13 NMR characterization on a basis of the observed flow activation
energies, are described in Table 8. A fifth polymer, called "E", was also
examined as a reference polymer because it was not expected to contain any
significant long chain branching as indicated by its flow activation energy
(see Table 8). Carbon-13 NMR data were obtained at a high field (50 MHz,
47 kG) to achieve improved sensitivity.

The 50 MHz C-13 NMR spectrum of polymer "A" is reproduced in Fig. 8. It is
nearly a classical representation of the spectrum antitipated for a polyethylene containing long chain branching. Only the five resonances expected
for linear polyethylene systems plus the α, and β and methine resonances
for long chain branches are observed. A simple inspection of the
intensities as compared to the end-group resonances indicates that fewer
than one polymer molecule out of three has a long chain branch.

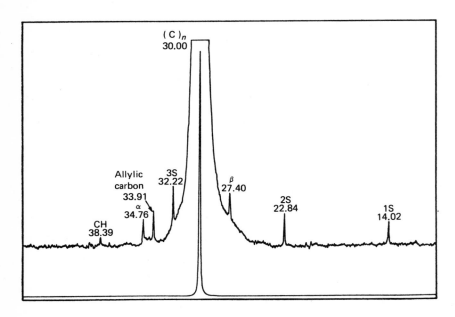

Fig. 8   C-13 NMR spectrum at 50 MHz of polymer A (75%
         in trichlorobenzene at 125°C, number of transients
         accumulated 45,300). Spectrum provided by courtesy
         of Nicolet Technology Corporation.

TABLE 8   Molecular Weights, Flow Activation Energies and
          Type of Catalyst for a series of Polyethylenes
          Examined for Long Chain Branching.

Polymer	M_w	M_n	E_a	Catalyst	Comonomer
A	159,000	19,100	8.0 kcal/mol	chromium based	none
B	224,000	13,600	8.6 kcal/mol	chromium based	none
C	148,000	17,900	9.3 kcal/mol	chromium based	none
D	226,000	8,500	9.6 kcal/mol	chromium based	hexene-1
E	172,000	32,400	6.1 kcal/mol	titanium based	none

Polymer "C", shown in Fig. 9, gives a more complex C-13 NMR spectrum than
observed for polymer "A" because four carbon side-chain branches are positi-
vely indicated even though no comonomer was added during polymerization.

From a quantitative viewpoint, it is evident that the relative intensities of the resonances from carbons associated with branches and end-groups can be compared to the intensity for the major methylene resonance, "n", at 30.00 ppm to determine branch concentrations and number average molecular weight or carbon number. The following definitions are useful in formulating the appropriate algebraic relationships:

n = intensity of the major methylene resonance at 30 ppm

$\bar{s}$ = average intensity for a saturated end-group carbon

a = the allylic carbon intensity at 33.9 ppm

C_{tot} = the total carbon intensity

$\bar{\alpha}$ = 1/2 $(\alpha + \beta)$ carbon intensities

N = average number of long chain branches per polymer molecule

$N + 2$ = average number of end-groups per polymer molecule

With the previous definitions, the polymer carbon number and number average molecular weight are given by:

$$\text{Carbon number} = C_{tot} \ (N+2)/(\bar{s} + a) \qquad (1)$$

$$Mn = 14 \times \text{Carbon number} \qquad (2)$$

For linear polymers where "N" is zero, the carbon number and number average molecular weight can be easily and reliably determined. For those polymers containing long chain branching, one must utilize the ratio of the carbon resonance intensities associated with branching to the end-group carbon resonance intensities to determine "N" as follows:

$$N = 2 \ \bar{\alpha} / \ (3(\bar{s} + a) - \bar{\alpha}) \qquad (3)$$

The number of long chain branches per ten thousand carbon atoms is similarly given by:

$$\text{Branches}/10,000 \ C = (\ (1/3 \ \bar{\alpha})/(C_{tot} \times 10^{4}) \qquad (4)$$

Equation 4 can be easily modified for the number of short chain branches (fewer than six carbons) per 10,000 carbons by replacing 1/3 ∞ with an intensity appropriate for one branch carbon from the short chain branch. The method discussed above, when applied to the C-13 NMR data from polymers A through E, gave the results listed in Table 9 for M_n, N and degree of branching.

Randall (82, 85) concludes that 13C NMR is a highly attractive method for characterizing polyethylenes. A serious drawback is not encountered even though branches six carbons in length and longer are measured collectively. The short branches are generally less than six carbons in length and truly long chain branches tend to predominate. On occasions there may be special exceptions for "intermediate" branch lengths, so independent rheological measurements should be sought as a matter of course. Nevertheless 13C NMR is a direct method, which possesses the required sensitivity to determine long chain branching in high density polyethylenes.

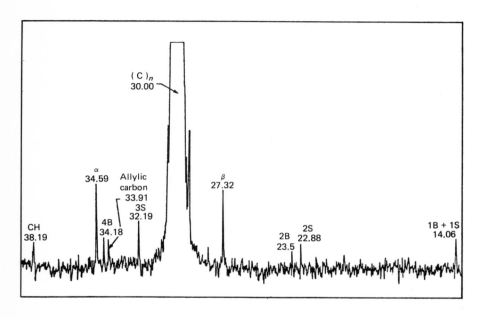

Fig. 9   C-13 NMR spectrum at 50 MHz of polymer C PE
         homopolymer (10% in trichlorobenzene at 125°C,
         number of transients accumulated 50,000).   Spectrum
         was provided courtesy of Varian Associates.

TABLE 9   Number Average Molecular Weight, Long Chain
          Branching, Short Chain Branching and End-Group
          Distribution for Polymers A, B·, C, D and E.

Polymer	C_6 Branches/ molecule	Branches/10,000		By 13C NMR M_n	By ion exclusion chromatography M_n(see Table 11)
A	0.28	(C_{6+})	1.8	21.700	19,000
B	0.29	(C_{6+})	2.1	18,680	13,600
C	0.23	(C_{6+})	1.4)	23,100	17,900
		(butyl)	5.5)		
D	-	(butyl)	6.4	10,300	8,500
E	-	(ethyl)	2.2	28,650	32,400

Electron Spin Resonance Spectroscopy

Ohnishe et al (104) have reported an electron spin resonance study of the
processes occurring during the 0.2 Mev γ ray radiation oxidation of polyethy-

lene.   Infrared spectroscopy was also used in this study. They showed that
in polyethylene three kinds of radicals are produced on irradiation: $-CH_2-CH-$
CH_2,(FR_{v1}), $-CH-CH=CH-(FR_{v11})$, and asymmetric singlet of the peroxide radical
which showed anisotropy. Reversible oxygenation of FR_{v11} took place even as
low as -113°C. Radiation oxidation of polyethylene proceeds mainly via
FR_{v11}, which produced eventually about 12 carbonyl groups and about 5 hydroxyl
groups in the course of post-oxidation, the trans-vinylene group decreasing
by an amount nearly equal to the amount of initial FR_{v11}. FR_{v1} played only
a minor role in the oxidation. An oxidation mechanism involving chain
reaction is proposed which suggests an intramolecular free back reaction of
hydrogen to the peroxide radical of FR_{v111}. FR_{v1} (polyenyl radicals .of
longer conjunctions) is oxidized to form carbonyl and hydroxyl groups,
although it does not give the spectrum of the peroxide radical. Polyenyl
radicals of shorter conjunctions were shown to react faster.

Hama et al (106) and Seiki and Takishita (107) carried out an electron spin
resonance study of free radicals produced in polyethylene by ultraviolet
light. Tsuji carried out an electron spin resonance study on radical conversions in irradiated low density polyethylene (108) and on radiation-induced
radical formation in polyethylene (109). Hori et al (110, 111) used electron
spin resonance to study radicals in polyethylene.

Seguchi and Tamura (112, 113) carried out electron spin resonance studies on
radiation graft copolymerization initiated by alkyl and alkyl radicals in
irradiated polyethylene.

Ono and Feii (105) have carried out electron spin resonance studies on
Ziegler-Natter type polyethylene catalyst systems.

The ESR spectrum of low density polyethylene after photolysis indicated
formation of the alkyl radicals $-CH_2CHCH_3$ (69). Tsuji (70) confirmed this
observation from photolysis; however, from ultraviolet and electron beam
irradiation, $-CH_2CHCH_2-$ radicals were found.

ESR measurements have been applied to studies of structural changes in high
density polyethylene from cyclic tensile stressing (71); of rotation of
peroxy radicals in polyethylene at 77-300°K (72) and of radicals produced by
fracture of polyethylene (73).

Browning et al (114) have investigated the application of electron spin
resonance spectroscopy to ultraviolet irradiated polyethylene.

Pyrolysis-Gas Chromatography

Brauer (137) has applied pyrolysis-gas chromatography to the elucidation of
the structure of polyethylenes of various types.

Cieplinski (138) pyrolysed polyethylene at 600°C - 700°C in a furnace and the
pyrolysates were analysed by programmed temperature gas chromatography.
Capillary columns coated with Carbowax 1540 or Apiezon L are used to examine
the pyrolysis products.The mechanism of pyrolysis of polyethylene has been
discussed by Wall (139).

Voigt (140) used a platinum filament type flash pyrolyser, (filament temperature 550°C). The decomposition products from up to 2 mg polymer were then
led directly on to the gas chromatographic column. Hydrogenation of the
pyrolysis products was not carried out in this work, Di-n-decylphthalate on
kieselguhr was used as the column packing.

In Fig. 10 is shown pyrograms obtained from Hostalen GC low pressure polyethylene. The heptene-1 and octene-1 peaks are suitable for identifying
polyethylene in mixtures. Identification is not always unambiguous, i.e.
in the majority of cases, one peak can correspond to several decomposition
products which may yield to separation only by selecting a different column
material from case to case.

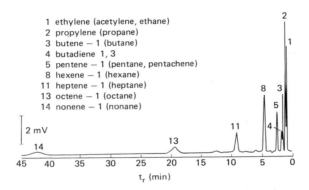

1 ethylene (acetylene, ethane)
2 propylene (propane)
3 butene – 1 (butane)
4 butadiene 1, 3
5 pentene – 1 (pentane, pentachene)
8 hexene – 1 (hexane)
11 heptene – 1 (heptane)
13 octene – 1 (octane)
14 nonene – 1 (nonane)

Fig. 10   Pyrolysis-gas chromatography of polyethylene
          (Hostalen GC)

Up to and including the peaks due to the heptenes, the retention times for
the pyrolysis products of polypropylene and polyethylene are more or less
identical, so that, qualitatively, they cannot be distinguished. The cleanly
defined heptene-1 peak, which indicates the presence of polyethylene, is,
however, characteristic.

The formation of decomposition products of polyethylene is a result of a
primary step in a statistical thermal breaking down of a chain:

$$\sim CH_2-CH_2-CH_2-CH_2 \sim \;\longrightarrow\; \sim CH_2-CH_2 \quad + \quad .CH_2-CH \sim$$

in which the free radicals formed can react further either by depolymerization
or by the transfer of hydrogen atoms. The first type of reaction leads to the
corresponding monomers, while the latter competes with the continuing decomposition of the chains. to form a spectrum of hydrocarbons with different chain
lengths. This spectrum contains, apart from saturated hydrocarbons and diolefins, olefins which, in their structure and quantitative distribution, are
dependent in a characteristic manner on the nature of the initial polymers as
might well be expected. The very marked depolymerization which takes place
with other polymers, such as polymethacrylates, polystyrene, and the like,
does not appear to play any major role in the cases of polyethylene or other
polyolefins, since the corresponding monomers do not exist in any considerable
excess as compared with the other decomposition products. As is to be expected, only unbranched fractions arising from statistical decomposition arise
in the case of polyethylene, which breaks down with reactions of the following type:

$$\sim CH_2-CH_2-CH_2 . \;+\; .CH_2-CH_2 \sim \rightarrow \sim CH_2-CH = CH_2 + CH_3-CH_2 \sim$$

or

$$\sim CH_2-CH_2-CH_2\sim \ + \ \ .CH_2-CH_2\sim \rightarrow \ \sim CH_2-CH-CH_2\sim \ + \ \ CH_3-CH_2\sim$$

$$\sim CH_2-CH_2-CH_2-CH_2-CH_2\sim \rightarrow \ \sim CH_2-CH = CH_2 \ \ + \ \ .CH_2-CH_2\sim$$

The very small amounts of branched molecules which are present even in low-pressure polyethylenes do not make themselves noticeable on the gas chromatogram under the conditions employed by Voigt (140). In this connection it should be pointed out that high-pressure polyethylenes produced completely identical gas chromatograms.

Kiran and Gilham (141) used a pyrolyzer with programming capability with a thermal conductivity detector and a mass chromatograph to study the degradation of polyethylene. This apparatus simplified identification of unknowns by providing molecular weights, quantities and retention times of constituents.

Van Schooten and Evenhuis (142, 143) applied their pyrolysis (at $500^{\circ}C$) hydrogenation-gas chromatographic technique to the measurement of short chain branching and structural details of three commercial polyethylene samples, a linear polyethylene, a Ziegler polyethylene of density 0.945 and a high density polyethylene of density 0.92. Details of the pyrograms are given in Tables 10 and 11.

TABLE 10   Relative Sizes of n-Alkane Peaks in Pyrograms of various Polyethylenes

Peak ratio	Linear polyethylene	Ziegler polyethylene	High pressure polyethylene
$n-C_4-n-C_7$	0.71	0.74	1.38
$n-C_5-n-C_7$	0.59	0.57	0.75
$n-C_6-n-C_7$	1.52	1.25	1.28
$n-C_8-n-C_7$	0.63	0.65	0.68
$n-C_9-n-C_7$	0.67	0.64	0.72
$n-C_{10}-n-C_7$	1.02	0.91	0.88
$n-C_{11}-n-C_7$	1.00	0.71	0.75

It was observed that the sizes of the iso-alkane peaks increase strongly with increasing amount of short chain branching. None of the n-alkane peaks in the Ziegler polyethylene pyrogram is significantly greater than the corresponding peak in the pyrogram of a linear polyethylene.

The pattern of the increased iso-alkane peaks in the Ziegler polyethylene pyrogram strongly suggests that these peaks are mainly due to ethyl side groups. This is in good agreement with the results of electron irradiation experiments. For the high pressure polyethylene Van Schooten and Evenhuis (142, 143) found that the n-butane peak of the pyrogram showed a clear increase in size, and the n-pentane peak a smaller, although probably significant, increase. The increases in the n-butane and n-pentane peaks are

Analysis of Plastics

TABLE 11   Iso-alkane Peaks in Polyethylene Pryrograms

Sample	Iso-alkane peak
Ziegler polyethylene	$(i-C_5,\ 3MC_5,\ (3MC_6),\ 3MC_7,$
	$(3MC_8,\ (3MC_9),$
High pressure polyethylene	$(iC_4,\ 2C_5,\ 2MC_5,\ 3MC_5,\ 2MC_6,\ 3MC_6$
	$(2MC_7-4MC_7,\ 3\ MC_7,\ 2MC_8-4MC_8,\ 3MC_8,$
	$(4MC_9-5MC_9,\ 2MC_9,\ 3MC_9,\ 4MC_{10}-5MC_{10},$
	$(2MC_{10}-4MC_{10},\ 3MC_{10}$

probably due to n-butyl and n-pentyl side groups, respectively, pentyl groups being much less numerous than butyl groups. However, from the pyrograms of the ethylene-butene, ethylene-hexene-1 and ethylene-octene-1 copolymers it is known that n-butyl side groups give a 2-methyl C_6 peak which is at least equal to the 3 methyl C_7 peak. In the high pressure polyethylene pyrogram, however, the 3 methyl C_7 peak, is more than three times as large. The main part of the 3 methyl C_7 peak, therefore, is probably due to ethyl side groups. This is in agreement with the sizes of the other 3-methylalkane peaks. It may be concluded that in high pressure polyethylene the short-chain branches are mainly ethyl and, for a smaller part, n-butyl groups, while also some n-pentyl groups may be present.

Van Schooten and Evenhuis (142, 143) concluded that the results obtained by pyrolysis-hydrogenation gas chromatography appear to be·in good agreement with those obtained by infrared and electron irradiation studies. They showed that the pyrograms of a linear polyethylene contained only very small peaks for branched and cyclic alkanes and very large n-alkane peaks. The largest peaks in the pyrogram are those for propane, n-hexane, n-heptane, n-decane and n-undecane, indicating important hydrogen exchange reactions followed by scission with the fifth (C_3 and $n-C_6$), ninth ($n-C_7$ and $n-C_{10}$) and thirteenth ($n-C_{11}$ and $n-C_{14}$) carbon atoms. Hydrogen transfer with the sixth carbon atom would account for the rather large n C_4 peak ($n-C_4$ and $n-C_7$), but this peak could also be due to intermolecular chain transfer reactions.

Pyrograms were also prepared for a range of polyethylene containing different amounts of short chain branching. (See Table 12 for details of samples). Previous work by high energy electron irradiation and mass spectrometry (144-146) has shown that the short branches in high pressure polyethylenes are mainly ethyl and n-butyl groups, but other short branches have also been supposed to be present (144).

The pyrograms obtained on these various polymers are shown in Figs. 11 and 12, ($n-C_6$ peak taken as reference). These pyrograms show marked differences which can be attributed to differences in short chain branching. The small amount of branching in Marlex 5003 and Ziegler polyethylenes is reflected only in the somewhat larger iso-alkane peaks, whereas the n-alkane pattern is practically the same as found for Marlex 50 (low branching). The Alkathene 2 and Lupolen H high pressure polyethylenes show, on the other hand, larger n-butane and n-pentane peaks, (Fig. 11). The iso-alkane peaks that show the

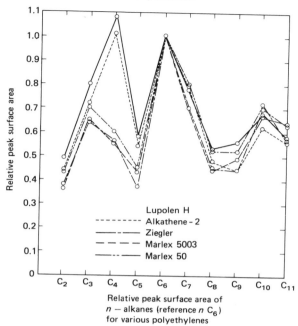

Relative peak surface area of
n — alkanes (reference *n* C$_6$)
for various polyethylenes

Fig. 11 Pyrolysis-gas chromatography carbon number
distribution (Relative peak surface areas of
n-alkanes, reference n C$_6$) of various
polyethylenes.

largest increase (for Alkathane 2 and Lupolen H) are the iso-pentane and 3
methyl alkane peaks (Fig. 12). The results in Fig. 11 and 12 clearly show
that the highest amount of branching is present in Lupolen H, the lowest in
Marlex 50. Assuming arbitrarily that these polymers, respectively, contain
24 and 1 short side chains/1000 carbon atoms and that the relative increase
of the n-butane and the iso-alkane peaks is linearly related with the amount
of branching, then the branching frequency of the other three samples can be
obtained by interpolation, (Table 12). These values are in good agreement
with those found in the literature. From the large increase in the n-butane
peak and the relatively small increase in the ethane peak it is concluded
that the two high pressure polyethylenes, (Alkathene 2 and Lupolen H),
contain mainly n-butyl side chains.

Eggertson and Tremoureux (147) have carried out a study of the effect of
heating rate on the composition of volatiles produced in the pyrolysis of
polyethylene. They state that in the technique of filament pyrolysis it is
customary to pyrolyze with a certain voltage to produce a specified maximum
temperature. It is presumed that precise setting of the voltage provides
sufficient control to reproduce pyrolysis conditions. However, in their
experience, it appears that the heating rate of the filament, rather than
its maximum temperature, is the primary factor governing reproducibility of
the product distribution. Poor control of the heating rate may well explain
the reported non-reproducibility of the filament, compared with the furnace
pyrolyzer. It is to be expected that variations in heating rate of the

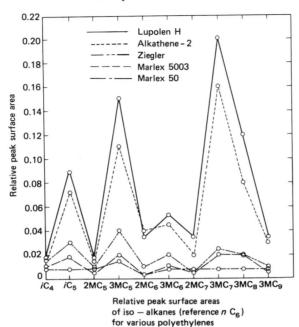

Relative peak surface areas
of iso — alkanes (reference n C_6)
for various polyethylenes

Fig. 12   Pyrolysis-gas chromatography carbon number
distribution (relative peak surface areas of
iso-alkanes, reference n C_6) of various
polyethylenes.

filament can result in different depths of cracking and different amounts of
secondary pyrolysis reactions.

The effect of heating rate on the pyrolysis patterns is illustrated by the
data given in Table 13, which were obtained by pyrolysis-hydrogenation-
gas chromatography of polyethylene. The patterns for the linear polyethylene are compared on the basis of C_1 plus C_2, and the non-normal hydrocarbons. The results are fairly consistent for heating rates of 70^oC per
second and lower. Thus, below a certain heating rate one does not need to
control the rate precisely. At 170^oC per second and above the amounts of
C_1 plus C_2 and non-normals are increased markedly.

The filament temperature programming technique has the following advantages:
(1) minimizes secondary pyrolysis reactions; (2) provides a means of dupli-
cating product distributions with different filaments; (3) offers a standard
method of thermal decomposition which is suitable for polymers of widely
varying thermal stabilities.

The use of known infrared assignments for polyethylene (148) as a model to
study chain branching (149-152) in reduced PVC has yielded variable results.

Quantitative estimates of the number of methyl groups are dependent upon the
extinction coefficient used (153). The method gives little information about
the nature of the branches in the polymer.

TABLE 12   Branching Frequency of Polyethylenes Estimated
           from Pyrogram

| | Branches/1000 carbon atoms (from peak surface area ratios) | | | |
Sample	n-butane	iso-pentane	3 methyl/pentane	3 methyl/heptane
Marlex 50 ex Phillips (very little branching -1 methyl group/1000 carbon atoms)	(1)	(1)	(1)	(1)
Marlex 5003 ex Phillips (polyethylene containing a little copolymerised butane-1)	0.5	4	2	2
Ziegler low pressure polyethylene (about 3-6 branches/1000 carbon atoms)	0.3	7	5	5
Alkathene 2 ex ICI high pressure polyethylene (20 -30 branches/1000 carbon atoms)	21	19	17	19
Lupolen H ex BASF high pressure polyethylene (20 -30 branches/1000 carbon atoms)	(24)	(24)	(24)	(24)

The γ-radiolysis work of Salovey (154), Harlan (155), and Kamath and Barlow (156) shows that low density polyethylene contains a significant amount of ethyl branches with lesser amounts of butyl branches. Salovey has indicated that ethyl branches may be more readily determined than butyl branches by this method, but concludes that even considering this difference, ethyl branches account for a large portion of the total branch content of low density polyethylene. ^{13}C NMR (164) and other techniques (165, 166) have been used for the determination of short chain branches in low density polyethylene.

Pyrolysis-hydrogenation-gas chromatography (157) has been demonstrated to be a particularly valuable technique for studying the branch structure of polyolefins (158, 159). Structural investigations (160, 161) on polyethylene by this technique have shown that methyl, ethyl and butyl branches can be distinguished and quantitative differences in the amounts of these branches can be determined in low density polyethylene, Ziegler polyethylene and Phillips polyethylene. Evidence that the products formed from the pyrolysis of polyethylene and ethylene-propylene copolymers agree with theoretically predicted products has also been published by Seeger, Exner, and Cantow (162, 163). Thus, a firm background relating pyrolysis products to chain branching in polyethylene has been established. Using pyrolysis-hydrogenation gas chromatography, Van Schooten concluded (159) that in low density polyethylene the short-chain branches are mainly ethyl with some butyl groups. In more recent work, Seeger and Barrall (161), using model copolymers as standards, determined the C_2:C_4 ratio to be about 2:1 in low density polyethylene.

Analysis of Plastics

TABLE 13   Effect of Filament Heating Rate on Pyrolysis Pattern

Sample	Heating Rate, °C/Sec.	Maximum Temp. °C	C_1 plus C_2	Area % (a) branched + cyclic $C_4 - C_{10}$	Ratio $1C_6/nC_6$
Filament					
Linear polyethylene	8	550	7	9	3
(Marlex 6009)	20	550	7	7	2
	20	800	7	5	-
	70	550	9	6	3
	170	650	25	8	6
	280	800	47	9	12
Furnace					
	-	550	12	13	8
	-	650	28	21	19
	-	800	52	83	73

			Area % MMA (b)
Filament			
Ethylene/Methyl-methacrylate	20	550	26
Copolymer	170	650	19
(24 %w MMA)	300	800	13
Furnace			
		550	7

(a)   Program-heated silicone column; peak areas normalized through C_{10}
(b)   Program-heated Carbowax column; peak areas normalized through C_{12}.

According to Wall et al (167) the pyrolysis of polyethylene proceeds by a radical chain mechanism. The products formed result from the process of random-chain cleavage, followed by intermolecular or intramolecular hydrogen abstraction. Hydrogen abstraction occurs preferentially at tertiary carbon atoms, and product formation results from homolysis of the carbon-carbon bond β to the radical site. The major products formed are the n-alkanes and the α , ω diolefins. The peaks between the triplets result from chain branching.

The effect of chain branching on pyrolysis of polyethylene may be viewed simply as promoting cleavage of the main polymer chain at carbon atoms α and β to the branch site (168-170). Cleavage of the branch from the main chain becomes more important with increasing branch length. The elegant work of Seeger, Exner and Cantow (169-170) with labelled ethylene-propylene copolymers on the relative amounts of α and β cleavage compared to statistical chain cleavage provides a foundation for discussion of the products arising from these reactions. They concluded that the probability of backbone cleavage at the branch sites is much greater than statistical chain cleavage and that α and β cleavage occur with equal frequency.

Since there are a number of different olefin products possible from cleavage
at a branch site, the determination of the type of chain branch is best
accomplished by pyrolysis hydrogenation-gas chromatography. On-line cataly-
tic hydrogenation of the pyrolysis products in the injection port of the
chromatograph converts all of the olefins to saturated hydrocarbons, therefore
reducing the number of possible products. Cleavage of the carbon-carbon bond
in the polymer backbone at the branch site (α cleavage) results in the formation of n-alkanes. Backbone cleavage at the bond β to the branch site
gives methyl alkanes. In the case of methyl branch sited, β cleavage gives
2-methylalkanes, and chain cleavage β to ethyl and butyl branches gives 3-
methyl- and 5-methylalkanes, respectively as discussed below (171). Satisti-
cal cleavage will produce a mixture of branched alkanes.

Additional proof of the above has been provided from studies on model ethylene-
1 butene copolymers and ethylene-1-hexene copolymers (172). Pyrolysis hydrogenation of these model polymers shows that the areas of the 3-methylalkanes
are increased relative to the areas of the corresponding 2-methylalkanes in
the programs of ethylene-1-butene copolymers, and that 5-methylalkanes are
indicative of butyl branching (171-172). Other indications of C_4 branches
are given by an increase in the relative amounts of 2-methylhexane (2-MC_6)
and 3-methylheptane (3-MC_7) and an increase in the relative amount of the
n-C_4 peak. However, the areas of 2-MC_6 and 3-MC_7 are equal in the case of a
C_4 branch, while for ethyl branches, the area of 3-MC_7 is enhanced relative
to 2-MC_6. Thus not only must individual peaks be interpreted for the determination of the type of branch, but the overall pyrolysis pattern must be
considered as well.

Seeger and Barrall (173) have described a system with flash pyrolysis and gas
chromatography including in-line hydrogenation which they applied to an
analysis of chain branching in polyethylene (density 0.916 to 0.962), high
density polyethylene and ethylene-butene-1 and ethylene-propylene copolymers.
A mass spectrometer was coupled to the outlet of the gas chromatographic
column. This system has been described by Michajlov et al (174, 175) and
Seeger et al (176). Pyrograms obtained on this equipment have proven to be
quite reproducible and this permitted the serious intercomparison studies
which are necessary for the determination of ethylene sequence distribution
in copolymers with propylene. Investigation of various degradation mechanisms give a more detailed picture of hydrocarbon polymer degradation.
These are: (a) primary random scission followed by hydrogen transfer stabilization and (b) decomposition by an intramolecular cyclization process via
transfer of radicals from the first to the fifth carbon followed by a chain
cleavage at the β -position. Importantly, an increased probability of
scission at the α - and β - position to a tertiary carbon atom was also observed by Seeger and Barrall (173) as contrasted to a linear polyethylene chain
(177). In copolymers of low propylene content it was possible to evaluate
with reasonable precision the number of tertiary atoms from the yield of
branched fragments (methylalkanes).

Seeger and Barrall (173) found that, for low concentration of branches, yield
and distribution of the isoalkanes is very sensitive to the type of polymerization used to form the polyethylene. A great number of the single-branched
isoalkanes were identified by comparing their retention time with the scale
of respective boiling points and by pyrolyzing well characterized ethylene-
propylene copolymers. The pyrograms of the ethylene-propylene copolymers
demonstrated that a C-C bond in an α - or β - position to a tertiary carbon
cleaves about two times more readily than a C-C bond in a linear chain. It
was further found that the branch length of the side chain is indicated by
the identity of the fragments produced, and the probability of formation of

these fragments is directly related to the possibility of a cyclic decomposition mechanism. From the high yields of 3- and 5- methyl-alkanes it was
possible to detect about 20 ethyl and 10 n-butyl branches per thousand in a
low density polyethylene. Figure 13 illustrates the excellent pyrograms
obtained by this technique for polyethylenes.

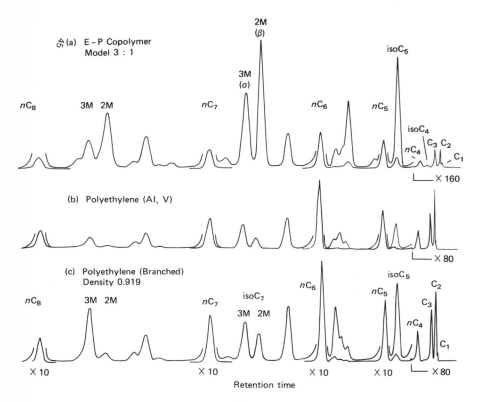

Fig. 13   Pyrograms of polyethylenes and copolymers.
          Column, 33 ft; program rate, 1°C/min;
          fragments C_1-C_8.

Seeger et al (178) have also applied their flash pyrolysis (at 600-800°C)
gas chromatography system with in-line hydrogenation of the fragments to
ultra thin films of polyethylene produced by plasma polymerization obtained
on a radio frequency glow discharge (180). Gas chromatographic separation
was carried out using a packed column (179).

Ahlstrom et al (181) applied the techniques of pyrolysis-gas chromatography
and pyrolysis-hydrogenation-gas chromatography to the determination of
branches in low density polyethylene. Their data indicate that high density
polyethylene contains mainly methyl branches. The relative areas of the
2-MC_6 and 2-MC_7 peaks in low density polyethylene were smaller than those
found for high density polyethylene. Large increases in the size of the 3-
methylalkane peaks in low density polyethylene indicates the presence of a
substantial amount of ethyl branches. This can be seen most clearly in Figs.

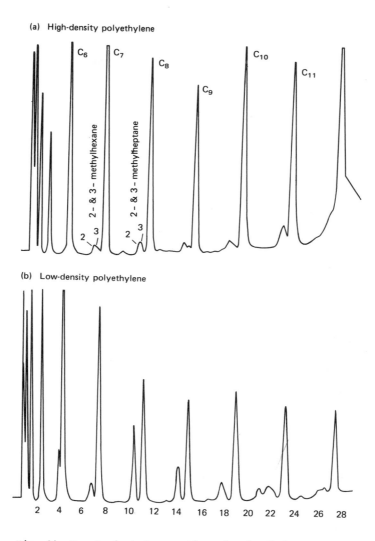

Fig. 14   Pyrolysis hydrogenation of polyethylenes
          (a) high density polyethylenes
          (b) low-density polyethylene
          Durapak column fragments C_1-C_{11}.

14 and 15 where the area of the 3-MC_7 peak is about two-thirds that of the
n-C^8 peak..

In related work, Michajlov, Zugenmaier and Cantow (179), observed by pyrolysis-hydrogenation-gas chromatography that beyond C_{10} the peak maxima differ
for low and high density polyethylene. They found that for low density
polyethylene peak maxima occur at C_{14} and C_{18} while for high density poly-

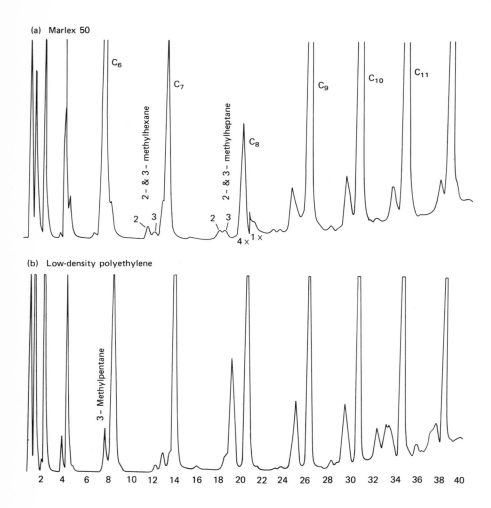

Fig. 15 Pyrolysis hydrogenation of Marlex 50 and LDPE;
(a) Marlex 50; (b) low-density polyethylene.
Durapak column fragments C_1-C_{11}.

ethylene peak maxima occur at C_{15}, C_{20-22} and beyond. Although Ahlstrom et al (181) did not extend their pyrograms beyond C_{20} their work also showed the C_{14}, C_{15} difference in peak maxima for these polymers. This difference appears to be characteristic and may be explained by an extension of the work of Tsuchiya and Sumi (182, 183).

From an examination of the products formed from the pyrolysis of high density polyethylene and polymethylene, Tsuchiya and Sumi (182, 183) proposed that the major hydrogen-abstraction reaction is due to an intramolecular cyclization. They proposed that, following initial radical formation at C_1,

successive intramolecular hydrogen abstractions occur along the chain resulting in the formation of new radicals at C_5, C_9 and C_{13} as shown in eqs.
(1) - (4).

$$R -- R \longrightarrow 2R. \tag{1}$$

$$CH_3 -(CH_2)_3 -CH-CH_2-R \longrightarrow \tag{(2)}$$

$$CH_3-(CH_2)_6-CH_2-CH-CH_2-R \tag{(3)}$$

$$CH_3-(CH_2)_6-CH_2-CH-CH_2-R \longrightarrow$$

$$CH_3-(CH_2)_{11}-CH-CH_2-R \tag{(4)}$$

Cleavage of the carbon-carbon bonds β to these macroradicals results in the
formation of increased amounts of C_6, C_{10} and C_{14} α -olefins and C_3, C_7,
and C_{11} n-alkanes over that which could be predicted from statistical-chain
cleavage.

Ahlstrom et al (181) propose that cyclic intramolecular hydrogen abstraction
should proceed further along the chain for less-branched high density polyethylene than for the more highly branched low density polyethylene. Thus,
at longer chain lengths, the relative contribution of intramolecular hydrogen abstraction to the products formed as a result of this reaction should
be greater for high density polyethylene than for low density polyethylene.

A careful examination of the data on pyrolysis in helium shows that this
difference in C_{14}, C_{15} peak maxima is due to a relatively greater increase in
the amount of C_{15} n-alkane formed from high density polyethylene than to an
increase in C_{15} α - olefin concentration. They showed that for both high
density polyethylene and low density polyethylene the C_{14} α - olefin is the
favoured product from decomposition of the C_{13} macroradical. However, due
to a relatively greater contribution of products from C_{17} marcoradical in
high density polyethylene the increase in the amount of C_{15} n-alkane and
C_{18} α - olefin formed is favoured relative to that in low density polyethylene.

From this these workers conclude that the cyclic intramolecular hydrogen
abstraction is more favourable at longer chain lengths for high density polyethylene and this explains the observed differences in peak maxima in the
pyrolysis of low density polyethylene and high density polyethylene.

Infrared Dichroism and X-Ray Diffraction

Infrared dichroism, (184-186) x-ray diffraction (186, 187, 191), and density

(192) and optical birefringence (186-188) have all been used as qualitative measures of orientation in polyethylene and other polymer films. In polymers which are partly crystalline, some of the methods suffer from certain defects. The x-ray method is capable only of giving information about crystalline portions of the film and measures the true orientation distribution. The measurement of optical birefringence compounds orientation effects in crystalline and amorphous areas. It is furthermore, sensitive to form birefringence (189). Infrared methods measure only an average so that any distribution will appear either circular (i.e. indistinguishable from unorientated) or elliptical even though the true distribution is lobed.

The percentage of crystallinity and the distribution of orientation in crystalline and amorphous regions are two of the basic properties of polyethylene films. It is highly desirable to have available rapid, accurate methods for measuring these. Infrared dichroism is particularly suitable for these measurements. Tobin and Carrano (190) have shown, such methods are readily set up, and form the basis for rapid, reliable determinations.

It is obvious that a highly orientated system will show larger dichroic ratios than a poorly oriented system. The question arises whether the orientation distribution may be deduced from measurements of dichroism. That is, one would like to know what fraction of the total number of dipole moment change vectors lies within a given solid angle, defined relative to axes fixed in a film sample. This cannot be determined from measurements made solely with the film plane perpendicular to the beam of incident radiation. However, by making certain assumptions, in particular one as to the form of the distribution function, one can determine the distribution of the projections of these vectors in the plane of the film. If additional measurements could also somehow be made edge-on, then the complete distribution could be determined. In view of these facts, the best approach is to treat the two-dimensional problem of orientation distribution, since this makes use of all the information available from practicable measurements.

Stein and coworkers (194) employed rheo-optical techniques of dynamic x-ray diffraction and birefringence in studies of orientation rates in low density polyethylene and of the nature of the α mechanical loss mechanism of polyethylene (195). The effects of stretching highly oriented films of polyethylene along the c-axis were studied by Ginzburg et al (196). Small-angle scattering experiments on extruded polyethylene films have aided in calculating local strains in spherulite regions (198). Roe and Gieniewski (199) have used small-angle x-ray diffraction to study scattering power and mass density as functions of degree of crystallinity in linear polyethylene.

X-ray studies of ethylene copolymers including scattering at varying degrees of annealing of polyethylene ionomers have been discussed by Marx and Cooper (200).

X-ray diffraction and electron diffraction have been used by Markova et al (201) to study short-range order in amorphous and crystalline polymers. The heating of a high density polyethylene melt was accompanied by a change of macromolecular packing from orthorhombic symmetry in the crystalline polymer to hexagonal packing in the melt.

Stein and coworkers (202) combined birefringence measurements with x-ray diffraction in studies of orientation in low density polyethylene permitting separation of the former into parts due to amorphous from those due to crystalline orientation.

Applications of small-angle x-ray scattering to semicrystalline polymers were described by Brown and coworkers (203) including types of distribution functions for layer thicknesses from polyethylene and poly(methylene oxides).

Takaynagi (204) has reported on x-ray diffraction results on molecular motion in polyethylene crystals.

Studies of ionomers by Marx and coworkers (205) revealed that the wide angle ionomer x-ray peak is a measure of the average distance between ion scattering sites. Small-angle scattering resulted from crystalline lamellae in ethylene ionomers.

Schultz and Long (206) described an apparatus for rapid energy scanning of small-angle x-ray scattering of polyethylene. The degree of crystallinity and the state of order in polyethylene was determined from wide-angle x-ray diffraction patterns.

Das Gupta and Noon (207) used high-field x-ray diffraction and infrared studies at low temperatures to show that the nucleation of small crystalline regions in polyethylene occurred owing to the removal of defects from the paracrystal line boundary.

Rybnikar (208) determined the crystallinity of branched polyethylene from x-ray diffraction patterns in which the diffraction curve corresponding to the amorphous fraction was obtained by linear extrapolation of the scattering intensity of the melt as a function of the temperature.

Shimamura and coworkers (209) measured the intensities of small-angle x-ray scatterings from hot drawn high density polyethylene. Low angle and wide-angle x-ray diffraction measurements were made on solution crystallized high density polyethylene.

Pae and coworkers (211) found that polyethylene with an initial chain folded morphology and orthorhombic crystal structure did not transform to any other phase prior to melting; however, a reversible orthorhombic and hexagonal phase transition is possible in an extended chain morphology formed on crystallization at high pressures. Small angle x-ray scattering and Raman characterization of the lamellar structure in solution crystallized polyethylene indicated that the lamellae consisted of a crystal centre and disoriented and dis-ordered surface layers of thickness approximately 12 Å (212).

A systematic error was observed by McRae and Maddams (213) for x-ray diffraction and transmission measurements on blends of polyethylene with amorphous atactic polystyrene unless the background incoherent scattering from the amorphous component over a wide Bragg angle range was defined.

Baczek (214) described small angle x-ray scattering studies of the deformation of polyethylene.

The use of high pressure x-ray diffraction for the study of linear polyethylene has been discussed by Rozkuszka (215).

Thermal Analysis Methods

Clampitt (216) has utilized differential thermal analysis to examine annealed samples of linear high pressure polyethylene blends. He showed that this

polymer could be resolved into three peaks. These peaks came at 115, 124,
and 134°C. The 115°C peak was associated with the high pressure polyethylene, whereas the 134°C peak was shown to be proportional to the linear
content of the system. Clampitt (217) also applied differential thermal
analysis to a study of the 124°C peak which he describes as the cocrystal
peak. His results appear to indicate that there are two classes of co-
crystals in linear-high pressure polyethylene blends with the linear component being responsible for the division of the blends into two groups. The
property of the linear component which is responsible for this division is
related to the crystallite size of the pure linear crystal.

A typical DTA curve of a linear-high-pressure polyethylene blend is shown
in Fig. 16. The peak is the cocrystal peak marked 2.

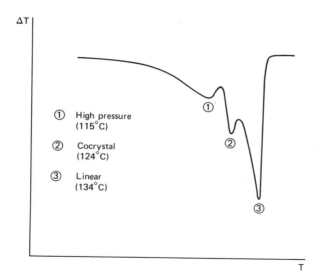

Fig. 16   Typical thermogram of a linear-high pressure
          polyethylene blend.

Holden (218) has used differential thermal analysis to carry out a study of
the effects of thermal history on polyethylene.

Techniques and apparatus for measuring oxidative stability of polyolefins
have been developed and simultaneous testing of a multiplicity of samples
is possible (219).

Bosch (220) has used differential analysis to identify polyethylene.

Evidence has been presented of the existence of morphologically different
crystal structures in polyethylene (221-223).

Differential scanning calorimetry has been used to monitor the cross linking
rate of polyethylene-carbon systems (224).

Spencer (225) used differential scanning calorimetry for the identification and quantitative analysis of multilayer thermoplastic packaging films containing nitrocellulose, polyethylene, cellophane.

Differential thermal analysis has been used to show the existence of morphologically different structures of polyethylene ionomers (226).

Schmitt (227) used mass spectrometry to evaluate gaseous products formed in thermal degradation of polyethylene and other polymers.

Fractionation

The time-consuming and laborious nature of fractionation procedures is illustrated well by the work of Nakajima (228) on the fractionation of polyethylene and its thermally degraded products involving extraction with boiling hydrocarbons with increasing boiling points between 45°C and 95°C and on the fractionation of polypropylene (229). It was necessary to extract the polymer using a Soxhlet apparatus with 17 different hydrocarbon fractions based on normal paraffins with different boiling temperatures in the range from 35°C to 135°C. The extreme laboriousness of such procedures is self evident.

Fractional extractions of polymers by the column technique is no less laborious. Two types of column extraction procedure are known. Gradient elution fractionation is achieved at a given temperature by making use of solvents with gradually increasing solvent power, or the increasing temperature fractionation is performed with a given solvent at increasing temperatures. According to the findings on column techniques by Wijga et al (230) the gradient elution method at a temperature (150°C) sufficiently near the melting point of the polymer separates fractions only according to molecular weight, whereas the increasing temperature method fractionates the polymer mainly according to tacticity. On the other hand, in the fractional extraction in a vapour-jacketed Soxhlet apparatus with boiling solvents with increasing boiling points, separation is considered to be conducted mainly by tacticity.

Kenyon et al (238) have described methods for the large scale elution fractionation of polyethylene. 500 g samples of polymer can be fractionated in 20 hours.

Kolke and Billmeyer (239) devised an analytical procedure for the solvent-gradient elution fractionation of polyethylene and applied it to both linear and branched polyethylenes. Number-average molecular weights derived from fractionation data are in good agreement with those measured directly.

Williamson and Cervenka (240) have studied gel permeation chromatographic separations of low and high density polyethylene and confirmed the Benoit-type universal calibration. Nikajima (241) has reported on a IUPAC sponsored study of linear polyethylene which involved gel permeation chromatography, osmotic pressure, infrared, melt, viscosity, and intrinsic viscosity measurements; average molecular weight on two samples were M_n = 10500-11000, M_w = 150000-165000 and M_n = 13600-18500, M_w = 40000-48000, respectively. Fractionations of high pressure polyethylene have been reported by El'darov et al (242) and Lovric (243, 244).

Peyrouset et al (245) fractionated linear polyethylene to a narrow molecular weight distribution.

Akutin (246) has shown that by a thermal precipitation method that the molecular weight distribution of low density polyethylene could be deduced from the precipitation curves using simple calculations.

Gianotti et al (247) have used preparative gel permeation chromatography to
obtain narrow molecular weight range fractions of low density polyethylene
and used these fractions to evaluate long chain branching in these polymers
by viscometric and light scattering measurements.

Molecular Weight Distribution

Taylor and Tung (249) have developed a rapid turbidimetric technique for the
determination of the molecular weight distribution of polyethylene which in a
way is similar to the thermal gradient preparative method in that it employs
polymer solubility as a function of temperature. A solution of solvent (α
chloronapthalene) and nonsolvent which contains a very low concentration of
polymer is slowly cooled. The high molecular weight species become insoluble
and separate out causing a small amount of turbidity. As the temperature
continues to decrease, increasing amounts of polymer are precipitated out
according to their molecular weight. Finally, a point is reached at which
even the lowest molecular weight species become insoluble in the solution.
At this point the turbidity is greatest, and ideally all of the polymer is
precipitated but remains in suspension as very fine particles. If the increase in turbidity is plotted against the decreasing temperature, a cumulative
plot is obtained which is similar to a cumulative weight per cent versus
molecular weight. The increase in turbidity is related to the cumulative
weight per cent and the molecular weight is related to the decrease in temperature

Fig. 17 shows a comparison of typical results from a stirred cell and a divided cell. The ordinate in this is in terms of Δ E, the decrease in millivolts in phototube output from the starting value.

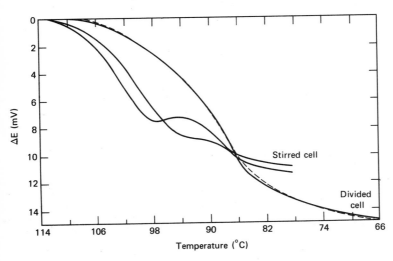

Fig. 17   Comparison of turbidity, temperature curves
from stirred and divided cells.

Taylor and Tung (249) discuss the application to their results to procedures described by Morey and Tamblyn (250) and Claesson (251) to convert the experimental turbidity into molecular weight distribution data and the difficulties they encountered in this work.

Gamble et al (253) used a photoelectric turbidimeter (Fig. 18) for measuring molecular weight distribution of poly- α-olefins and ethylene-propylene copolymers. This instrument measures changes in turbidity as a function of temperature (Fig. 19).

Fig. 18   Photoelectric turbidimeter.

A parameter designated as S was chosen by Gamble et al (253) to be the difference in temperature between points representing 20% of the maximum turbidity and 50% of the maximum turbidity as suggested by Taylor and Tung (249). This portion of the curve is essentially linear.

For measuring polydispersity the parameter S is correlated with a parameter determined from the Wesslau equation (254).

Because of the fairly narrow working temperature range, 80-25°C, of the solvent-non-solvent mixture used (heptane-n-propanol) it was necessary to study the effect of polymer concentration to provide a suitable, turbid system. Curves in Fig. 20 demonstrate the effect of changing the concentration of ethylene-propylene rubber without changing the concentration of the non-solvent.

Chiang (255) has investigated on the intrinsic viscosity-molecular weight relationship for polyethylene.

Schreiber and Walman (256) studied the effect of temperature on molecular weight measurements in polyethylene.

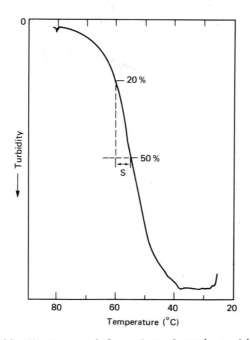

Fig. 19 Chart record from photoelectric turbidimeter
        X-Y recorder, Mw/Mn of polymer = 8.0.

De la Cuesta and Billmeyer (257) in their studies of the molecular structure
of polyethylene have reported on the measurement of intrinsic viscosities of
polyethylene solutions.

Das N Palit (258) have carried out viscosity studies of polyethylene in a
    solvent mixture composed of two non-solvents.

Chien has investigated the molecular weight distribution of polyolefins (259).

The National Bureau of Standards supply standard polyethylenes suitable for
molecular weight calibration purposes (260).

Maley (261) and Starck (263), respectively have applied gel permeation
chromatography and size exclusion chromatography to the determination of the
molecular weight distribution of polyethylene.

Platonov et al (262) measured the molecular weight distribution and branching
of polyethylene using rapid sedimentation and viscometry.

Gel permeation chromatography in combination with viscometry and light scattering have found application in the study of branching in polyethylene (263).

Viscometric determination of the molecular weight of polyethylene have been
reviewed by Lanikova et al (264).

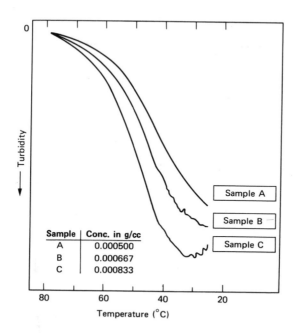

Fig. 20 Effect of polymer concentration on turbidimetry.
Ethylene-propylene rubber.

Ross and Shank (265) used infrared detectors in molecular weight measurements
on polyethylene fractions with a precision better than ± 2.5% by gel permeation chromatography.

Autoradiography

Autoradiography following neutron irradiation is a useful technique for
assessing the distribution of impurities such as chlorine, sodium, titanium
and aluminium remaining as catalyst remnants in high density polyethylene
and in polypropylene thin sections or film.

Irradiation of plastic films with thermal neutrons in a nuclear reactor
yields radioisotopes by the following reactions:-

	Half life	Cross-section
Al^{27} (n.γ) Al^{28}	2.3 minute	0.21 Barns
Cl^{37} (n.γ) Cl^{38}	37 minutes	0.14 Barns
Na^{23} (n.γ) Na^{24}	15 hours	0.54 Barns

These radioisotopes emit both β and γ - rays and can be detected by
autoradiography. The cross sections of other trace elements likely to be
present in the polymer, (Ti, K, Fe), are too small to produce significant

quantities of radioisotopes during short (5 minute) irradiations.

To study the distribution of chlorine in the films samples were irradiated in
a similar way for a longer period (30 minutes) and was autoradiographed after
allowing the aluminium to decay. Figure 21 is an illustration of the type of
results achievable by this technique. It should be noted that the 5 minute
irradiation activated the shorter-lived elements (Al, Cl) and would not
induce significant sodium activity in samples of low sodium content

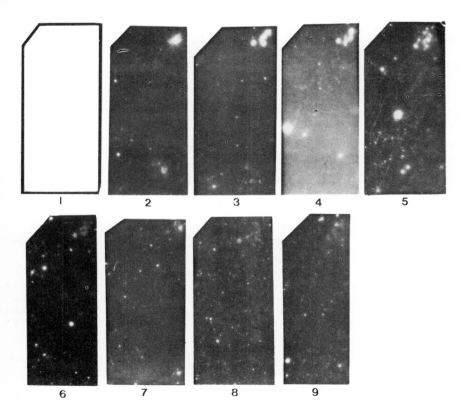

Fig. 21 Autoradiographs, metals residues in polyolefins.

Miscellaneous

Molecular relaxation processes in polyethylene have been studied by radio-
thermoluminescence (266).

Looyenga (267) has discussed the relation between refraction index and density of polymer solutions, including specific refraction data on polyethylene
and polystyrene.

Stein (268) reported on rheooptical and dielectric studies of the α -loss
region of low density polyethylene, relating observations to slipping of
crystalline lamellae.

Kavak Chuck et al (269) have shown that gamma radiation (sterilization dosages and higher) of high pressure polyethylene gave predominantly ketones at dosages equal to or less than 40 Mrad, and alcohols, ketones and acids at dosages 40 to 100 Mrad.

Crewthen et al (270) have carried out a light scattering study of linear polyethylenes.

Olsen and Osteraas (271) examined sulphanated polyethylene surfaces using frustrated multiple internal reflection spectroscopy.

Thermoluminescence studies have been reported for polyethylene (272).

Onagi (273) has reviewed rheooptical studies by infrared dichroism on polyolefins and monoolefin block copolymers.

The random coil configuration and the radius of gyration of polyethylene, employing hydrogen and deuterium-tagged polymers has been determined by Schelten and coworkers (274).

Bergmann (275) has reported studies of molecular motions in partially crystalline polymers such as polyethylene.

Ballard et al (276) examined the chain configuration of pressure crystallized polyethylene using neutron scattering and interpreted the results in terms of a chain folding mechanism in which a molecule was bounded by the surfaces of a lammellar block.

Light scattering and viscosity have been used to characterize the long chain branching of polyethylene in α-chloronapthalene (277).

Electron scanning chemical analysis has been found to be suitable to establish a comprehensive picture of the early stages of the fluorination process of polyethylene (278, 279).

King et al (280) discussed small angle neutron scattering on poly-(deutero-ethylene) dissolved in polyethylene.

The factors governing molecular aggregation in polyethylene-d_4-polyethylene blends have been studied by small angle neutron scattering (281).

Picot et al (282) discussed neutron small angle scattering experiments on hot-stretched and annealed tagged polyethylene films in terms of coil deformation.

The ultraviolet absorption spectrum of polyethylene has been discussed by McCubbin and Weeks (283) who commented on the appearance of a small peak at about 200 nm which appears to be intrinsic to the polymer and not due to the presence of impurities.

Studies of the morphology of linear polyethylene carried out by Anderson (284) have revealed the occurrence of a fractionation of molecular weights during crystallization or annealing. In isothermally bulk-crystallized samples, the different molecular weight species form different types of lamellae.

Schreiber and Bagley (285) have reported a theoretical treatment on the use of the Newtonian Melt Viscosity at 190°C of polyethylene as an index of long chain branching. They developed the relationship between Newtonian melt

viscosity η_0 and the weight- average molecular weight, M_w for polyethylenes
by use of melt viscosities obtained from capillary viscometry and molecular
weights from intrinsic viscosity and bulk flow measurements in addition to
light-scattering determinations. The work of Schreiber and Bagley (285)
indicates the necessity of using polyethylenes which have been stabilized
against the effects of thermally induced changes if meaningful values of
Newtonian viscosity are to be obtained.

The mechanism of formation of internal double bonds in polyethylene
following irradiation by fission of two hydrogen atoms from contiguos
carbon atoms has been established by Slovokhotova et al (286).

Bursfield (287) has used 14C labelled triethylaluminium cocatalyst in his
measurements of the number of active centres in the Ziegler-Natta polymerization.

Braun and Guilett (288) have used inverse gas chromatography to determine
the crystallinity of low and high polyethylene waxes.

Kato (134) studied the formation of carbonyl functions on polyethylene film
treated with chromic acid. Changes in amount of hydrazones formed following
reaction with 2,4-dinitrophenylhydrazine were followed by ultraviolet spectra
and compared to changes in wettability of the film with water.

The effect of electrical discharge in the atmosphere on structural changes in
polyethylene films have been studied by Guseinov and coworkers (135). Ozonides were formed down to a 12- μm depth and carbonyl groups down to 4 μm.
Some low molecular weight compounds containing $RONO_2$ groups formed on the
film surface.

LePoidevin (136) showed that the oxidation of stabilized polyethylene in
aqueous solution occurred in four steps; the first involving the formation
of aldehydes and/or acids followed by a decarbonylation-decarboxylation
reaction; the second step involved ester formation and was confined to
short polymer chains; the final step involved the rapid formation of acids
and ketones.

Viscometric measurements have been made on polyethylene in a cosolvent
mixture such as xylene-carbon disulphide (128) in studies of photooxidative
degradation in the presence and absence of stabilizers (129).

Miscellaneous analytical methods have been described for automatic determination of oxygen absorption (130) in polyethylene.

The depolarization thermocurrent has shown promise as a useful tool for
examination of transitions and relaxations in many polymers. It has been
used for study of radiation-induced oxidative degradation of polyethylene
(131).

Determination of Water

Conventional weight loss methods for determining water in polymers have
several disadvantages not the least of which is that any other volatile
constituents of the polymer such as solvents, dissolved gases, volatile
substances produced by decomposition of the polymer or additives therein
are included in the determination. The method described below has the
advantage that it is absolutely specific for water. In addition, by contr-
olling the temperature of the sample, even in the extreme by temperature

programming the sample, information can be obtained regarding the rate of release of water from the polymer and its dependence on sample temperature:

Semi automated Karl Fischer titration method for determination of water in polymers.

The instrument operates by means of a standard Karl Fischer dead-stop indicator circuit, with a moving coil relay replacing the microammeter. The moving coil relay operates as a switch, in conjunction with a second relay, to deactivate (at the end-point) and to activate (when water is present) an automatic titrant dispenser, which dispenses a given volume of titrant into the titrand each time it is activated, (Figs. 22 & 23 and Table 14).

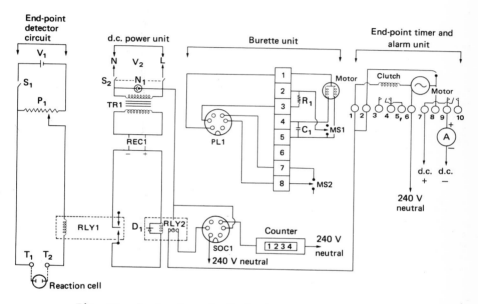

Fig. 22   Semi-automatic Karl Fischer water apparatus, Burette valve construction.

Traces of water in polyethylene have been determined by devolatilizing the polymer followed by gas chromatography (289).

Determination of Volatiles

A method involving heating of the sample followed by gas chromatography has been described for the determination of lower alkanes and aromatic hydrocarbons in polyethylene (290, 291). Toluene and ethyl acetate have been determined by a procedure involving vacuum treatment of the polymer followed by gas chromatography. Alkanes have been determined by devolatilizing the polymer followed by gas chromatography (289).

Analysis of Plastics

TABLE 14 Components List
(Instrument assembly by ICAM Ltd. Northop,
Mold, Flintshire)

Item	Description	Manufacturer
S1	Switch SPDT)	
	)	
S2	Switch SPDT)	
	)	
T1 and T2	Insulated terminals)	
	)	
N1	Panel neon clear 240V)	Radiospares Limited
	)	P.O. Box 2BH,
TR1	Transformer Hygrade 240V)	4-8 Maple Street,
	50 cycles 2 x 6.3V)	London, W.1.
	)	
REC1	Rectifier Rec 20)	
	)	
RLY2	Relay Type 1 12V d.c.)	
	120 ohms)	
	)	
D1	Diode 10 DE Type REC50A)	
P1	Potentiometer Model A 10 turn	Beckman, Glenrothes,
	500 ohms and Duo-Dial Model RB	Scotland.
RLY1	S170 d.c. relay make at 90	Sangamo Weston
	micro-amps resistance 3300	Enfield, Middlesex.
	ohms Spec S170/1/457	
PL1 and SOC1	6-pin plug and socket Part No.	A.F. Bulgin, Barking,
	P194	Essex.
MS1 and MS2	Microswitch Type HA1	Crouzet Ltd.,
		Brentford, Middlesex.
V1	1.5V d.c. battery	
V2	240V a.c. supply	
A	Audible alarm Bleeptone	A.P. Beeson Ltd.,
	12V d.c.	Hove, Sussex.
Burette Unit	Fisons automatic dispenser	Fisons Ltd.,
	includes R1 4700 ohm resist-	Loughborough.
	ance and C1 0.4 uF condenser	
PTFE valve block		ICAM Ltd. Northop,
		Mold, Flintshire.
Counter	Reset vending counter Part No.	Veeder-Root, Croyden
	KK1441	
Timer	Chronoset CF 0-36 minutes	Technical Representations
	direct clutch model	Ltd. Stockport, Cheshire.
Case	Type DA 40168	Bedco Ltd., Harpenden,
		Herts.

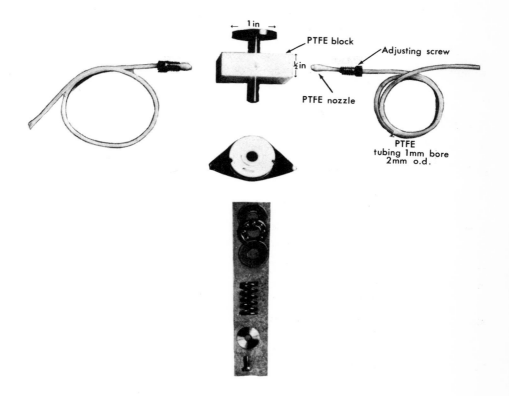

Fig. 23 Semi-automatic Karl Fischer water apparatus,
Burette valve construction.

Additives

Antioxidants in polyethylene (see also polypropylene)

Gas Chromatography

Antioxidants; uv absorbers, lubricants, antistatic agents optical brighteners	solvent extraction, glc	292,293
2,6 di-t-butyl-4-methylphenol	solvent extraction, glc	294
butylated hydroxy anisole	solvent extraction, glc	295
butylated hydroxy toluene	solvent extraction, glc	295
2,6-butyl-4-methylphenol 2,6 di-t-butyl-4-methylphenol p-t-butylphenol	solvent extraction, glc	296

Topanol CA (1,1 3 tris(2-methyl-4
hydroxy-5-t butylphenyl) butane)
Polygard, (tris(nonyl phenyl) phosphite)
Inganox 1010 (pentaerythritol tetra-3(3,5-
di-t-butyl-4-hydroxyphenyl propionate)
Inganox 1076 (Octadecyl 3(3,5 di-t-butyl-
4-hydroxyphenyl) propionate)
Santonox R (4,4 thiobis (6-t-butyl-m-
cresol))
Annulex PBA15 (1,1, di (3-t-butyl-4-
hydroxy-6-methylphenyl) butane)
Nonox DCP (2,2, bis (3-methyl-4-hydroxy-
phenyl) - (propane)
Nonox-WSP (bis (2-hydroxy-5-methyl-3-
(methyl-cyclohexyl) phenyl) methane)
Ionox 330 (1,3,5 trimethyl-2,4,6-tris
(3,5 di-t-butyl-4 hydroxybenzyl) benzene)

Visible Spectrophotometric methods

Bis-(3,5-di-t-butyl-
4-hydroxy phenol) methane

Ionol (2,6 di-tert-
butyl-p-cresol)

phenolic antioxidants
including dicresylol
propane and Santonox R
(4-4, -thio-bis-(-3-
napthyl-6-tert butyl-
phenol)
phenolic antioxidants
including Succonox 18,
butylated hydroxy toluene,
Ionol, (2,6-di-tert butyl
-p-cresol), Nonox CI
(N-N, di-βlnapthyl-p-
phenylenediamine,
Santonox R (4,4 thio-bis-
3 methyl-6-tert-butylphenol)
phenolic antioxidants
including Agerite Alba
(hydroquinone monobenzyl
ether), Agerite Spar
(styrenated phenol),
Agerite Superlite (polyalkyl
polyphenol)
Antioxidant 425, i.e.
(2,2' methylene-bis (6-
tert-butyl-4 methyl phenol)),
Antioxidant 2246, i.e.
(2,2'-methylene-bis (6-
tertlbutyl-4-methyl phenol)),
Deenax, (2,6, Di-tert-butyl-
p-cresol),
Ionol, (2,6-di-tert-butyl-
p-cresol),
1-Naphthol,
2-Naphthol,
Naugawhite, (alkylated phenol),
Nevastain A, (not disclosed),
Nevastain B, (not disclosed),
Nonyl phenol
p-Phenyl phenol,

Polygard, (Tris (nonylated
phenyl)-phosphite)
Santovar A, (2,5-Di-tert-
amyl-hydroquinone)
Santovar O, (2,5-Di-tert-

finish using αα'
diphenyl-β- picryl
-hydrazyl

solvent extraction 312
oxidation of alka-
line solution and
spectrophotometric
evaluation
cyclohexane extrac- 313
tion, visible spectro-
photometric finish
(low levels)
toluene extraction, 314
coupling with
sulphanilic acid
spectrophotometric
finish
(low levels)
solvent extraction, 315
reaction with
ethanolic $FeCl_3$ to
produce Fe^{2+} which is
estimated spectrophoto-
metrically using 2-2'
dipyridyl (low levels)

methanol or ethanol 316
extraction coupling
with diazolized p-
nitroanaline- spectro-
photometric finish

316
continued

butyl-hydroquinone)
Santowhite crystals, (4,4'-Thio-bis-
(6-tert-butyl-2-methylphenol))
Santowhite MK (reaction product of
6-tert-butyl-m-cresol and SCl_2)
Santowhite powder, (4,4'-Butylidene-
bis (3-methyl-6-tert-butylphenol))
Solux, (N-p-Hydroxyphenyl-morpholine)
Stabilite white powder, (not disclosed)
Styphen 1, (Styrenated phenol)
Wingstay S, (Styrenated phenol)
Wingstay T, (a hindered phenol)
amine antioxidants e.g.   solvent extraction   315
Nonox CI (N,N' di-2-napthyl-p-   spectrophotometric
phenylene diamine)   finish using dipyridyl
Succonox 18 (N-stearoyl-p-aminophenyl)
Nonox CI   toluene extraction,   317, 318
                                             methanol precipi-
                                             tation of polymer,
                                             spectrophotometric
                                             determination after
                                             reaction with
                                             hydrogen peroxide.

Ultraviolet methods

Phenolic type
Topanol OC (2,6 di-t-butyl-4-methylphenol)   solvent extraction,   319
Borox M (bis-(3,5-di-t-butyl-4-hydroxy   thin-layer chromato-
phenyl methane))   graphy, then ultra
                                             violet spectroscopy

Ionox 330 (1,3,5, trimethyl-2, 4, 6 tris
(3,5 di-t-butyl-4-hydroxy benzyl) benzene)
ditto   solvent extraction   319, 320
                                             oxidation in PbO_2 or
                                             nickel peroxide -
                                             ultra violet spectro-
                                             scopy in neutral and
                                             alkaline solution ,
p-methoxyphenol 4,4' methylene bis-   solvent extraction,   321
(2,6-di-tert-butyl-phenol)) and   ultraviolet spectro-
Santonox R, (4,4'-thio-bis-(6-tert-butyl-   scopy (using batho-
m-cresol))   chromic shift in
                                             sodium hydroxide
                                             solution)

Ionol (2,6 di-tert-butyl-p-cresol)   chloroform extraction-322, 323
Santonox R, (4,4'-thio-bis-6-tert-   ultraviolet spectro-. 324, 325
butyl-m-cresol)
Irganox 1010   ultraviolet spectro-   326
                                             scopy
Ionol (2,6 di-tert-butyl-p-cresol)   solvent extraction -
Santonox R, (4,4'-thio-bis (6-t-   ultraviolet spectro-
butyl-m-cresol)   scopy solvent extraction

Fluorescence and Phosphorescence methods

Phenyl-napthylamines   solvent extractions   327
                                             fluorescence method

Agerite D (polymeric dihydroxy) -quinone) phenyl-2-napthylamine	solvent extractions fluorescence method	328
Santonox R, (4,4'-thio-bis-(6- tert-butyl-m-cresol)) phenyl-2-naphthylamine	solvent extractions phosphorescence method	328

Spectroscopic techniques

dilaurylthio-dipropionate	infrared spectroscopy of polymer film	329
Ionol (2,6 di-t-butyl-p-cresol) Santonox R (4,4'-thio bis-(6'- t-butyl-m-cresol)	iso-octane extraction infrared spectroscopy	330
phenolic antioxidant	solvent extraction, various finishes including spectro- photometry, NMR analysis and mass spectrometry	331, 332, 333, 334, 335, 336, 337
antioxidants	extraction with chloroform or hexane or diethyl ether or toluene or carbon disulphide-iso octane mixtures or cyclo- hexane or water followed by various instrumental finishes	338, 339, 340, 341 342, 343, 344, 345, 346, 347

Thin-layer and paper chromatography

antioxidant	solvent extraction, paper chromatography using chromogenic reagents	348, 349, 350, 351, 352, 353,
Amine antioxidant	solvent extraction, thin-layer chromato- graphy	354, 355, 356, 357
antioxidants	solvent extraction, thin-layer chromato- graphy	358, 359, 360, 361- 369
Ionol (2,6 di-t-butyl-p- cresol) Santonox R (4,4'-thio-bis-(6'-t- butyl-m-cresol))	hexane extraction, ethanol solution of residue examined by thin-layer chromato- graphy	370, 371
Ionol (2,6 di-t-butyl-p-cresol 2,4,6 tri-t-butyl phenol	solvent extraction, thin-layer chromato- graphy	372
Nonox range of antioxidants (amine and phenolic types)	solvent extraction, thin-layer chromato- graphy	373
Butylated hydroxy toluene "2246" Santofex range, (hydroquinone type) Superlite Polygard tris (2,3 trinonylphenyl) phosphite		

Irganox 1076
4,4' butylidene, (2-t-butyl-5-
methyl) phenol

4,4' thio bis (6-t-butyl-m-cresol)
pentaerythritol tetrabis 3,5, di-
t-butyl-4-hydroxyhydrocinnamate
2,2' methylene bis(4-methyl-6-t
butyl phenol)
octadecyl (3,5-di-t-butyl-4-
hydroxyphenyl) acetate
Ionol, (2,6 di-t-butyl-p-cresol)

n-heptane extraction   374
thin-layer chromato-
graphy

Column chromatography

antioxidants

solvent extraction,   375, 376,
column chromatography 377
on silica gel using
a range of solvents
and various detectors

phenolic and amine
antioxidants

solvent extraction,   378
high performance
liquid chromatography

Dilaurylthiodipropioate

Gel permeation chrom- 379
atography

Santowhite, (4,4' butylidene-bis-
(6-t-butyl-m-cresol))
Butylated hydroxy toluene
Santonox R, (4,4' thio bis(6-t-
butyl-m-cresol))

Chloroform extraction 380
column chromatography,
successive elution with
chloroform (removes
BHT) 1:10 aqueous
methanol (removes
Santowhite and
Santonox R) Eluates
analysed by ultra-
violet spectroscopy

Butylated hydroxytoluene
Santonox R (4,4" thio bis (6-t-butyl
-m-cresol)
Irganox 1076
CAO 14

freeze grinding,   381
diethyl ether extrac-
tion, column chroma-
tography, with
refractive index and
ultraviolet detectors
at column outlet

antioxidants

solvent extraction,   414
column chromatography

Miscellaneous methods

2-t-butyl-4-methylphenol
2,6 di-t-butyl-4-methylphenol
2,2'-bis (4-methyl-6-butylphenyl)
methane

solvent extraction,   382
bromometric estimation

Santonox R (4,4' thio bis(6-t-
butyl-m-cresol))

ditto   383

Dilaurylthio-dipropionate

solvent extraction,   384
polarography

diorgano-sulphide
tert phosphite types
distearyl thiodipropionate

oxidatives with m-   385
chloropenoxybenzoic
acid, unreacted oxidant

dilauryl thiodipropionate   estimated iodimet-
4,4' thio bis (6-tert butyl-m-cresol   rically
1'1' thio bis (2-naphthol)
triphenyl phosphite
triethylphosphite
tri-p-tolyl phosphite
tris (dinonyl phenyl) phosphite
2,2' thio bis (-6- tert-butyl-p-cresol)
triisopropylphosphite

The concentration of antioxidants in polyethylene has been determined by a
thermogravimetric technique based on measuring the oxidation induction time
(132, 133). By this method the exudation rate, solubility, and diffusion
coefficient of antioxidants in polyethylene could be determined.

Ultraviolet absorbers in polyethylene (see also polypropylene)

Thin-layer chromatography

benzobenone and salicylic   solvent extraction,   386
acid types   thin-layer chromato-
                                 graphy
benzophenone type   solvent extraction,   387
salicylate type   thin-layer chromato-
                                 graphy
benzotriazole type   for quantitative
                                 estimation
salicylate type
substituted acryonitriles
onganonickel type
ultraviolet absorbers   solvent extraction,   388-397
and optical brighteners   thin-layer chromato-
                                 graphy
optical whiteners   solvent extraction,   398
                                 thin-layer chromato-
                                 graphy

Miscellaneous methods

Tinuvin P (2'hydroxy 5'-methylphenylbenzo   Toluene extraction,   399
-triazole))   glc
Tinuvin 326 (2-(2' hydroxy-3'-t-butyl
5' methylethylphenyl-3 chlorobenzo-
triazole))
Tinuvin 327 (2-(2' hydroxy-3'-5'-di-
6-butyl(phenyl)-3-chlorobenzotriazole))
Cyasorb UV531 (2 hydroxy-4-n octoxy
benzophenone)
benzophenone type   solvent extraction,   400
                                 potentiometric
                                 titration with sodium
                                 methoxide in dimethyl-
                                 formamide
antioxidants   solvent extraction,
                                 column chromatography 415

Methods involving solvent extraction of the polymer followed by thin-layer
chromatography have been described (406, 408-413, 573, 605, 612) for the
determination of antistatic agents in polyethylene. Gas chromatography has
also been used to determine antistatic agents in polyethylene extracts (401-
404).

Antiozonants in Polyethylene

Methods have been described for determining antiozonants, based on thin-layer
chromatography of a solvent extract of the polymer (405-411).

POLYETHYLENE COPOLYMERS AND TERPOLYMERS

ETHYLENE-PROPYLENE COPOLYMERS

Chemical

Narasaki et al (416) used spectrophotometry to compare the recovery of phosphorus in organic additives via wet ashing and oxygen bomb and acid digestion of the residues. Both decomposition procedures were satisfactory on
phosphorus alone but, in an ethylene-propylene copolymer system, the latter
method gave a considerably higher recovery of phosphorus at about the 100-
ppm level.

Measurement of Ethylene-Propylene Ratios and other Structural Features

Processes for the manufacture of ethylene propylene polymers can produce
several distinct types of polymers. Polymer might consist of mixtures of
the following types of polymers:-

   i)   Physical mixture of ethylene homopolymer and propylene copolymer,

        E-E-E-E-E-   P-P-P-P

  ii)   Copolymers in which the propylene is blocked, e.g.

        E-E-P-P-P-P-P-E-E-E E-E-E-P-P-P-

 iii)   Copolymers in which the propylene is randomly distributed, e.g.

        -E-P-E-P-E-P-E-P   (alternating e.g. pure cis 1:4 polyisoprene

or   -E-E-E-P-E-E-E-E-P-P-E E-E E

  iv)   Copolymers containing random (or alternating) segments together with
        blocks along the chains, i.e. mixtures of (iii) and (ii) (random and
        block) or (iii) and (ii), (alternating and block).

   v)   Containing tail to tail propylene units in propylene blocks:

$$
\begin{array}{llll}
CH_3 & CH_3 & CH_3 CH_3 & CH_3 \\
| & | & |\ \ | & | \\
-CH-CH_2-CH_2-CH-CH_2-CH_2-CH_2-CH-CH-CH_2-CH_2-CH_2-CH_2-CH-
\end{array}
$$

i.e. head to head and tail to tail addition giving even numbered sequences of methylene groups.

Various methods are available for determining the % of ethylene and of propylene and of the ethylene propylene ratio in these polymers and these methods are discussed in further detail in this section.

In fact, the problem is somewhat more subtle than expressed above and the type of measurements that are commonly required are listed below:

(a) Determination of total precent propylene in the above polymers, regardless of the manner in which the propylene is bound.

(b) Determination of total precent ethylene in the above polymers, regardless of the manner in which the ethylene is bound.

(c) Determination of proportions of total propylene content of polymer which is blocked and that which is randomly distributed along the polymer chain.

Several possible techniques are available for carrying out these analyses, and these are now discussed.

(a) Pyrolysis of sample followed by gas/liquid chromatography of pyrolysis products.

(b) Direct infra-red spectroscopy of polymer, either at room temperature or at elevated temperatures.

(c) Pyrolysis of sample followed by infra-red spectroscopy.

(d) Nuclear magnetic resonance spectroscopy.

Pyrolysis-Gas Chromatography

In one of the earliest references to this technique Bua and Manaresi (417), described an analytical procedure for ethylene-propylene copolymers, wherein depolymerization is accomplished at 470-520°C. This procedure produces a mass spectrum essentially of C_2 and C_3 monomers, extending to about mass 56. An analytical method was developed from such spectra.

High vacuum pyrolysis at 400°C of ethylene-propylene copolymers carried out by Van Schooten et al (418), followed by trapping of released volatiles in dry ice-acetone produced a mixture of α olefins and α'olefins. The most volatile fractions, collected in dry ice-acetone were hydrogenated to saturated hydrocarbons, which were analysed by gas-liquid chromatography. For samples of copolymer prepared using either titanium trichloride or vanadium oxychloride catalysts the chromatogram of this fraction showed peaks of 2,4-dimethylheptane, 2-methylheptane, 4-methylheptane, 2,4-dimethylhexane, 3-methylhexane and 2-methylhexane, but only in the chromatogram of the volatile fraction from the copolymer produced using vanadium was a peak of 2,5-dimethylhexane found. This is an indication that polymers prepared with a catalyst containing vanadium oxychloride contains methylene sequences of two units between branches. Van Schooten et al (418) conclude that ethylene-propylene copolymers prepared with vanadium-containing catalysts, especially those with $VOCl_3$ or $VO(OR)_3$, have methylene sequences of two and four units. The detection of methylene sequences of two units is a strong indication that head-to-head orientation of the propylene units may occur with certain catalysts of the Ziegler-Natta type.

Voigt (419-421) employed the apparatus shown in Fig. 24 attached directly to
the gas inlet of the gas chromatograph for the examination of ethylene-
propylene copolymers. Figure 24 shows the Pyrex glass pyrolysis cell b loc-
ated in the oven a, the carrier gas entering the cell from above, and bring-
ing with it the pyrolysate formed on the red-hot platinum wire c on leaving
the cell and passing through the gas inlet and a heated feed line to the gas
chromatograph. d is the junction of a chromel-alumel thermocouple, the lead
wires of which are embedded in a Ni-Cr-Fe alloy tube filled with magnesium
oxide. Like the platinum wire of the heating spiral, this tube has a diameter of only 0.5 mm.

Specimens weighing about 2 mg were heated for 18 seconds up to a maximum
temperature of 550°C. During the pyrolysis, the carrier gas flushes the
products of pyrolysis into the separating column. The pyrolysis cell is cut
out of the carrier gas flow 1 min after starting the pyrolysis.

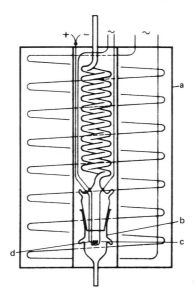

Fig. 24   Pyrolysis apparatus (a) jacket, (b) pyrolyser,
          (c) platinum filament, (d) thermoelement.

Voigt (419-421), reached the following conclusions regarding pyrolysis-gas
chromatography of ethylene-propylene copolymers and the two homopolymers.

(a)   Only unbranched molecules arising from statistical decomposition of the
      polymer arise in the case of polyethylene. The very small amount of
      branched molecules present even in low pressure polyethylenes were not
      detected on the pyrogram (See Fig. 25 and Table 15). High pressure
      polyethylenes produced identical pyrograms.

(b)   In the case of polypropylenes (Fig. 26 and Table 15) no straight chain
      decomposition products with a length greater than C_5. The larger
      decomposition products all contain methyl branches, or 2.4 dimethyl
      branches. The considerable quantity of 2:4 dimethyl heptene is part-

-icularly noticeable. This substance can, to a certain extent, be regarded as tri-meric propylene and can be designated as the major product
of polypropylene decomposition.

(c)   Block copolymers give rise to the same pyrogram as do the corresponding
       mixtures containing the same proportions of homopolymers, (compare
       Figs. 27 and 28 and for interpretation of peaks see Table 15). Thus,
       the repeating propylene units in block copolymers behave like propylene
       homopolymer.

(d)   Different considerations apply in the case of random ethylene/propylene
       copolymers. Repeating propylene units are absent in these polymers,
       thus lower concentrations of 2:4 dimethyl heptene (propylene trimer)
       would be expected to be present in the pyrolysate of a random copoly-

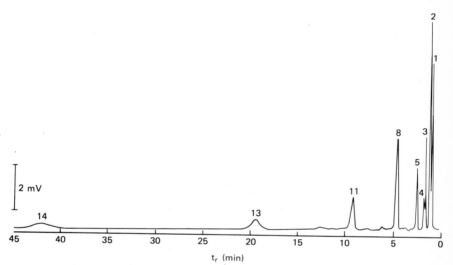

Fig. 25   Pyrolysis gas chromatograph of low pressure
           polyethylene (Hostalen 9C). For the identification
           of the numbered peaks see Table 27.

Comparison of the pyrogram of a block copolymer containing 27% propylene
(Fig. 28) with that of a random copolymer containing 43% propylene (Fig. 29)
confirms this view. In Fig. 27 the characteristic 18 PP peak for consecutive
polypropylene units is completely absent. Apparently, the chances of a tri-
meric propylene unit being formed from a statistically arranged sequence of
ethylene and propylene molecules is extremely small. On the other hand, the
polypropylene peak 13 PE and the polypropylene peak 14 PP are very well defined in the pyrogram of the statistical copolymer (Fig. 29). The suppression
of peak 18 PP is highly characteristic to the statistical nature of the distribution of ethylene and propylene units in ethylene/propylene copolymers.

Voigt (419-421) described methods for the determination of the propylene content of block copolymer based on the estimation of the ratio of the areas
18 PP/11 PE, (see Table 27) and for the determination of the propylene

Analysis of Plastics

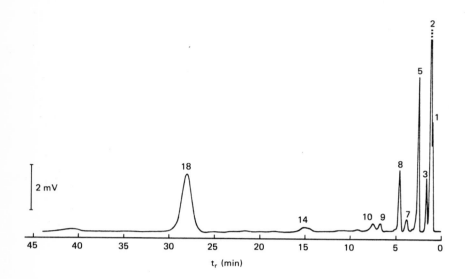

Fig. 26   Pyrolysis gas chromatogram for polypropylene,
(Hostalen PPN).   For the identification of
numbered peaks see Table 27.

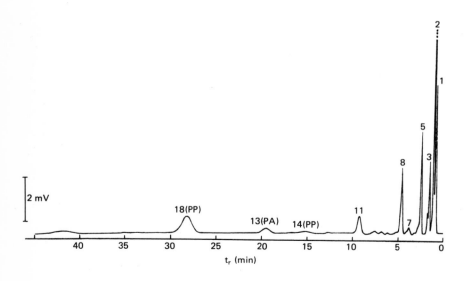

Fig. 27   Pyrolysis gas chromatogram of a mixture of 70%
w/w polyethylene, 30% w/w polypropylene.   (18 PP
is 2.4 dimethyl heptane-1, 14 PP is 4 methyl
heptane 1 and 13 PE is octene -1.).

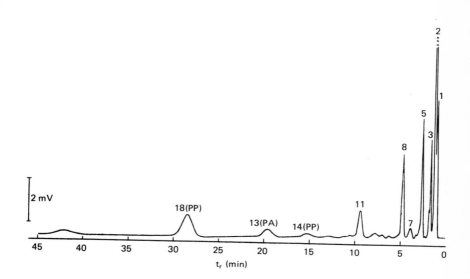

Fig. 28 Pyrolysis gas chromatograph for a block copolymer consisting of 73% w/w ethylene and 27% propylene (18 PP is 2.4 dimethyl heptene-1, 13 PE is octene -1 and 14 PP is 4 methyl heptene).

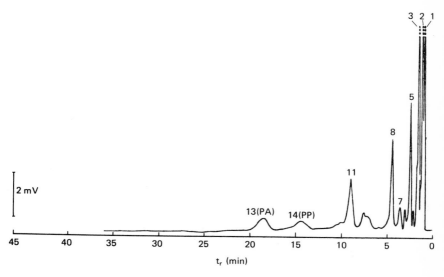

Fig. 29 Pyrolysis gas chromatogram of an ethylene/propylene rubber, a statistical copolymer containing 43% w/w propylene.

TABLE 15 Decomposition Products Identified in the Pyrogram
          of Polyethylene and Polypropylene

Polyethylene (See Fig. 25)   Polypropylene (See Fig. 26)

Peak No.		Compound	Peak No.	Compound
1 PE		Ethylene	1 PP	Ethylene
2 PE		Propylene	2 PP	Propylene
3 PE	Butene-1		3 PP	Isobutene
4 PE		Butadiene	4 PP	
5 PE		Pentene-1	5 PP	Pentene
6 PE			6 PP	
7 PE			7 PP	4-methyl pentene (1) or (2)
8 PE		Hexene-1	8 PP	2-methyl pentene (1)*
9 PE			9 PP	2-4-dimethyl pentene (1)
10 PE			10 PP	C_6 diolefin
11 PE		Heptene-1	11 PP	
12 PE			12 PP	
13 PE		Octene-1	13 PP	
14 PE		Nonene-1	14 PP	4-metnyl heptene-1 or (2)
15 PE			15 PP	
16 PE			16 PP	
17 PE			17 PP	
18			18 PP	2-4-dimethyl heptene (1) or (5) or (6)

* or 2-methyl pentadiene (1.4).

content of random copolymers (based on a plot of the ratio 14 PP/13 PE, i.e.
4-methylheptene/octene-1 (See Table 15).

Brauer (422), applied pyrolysis gas chromatography to the elucidation of the
structure of ethylene-propylene copolymers and found a relationship between
peak areas on the pyrogram and weight % propylene in the copolymer.

The application of pyrolysis-gas chromatography to the analysis of ethylene-
propylene copolymers has also been investigated by several other workers
(423-426).

Some of the classic work on the pyrolysis-gas chromatography of ethylene-
propylene copolymers was reported by Van Schooten and Evenhuis (427, 428,
431, 433, 434); In their earlier work the sample (20 mg.) in a platinum
dish was submitted to controlled pyrolysis in a stream of hydrogen as carrier gas. The pyrolysis products were then hydrogenated at 200ºC by passing
through a small hydrogenation section containing 0.75% platinum on 30/50
mesh aluminium oxide. The hydrogenated pyrolysis products are then separated
on a squalane on fireback column and the separated compounds detected by a
katharometer. Under the experimental conditions used in this work only
alkanes up to C_9 could be detected. These workers examined polyethylenes and
polypropylenes of different origins and ethylene-propylene copolymers containing between 0% and 100% propylene. They also examined hydrogenated polyisoprene.

Pyrograms for polyethylene, polypropylene and an ethylene/propylene copolymer
are shown in Fig. 30. Tentative identifications are made of most of the
hydrogenated pyrolysis products. Comments on the chromatograms are made in
Table 16. Some peaks are nearly independent of propylene content. Others,

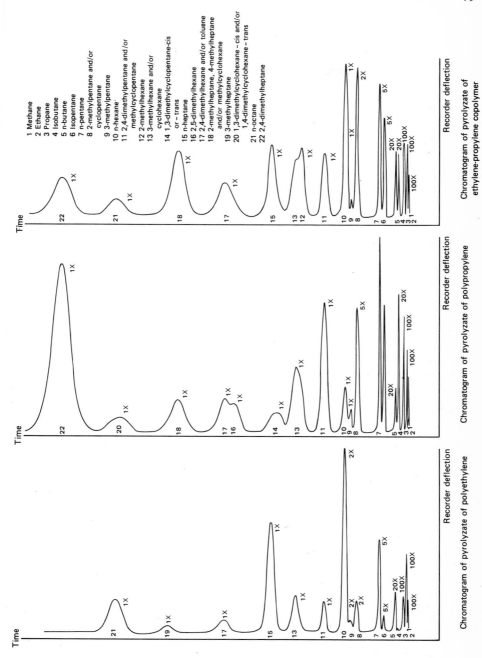

Fig. 30   Gas chromatograms of pyrolysates of polyethylene,
polypropylene and ethylene-propylene copolymer.

TABLE 16   Comments on Pyrolysis Products obtained from Various Polyolefins

Polyethylene		Polypropylene		Polyethylene/Polypropylene copolymer
Major	Minor	Major	Minor	
n-alkanes predominant, viz: ethane, propane, butane, pentane, hexane, octane.	Small amounts of 2-methyl and 3-methyl compounds present (iso-pentane and 3-methyl pentane indicative of short chain branching).	Branched alkanes predominate peaks of branched hydro-carbons appear in a regular pattern, e.g. 2-methyl and 3-methyl and 2:4 dimethyl configurations. Part-icularly noticeable are the large peaks due, 2:4 dimethyl pentane and 2:4 dimethyl heptane (almost absent in polyethylene pyrolysate).	n-paraffins present in decreasing amounts up to and including a small amount of n-hexane, (possibly n-heptane and n-octane present but these peaks might be obscured by cyclic components).	Both branched and n-alkanes present 2-methyl and 3-methyl and 2:4 dimethyl configuration present 2:4 dimethyl heptane and 2:4 dimethyl pentane concen-trations less than amounts found in polypropylene.

however, vary with propylene content, e.g. iso-butane, 2 methyl pentane, n-hexane, n-heptane, n-octane, 2.4 dimethyl heptane. Excellent graphs are obtained relating bound propylene content of the copolymer with the peak areas of the last four compounds in this list.

A problem in the analysis of ethylene/propylene copolymers is to distinguish between 'blocked' propylene units and 'random' propylene units (for definition see start of Section)

The two extremes are:-

(a)   a polymer in which the blocks of ethylene units and propylene units are so long that, upon pyrolysis, it resembles a physical mixture of two homopolymers.

(b)   a completely alternating copolymer for which hydrogenated polyisoprene can be taken as a model substance, i.e. -E-P-E-P-E-P-E-

Van Schooten and Evenhuis prepared pyrograms for the three following polymers:

     i)   1:1 mixture of polyethylene and polypropylene
    ii)   1:1 ethylene/propylene copolymer
   iii)   Hydrogenated polyisoprene

This comparison showed that the pyrogram of the 1:1 mixture of polyethylene and polypropylene differs significantly from that obtained by assuming additivity only for the peaks up to the butanes, in which region the reproducibility of the gas/liquid chromatographic analysis was rather poor.

The copolymers gave pyrolysis spectra slightly different from both hydrogenated polyisoprene and the 1:1 mixture. Different from the copolymer 1:1 mixture are the methane, ethane, propane and isobutane peaks, but also the 2 methyl hexane and 2 methyl heptane peaks, which are higher, and the 2:4 dimethyl pentane and 2:4 dimethyl heptane peaks, which are lower.

In the earlier method discussed above the polymer was pyrolysed in a platinum dish after which the hydrogenated pyrolysis products were analysed by gas chromatography using a katharometer detector. This procedure enabled Van Schooten and Evenhuis to analyse volatile polymer decomposition products up to C9. Using this procedure, however, a large sample (20-30 mg.) was required, which favoured the occurrence of consecutive and side reactions of the primary pyrolysis products. In their improved method, (429, 430) discussed below, Van Schooten and Evenhuis pyrolysed a much smaller polymer weight (0.4 mg.) in a stream of hydrogen on an electrically heated nichrome filament at 500°C and lead the pyrolysis products onto a hydrogenation catalyst to convert all products to saturated hydrocarbons. Operation with such small samples was made possible by using a sensitive flame ionisation G.L.C. detector. This and the use of programmed heating of the Apiezon N/firebrick G.L.C. column extended the range of detectable volatile products from C9 to iso C13 with considerable improvement in resolution of the pyrograms especially of the more volatile components.

These modifications of experimental technique minimized the occurrence of secondary reactions, with the result that the relative amounts of lowest molecular weight fragments are considerably reduced. This is illustrated in Figs. 31 and 32 where peak surface areas are given for the n-alkanes in the pyrograms of polyethylene and polypropylene as obtained by the old and the new pyrolysis apparatus. The modified technique, therefore, provides a much more

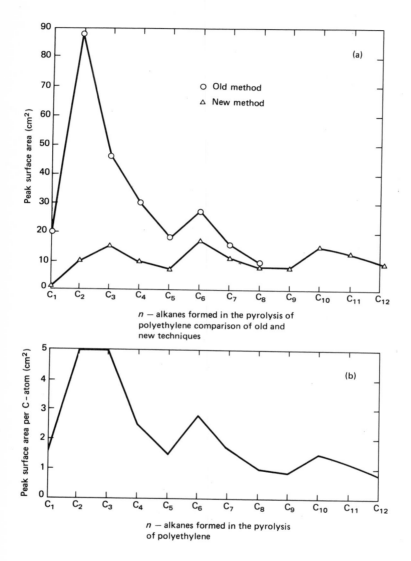

Fig. 31   Peak surface area-n-alkane distributions of
          poly ethylene.

reliable picture of the primary pyrolysis reactions, and enables some conclusions to be drawn regarding primary reaction mechanisms.

Van Schooten and Evenhuis (429, 430), also applied their modified pyrolysis-
gas chromatographic technique using the nichrome filament at 500°C to ethylene-propylene copolymers and to other copolymers (e.g. ethylene-butene-1,
propylene-butene-1 copolymers and to partly unsaturated ethylene-propylene

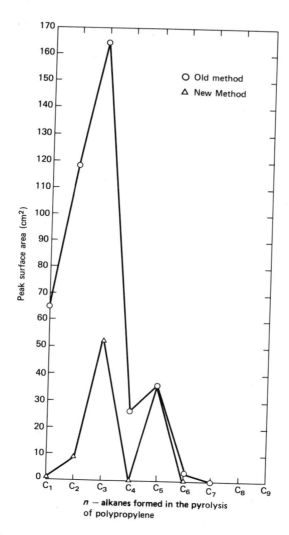

Fig. 32 Peak surface area-n-alkane distributions of
         polypropylene

terpolymers). The method they developed provides information both about the
composition and about the structure, degree of alteration, head-to-head or
tail-to-tail arrangement, etc., of copolymers. They describe quantitative
methods for ethylene-propylene copolymers and for ethylene-propylene-dicyclopentadiene terpolymers.

Van Schooten and Evenhuis (427, 428) applied their pyrolysis-hydrogenation-
gas chromatography technique to the quantitative determination of the ethylene-propylene copolymer, an analysis which presents difficulties in solvent

solution-infrared methods, especially with samples that are only partly soluble in suitable solvents such as carbon tetrachloride. Since the pyrogram
of polyethylene consists almost exclusively of normal alkanes and that of
polypropylene of iso-alkanes the ratio of the peak heights of a n-alkane to
an iso-alkane is a good measure of the copolymer composition. The ratio
$n-C_7/(2 \text{ methyl } C_7 + 4 \text{ methyl } C_7)$ was a good measure of ethylene-propylene
ratio in copolymers (Fig. 33).

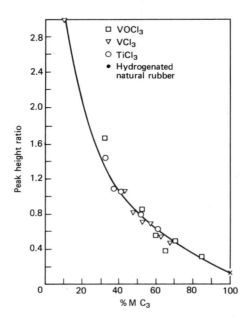

Fig. 33 Ethylene-propylene copolymers. Peak height ratio
         $n-C_7/(2MC_7 + 4MC_7)$ as a function of propylene
         content.

Van Schooten and Evenhuis (435), prepared gas chromatograms of hydrogenated
pyrolysis products of polyethylene, polypropylene, ethylene-propylene copolymers and hydrogenated polyisoprene. These indicate a high degree of
alternation in the ethylene-propylene copolymers. They identified most
peaks in the chromatograms and ascribed to a single component, some to two
or three different iso-alkanes or cycloalkanes.

A survey of the relative concentrations of the pyrolysis products is presented in Figs. 34 and 35, where peak areas have been given for the pyrolysis
products of polyethylene, polypropylene, hydrogenated polyisoprene and for a
copolymer containing about 50%m ethylene.

A cursory examination of these figures shows that, apart from a few minor
differences, the pyrolysis spectra of ethylene-propylene rubber of 50%m C_2
and hydrogenated polyisoprene are very similar. This would indicate a high
degree of alternation in the ethylene-propylene copolymer, because presumably the spectrum of a C_2/C_3 copolymer having poor alternation resembles

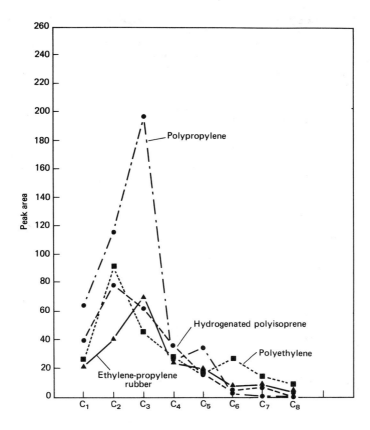

Fig. 34   Peak area of n-alkanes obtained after pyrolysis
          of some polymers.

that of a 1:1 mixture of polyethylene and polypropylene.   The same conclusion
was drawn from the size of the 2,4-dimethylheptane peak.

The main features of the pyrolysis spectra of polyethylene, polypropylene
and hydrogenated polyisoprene can be interpreted on the basis of simultaneous
breakage of carbon-carbon bonds in the main polymer chain.   For a more detailed interpretation of the pyrolysis spectrum it is necessary to assume a
series of radical-chain reactions in which intramolecular hydrogen abstraction plays an important role.   In this matter the occurrence of a number of
fragmentation products can be reasonably explained.

For example, by assuming hydrogen transfer with the fifth carbon atom it is
possible to explain:

(1)   the maximum at n-hexane in Fig. 34 for polyethylene;
(2)   the maximum at n-pentane in Fig. 34 for polypropylene;
(3)   the large amounts of 2-methylheptane and 3-methylhexane in the
      chromatogram of hydrogenated polyisoprene.

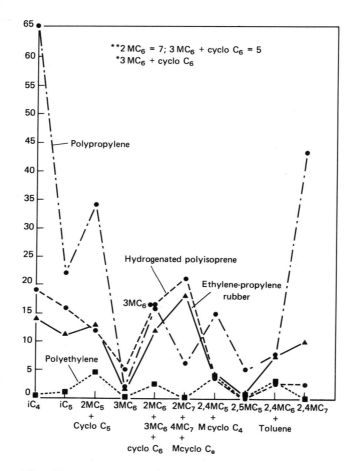

Fig. 35   Peak area of isoalkanes obtained after pyrolysis
          of some polymers.

(4)   the much larger peak for 2,4-dimethylheptane than for 2,4-dimethylpen-
      tane in the chromatogram of polypropylene;

Eggertson et al (436), have discussed an apparatus for the pyrolysis of ethylene-propylene copolymers, as well as polypropylene, polyethylene and blends
of these two homopolymers, a noteworthy feature of which is the provision for
rapid and convenient evaluation of complex pyrolysis patterns by tape-recor-
der computer integration of the gas chromatographic peaks. In order to simplify the chromatograms and to make possible the mass spectrometric identification of fractions, the pyrolysis products were hydrogenated in the gas
chromatographic column during the separation. The hydrogenation is accomplished on a one-foot column packed with platinum dioxide catalyst impregnated
on SE30 (silicone rubber) coated Chromosorb. This column is placed just
ahead of the separation column. Olefins appear to be completely hydrogenated
in this system while aromatics are not affected.

Eggertson (436) showed that specimens of fractionated ethylene-propylene rubber polyethylene-ethylene propylene rubber-polyethylene block copolymers had C_2 contents of between 67 and 82%.

Infrared Spectroscopy

The infrared absorption of ethylene copolymers in the 700-850 cm^{-1} region can provide information about their sequence distributions. Studies on model hydrocarbons (438-441) have shown that the absorption of methylene groups in this region is dependent on the size of methylene sequences in the compounds. Methylene absorptions observed in this region and their relation to structures occurring in ethylene copolymers are shown in Table 17.

TABLE 17   Methylene Absorption and Copolymer Structure
           Responsible for Absorption in Ethylene Copolymers

$-(CH_2)_n-$	cm^{-1}	Sequence	
$(CH_2)_1$	815	$\sim CH_2CHRCH_2CHRCH_2CHR \sim$	(2)
$(CH_2)_2$	751	$\sim CH_2CHRCH_2CH_2CHRCH_2 \sim$	(3a)
		or	
		$\sim CHRCHRCH_2CH_2CHRCHR \sim$	(3b)
$(CH_2)_3$	733	$\sim CH_2CHRCH_2CH_2CH_2CHR \sim$	(4)
$(CH_2)_4$	726	$\sim CH_2CHR(CH_2CH_2)_2CHRCH_2 \sim$	(5)
		or	
		$\sim CHRCHR(CH_2CH_2)_2CHRCHR \sim$	(5b)
$(CH_2)_4$	722	$\sim CH_2CHR(CH_2CH_2)_nCH_2CHR \sim$	(6)

The absorption at 724 and 731 cm^{-1} of several ethylene-propylene copolymers have been assigned (442) to structures (6) and (4), respectively (Table 17). The relative intensities observed were qualitatively consistent with sequence distributions calculated for the copolymers, assuming reactivity ratios of 7.08 and 0.088 for ethylene and propylene, respectively. Veerkamp and Veermans (443) developed a technique to measure the intensities of these absorptions accurately. By assuming similar extinction coefficients for the two bands, the ratio of methylene units in $(CH_2)_3$ and larger methylene sequences in a number of copolymers was determined. The results agreed reasonably well with theoretical values based on Natta's calculations.

The presence of $(CH_2)_2$ sequences in ethylene-propylene copolymers has been considered in studies of their structure (444). Such sequences could result from the tail-head incorporation of propylene units into the copolymers (Structure 3a, Table 17).

Several methods have been reported for determining propylene in ethylene-propylene copolymers. The basis for calibration of many of these methods is the work published by Natta (445), which involves measuring the infrared absorption of the polymer in carbon tetrachloride solutions. The absorption at 7.25 microns, presumably due to methyl vibrations, is related to the propylene concentration in the copolymer. In some cases the dissolution of

copolymers with low propylene content or some particular structures is difficult (446, 447). Moreover, Natta's solution method was calibrated against
his radiochemical method (445), for which the precision of the method was
not stated; a considerable amount of scatter is evident in the data presented. Typical methods that have used Natta's solution procedure (445) for
calibration are described in publications by Wei (448) and Güssl (449).
These infrared methods avoid the solution problems by employing intensity
measurements made on pressed films. The ratio of the absorption at 717 cm^{-1}
to that at 1149 cm^{-1} is related to the propylene content of the copolymer.
Some objections (450, 451) to the use of solid films have been raised because
of the effect of crystallinity on the absorption spectra in copolymers with
low propylene content. These film methods are reliable only over the range
of 30 to 50 mole % propylene.

Bau and Manaresi (452) also calibrated a mass spectrometric method by Natta's
solution method (445). The propylene concentration was reported to be related to a certain ratio of mass peaks measured in the mass spectra of the
pyrolysis products of the copolymers. An approximate calibration could be
established with homopolymer mixtures of polyethylene and polypropylene.
Drushel and Iddings (446, 447) reported a method calibrated against C^{14} labelled copolymers. They found that the ratio of absorbances of two bands in
the C-H stretching region or two bands in the C-H wagging region could be
used to measure the propylene content of pressed films.

Natta et al (453), determined the degree of alternation of ethylene and propylene units in ethylene propylene copolymers from the infrared spectrum,
using peaks at 13.35, 13.70 and 13.83 μ (750, 731 and 724 cm^{-1}), the one at
13.70 μ being attributed to a sequence of three methylene groups between
branch points, presumably due to the insertion of one ethylene between two
similarly oriented propylene molecules.

In order to check the correctness of this interpretation Van Schooten et al
(459), examined the infrared absorption spectra between 13 and 14 μ of
ethylene-propylene copolymers prepared with various catalyst systems and
compared them with the spectra of some model compounds, namely 2,5-dimethyl-
hexane, 2,7-dimethyloctane, 4-methylpentadecane, 4-n-propyl-tridecane, polypropylene, polyethylene, polybutene-1 and hydrogenated polyisoprene, the
last being considered as an ideally alternating ethylene-propylene copolymer.
They supported the conclusions drawn from the spectroscopic investigation by
the results of pyrolysis-gas chromatography experiments.

The infrared absorption bands between 13 and 14 μ (700 and 770 cm^{-1}) of the
various samples are given in Fig. 36. These bands are due to rocking modes
of the CH$_2$ groups and their frequency depends on the length of the CH$_2$ sequences. In 2,5-dimethylhexane, hydrogenated polyisoprene and 2,7-dimethyl-
octane there are two, three and four CH$_2$ groups between branches, respectively, corresponding to peaks in the infrared spectra at 13.28 μ (754 cm^{-1})
($\mu\epsilon_{spec.}$ = 0.023 l./g.cm), 13.53 μ (740 cm^{-1}), ($\epsilon_{spec.}$ = 0.058 l./g.cm) and
13.75 μ (729 cm^{-1}) ($\epsilon_{spec.}$ = 0.042 l./g.cm). Amorphous and crystalline
polypropylene do not show any absorption at all in this region, whereas crystalline polyethylene shows two peaks at 13.65 and 13.85 μ (734 and 723
cm^{-1}). The latter peak is also found in liquid low molecular weight hydrocarbons (454) and in amorphous polymers containing long sequences of CH$_2$
groups (454); the 13.65 μ peak is a crystalline band, attributed to the CH$_2$
rocking in the polyethylene crystal (454-456). From the spectra of an amorphous butene-1 polymer it was concluded that ethyl side-groups give rise to
a peak at around 13.1 μ (765 cm^{-1}) and from the spectra of 4-methylpenta-
decane and 4-n-propyltridecane that n-propyl groups absorb at 13.53 μ

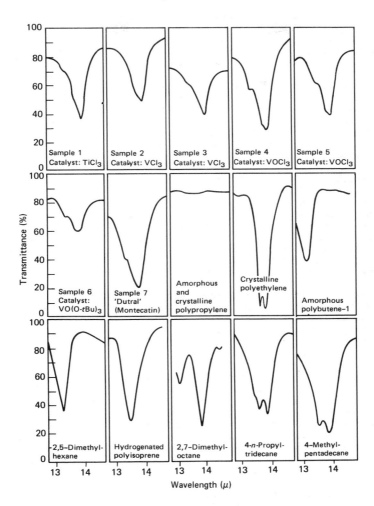

Fig. 36   Infra-red spectra between 13 and 14μ of C_2-C_2
          copolymers and model compounds.

(740 cm^{-1}), proving that the shoulder at 13.3 μ found in some polymers is not
due to ethyl or n-propyl end-groups. A survey of the various peaks and
shoulders present in the spectra is given in Table 18.

In the spectra of all copolymers, except sample 6, they found a sharp peak at
13.80 to 13.85 μ which must be ascribed to sequences of CH_2 groups longer
than four, i.e. sequences of more than two ethylene units (454).

Significant differences are, however, observed between the various spectra in
the wavelength region from 13.0 to 13.7 μ . None of the copolymer spectra
shows a clear shoulder at 13.53 μ , where hydrogenated polyisoprene shows
maximum absorption.

TABLE 18   Peaks and Shoulders between 13 and 14 μ (700 and 770 cm^{-1}) in Infrared Spectra of C_2-C_3 Copolymers

Wavelength, μ

Sample	13.8 to 13.9	13.7 to 13.8	13.6 to 13.7	13.50 to 13.60	13.0 to 13.35
2,5 Dimethylhexane	-	-	-	-	Peak at 13.28
Hydrogenated polyisoprene	-	-	-	peak at 13.53	-
2,7-Dimethyloctane	-	peak at 13.73	-	-	-
Amorphous butene-1 polymer	-	-	-	-	peak at 13.12
4-n-Propyltridecane	peak at 13.87	-	-	peak at 13.57	-
4-Methylpentadecane	peak at 13.85	-	-	peak at 13.53	-
Cryst. polypropylene	-	-	-	-	-
Cryst. polyethylene	peak at 13.85	-	peak at 13.65	-	-
Samp. 1 TiCl$_3$ catalyst	peak at 13.86	-	-	-	Vague Shoulder
Samp. 2 VCl$_3$ catalyst	peak at 13.85	-	-	Vague Shoulder around 13.60	-
Samp. 3 VCl$_3$ catalyst	peak at 13.86	-	-	-	Shoulder at 13.3
Samp. 4 VOCl$_3$ catalyst	peak at 13.85	Vague Shoulder at 13.7	-	-	Pronounced Shoulder at 13.3
Samp. 5 VOCl$_3$ catalyst	peak at 13.86	Shoulder at 13.75	-	-	Shoulder at 13.35
Samp. 6 VO(O-t-Bu)$_3$ catalyst	-	Broad peak at 13.75	-	-	peak at 13.35
Samp. 7 Dutral	peak at 13.82	-	-	-	Pronounced Shoulder at 13.?

This means that in all the samples the content of $(CH_2)_3$ sequences is low. There are, however, several samples showing a pronounced shoulder at ca. 13.3 μ viz. samples 3, 4, 5, 6 and 7 (Table 18). This shoulder must probably be assigned to sequences of two CH_2 groups between branch points (cf. spectrum of 2,5-dimethylhexane). Of these samples two have been prepared with a VOCl$_3$-containing catalyst and one with a VO(O-t-Bu)$_3$-containing catalyst. Only a few of the samples prepared with a catalyst containing VCl$_3$, (No. 3), showed the shoulder at 13.3 μ . This shoulder was always observed in samples which had been prepared with a catalyst obtained from VOCl$_3$ or VO(OR)$_3$. It might well be that this band arises from the presence of $(CH_2)_4$ sequences. This holds for samples 4, 5 and 6, which were prepared with catalysts containing VOCl$_3$ or VO(O-t-Bu)$_3$ and which, as we have seen, also display absorption bands that indicate the presence of C_2 sequences. None of

the copolymer spectra obtained thus far indicates the presence of the structure.

$$
\begin{array}{ccc}
\text{H} & & \text{H} \\
| & & | \\
\sim\text{C} & - & \text{C} \sim \\
| & & | \\
\text{H}_3\text{C} & & \text{CH}_3
\end{array}
$$

which would give rise to absorption at 8.9 μ (1,125 cm^{-1}). This means that
head-to-head orientation of propylene occurs only after addition of an ethylene unit. Experiments to obtain propylene homopolymers with the vanadyl-
containing catalysts at best gave poor yields and sometimes no polymer what-
soever. Where polymer was obtained this possessed the normal head-to-tail
structure.

Ciampelli et al (457), have developed two methods based on infrared spectroscopy of carbon tetrachloride solutions of the polymer at 7.25, 8.65 and
2.32 μ for the analysis of ethylene-propylene copolymers containing greater
than 30% propylene. One method can be applied to copolymers soluble in solvents suitable for the infrared analysis; the other can be applied to the
insoluble ones. The absorption band at 7.25μ due to methyl groups is used in
the former case, whereas the ratio of the band at 8.6 μ to the band at 2.32μ
is used in the latter. Infrared spectra of polymers containing between 55.5
and 85.5% ethylene are shown in Fig. 37.

This approach has also been discussed by Bucci and Simonazzi (458). This
method is based on measurements of absorption of solutions of the polymer at
1.692 μ (CH$_3$ groups, 1.764 μ (CH$_2$ groups), and is applicable to polymers containing as high as 52 mole % propylene.

Van Schooten et al (459), have shown that ethylene-propylene copolymers prepared with VOCl$_3$- or VO(OR)$_3$-containing catalysts not only contain odd-numb-
ered methylene sequences as was expected, but also sequences of two units.
Some indications for the presence of methylene sequences of four units were
found. They (460), also investigated copolymers with much lower ethylene
contents and pure polypropylenes prepared with catalysts consisting of VOCl$_3$
or VO(OR)$_3$ and alkylaluminium sesquichloride.

The infrared spectra of the ethylene-propylene copolymers containing 15 to 30
mole per cent of ethylene show little or no absorption above 13.70 μ (Fig.
38). This means that very few methylene sequences of four or more units are
present. The absorption peaks at 13.30 and 13.60 μ reveal the presence of
sequences of two and three units respectively. Thus, these copolymers, unlike those previously described, contain nearly all of the ethylene in isolated monomer units.

The earlier work (459), indicated that polypropylenes obtained with vanadyl
catalysts possessed the normal head-to-tail structure. More detailed examination, however, showed that the amorphous fractions isolated from these poly-
propyelenes do show absorption at 13.3 μ pointing to the presence of methylene sequences of two units, which can only mean that tail-to-tail arrangement
of propylene units does occur (Fig. 39). A very small absorption peak at
13.3 μ was also found in the spectrum of the crystalline fractions (Fig. 39).
Polypropylenes prepared with catalysts based on VCl$_3$ show only the normal
head-to-tail arrangement in both amorphous and crystalline fractions as do
polymers prepared with TiCl$_3$-catalysts.

Analysis of Plastics

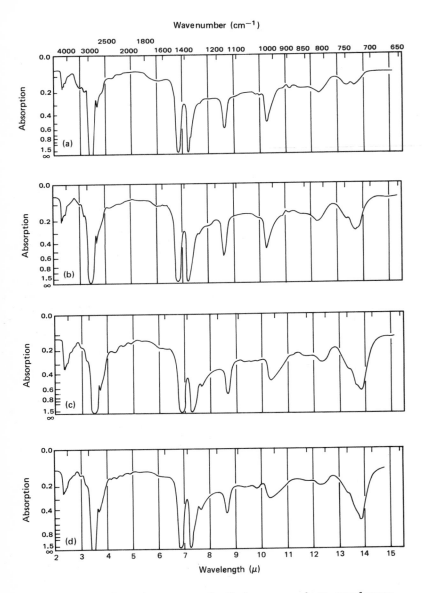

Fig. 37 Infrared spectra of ethylene-propylene copolymers
of various compositions
A   85.5% polyethylene
B   74.0%   "
C   65.9%   "
D   55.5%   "

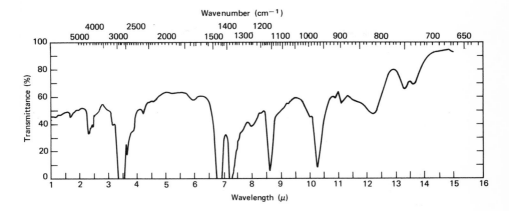

Fig. 38   Infrared spectrum of ethylene-propylene copolymer
          containing 85 mole per cent propylene.

The amount of propylene units coupled tail-to-tail was estimated to range
from about 5 to 15 per cent for the amorphous and from about 1 to 5 per cent
for the crystalline fractions.   The amount of propylene units in tail-to-
tail arrangement was calculated from spectra of thin films by comparing the
ratio of the absorbances at 13.30 and 8.65 μ (methyl band) in the spectra of
the polypropylenes with the ratio of the absorbances at 13.60 μ and 8.65 μ
in the spectrum of hydrogenated natural rubber.   This implies the assumption
that the absorbance per CH_2 group is the same at 13.30 μ for $(CH_2)_2$ sequences
as at 13.60 μ for $(CH_2)_3$ sequences.

The differences in amount of tail-to-tail coupled units between crystalline
and amorphous fractions are to be expected since every head-to-head and tail-
to tail configuration disturbs the regularity of the isotactic chain.   In the
polypropylenes every tail-to-tail configuration must necessarily be accompanied by a head-to-head coupling.   This would be expected to show up in an
absorption peak at 8.8 to 9.0 μ , characteristic of the structure

$$
\begin{array}{ccc}
H & & H \\
| & & | \\
\sim C & - & C \sim \\
| & & | \\
CH_3 & & CH_3 \\
\end{array}
$$

which is also found in hydrogenated poly-2,3-dimethylbutadiene, used as a
model compound and in alternating copolymers of ethylene and butene-2 (461).
In the polypropylenes examined by Van Schooten and Mostert (460), and also in
the copolymers they did indeed find an absorption band near 9 μ , although it
is much less sharp than in the model compound.

All spectra containing the 13.3 μ peak show a further small band at 10.9 μ ,
which is also found in the spectrum of poly-2,3-dimethylbutadiene.   The
spectrum of the amorphous alternating copolymer of ethylene and butene-2,
published by Natta (462), shows a band at about 10.8 μ .   The fact that the
polypropylenes prepared with $VOCl_3$- or $VO(OR)_3$-containing catalysts show tail-

to-tail arrangement means that tail-to-tail coupling of propylene units may
also occur in the ethylene-propylene copolymers. However, because the content of $(CH_2)_2$ sequences in the copolymers is much higher than in the polypropylenes prepared with the same catalysts, a large part of these sequences
most likely stems from isolated ethylene units between two head-to-head
oriented propylene units, their relative amount depending on the ratio of
reaction rates of formation of the sequences.

$$
\begin{array}{cccccccc}
| & | & | & | & \\
+C-C+C-C+C-C+ \\
| & | & & | & | \\
& C & & & C \\
& | & & & | \\
& & +C-C+C-C+ \\
& & | & | & | \\
& & C & C
\end{array}
$$

and

Veerkamp and Veermans (462), neglected this possibility in their determination of the fraction of isolated ethylene units.

Natta (464), used a radiochemical method to determine composition and Stoffer
and Smith (465), published a procedure for preparing standards by liquid
scintillation counting for use in calibrating infrared methods for the analysis of ethylene-propylene copolymers.

Drushel and Iddings (463), used carbon [14]-labelled standards (ethylene tagged)
prepared by a modification of the procedure of Stoffer and Smith (465).
These standards were used for calibration purposes in the C-H stretching,
bending, rocking, and wagging region of the infrared spectrum and in the combination region of the spectrum. In each case, a ratio of band intensities
was used which eliminates the necessity of measuring film thickness. Not
only does the ratio technique provide a greater degree of precision, as contrasted to the use of film thickness, but the derived working curves may be
applied to other instruments with an acceptable degree of accuracy. Advantages and disadvantages in the use of each specific region of the spectrum is
discussed by Drushel and Iddings (463).

Methyl and methylene rocking frequencies and intensities were interpreted by
these workers and used to elucidate the degree of randomness in the addition
of both ethylene and propylene to the copolymer. For interpretation, use was
made of the intensity ratio of the 1150-cm^{-1} and 968-cm^{-1} methyl group bands
and the intensity ratio of the 745-cm^{-1} and 720-cm^{-1} methylene group bands.

Liang, Lytton and Boone (473) on the basis of frequencies, relative intensities, polarization properties, and effects on deuteration of polypropylene,
have tentatively assigned the 968-cm^{-1} band in polypropylene to the methyl
rocking mode mixed with CH_2 and CH rocking vibrations. From this assignment,
perhaps it is a change in the magnitude of the mixing with the CH_2 and CH
modes in the case of the isolated propylene units in the copolymer which
gives rise to a more diffuse absorption band at 968-cm^{-1}. In this connection,
Liang and Watt (474), prepared nonstereospecific polypropylene in which the
968-cm^{-1} and 1150-cm^{-1} bands are either missing or very weak. Also, the
occurrence of head-to-head arrangements of contiguous propylene units would
be expected to influence the behaviour of the methyl rocking mode and any
mixing which may be involved.

For the purpose of polymer characterization the ratio of absorbances at 968
and 1150 cm^{-1} should provide a measure of the degree of randomness with respect to the introduction of propylene units into the copolymer. On the basis
of the initial observations, the ratio $A_{968\ cm^{-1}}/A_{1150\ cm^{-1}}$ should decrease

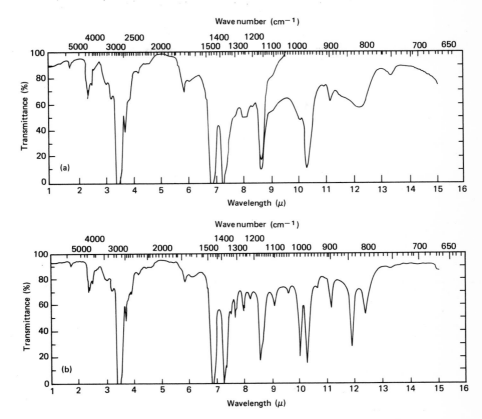

Fig. 39   Infrared spectrum of polypropylene prepared with
          vanadyl-based catalyst: (a) amorphous part.
          (b) crystalline part.

as the randomness increases. In addition, an increase in ethylene content is
expected to increase the probability of producing propylene units isolated by
ethylene units. For comparison of the 968-cm^{-1} to 1150-cm^{-1} absorbance ratio
with other data given below, correction for the effect of composition is made
by multiplying by the ratio mole % C_2/mole % C_3, the molar ratio of ethylene
to propylene in the copolymer. Relationships between this modified absorbance ratio and other established physical or spectral properties indicative
of randomness (or conversely, block polymer) are discussed below.

The temperature of incipient melting (or crystallization) obtained from
stress-temperature relationships has been used to estimate the extent of
"tacticity" or block copolymerization (475-477). The original relationship
devloped by Flory (478) for copolymers, which was extended by Coleman (477),
described the relationship between the melting temperature of the copolymer
(T_m) and the homopolymer (T_m^o) to the probability (p) of the appearance of
contiguous monomer units in the copolymer.

$$1/T_m - 1/T_m^o = -(R/\Delta H \mu) \ln p \tag{1}$$

Therefore, as the temperature of incipient melting increases, the ratio $A_{968 \text{ cm}^{-1}}/A_{1150 \text{ cm}^{-1}}$ would be expected to increase (see Fig. 40). This ratio $A_{968 \text{ cm}^{-1}}/A_{1150 \text{ cm}^{-1}}$ should correlate with the number of isolated ethylene groups in the copolymer

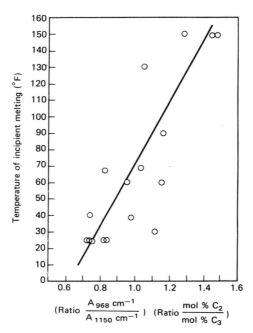

Fig. 40   Relationship between the temperature of incipient
          melting and the ratio of -CH$_3$ bands sensitive
          to the number of isolated propylene units.

Veerkamp and Veermans (479), have reported on the differential measurement of $(CH_2)_2$ and $(CH_2)_3$ units at 745 cm^{-1} and 630 cm^{-1}, respectively, in ethylene-propylene copolymers. The band at 720 cm^{-1} was assigned to units of 5 or more CH$_2$ units. Thus, the absorbance ratio $A_{745 \text{ cm}^{-1}}/A_{720 \text{ cm}^{-1}}$ provides a measure of the number of isolated ethylene units, and when corrected for the effect of composition (by multiplying by the ratio mole % C$_3$/ mole % C$_2$) is related to the inverse of the ratio $A_{968 \text{ cm}^{-1}}/A_{1150 \text{ cm}^{-1}}$ times the ratio mole % C$_2$/mole % C$_3$. The more random the copolymer, as indicated by the absorbance ratio $A_{745 \text{ cm}^{-1}}/A_{720 \text{ cm}^{-1}}$ for the relative number of isolated ethylene units, the more diffuse the 968-cm^{-1} band becomes.

In summary, the appearance of contiguous vs. isolated monomer units may be studied by the spectral characteristics of the methyl and methylene group rocking and wagging bands. The C-H stretching frequencies are useful for measurement of copolymer composition when spectrophotometers of sufficient resolution are available. Large absorptivities provide the sensitivity to examine extremely small samples from fractionation studies. Use of this region is more convenient since quartz disks may be used to support the cast film. A typical infrared spectrum of a cast film of ethylene-propylene

copolymer, showing the base line is reproduced in Fig. 41. The ratio between
the CH_3 and CH_2 asymmetrical C-H stretching band was not used because of the
large difference in relative intensities. The ratio between the asymmetrical CH_3 and the symmetrical CH_2 bands, which have nearly the same relative
intensities, produced more reliable results. Resolution of the CH_3 from the
CH_3 asymmetrical as well as the symmetrical bands must be satisfactory as no
curvature of the calibration curve was seen.

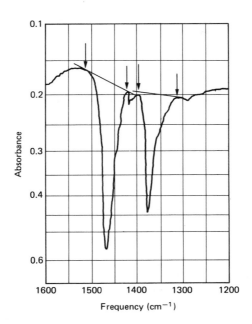

Fig. 41   Typical infrared spectrum of an ethylene-propylene
          copolymer in the C-H bending region of the spectrum
          Base line points shown by arrows
          Grating to prism changeover at 1416 cm^{-1}

If the CH_3 symmetrical band was not resolved from the CH_2 band, a change in
the slope of the curve at high propylene contents would have been observed.

Drushel and Iddings (463), also studied the near infrared spectrum of ethylene-propylene copolymers in the combination region of the spectrum where the
absorptivities are much lower. A typical spectrum of the copolymer is shown
in Fig. 42. Assignment of the specific frequencies in the C-H stretching and
bending regions which could give rise to these combination bands was not
made.

Bucci and Simonazzi (480), reported the use of overtone bands near 5882 cm^{-1}
(1.7 μ), but resolution of the CH_3 from the CH_2 bands was not as good as
found for the combination region discussed above.

Bucci and Simonazzi (485), have investigated the assignment of infrared bands
of ethylene propylene copolymers in the spectral region 900-650 cm^{-1}. They

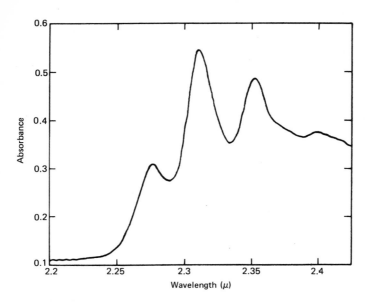

Fig. 42   Typical near-infrared spectrum of an ethylene-
          propylene copolymer.

also calculated absorbances at various frequencies and attempted a numerical
evaluation of the distribution of monomeric units in the copolymer. The
spectra of ethylene, propylene copolymers in this frequency range show peaks
or shoulders at 815, 752, 733 and 722 cm^{-1} whose intensity changes with composition (Fig. 43). For the assignment of these bands they examined spectra
of several model compounds which contain $(-CH_2-)_n$ .sequences with different
values of n, (Table 19). In the same table they compare their results with
those given in the literature (481-484).

On the basis of this comparison they assign the bands at 815, 752, 733 and
722 cm^{-1} to the $(-CH_2-)_n$ sequences with n = 1, 2, 3 and 5 or more, respectively.

This assignment does not agree with that proposed by Van Schooten and coworkers.

Bucci and Simonazzi (485), think that the difference lies in the fact that
these authors assigned the band at 733 cm^{-1} to the sequence $(-CH_2-)_4$. Bucci
and Simonazzi were not able to detect in ethylene-propylene copolymer any
absorption at 725 cm^{-1} where the sequence $(-CH_2-)_4$ should occur.

Bucci and Simonazzi recorded the spectrum of squalane compensated with 2, 6,
10, 14-tetramethylpentadecane. They did find a band due to $(-CH_2-)_4$ at 726
cm^{-1} (Fig. 44).

In Table 20 are listed the sequences and absorption frequencies and absorbtivities (Kvi) for various model compounds.

TABLE 19

Sequences and compounds	Frequency, cm^{-1}			
	Bucci and Simonazzi	McMurry-Thornton	Van Schooten et al	Natta et al
$(-CH_2-)_1$	815	785-770		
Atactic polypropylene	815			
$(-CH_2-)_2$	752	743-734		
2,5-Dimethylhexane			753	
Ethylene-cis-butene-2 copolymer	752			752
Hydrogenated poly-2,3-dimethyl-butadiene			752	
$(-CH_2-)_3$	733-735	729-726		
2,6,10, 14-Tetramethyl-pentadecane	735			
Squalane	735			
Hydrogenated natural rubber	735		739	
$(-CH_2-)_4$	726	726-721	728	
Squalane compensated with 2,6, 10, 14-tetramethylpentadecane				
$(-CH_2-)_5$ or more	722-721	724-722	722	
n-Heptane	722			
n-Decane	721			
n-Nonadecane	721			

TABLE 20

Sequences	Absorption frequency cm^{-1}	Model compounds	Absorptivities for CH$_2$ (Kvi) cm^{-1}mol^{-1}cm^{-1}ml
$(-CH_2-)_1$	815	Atactic polypropylene with head-to-tail linking	0.6 0.10x10^4
$(-CH_2-)_2$	752	Atactic polypropylene with a definite amount of head-to-head linking[a]	1.0 0.15x10^4
$(-CH_2-)_3$	733	Hydrogenated natural rubber, squalane, 2,6,10,14-tetra-methylpentadecane	1.2 0.20x10^4
$(-CH_2-)_5$	722	Linear C$_{10}$-C$_{19}$ hydrocarbons	1.2 0.25x10^4

a Determined from the decrease of intensity of the 815 cm^{-1} band.

Using Dn (the absorbtivity of the bands due to various frequencies) and Kvi (absorbtivities for CH$_2$) Bucci and Simonazzi (485) calculated for various samples the CH$_2$ content (in grams/100 g) corresponding to each band at frequency v$_i$. These results are listed in Figure 45 where the quantity G$_n$ is defined as G$_n$ = Dn/Kvi.

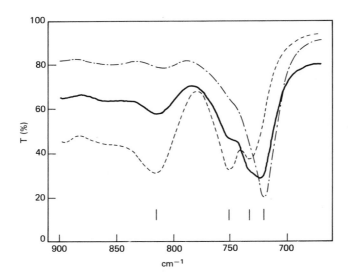

Fig. 43 Infrared spectra (900-670 cm⁻¹) of ethylene-
propylene copolymers at various compositions:
(-.-.-25%(3 mol.)); (_____50%(3 mol.);
(----75% (3 mol.)).

Lomonte and Tirpak (486) have developed a method for the determination of per cent ethylene incorporation in ethylene-propylene block copolymers by infra-red spectroscopy. The method does not involve previous fractionation of the polymer. Standardization is done from mixtures of the homopolymers. Both standards and samples are scanned at 180°C in a spring-loaded demountable cell. The standardization was confirmed by the analysis of copolymers of known ethylene content prepared with C^{14}-labelled ethylene. By comparison of the infrared results from the analyses performed at 180°C and also at room temperature, the presence of ethylene homopolymer can be detected. These workers derived an equation for the quantitative estimation of per cent ethylene present as copolymer blocks.

The method distinguishes between true copolymers and physical mixtures of copolymers. This method makes use of a characteristic infrared rocking vibration due to sequences of consecutive methylene groups. Such sequences are found in polyethylene and in the segments of ethylene blocks in ethylene-propylene copolymers. This makes it possible to detect them at 730 and 720 cm⁻¹. There are bands at both these locations in the infrared spectrum of the crystalline phase but only at 720 cm⁻¹ in the amorphous phase. The ratio of these two bands in the infrared spectrum of a polymer film at room temperature is a rough measure of crystallinity. As seen by this ratio, the infra-red spectra of the copolymers show varying degrees of polyethylene type crystallinity, dependent on ethylene concentration and method of incorporation. It is this varying degree of crystallinity which allows the qualitative detection of ethylene homopolymer in these materials. A calibration curve of absorbance at 720 cm⁻¹ versus ethylene was made from known mixtures for both hot and the cold runs. Both plots resulted in straight lines from which the following equations were calculated.

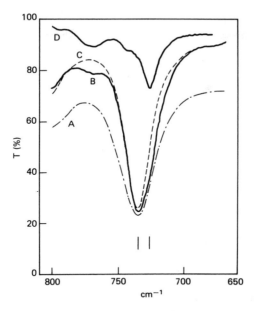

Fig. 44   Infrared spectra of model compounds: (A) hydrogenated natural rubber; (B) squalane; (C) 2,6,
10,14-tetramethylpentadecane; (D) squalane
compensated with 2,6,10,14-tetramethylpentadecane.

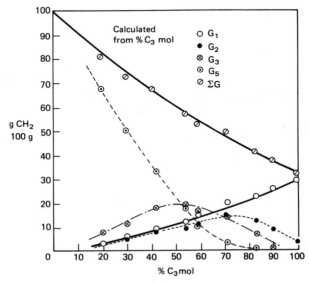

Fig. 45   Distribution of $(-CH_2-)_n$ sequences as a function
of molar composition.

Analysis of Plastics

% Ethylene at 180°C   = $A/(0.55b)$

where A = absorbance measured at 720 cm^{-1} and b = thickness of wire spacer in centimeters and

% Ethylene at room temperature   = $A/(3.0b)$

A series of ethylene-propylene block copolymers prepared with C^{14}-labelled ethylene was analysed for per cent ethylene incorporation by radiochemical methods. These samples when scanned at 180°C gave results which checked with the radiochemical assay quite well. However, when the cooled samples were scanned, the results from the cold calibration were low in comparison with the known ethylene content. These data are shown in Table 21.

TABLE 21 C^{14} Labelled Ethylene-Propylene Copolymers

		Ethylene, %	
Sample No.	Radiochemistry	Hot infrared scan	Cold infrared scan
3401	2.4	2.9	0.9
3402	4.0	3.65	1.3
3403A	22.4	20.7	14.7
3403B	24.5	22.2	15.3
3404	12.4	13.0	7.1
3405	14.0	14.1	7.5

A pair of samples were prepared in which the active sites on the growing propylene polymer were eliminated by hydrogen before addition of ethylene. Practically identical values for per cent ethylene incorporation were calculated for both the hot and the cold scans. These data are shown in Table 22. Lomonte and Tirpak (486) discuss the implications of these findings.

TABLE 22   Samples with Hydrogen-Reduced Active Sites

Sample No.	Ethylene %	
Sample No.	Hot infrared scan	Cold infrared scan
1487	5.0	5.3
1553	7.1	6.9

Tosi and Ciampelli (488) have reviewed the literature on the infrared analysis of ethylene-propylene copolymers and concluded that none of the methods described satisfactorily solved the problem of determining the composition of ethylene copolymers containing less than 20% of propylene.

Tosi and Simonazzi (487) have described two infrared methods for the evaluation of the propylene content of ethylene rich ethylene-propylene copolymers. The first is based on the ratio between absorbances of the 7.25 μ band and the product of the absorbances by the half width of the 6.85 μ obtained on diecast polymer film at 160°C and the second method is based on measurements at 1.69 μ and 1.76 μ in the near infrared.

The 7.25 μ band is also employed in the method proposed by Corish and Tunnicliffe (489) which is likely the one of widest applicability for C_2-C_3

The calibration curve, based on a series of standard copolymers prepared with ^{14}C-labelled either ethylene or propylene as shown in Fig. 46 is obtained by plotting the $A_{7.25}/A_{6.85}$ ratio against the C_3 weight fraction.   To improve precision Tosi and Simonazzi (487) ratioed $A_{7.25}\mu$ to the product of the absorbance of the 6.85 µ band by its half-width.   For the sake of simplicity the latter quantity was expressed in wavelengths instead of wavenumbers: $\Delta\lambda$ ½ $A_{6.85}$ is approximately 0.10 µ at room temperature and 0.14 at 160°C.

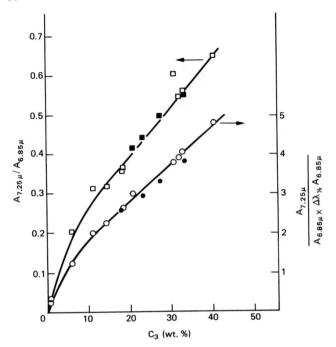

Fig. 46   Calibration curves for the determination of C_3
in high-C_2 copolymers. Full points refer to
model hydrogenated poly-3 methyl alkenamers).

Table 23 lists the calibration data of Tosi and Simonazzi (487).

Tosi and Simonazzi (487) compared the upper curve of Fig. 46 to calibration curves obtained by other workers for C_2-C_3 copolymers based on the $A_{7.25}/A_{6.85}$ ratio.   This comparison when the C_3 content is not too low should hold irrespective of the recording temperature: in fact they found only minor differences in the $A_{7.25}/A_{6.85}$ ratio at room temperature and at 160°C (as seen in Table 24).

However, as seen in Table 24, data obtained by various workers for a range of ethylene-propylene copolymers do differ strongly, and it is rather difficult to account for such differences, which are much too large to be justified only by the different choice of base lines.   The variety of catalyst systems employed to synthesize copolymers should not affect the ratio.   The type of spectrophotometer and the settings adopted to record the spectra probably exert an appreciable influence on the $A_{7.25}/A_{6.85}$ ratio.

Analysis of Plastics

TABLE 23 Calibration Points for the Composition Analysis
of C_2-C_3 Copolymers (Spectra Recorded at 160°C)

Sample	C_3 (wt-%)[a]	$A_{7.25}/A_{6.85}$	2 (%)[b]	$\dfrac{A_{7.25}}{A_{6.85}}$ x	2 (%)[b]
3137-25	40.5	0.652 ± 0.011	1.7	4.80 ± 0.20	4.1
3629-48	32.9	0.561 ± 0.053	9.5	4.07 ± 0.06	1.5
3629-44	32.0	0.548 ± 0.032	5.8	3.92 ± 0.06	1.5
3137-38	30.7	0.606 ± 0.024[c]	4.0	3.80 ± 0.07	1.9
3137-31	20.7	0.416 ± 0.018	4.2	3.02 ± 0.05	1.7
3297-55	18.2	0.365 ± 0.037	10.0	2.62 ± 0.09	3.4
3297-59	18.0	0.357 ± 0.015	4.3	2.59 ± 0.07	2.7
3274-53	14.2	0.318 ± 0.015	4.8	2.25 ± 0.04	1.8
Blend No. 1	11.0	0.315 ± 0.022	6.8	2.00 ± 0.13	6.5
Blend No. 2	6.0	0.204 ± 0.013	6.4	1.25 ± 0.06	5.0
Polyethylene	0.7	0.036 ± 0.007	18.5	0.28 ± 0.05	20.3

[a] Radiochemical analysis

[b] the standard deviation has been multiplied by the Student's $t_{95\%}$ coefficient
to the number of replications (on the average 5) for each calibration
point.

TABLE 24 Comparison of Calibrations Based on the
$A_{7.25}/A_{6.85}$ Ratio

C_3 (wt-%)	$A_{7.25}/A_{6.85}$				Toni & Simonazzi (487)
	Ref. 490	Ref. 491	Ref. 492	Ref. 494	
10	0.21	0.12	-	0.15	0.264
20	0.35	0.24	-	0.28	0.395
30	0.47	0.38	-	0.41	0.521
40	0.60	0.52	0.44	0.53	0.647
50	0.70	0.68	0.53	0.65	-

Earlier, Bucci and Simonazzi (495) had proposed a near infrared method involving measurements at 1.69 μ and 1.76 μ to determining propylene in ethylene-
propylene copolymers containing higher concentrations of propylene,/ above
25% by weight. Tosi and Simonazzi (496) investigated the applicability of
this method to copolymers containing less than 25% propylene.

They feared that this method could fail to give reproducible results below
25% C_3 by weight, since at these compositions an accurate measurement of the
1.69 μ band absorbance becomes very difficult, there being only an impercep-

tible shoulder on the side of a steep band, and the position where to make
the absorbance reading better being determined by cutting the spectrum with
a vertical line at a fixed distance from the 1.76 μ reference band than by
eye. Fortunately, with a high resolution instrument such as the Cary model
14 spectrophotometer, the near-infrared method turned out to be applicable
also to copolymers with a molar C_3 content smaller than ca. 15%. In fact,
extrapolation of the calibration curve formerly obtained from standards with
higher C_3 content produced fairly accurate results for five labelled copolymers, also mechanical mixtures of C_2-C_3 copolymers with polyethylene, and
hydrogenated poly-3-methyl alkenamers. This agreement also proves that the
method is barely influenced by C_2 crystallinity. A typical spectrum of a
copolymer containing 20.7% w/w C_3 is shown in Fig. 47.

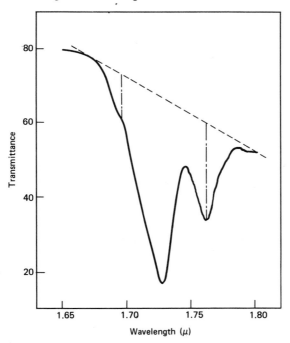

Fig. 47   Near-infrared spectrum (recorded at room
          temperature) of ethylene-propylene copolymer
          containing 20.7 wt% propylene.

Brown et al (497) showed that pyrolysis of ethylene-propylene copolymers at
450°C produces derivatives that are rich in unsaturated vinyl and vinylidene
groups, similar to the pyrolysis of natural rubber and styrene-butadiene
rubber mixtures (500), which produces vinyl groups derived from the butadiene part of the molecule and the vinylidene groups from the methyl branches
of the isoprene units. This unsaturation exhibits strong absorption in the
infrared region. The ratio of the absorption of the vinyl groups to that of
the vinylidene groups varies with the mole fraction of propylene in saturated ethylene-propylene copolymers (498). Making use of this ratio, they
developed an analytical method for determining propylene in both raw and
vulcanized ethylene-propylene copolymers. The vinyl group absorbs at about
909 cm^{-1} and the vinylidene at about 889 cm^{-1} (499, 498).

Fig. 48 shows some typical spectra obtained on the pyrolyzates in the region
of 950 cm^{-1} to 850 cm^{-1}. The values of the ratio, R (X100) range from 9.977
to 0.0290, respectively, for 0 to 100 mole % propylene for the raw samples
and from 5.440 to 0.0431, respectively, for 10 to 100 mole % propylene for
the vulcanized samples. The common logarithm of the ratio, R, can be represented by a linear function of the mole % of propylene in the copolymer.

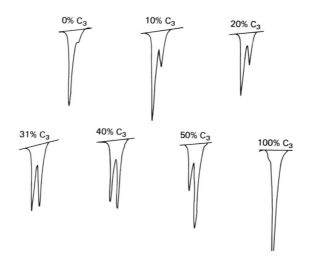

Fig. 48   Typical infrared spectra of pyrolyzates obtained
          from raw ethylene-propylene copolymers. The
          position of the absorption peaks from left to
          right are near 909 cm^{-1} and 889 cm^{-1} respectively,
          for each pair and the nominal compositions are
          indicated in mole % propylene.

Table 25 lists the results, expressed as common logariths of 100 R, for
unvulcanized samples.

Raman and infrared spectra of ethylene-propylene copolymers indicated that
overall amounts of propylene altered crystallinity in blocks of ethylene
(501). Tosi (502) shows that experiment agreed with theory in ethylene-
propylene copolymer composition and intensity of CH_2 rocking vibrations of
methylene sequences. Popov and Duvanova (505) used a comparable approach to
analysis of ethylene-propylene block copolymers. In addition to infrared
spectroscopic analysis for ethylene-propylene-diene terpolymers, Morimoto
and Okamoto (506) used solubility tests and electron microscopy to character-

TABLE 25 Log_{10} (100 R) for Polymer Pyrolyzates
(Raw Samples)

Log_{10} (100R) at various propylene concentrations, mole %

0	10	20	31	40	50	100
2.827	2.584	2.309	2.041	1.931	1.620	0.695
2.840	2.525	2.309	2.048	1.896	1.614	0.743
2.946	2.604	2.318	2.060	1.940	1.592	0.596
2.999	2.587	2.376	2.047	1.908	1.596	0.580
2.996	2.552	2.238	2.097	1.886	1.589	0.542
2.954	2.568	2.327	2.055	1.896	1.588	0.432
2.989	2.562	2.315	2.063	1.902	1.589	0.542
2.951	2.578	2.301	2.048	1.916	1.582	0.461
2.897	2.567	2.340	2.053	1.933	1.620	0.591
2.964	2.561	2.362	2.068	1.904	1.588	0.658
2.9365	2.5687	2.3193	2.0579	1.9114	1.5979	0.5840 Av.

ize organic solvent insoluble fractions. Seeger and Exter (507) and Seno and Tsuge (508) have carried out microstructure studies on sequence distributions in ethylene-propylene copolymers.

Nuclear Magnetic Resonance Spectroscopy

Carbon 13 NMR has proved to be an excellent technique for analysis of sequence distributions and comonomer contents in ethylene-propylene copolymers (509-517). These analyses are particularly straightforward if one of the monomer units is present at a level of 95% or greater because the other monomer will then occur primarily as an isolated unit.

Porter (509) has applied NMR spectroscopy to the determination of branching in polyolefins. It is significant that he quoted no results for branching in high density polyethylene and ethylene-propylene copolymers. It is believed that quantitative data cannot be obtained for copolymers containing less than 10% polypropylene by this method. The NMR method is applicable, however, to propylene-ethylene copolymers where the propylene content is greater than 50%. The investigation of the chain structure of copolymers by ^{13}C-NMR has several advantages because of the large chemical shift involved. Crain et al (514) and Cannon and Wilks (515) have demonstrated this for the determination of the sequence distribution in ethylene-propylene copolymers. In these studies, the assignments of the signals were deduced by using model compounds and the empirical equation derived by Grant and Paul (517) for low molecular weight of linear and branched alkanes. Little work has been reported on the quantitative analysis of polymers by ^{13}C-NMR spectroscopy (516) by use of the continuous wave or pulse Fourier transform technique (516). Tanaka and Hatada (518) studied the accuracy and the precision of such intensity measurements in ^{13}C-NMR spectra and elucidated the distribution of monomeric units in ethylene-propylene copolymers. Ethylene-propylene copolymers can contain up to four types of sequence distribution of monomeric units. These are propylene to propylene (head-to-tail) and head-to head) ethylene-propylene and ethylene to ethylene. These four types of sequence and their average sequence lengths of both monomer units are measured by the Tanaka and Hatada (518) method. Measurements were made at 15.1 MH_3. Assignments of the signals were carried out by using the method of Grant and Paul (517) and also by comparing the spectra with those of squal-

ane, hydrogenated natural rubber, polyethylene and atactic polypropylene.
The accuracy and the precision of intensity measurements, that is, the deviation from the theoretical values and the scatter of the measurements, respectively, were checked for the spectra of squalane and hydrogenated natural
rubber and were shown to be at most 12% for most of the signals.

Randall (519) has examined the branching content for ethylene-propylene copolymers containing predominantly ethylene but, until the work of Paxton and
Randall (520), no method had been reported for corresponding measurements on
copolymers containing predominantly propylene. This method utilizes infrared spectroscopy at 732 cm^{-1} and 13C NMR. Paxton and Randall (520) used 13C
NMR to provide reference standard for the less time-consuming infrared measurements. Their work shows that an excellent correlation, which can be used
for quantitative analysis, is obtained between 13C NMR and infrared results
on a series of ethylene-propylene copolymers containing between 97 and 99%
propylene.

As shown by a typical example in Fig. 49 each 13C NMR spectrum was recorded
with proton noise-decoupling to remove unwanted 13C-^{1}H scalar couplings. No
corrections were made for differential nuclear Overhauser effects (NOE) since
constant NOEs were assumed (523, 524) in agreement with previous workers
(509-512). Constant NOEs for all major resonances in low ethylene content
ethylene-propylene copolymers have been reported (525).

The 13C NMR spectrum of an ethylene-propylene copolymer, containing approximately 97% propylene in primarily isotactic sequences, is shown in Fig. 49.
The major resonances are numbered consecutively from low to high field. Chemical shift data and assignments are listed in Table 26. Greek letters are
used to distinguish the various methylene carbons and designate the location
of the nearest methine carbons.

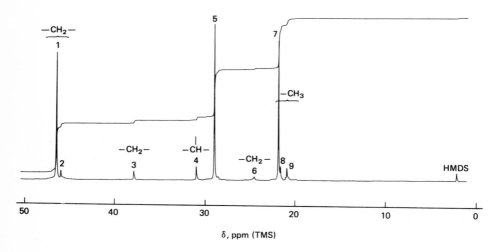

Fig. 49 A 13C NMR spectrum at 25.2 MHz and 125°C of a 97/3
         propylene-ethylene copolymer in 1,2,4-trichloro-
         benzene and perdeuterobenzene. The internal
         standard copolymer in 1,2,4-trichlorobenzene and
         perdeuterobenzene.

TABLE 26   Observed and Reference ^{13}C NMR Chemical Shifts in ppm for Ethylene-Propylene Copolymers and Reference Polypropylenes with Respect to an Internal TMS Standard

Line	carbon	3/97 E/P[a]	E/P (Ray et al(512))	97/3[a] E/P	sequence assignment	Ref. cryst. Ref 523	PP amorphous Ref 523
1	$\alpha\alpha$-CH$_2$	46.4	46.3		PPPP	46.5	47.0-47.5 r 46.5 m
2	$\alpha\alpha$-CH$_2$	46.0	45.8		PPPE		
3	$\alpha\alpha$-CH$_2$	37.8	37.8		PPEP		
4	CH	30.9	30.7		PPE		
5	CH	28.8	28.7		PPP	28.5	28.8 mmmm 28.6 mmmr 28.5 rmmr 28.4 mr+rr
6	$\beta\beta$ CH$_2$	24.5	24.4		PPEPP		
7	CH$_3$	21.8	21.6	P	PPPPP	21.8	21.3-21.8mm 20.6-21.0mr 19.9-20.3rr
8	CH$_3$	21.6	21.4		PPPPE		
9	CH$_3$	20.9	20.7		PPPEP		
	CH$_3$		19.8	19.8	EPE		
	CH		33.1	33.1	EPE		
	α-CH$_2$			37.4	EPE		
	β-CH$_2$			27.3	EPE		
	-CH$_2$)$_n$-		29.8	29.8	EEE		

[a]As measured by Paxton and Randall (520)

Paxton and Randall (520) in their method use the reference chemical shift data obtained on a predominantly isotactic polypropylene and on an ethylene-propylene copolymer (97% ethylene). They concluded that the three ethylene-propylene copolymers used in their study (97-99% propylene) contained principally isolated ethylene linkages. Knowing the structure of their three ethylene-propylene copolymers, they used the 13C NMR relative intensities to determine the ethylene-propylene contents and thereby establish reference copolymers for the faster infrared method. After a detailed analysis of the resonances Paxton and Randall (520) concluded that methine resonances 4 and 5 (Table 26) gave the best quantitative results to determine the comonomer composition. The composition of the ethylene-propylene copolymers were determined by peak heights using the methine resonances only. In no instance was there any evidence for an inclusion of consecutive ethylene units. Thus, composition data from 13C NMR could now be used to establish an infrared method based on a correlation with the 732 wavenumber band which is attributed to the rocking mode, γ r, of the methylene trimer, -(CH$_2$)$_3$- .

Randall (527) has developed 13C NMR quantitative method for measuring ethylene propylene mole fractions and methylene number average sequence lengths
in ethylene-propylene copolymers. He views the polymers as a succession of
methylene and methyl branched methine carbons as opposed to a succession of
ethylene and propylene units. This avoids problems associated with propylene inversion and comonomer sequence assignment. He gives methylene sequence distributions from one to six and larger consecutive methylene carbons
for a range of ethylene-propylene copolymers and uses this to distinguish
copolymers which have either random, blocked or alternating comonomer sequences.

Natural abundance ^{13}C NMR at 22.6 and 39.9 MHz and nuclear Overhauser enhancements were used by Sanderson and Kamorski (528), to study the stereochemical configuration in a low ethylene content ethylene-propylene copolymer.
No evidence was found for a mixture of meso and racemic configurations in
the polypropylene chain across an inserted ethylene unit.

Ray et al (529) determined comonomer content in isotactic ethylene-propylene
copolymers, complete diad and triad content as well as partial tetrad and
pentad distributions using ^{13}C NMR.

Proton Magnetic Resonance Spectroscopy

Dudek and Buesche (530) have suggested a conformational analysis by proton
magnetic resonance spectroscopy of ethylene-propylene rubbers.

Proton magnetic block copolymers has been used to characterize ethylene-
propylene block copolymers, containing from zero to 40% ethylene, either as
homopolymer or copolymer blocks, Porter (531). The test is independent of
tacticity and provides qualitative information on copolymer sequencing and
propylene chain structure. The analysis was developed using a series of
standard reference polymers synthesized to contain various ratios of C^{14-}
tagged ethylene and propylene. All measurements were made on 10% polymer
solutions in diphenyl ether. Analyses are accurate to about $\pm$ 10% at higher
ethylene concentrations. The method is sensitive, with less precision, to
below 1% for ethylene either as blocks or homopolymer. Porter (531, 532)
found that resolution improves with increasing temperature and with decreasing polymer concentration.

Fig. 50 shows the PMR spectrum of hot n-heptane solution of atactic polypropylene containing up to 40% syndiotactic placement, and by Natta definition may be called stereoblock (531). The spectrum is inherently complex,
as a first-order theoretical calculation indicates the possibility of at
least 15 peaks with considerable overlap between peaks because differences
in chemical shifts are about the same magnitude as the splitting due to spin-
spin coupling.

The largest peak, at high field in Fig. 50 represents pendant methyls in propylene units. It is characteristically split by the tertiary hydrogen. By
area integration, about 20% of the nominal methyl proton peak is due to
overlap of absorption from chain methylenes. This overlap is consistent
with a reported syndiotactic triplet, two peaks of which are close to A and
B in Fig. 50 and a third peak which falls with the low field branch of the
methyl split (532). The absence of a strong singlet peak in the methylene
range indicates the virtual absences of "amorphous" polymer in the atactic
polypropylene shown in Fig. 50 (531, 532), which could possibly be due to
head-to-head and tail-to-tail units. The low field peak represents the
partial resolution of tertiary protons which are opposite the methyls on the
hydrocarbon chain.

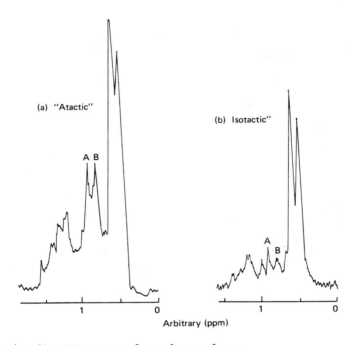

Fig. 50   PMR spectra for polypropylenes

Fig. 50 also shows a spectrum for an isotactic polypropylene, > 95% isotactic by solubility (< 5% soluble in boiling heptane).   This spectrum has the
same general character as the atactic polypropylene.   The important difference is a marked decrease of peak intensity in the chain methylene region.
This decrease is caused by extensive splitting and the difference in chemical shifts for the nonequivalent methylene hydrogens in isotactic environments.   This is in accord with the study of Stehling (526), on deuterated
polypropylenes, which indicates that much of isotactic methylene absorption
is "buried" beneath the methyl and tertiary hydrogen peaks.   The fractional
area in the nominal methylene region of the spectrum is thus sensitive to
the number of isotactic and syndiotactic dyads and, therefore, may be used
as a measure of polypropylene tacticity.

Fig. 51 shows the spectrum of a physical mixture of a polypropylene and
linear polyethylene.   The low field absorption in Fig. 51, characteristic of
aromatic hydrogens, is due to the polymer solvent, diphenyl ether, which was
used throughout.   Polymer concentrations in solution can be readily calculated from the ratios of peak areas adjusted to the same sensitivity.   The
superpositions of spectra that were obtained separately on the homopolymers
show the same pattern, with different intensities, as spectra on physical
mixtures.   The polyethylene absorption falls on the peak marked A of the
chain methylene complex in polypropylene.   Peak B, also due to chain methylenes in polypropylene, is resolved in both spectra in Fig. 51.   Fig. 51 also
gives the spectrum of an ethylene-propylene block copolymer.   The ethylene
contribution again falls on peak A.

Analysis of Plastics

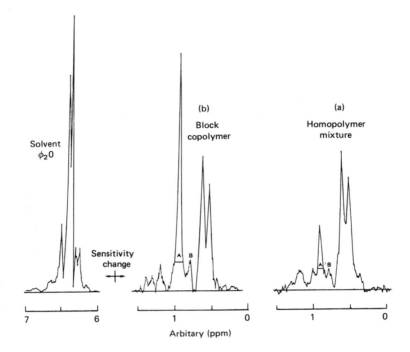

Fig. 51   PMR spectra for polymers of ethylene and propylene

Various workers have developed analyses for physical mixtures and block
copolymers based on the ratio of the incremental methylene area to the total
polymer proton absorption. This concept has been tested by Barrall et al
(533), using PMR analyses on a series of physical mixtures and block copolymers synthesized with C^{14}-labelled propylene and others with C^{14}-labelled
ethylene. A most important feature of this analysis is that the methylene
peaks A and B have virtually the same relative heights in polypropylenes with
a variety of tacticities (note Fig. 50). This is also true for PMR spectra
given by Satoh and others for a tactic series of polypropylenes (534).. This
suggests that PMR analyses for ethylene are independent of tacticity since
the area increment of peak A above peak B has been used for analysis.

Qualitative polypropylene tacticities can be estimated by PMR not only for
homopolymers (Fig. 50), but also in the presence of polyethylene and ethylene copolymer blocks. The relative heights of the peaks for secondary and
for tertiary hydrogen in Fig. 51 indicates that the polypropylene in the
copolymer and the physical mixture is dominantly isotactic.

Table 27 gives analyses carried out by Porter (531) for a series of physical
mixtures made up with tagged polypropylene standardized by radiocounting.
The three sets of values are in good accord.

Side Chain Analysis of Ethylene -   Olefin Copolymers

Kamath and Barlow (535) showed by a study of the vacuum radiolysis products

TABLE 27   Analyses of Linear Polyethylene-Polypropylene
            Physical Mixtures

Polyethylene, wt.-%

Sample	Made up	PMR	Tracer
6	5.0	4.6	4.2
7	9.0	8.6	7.6
5	17.0	16.7	16.3
8	34.0	31.1	31.0

of a series of ethylene α -olefin copolymers and ethylene homopolymers that
if a correction is applied, to take into account the fragments arising from
scission at chain ends, the remaining products can be quantitatively accounted for as entirely due to scission of side branches introduced onto the
back-bone chain by the α -olefin comonomer. The cleavage of branches takes
place, for all practical purposes, exclusively at the branch points at which
the branches are attached to the backbone chain. The same data together
with similar radiolysis data of poly(3-methyl pentene-1) and poly(4-methyl
pentene-1) further showed that all branches cleave with equal efficiency,
regardless of their length. Radiolysis does, therefore, provide a reliable
and convenient tool for the quantitative characterization of high-pressure
polyethylene with regard to the unique short-chain branching distribution
that is characteristic of each.

The results obtained in a series of irradiations of ethylene- α -olefin
copolymers containing about 4 mole % comonomer are shown in Table 28.

TABLE 28   Radiolysis Products of Ethylene   -Olefin
            Copolymers

(U235 radiation source)

G value x 10^{2}*

Copolymers	CH_4	C_2H_6	C_3H_8	$i-C_4H_{10}$	$n-C_4H_{10}$	$n-C_5H_{12}$	$n-C_6H_{14}$
Ethylene-propylene	1.7	0.1	0.1	-	0.03	-	-
Ethylene-butene-1	0.2	1.5	0.1	0.1	0.2	0.03	-
Ethylene-pentene-1	0.2	0.3	2.1	0.02	0.05	-	-
Ethylene-hexane-1	0.4	0.3	0.2	0.03	1.2	0.02	-
Ethylene-octene-1	0.2	0.3	0.1	-	0.1	0.1	1.2
Linear polyethylene	0.2	0.3	0.2	0.03	0.06	0.05	-

*G value defined as the number of molecules of the particular product
produced per gram of sample per 100 e.v. of incident radiation dose. This is
calculated, with 10% accuracy, from the gas chromatographic analysis data of
the products, weight of the polymer sample irradiated, and radiation dose to
which the sample had been subjected.

It was concluded by Kamath and Barlow (535) that during radiolysis of polyethylene there also takes place a certain amount of random scission at chain
ends in addition to the cleavage of branches, and that the observed extraneous hydrocarbons are simply the products derived from this fragmentation at

the chain ends. It is quite probable that a portion of the extraneous hydrocarbons is derived from scission of stray branches that might have been
introduced on the chains by stray impurities during polymerization, but the
fact that one also observes a consistent decrease in the total amount of
these extraneous hydrocarbons derived from the homopolymer with an increase
in its molecular weight leads to the conclusion that the random scission at
chain ends is their main cause. Obviously, then, if one makes an appropriate
allowance for these radiolysis fragments derived from chain ends, then the
only significant paraffin left in the radiolysis products of each copolymer
is that corresponding to the branch introduced on the polyethylene back-zone
by the comonomer.

Since it is generally agreed that high-pressure polyethylene actually contains a variety of short branches and not just one type of branch, it is
obvious that, if one wants to translate the hydrocarbon analysis of its
radiolysis products into quantitative branch-type analysis, one will also
need accurate information on the relative efficiency of the scission of different branch types. To obtain this information Kamath and Barlow (535) irradiated some additional ethylene α -olefin copolymers, differing significantly in their comonomer content and the hydrocarbons in their radiolysis
products were analyzed as before. Table 29 lists these results. The efficiency of scission is calculated from the known comonomer content (methyl
group analysis) and the observed G values of the principal hydrocarbon after
application of the appropriate correction against the small chain-end fragmentation. It is clear from the data that all branches of up to 6 carbon
atoms or more break off with equal efficiency and that the branch length per
se exerts little or no effect on the ease of scission.

TABLE 29   Efficiency of Radiation Scission of Short Branches

				G value x 10^2				Average G x 10^2
Copolymers	CH_3 per 1000 CH_2	CH_4	C_2H_6	C_3H_8	$n-C_4H_{10}$	$n-C_6H_{14}$		CH_3 per 1000 CH_2
Ethylene-propylene	41.9	3.4						0.074
	26.4	1.7						
Ethylene-butene-1	28.7		2.0					
	21.6		1.5					0.072
	4.1		0.3					
Ethylene-pentene-1	25.6			2.1				0.072
	17.1			1.1				
Ethylene-hexene-1	23.5				1.2			0.073
	7.8				0.7			
Ethylene-octene-1	19.1					1.3		
	15.1					1.2		0.074
	11.6					0.9		

In Table 30 are presented results obtained in the radiolysis of high pressure polyethylenes of various densities. Noteworthy is the disparity between
the total number of all branches, as determined from the radiolysis data,
and the methyl group content, as derived by the infrared method. This may
be attributed to the fact that the usual infrared method of determining the
methyl group content, based on the absorbance at 7.25 μ , does not necessarily count all methyl groups. For example, if two methyl groups are attached to one carbon atom, as in polyisobutylene, the characteristic methyl

absorption at 7.25 splits, giving two bands at 7.20 μ and 7.40 μ respectively (536) which consequently are lost in the methyl group determination procedure.

In line with the observations reported by earlier workers, the two most pop-
ulous branches, according to the radiolysis method, are ethyl and n-butyl,
which occur in the ratio 2:1, as observed by others (537).

TABLE 30   Distribution of Short-Chain Branches in High
Pressure Polyethylene

| Resin No. | Density, g./cc. | Branches per 1000 methylene units[b] | | | | | | Total branches per 1000 methylene | Methyl[a] group content per 1000 methylene |
		-CH$_3$	-C$_2$H$_5$	-C$_3$H$_7$	i-C$_4$H$_9$	-n-C$_4$H$_9$	-C$_5$H$_{11}$		
A	0.934	2.9	11.8	1.7	0.3	5.9	1.6	24.2	18.8
B	0.929	2.5	15.2	2.1	0.3	8.5	2.0	30.6	24.0
C	0.924	3.8	16.7	2.2	-	9.9	1.8	34.4	30.9

[a]Determined by infrared analysis

[b]Calculated from G values of isolated hydrocarbons assuming scission efficiency of 0.073 x 10^2 per 1000 carbon atoms. As short chain branches out
number chain ends by 30;1, no correction for the fragmentation products at
chain ends was made.

X-Ray Diffraction

High-angle x-ray diffraction studies of atactic poly(1-methylhexamethylene),
poly(1-methyloctamethylene), and poly(1-methyldecamethylene), and poly(1-dodecam-
ethylene) have revealed a partial intra- and intermolecular order suggesting
that these copolymers may be considered as ethylene-propylene copolymers containing along the chains 2, 3, 4 and 5 ethylene units for every propylene
unit respectively (538).

Fractionation

Kenyon et al (539) have described methods for the large scale elution fractionation of ethylene-propylene copolymers. 500 g samples of polymer can be
fractionated in 20 hours. Ogawa (540) determined the molecular weight of
ethylene-propylene copolymers using gel permeation chromatography.

Fractionations of ethylene-propylene copolymers have been reported by Gol'
denberg et al (541) and Ogawa and coworkers (542, 543).

Ogawa and Inaba (544) compared a computer simulation of the solution fractionation of ethylene-propylene copolymers with experimental results and
showed that the products should be classified into five types: ethylene-
propylene copolymer, polyethylene-polypropylene blends, polypropylene-copoly-
mer blends, polyethylene-copolymer blends, and propionate copolymers on the
sequence distribution of vinyl acetate units.

Sitnikova et al (545) fractionated ethylene-propylene copolymer blends.

Molecular Weight

Taylor and Graham (546) have described a dual beam turbidmetric photometer
which they claim has distinct advantages over the earlier single beam instruments (547, 548) and have applied it to the determination of the molecular
weight distribution of polyethylene,polypropylene, ethylene-propylene rubbers
and other polymers (549).

Miscellaneous

Bushick (550) determined monomer reactivity ratios in ethylene-propylene
copolymers. Semura (551) carried out swelling measurements of ethylene
copolymers crosslinked with dicumyl peroxide.

ETHYLENE-PROPYLENE DIENE TERPOLYMERS

Chemical Methods

Polymerization of ethylene and propylene results in a saturated copolymer.
In order to vulcanize this rubber, some unsaturation has to be introduced.
This is commonly done by adding a few percent of a non-conjugated diene (termonomer) such as dicyclopentadiene, 1,4-hexadiene, or ethylidene norbornene
during the polymerization. Since only one of the double bonds of the diene
reacts during polymerization, the other is free for vulcanization. The
amount of unsaturation left in the ethylene propylene diene terpolymer is of
great interest because the vulcanization properties will be affected.

The iodine monochloride method has been used for a variety of polymers.
These polymers include those which are highly unsaturated, such as polybutadiene and polyisoprene (552-555) and polymers having low unsaturation such
as butyl rubber (556) and ethylene propylene diene terpolymer. Considerable
work has been done investigating the side reactions of iodine monochloride
with different polymers (556). These side reactions are substitution and
splitting out rather than the desired addition reaction.

Infrared Spectroscopy

Infrared spectroscopy has been used for the determination of unsaturation in
ethylene-propylene-diene terpolymers (557). Determination of extinction coefficients for the various terpolymers is required if quantitative work is
to be done.

NMR Spectroscopy

Sewell and Skidmore (558) used time averaged NMR spectroscopy at 60 MC/sec
to identify low concentrations of nonconjugated dienes introduced into ethylene-propylene copolymers to permit vulcanization. Although infrared spectroscopy (559) and iodine monochloride unsaturation methods (560) have been
used to determine or detect such dienes, these two methods can present difficulties. Identification of the incorporated third monomer is not always
practical in infrared spectroscopy at the low concentrations involved, and
in the high-resolution NMR spectra of these terpolymers the presence of
unsaturation is not usually detected, since the signals from the olefinic
protons are of such low intensity that they become lost in the background
noise. The spectra obtained by time averaged NMR are usually sufficiently
characteristic to allow identification of the particular third monomer incorporated in the terpolymer. Moreover, as the third monomer initially contains

two double bonds, differing in structure and reactivity, the one used up in copolymerization may be distinguished from the one remaining for subsequent use in vulcanization. Therefore information concerning the structure of the remaining unsaturated entity may be obtained. Table 31 shows the chemical shifts of olefinic protons of a number of different third monomers in the copolymers.

TABLE 31

Third monomer	Chemical shift, Υ
Cyclooctadiene 1,5	4.55
Dicyclopentadiene	4.55
1:4 Hexadiene	4.7
Methylene norbornene	5.25 and 5.5
Ethylidene norbornene	4.8 and 4.9

The cyclooctadiene and dicyclopentadiene terpolymers have olefinic protons with the same chemical shift, 4.55 Υ , and so these cannot be differentiated by this technique but may be distinguished by the use of iodine monochloride. The hexadiene type of terpolymer may be identified by its olefinic resonance at 4.7 Υ . These three monomers have what appears as a single olefinic resonance in the terpolymer. On the other hand, the two norbornadiene types of monomer each show two characteristic resonances. In the methylene nor-bornene terpolymer the olefinic resonances arise from two protons, each giving a separate signal, whereas in the ethylidene norbornene terpolymer there is only one proton, the signal of which appears as a doublet. In view of these considerations it is more difficult to detect the olefinic resonance in the latter instance.

Altenau et al (561) applied time averaging NMR to the determination of low percentages of termonomers such as 1:4 hexadiene, dicyxlopentadiene and ethylidene norbornene in ethylene-propylene termonomers. They compared results obtained by NMR and the iodine monochloride procedure of Lee et al (562). The chemical shifts and splitting pattern of the olefinic response were used to identify the termonomer. Infrared spectroscopy was also app-lied to the identification of the termonomer

Fig. 52 shows the time averaged NMR spectra of ethylene-propylene terpolymers containing various dienes.

Sewell and Skidmore (563) also showed that the termonomer can be identified by time-averaged NMR spectroscopy. They noted that very accurate chemical shifts cannot be determined in time-averaged NMR spectra because of the possibility of magnetic field drift during the scanning time.

Altenau et al (561) also prepared the infrared spectra of 5 thou films of terpolymers of ethylene-propylene with 1,4-hexadiene, dicyclopentadiene and ethylidene neobornene polymer.

Fig. 53 shows infrared spectra of terpolymers containing each of the above mentioned termonomers. Ethylidene norbornene has a broad band at 12.35 μ ; dicyclopentadiene has a broad band at 14.3 μ and a weak band at 110.55 μ ; and 1,4-hexadiene has a strong sharp band at 10.35 μ and a weak band at 11.25 μ. Although these termonomers have absorptions at other wavelengths, they found the above wavelengths to be the most useful for identification.

118   Analysis of Plastics

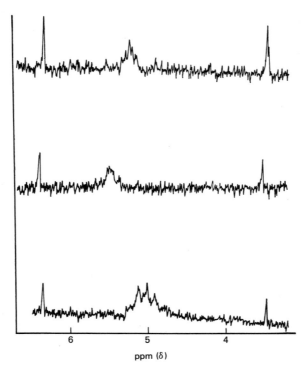

ppm (δ)

Fig. 52 Time-averaged NMR spectra of EPDM containing one
        of the following termonomers
        upper spectrum: 1,4-Hexadiene
        middle spectrum: Dicyclopentadiene
        lower spectrum: Ethylidene norbornene

Table 32 compares the amount of termonomer found by the NMR method of
Altenau et al (561) and by iodine monochloride procedures (554, 562). The
termonomers were identified by NMR and infrared.

Table 32 shows that the data obtained by the NMR method agree more closely
with the Lee, Kolthoff and Johnson iodine monochloride method (562) than
with the iodine monochloride method of Kemp and Peters (554). The difference
between the latter two methods is best explained on the basis of side reactions occurring between the iodine monochloride and polymer because of branching near the double bond (562). The reason for the difference between the
NMR and Lee, Kolthoff and Johnson (562) methods is not clear. · The reproducibility of the NMR method was ± 10 to 15%.

Pyrolysis-Gas Chromatography

Pyrolysis gas chromatography has been used by many workers to determine the
overall composition of ethylene propylene diene terpolymers (564). In atte-
mpting to determine the third component in ethylene propylene diene terpolymers, difficulties might be anticipated, since this component is normally
present in amounts around 5 weight per cent. However, dicyclopentadiene was

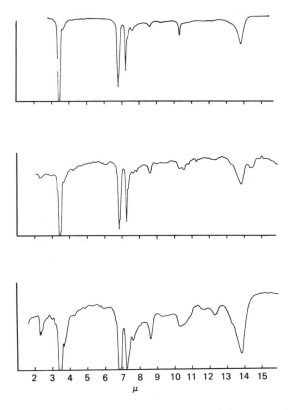

Fig. 53   Infrared spectra of EPDM containing one of the
          following termonomers
          Upper spectrum: 1,4-Hexadiene
          Middle spectrum : Dicyclopentadiene
          Lower spectrum : Ethylidene norbornene

identifiable in ethylene propylene diene terpolymers even when the amount
incorporated was so small it was of no practical value.

Van Schooten and Evenhuis (565, 566) applied their pyrolysis-(500ºC) hydrogenation-gas chromatographic technique to unsaturated ethylene-propylene
copolymers, i.e. ethylene-propylene-dicyclopentadiene and ethylene-propylene
norbornene terpolymers. The pyrograms summarized in Table 33 show that very
large cyclic peaks are obtained from the unsaturated ring: methyl cyclo
pentane is found when methyl-norbornadiene is incorporated; cyclo-pentane ·
when dicyclopentadiene is incorporated; methylcyclohexane and 1,2 methyl
cyclohexane when the addition compounds of norbornadiene with, respectively,
isoprene and dimethylbutadiene are incorporated; and methylcyclopentane when
the dimer of methylcyclopentadiene is incorporated. The saturated cyclopentane rings present in the same ring systems in equal concentrations, however,
give rise to peaks which are an order of magnitude smaller. Obviously,
therefore, the peaks which stem from the termonomer could be used to determine

TABLE 32   Determination of Termonomer in Ethylene-Propylene
           Diene Terpolymers - Comparison of Methods.
           (Data shown as Weight per cent Termonomer)

NMR	Lee (562) Kolthoff & Johnson	Kemp & Peters (554)	Termonomer
7.3	5.0		Dicyclopentadiene
1.1	1.6		1,4-Hexadiene
5.7	5.9	9.0	Ethylidene norbornene
2.8	3.6	4.5	Ethylidene norbornene
1.7	2.3	4.8	Ethylidene norbornene
4.6	5.4	6.0	Ethylidene norbornene

its content if a suitable calibration procedure could be found.

Van Schooten and Evenhuis (565, 566) subjected a number of terpolymers containing dicyclopentadiene and having different amounts of unsaturation, to
pyrolysis-GLC analysis and plotted the height of the characteristic peaks
(or the ratio of the heights of these peaks to the height of the n-C peak)
against unsaturation measured by ozone absorbtion (567). In Fig. 54 these
relationships are given for the cyclopentane peaks. Similar curves were
found for the methylcyclopentane or ethylcyclopentane peaks.

ETHYLENE-BUTENE COPOLYMERS

Infrared Spectroscopy

The infrared absorption of ethylene copolymers in the 700-850 cm^{-1} region can
provide information about their sequence distribution. Studies on model
hydrocarbons (568-570) have shown that the absorption of methylene groups in
this region is dependent on the size of methylene sequences in the compounds.
Methylene absorptions observed in this region and their relation to structures accuracy in ethylene-propylene copolymers are shown in Table 34.

In copolymers of ethylene with monomers such as 2-butene containing structures 3b and 5b absorption at 751 cm^{-1} can provide a measure of the number of
ethylene units centred in butene-ethylene-butene triads. Indeed Natta and
coworkers (571) cited the intense absorption shown by such copolymers at
751 cm^{-1} (13.29μ) as evidence for their highly alternating structure.

Brown (575) has developed a radiochemical method for the determination of
the composition of ethylene-butene-1 copolymer gels in the 0.6 to 7.3% butene
range, i.e. 1.5 to 20 ethylene branches/1000C. Ethylene-butene-1 copolymers
containing up to 4% molar 4-^{14}C butene-1 were prepared and the radioactive
^{14}C contents measured by scintillation counting. Standards prepared by this
method are suitable for the calibration of the more rapid infrared method
which involves measurements of the characteristic absorption of the ethyl
branches at 13 μ (576). Absorbance at 13   is directly proportional to the
concentration of ethyl branches up to 10 per 1000C

Methyl groups in ethylene-α-olefin copolymers such as in ethylene-butene-1
have been estimated from the relationship between absorbance at 1378 cm^{-1} and
1369 cm^{-1} (572).

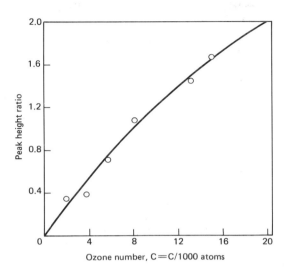

Fig. 54   Ethylene-propylene-dicyclopentadiene rubbers.
Peak height ratio CyC_5/n-C_4 versus ozone number.

Yur'eva et al (573) have developed a method for the determination of the composition of 1-butene-ethylene-propylene copolymers using mechanical mixtures of
1-butene-propylene copolymers and ethylene-propylene copolymers for calibration.

Popov and Duvanova (574) determined α-olefins in 1-butene-ethylene copolymer
by determining the concentration of methyl groups and introducing a correction
for the concentration of terminal methyl groups in the main polymer chain.

Pyrolysis Gas Chromatography

Neumann and Nadeau (578) applied pyrolysis gas chromatography to the examination of ethylene-butene copolymers. Pyrolyses were carried out at 410°C in an
evacuated gas vial and the products swept into the gas chromatograph. Under
these pyrolysis conditions, it is possible to analyze the pyrolysis gas components and obtain data within a range of about 10% relative. The peaks observed on the chromatogram were methane, ethylene, ethane, combined propylene and
propane, isobutane, 1-butene, n-butene, trans-2-butene, cis-2-butene, 2-methyl-
butane, and n-pentane. A typical pyrolysis chromatogram for polyethylene is
shown in Fig. 55.

Figure 56 shows the relationship between the amount of ethylene produced on
pyrolysis and the amount of butene in the ethylene-butene copolymer, which was
determined by an infrared analysis for ethyl branches. (579) The y intercept
of 16.3% ethylene should represent that amount of ethane which would result
from a purely linear polyethylene. An essentially unbranched Phillips-type
polyethylene polymer yielded 14.5% ethylene, which is fairly close to the predicted 16.3%.

As one would expect, ethyl branches will increase with the amount of butene

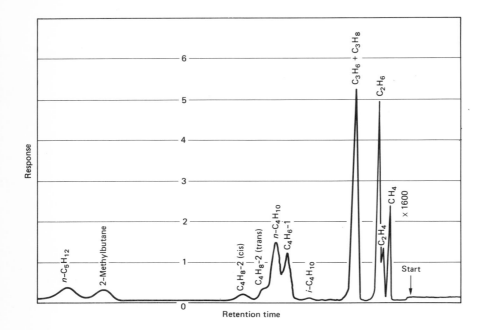

Fig. 55   A typical polyethylene pyrolysis chromatogram.

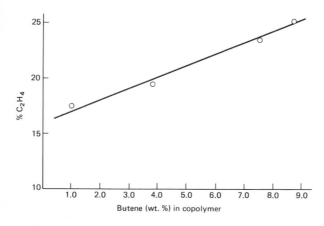

Fig. 56   Per cent C_2H_6 as function of butene content of
          ethylene-butene copolymers.

copolymerized with the ethylene. This is in agreement with the work of Madorsky
and Straus (581) who found that the thermal stability and breakdown products
obtained on pyrolysis can be related to the strengths of the C-C bonds in the

Table 33   Characteristic Peaks in the Pyrograms of a Number of Unsaturated Terpolymers.

Sample No.	1	2	3	4	5	6	7	8
% mC$_3$	32	n.d.	34	34	32	35	n.d.	32
Unsaturation, C=C/1000C	0	0.8 Norbornene	≤ 0.5	8	8.1 dicyclopentadiene	9.5	13	4.5
Termonomer	none							(dimer-of methyl cyclopentadiene)
Peak height cm, n-C$_4$	43	43.5	45	38	26.5	37	47	34
Peak height cm, cyclopentane	0.0	6.5	4.1	2.2	28.7	2.1	2.8	1.2
Peak height cm, methylcyclopentane	0.6	1.2	1.4	45	2.1	2.4	2.5	9.7
Peak height cm, cyclohexane	0.35	0.5	0.5	0.6	0.9	0.7	0.5	0.5
Peak height cm, methylcyclohexane	0.6	0.7	0.7	1.3	0.9	20.6	1.1	0.7
Peak height cm, ethylcyclopentane	0.2	0.5	0.5	3.1	1.1	-	1.2	0.4
Peak height cm, n-propylcyclopentane	0.0	0.15	0.1	0.15	0.5	0.1	0.15	1.7
Peak height cm, 1,2-methylcyclohexane	0.0	0.0	0.0	0.5	0.0	0.0	17.8	0.0

n.d.=not determined.
*=Intake: ca. 16 mmole per 1000 C atoms polymer.

polymer chain - i.e. secondary >tertiary >quaternary. Boyle, Simpson and Waldron (580) have shown by irradiation of hydrocarbon polymer that the major constituents of the paraffins formed are those corresponding in both carbon number and skeletal structure to the side chains in the hydrocarbon polymer.

Van Schooten and Evenhuis (565, 566) applied their pyrolysis-(at 500°C)hydrogenation-gas chromatographic technique to ethylene-butene copolymers prepared with a titanium trichloride triethylaluminium catalyst. Ethylene-butene-1 copolymers prepared by copolymerizing ethylene and butene-1 should give pyrograms with two types of peaks. On the one hand, peaks are expected which are characteristic of long sequences of one monomer, as found in the homopolymer pyrograms, and on the other, peaks which are typical of an alternating ethylene-butene copolymer - in fact exactly the same is found for ethylene-propylene copolymers. Assuming a close analogy between the pyrogram of an alternating ethylene-butene copolymer and that of an alternating ethylene-propylene rubber the characteristic peaks in the former pyrogram would be the 3-ethyl C$_6$ and 3-methyl C$_9$ peaks. These peaks could therefore in principle be used to find differences in degree of alternation between various types of ethylene-butene copolymers. The 3-ethyl C$_6$ peak, however, was found to coincide with the 3-methyl C$_7$ peak, which is

Table 34 <u>Methylene and Copolymer Structures Responsible
for Absorption in Ethylene copolymers</u>.

$(CH_2)_n$	cm^{-1}	sequence	
(CH_2)	815	$\sim CH_2CHRCH_2CHR\ CH_2\ CHR \sim$	(2)
$(CH_2)_2$	751	$\sim CH_2\ CHR\ CH_2\ CH_2\ CHR\ CH_2\sim$	(3a)
		or $\sim CHR\ CHR\ CH_2\ CH_2\ CHR\ CHR\sim$	(3b)
$(CH_2)_3$	733	$\sim CH_2\ CHR\ CH_2\ CH_2\ CH_2\ CHR\sim$	(4)
$(CH_2)_4$	726	$\sim CH_2\ CHR\ (CH_2\ CH_2)_2\ CHR\ CH_2\sim$	(5a)
		or $\sim CHR\ CHR\ (CH_2\ CH_2)_2\ CHR\ CHR\sim$	(5b)
$(CH_2)\ >4$	722	$\sim CH_2\ CHR\ (CH_2\ CH_2)_n\ CH_2\ CHR\sim$	(6)

characteristic of long butene sequences, and so only the 3-methyl C$_9$ peak remains.

Another polymer which can be regarded as an ethylene-butene copolymer is obtained by hydrogenation of polybutadienes. Each butadiene segment in the 1,4 arrangement corresponds, after hydrogenation, to two ethylene units and a butadiene in 1,2 configuration to a butene unit, long sequences of 1,4 arrangement therefore give long ethylene sequences and long 1,2 sequences give long butene sequences. An alternating 1,4-1,2 butadiene polymer will have methylene sequences of five units, and a random 1,4-1,2 distribution will have isolated methylene units and sequences of 5, 9, 13 and more, ethylene units; sequences of three methylene units, characteristic of an alternating ethylene-butene copolymer, will be absent, however.

On applying the rules which they have found to hold for the pyrograms of poly-α-olefins to the hydrogenated alternating 1,4-1,2 polybutadiene, Van Schooten and Evenhuis (565, 566) expected the pyrogram of this polymer to contain a rather large n-C$_8$ peak and a very small 3-methyl C$_9$ peak, in addition to the other peaks found for ethylene-butene copolymers.

$$\{(CH_2)_5 - \underset{\underset{C_2H_5}{|}}{CH}\}_n$$

They pyrolyzed samples of both types of ethylene-butene-1 copolymers. A survey of the areas of the main peaks is given in Fig. 57 for a copolymer. Peak heights are given as a function of butene content in Figs. 58 and 59 for directly prepared copolymers. Differences in the heights of the respective peaks for the two types of copolymers are only slight. An exception, however, was the considerable difference found between the 3-methyl C$_9$ peaks (see Fig. 60) as was only to be expected, considering that this peak is characteristic of butene-ethylene-butene (b-e-b) sequences:

$$\sim C - C - C \mid C - C - C - C - C - C \mid$$
$$\underset{C_2}{|} \qquad \underset{C_2}{|} \qquad \underset{C_2}{|}$$

Van Schooten and Evenhuis (582) went on to demonstrate the presence of very low levels of ethyl side groups in their elucidation of the microstudies of ethylene-butene-1 copolymers. By the use of larger amounts of sample (1 mg instead of 0.4 mg) and a high sensitivity of the detection system, branched alkane peaks, which are small compared to the n-alkane peaks, are enlarged considerably. Thus a demonstration was given of significant differences in the sizes of these branched-alkane peaks and in the microstructure of the

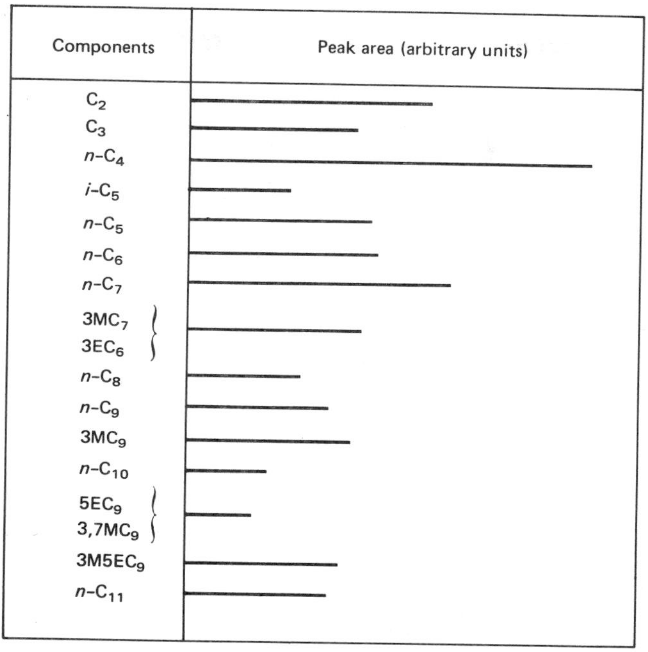

Components	Peak area (arbitrary units)
C_2	
C_3	
$n\text{-}C_4$	
$i\text{-}C_5$	
$n\text{-}C_5$	
$n\text{-}C_6$	
$n\text{-}C_7$	
$3MC_7$ \} $3EC_6$	
$n\text{-}C_8$	
$n\text{-}C_9$	
$3MC_9$	
$n\text{-}C_{10}$	
$5EC_9$ \} $3,7MC_9$	
$3M5EC_9$	
$n\text{-}C_{11}$	

Fig. 57   Areas of the main peaks in the pyrogram of an
ethylene-butene-1 copolymer containing 50% of
butene-1.

polymers mentioned.

As can be seen in Fig. 61, the pyrogram of the ethylene-1-butene copolymer
shows significant enlargements - compared with that of a linear polyethylene -
of the 3-methylheptane, 3-methyloctane and 3-methylnonane peaks, which can be
related to the presence of ethyl side groups. Butyl side groups in ethylene/-
1-hexene copolymers give rise to a considerable increase in the isobutane, iso-
pentane, 2-methylhexane, 3-methylheptane, 5-methylnonane and the 5-methyldec-
ane peaks. As could be expected, the enlarged 2-methyloctane, 3-methylnonane
and 4-methyldecane peaks correspond with the presence of hexyl side groups in
the ethylene/1-octane copolymer.

Ethylene-butene copolymers give a number of fragments characteristic of mono-
mer sequences containing one of the monomers only: $n\text{-}C_6$, $n\text{-}C_{10}$ and $n\text{-}C_{11}$ for
ethylene sequences, and $n\text{-}C_4$, $n\text{-}C_7$, $i\text{-}C_5$, 3-methyl C_7 and 3-methyl-5-ethyl C_9
for butylene sequences. These peaks were comparable in size for both directly
prepared C_2/C_4 copolymers and hydrogenated polybutadiene of the same composition. There are, however, large differences in size between the 3-methyl C_9
peaks which stem from $C_4C_2C_4$ sequences. The latter are absent in hydrogenated
polybutadienes, because a 1,2-1,4-1,2 sequence in polybutadiene yields a C_4C_2
C_2C_4 sequence.

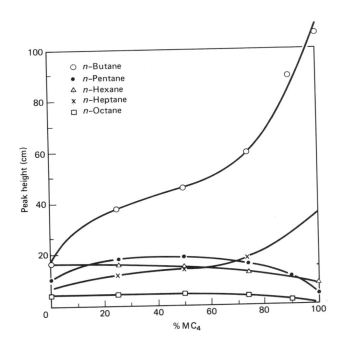

Fig. 58 Ethylene-butene-1 copolymers. Peak heights of n-
         alkanes as functions of butene content.

Seeger and Barrall (583) have investigated the applicability of pyrolysis-hydrogenation-gas chromatography to the elucidation of side chain branching in
ethylene-butene-1 copolymers. In Fig. 62 are shown the single branched fragment pattern up to ten carbon atoms obtained for this copolymer compared with
those obtained for Ziegler (linear) polyethylene and low density (0.918 g/cc)
polyethylene. It is possible to measure not only the kind of branching present
but also to identify the polymer from this simple series. The distribution of
isomers varies significantly with the type of branching. In the copolymer with
1-butene (ethyl branches) the 3-methyl isomer is dominant. The low density
polyethylene (0.918 g/cc) shows a high 3-methyl peak as well as a high 5-meth-
yl and 4-methyl peak yield. This indicates that both ethyl and butyl branches
are present in the polyethylene material. Such evidence confirms former suggestions, mainly on the basis of infrared measurements, that low density polyethylene has branches which are the result of certain intramolecular transfer
reactions during high-pressure polymerization.(584) For Ziegler-type polyethylene (Fig. 62b) there is no immediate correlation of the fragment yield with
a certain length of side chain branch. Ethyl branches may be the cause of the
3-methyl C_{10} peak. Butyl branches seem to be effectively lacking since the 5-
methyl peak has almost disappeared. On the basis of boiling points the main
peaks are probably 4-ethyl and 4-propyl isomers. A statistical length distribution of long branches seems to be present in Ziegler-type polyethylene. This
type of branching may be present in the ethylene portions of copolymers (Fig.
62a) and even in low-density polyethylene (Fig. 62).

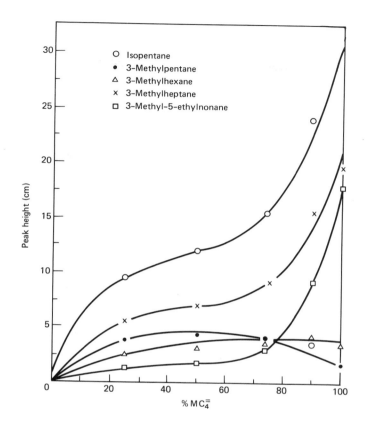

Fig. 59   Ethylene-butene-1 copolymers. Peak heights of
isoalkanes as functions of butene-1 content.

By using the copolymers with 1-butene (about five ethyl branches per 1000 carbon atoms) as standards the number of branches in the low density polyethylene may be estimated. The concentration of branches in these model copolymers
could be verified by pyrolysis chromatography. For a direct comparison of the
amplitudes, 50% has to be added in Fig. 62c, 20% has to be subtracted from
Fig. 62b, while in Fig. 62a the sensitivities are approximately the same.
From such approximations the low density polyethylene in Fig. 62c contains
approximately 20 ethyl and about 10 n-butyl branches for each 1000 carbon
atoms. In addition, some long side chains may be present. In the Ziegler polyethylene with a density of 0.962 g/cc(Fig. 62b), less than three ethyl branches
per 1000 carbon atoms are present; thus long branches seem to predominate.
Such data are in good agreement with those usually obtained from infrared
measurements. (584)

ETHYLENE-HEXENE AND ETHYLENE-OCTENE COPOLYMERS

Pyrolysis-Gas Chromatography

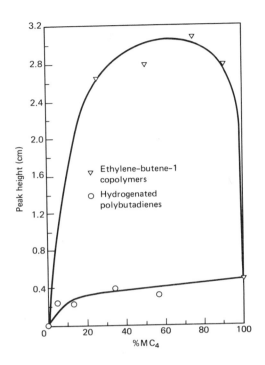

Fig. 60   Height of 3MC$_9$ peak as a function of butene-1
          content.

Short chain branching in polyethylene has been the subject of several publications (585-591) and has been studied by infrared spectroscopic techniques
and by a combination of high-energy electron irradiation and mass spectrometry (587, 588). This latter approach has shown that the short branches in high-
pressure polyethylenes are mainly ethyl and n-butyl groups, but other short
branches have also been supposed to be present.

Van Schooten and Evenhuis (592) modified their flash pyrolysis techniques to
the identification of the structures of copolymers of ethylene and small quantities of butene-1, hexene-1 and octene-1. The 20-30 mg pyrolysis samples
used in the original method were reduced to 1 mg in order to obtain a high
sensitivity for the branched alkane pyrolysis products without the occurrence
of confusing side reactions.

As can be seen in the figure in Fig. 63 using the 1 mg sample, the pyrogram
of the ethylene/1-butene copolymer shows significant enlargements - compared
with that of a linear polyethylene - of the 3-methylheptane, 3-methyloctane
and 3-methylnonane peaks, which can be related to the presence of ethyl side
groups. Butyl side groups in the ethylene-1-hexene copolymers give rise to a
considerable increase in the isobutane, isopentane, 2-methylhexane, 3-methyl-
heptane, 5-methylnonane and the 5-methyldecane peaks. As could be expected,
the enlarged 2-methyloctane, 3-methylnonane and 4-methyldecane peaks correspond with the presence of hexyl side groups in the ethylene-1-octene copoly-

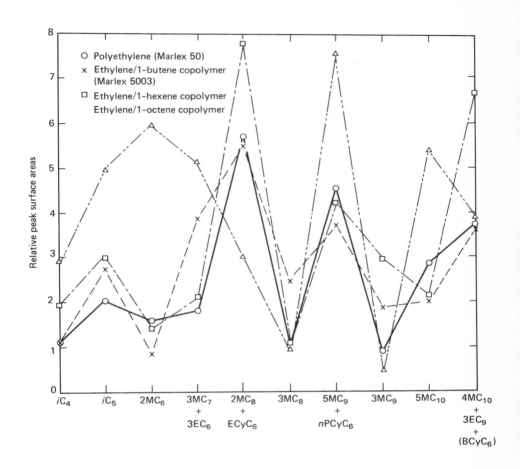

Fig. 61   Relative peak areas of isoalkanes obtained after
          pyrolysis of linear polyethylene and of copolymers
          of ethylene and 1-butene, 1-hexane and 1-octane,
          respectively.

mer.

In further work Van Schooten and Evenhuis (565) reported on the application
of their pyrolysis (at 500°C)-hydrogenation-gas chromatographic technique to
the measurement of short chain branching and structural details in polyethylenes and modified polyethylenes which contain small amounts of comonomer (about
ten per cent by weight of butene-1, hexene-1 and octene-1). A survey of the
isoalkane peaks of the pyrograms of these polymers with their probable assignment is given in Table 35. The effect of the comonomer on the relative sizes
of the n-alkane peaks is given in Table 36. The peaks stemming from the total
side group appear to be somewhat increased in intensity (n-C_4 peak in the
polyethylene-hexene pyrogram, n-C_6 peak in polyethylene-octene pyrogram). The
polyethylene-octene pyrogram also shows that the peak stemming from the n-alk-

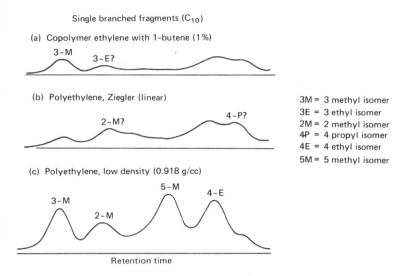

Single branched fragments (C$_{10}$)

(a) Copolymer ethylene with 1-butene (1%)

3-M
3-E?

(b) Polyethylene, Ziegler (linear)

2-M?
4-P?

3M = 3 methyl isomer
3E = 3 ethyl isomer
2M = 2 methyl isomer
4P = 4 propyl isomer
4E = 4 ethyl isomer
5M = 5 methyl isomer

(c) Polyethylene, low density (0.918 g/cc)

5-M
4-E
3-M
2-M

Retention time

Fig. 62   Alkane types produced on pyrolysis of polyolefins
and copolymer. Single branched fragments (C$_{10}$).

ane with one carbon atom less is somewhat enlarged.

This suggests that chain scission may occur both at the α- and at the β- C-C
bond. Only the latter type of scission will produce methyl-substituted isoalkanes, e.g. after intramolecular hydrogen transfer. A list of isoalkanes
that may be expected to be formed in this way, is compared in Table 37 with a
list of those that have been found to be significantly increased in size in
the copolymer pyrograms.

Seeger and Barrall (583) have investigated the applicability of pyrolysis-hydrogenation-gas chromatography to the elucidation of side chain branching in
ethylene-hexene-1 copolymers. In Fig. 64 are shown the single branched fragment pattern up to ten carbon atoms obtained for this copolymer compared with
those obtained for Ziegler (linear) polyethylene and low density (0.918 g/cc)
polyethylene.

It is possible to measure not only the kind of branching present but also to
identify the polymer from this simple series. The distribution of isomers varies significantly with the type of branching. In the copolymer with 1-hexene
(butyl branches) the 5-methyl and 4-methyl isomers are dominant. The low-density polyethylene (0.918 g/cc) shows a high 3-methyl peak as well as a high 5-
methyl and 4-methyl peak yield. This indicates that both ethyl and butyl branches are present in the polyethylene material. Such evidence confirms former
suggestions, mainly on the basis of infrared measurements, that low density
polyethylene has branches which are the result of certain intramolecular transfer reactions during high-pressure polymerization. (584)

ETHYLENE ACRYLATE COPOLYMERS

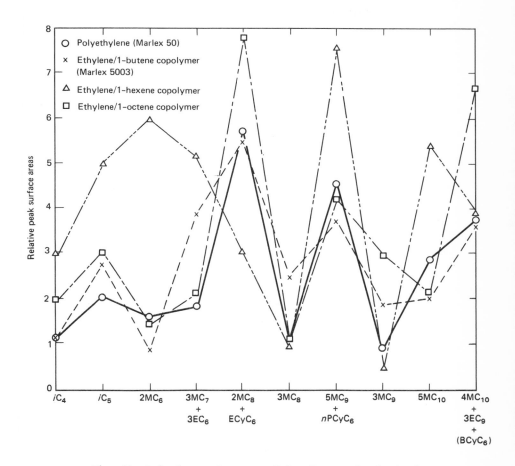

Fig. 63 Relative peak areas of isoalkanes obtained after
         pyrolysis of linear polyethylene and of copoly-
         mers of ethylene and 1-butene, 1-hexene and 1-
         octene, respectively.

NMR Spectroscopy

Porter et al (593) analyzed a series of ethylene polymers and copolymers by
high resolution nuclear magnetic resonance spectroscopy. Various ester-contain-
ing monomers- e.g. ethyl acrylate and vinyl acetate- copolymerized with ethylene were identified and determined. NMR analyses compare well with copolymer
determinations by pyrolysis-gas chromatography and by neutron activation analyses for oxygen. These workers showed that for certain ethylene homopolymers,
NMR can provide a measure of molecular weight and chain branching. Measurements were made on 10% polymer solutions in diphenyl ether at elevated temperatures (593). Resolution generally improves with increasing temperature and
decreasing polymer concentration in the diphenyl ether solvent. Peaks for CH_2O
protons sometimes appear better resolved below 200°C, although the total peak

Analysis of Plastics

Table 35 Areas of the Main Iso-Alkane Peaks in the Pyrograms of Linear Polyethylene, Ethylene-Butene-1,
Ethylene-Hexene-1 and Ethylene-Octene-1 Coplymers.

Component	Peak area in arbitrary units			
	Linear PE	Ethylene -Butene-1	Ethylene -Hexene-1	Ethylene -Octene-1
$i\text{-}C_4$	10	20	37	30
$i\text{-}C_5$	18	47	62	46
$2MC$	9	14	18	20
$3MC_5\text{-}CyC_5$	31	54	46	76
$2MC_6$	14	13	74	22
$3MC_6$	13	33	24	26
$2MC_7\text{-}4MC_7$	16	17	25	27
$3MC_7\text{-}3EC_6$	16	67	64	32
$2MC_8\text{-}4MC_8\text{-}ECyC_6$	50	44	38	119
$3MC_8$	9	39	12	17
$4MC_9\text{-}5MC_9\text{-}4EC_8$ } $i\text{-}PCyC_6\text{-}BuCyC_5$	11	21	51 94	36
$2MC_9\text{-}n\text{-}PCyC_6$	40	35	40	65
$3MC_9$	8	27	6	45
$4MC_{10}\text{-}5MC_{10}\text{-}secBuCyC_6$	25	23	67	32
$2MC_{10}\text{-}4MC_{10}\text{-}n\text{-}BuCyC_6$	33	30	47	102
$3MC_{10}$	15	32	14	34

Table 36 Relative Sizes of n-Alkane Peaks for Polyethylene and for Ethylene-Butene-1, Ethylene-hexene-
1 and Ethylene-Octene-1 Copolymers.

Peak Ratio	Marlex 50	Ethylene -Butene-1	Ethylene -Hexene-1	Ethylene -Octene-1
$n\text{-}C_4\text{-}n\text{-}C_7$	0.71	0.86	1.15	0.86
$n\text{-}C_5\text{-}n\text{-}C_7$	0.59	0.69	0.61	0.91
$n\text{-}C_6\text{-}n\text{-}C_7$	0.52	1.66	1.62	1.88
$n\text{-}C_8\text{-}n\text{-}C_7$	0.63	0.58	0.66	0.79
$n\text{-}C_9\text{-}n\text{-}C_7$	0.67	0.63	0.62	0.75
$n\text{-}C_{10}\text{-}n\text{-}C_7$	1.02	0.93	0.99	1.05
$n\text{-}C_{11}\text{-}n\text{-}C_7$	1.00	0.82	0.90	0.92

area remains nearly constant.

Figure 65 shows NMR spectra for an ethylene-ethyl acrylate copolymer. Spectra
indicate clearly both copolymer identification and monomer ratio. A distinct
ethyl group pattern (quartet, triplet) with the methylene quartet shifted
downfield by the adjacent oxygen is observed. The oxygen effect carries over
to the methyl triplet which merges with the aliphatic methylene peak. No other
ester group would give this characteristic pattern. The area of the quartet
is a direct and quantitative measure of the ester content. All features in
Fig. 65 are consistent with an identification of an ethylene-rich copolymer
with ethyl acrylate. Ethyl acrylate contents obtained by NMR (6.0%) agreed
well with those obtained by PMR (6.2%) and neutron activation analysis for
oxygen (6.1%).

Pyrolysis-Gas Chromatography

Table 37 Expected and Observed Increased Iso-Alkane Peaks in Copolymer pyrograms.

Ethylene-Butene-1		Ethylene-Hexene-1		Ethylene-Octene-1		Number of Backbone C Atoms in Fragment
Expected	Observed	Expected	Observed	Expected	Observed	
-	-	-	$i\text{-}C_4$	-	$i\text{-}C_4$	-
-	-	-	$i\text{-}C_5$	-	$i\text{-}C_5$	-
$i\text{-}C_5$	$i\text{-}C_5$	$2MC_6$	$2MC_6$	$2MC_8$	$2MC_8$	3
$3MC_5$	$3MC_5$	$3MC_7$	$3MC_7$	$3MC_9$	$3MC_9$	4
$3MC_6$	$3MC_6$	$4MC_8$	-	$4MC_{10}$	$4MC_{10}$	5
$3MC_7$	$3MC_7$	$5MC_9$	$5MC_9$	$5MC_{11}$	-	6
$3MC_8$	$3MC_8$	$5MC_{10}$	$5MC_{10}$	$6MC_{12}$	-	7
$3MC_9$	$3MC_9$	$5MC_{11}$	-	$7MC_{13}$	-	8

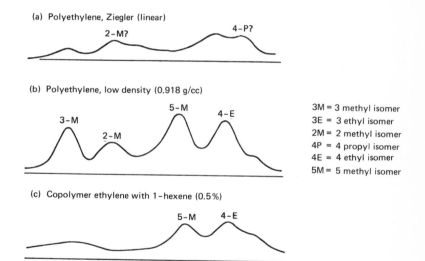

(a) Polyethylene, Ziegler (linear)
2-M? 4-P?

(b) Polyethylene, low density (0.918 g/cc)
3-M 5-M 2-M 4-E

3M = 3 methyl isomer
3E = 3 ethyl isomer
2M = 2 methyl isomer
4P = 4 propyl isomer
4E = 4 ethyl isomer
5M = 5 methyl isomer

(c) Copolymer ethylene with 1-hexene (0.5%)
5-M 4-E

Retention time

Fig. 64 Single branched fragments(C_{10}). Conditions as in Fig. 62.

Bombaugh et al (632) reported that the amount of methanol formed on pyrolysis of ethylene-methyl acrylate copolymers increases with the block character of the copolymers. The relative amounts of methanol and methyl acrylate formed on pyrolysis proved useful for distinguishing copolymers from homopolymer blends. Copolymers indicated by this technique as having block structures were shown to be partly crystalline by differential thermal analysis; random copolymers of the same composition were non-crystalline. Furthermore, the amount of methyl methacrylate regenerated when ethylene-methyl methacrylate copolymers were pyrolyzed decreased as the number of ethylene-methyl methacry-

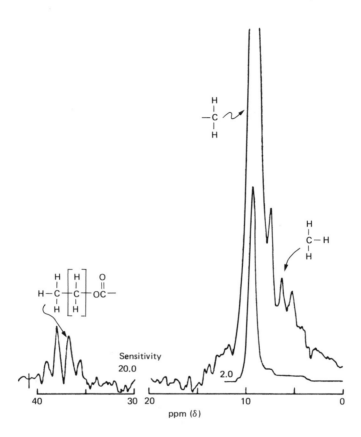

Fig. 65   NMR spectra. Poly(ethylene-ethyl acrylate).

late junctions in the copolymers increased. The results obtained in this study
are presented in Fig. 66.

The pyrolysis-gas chromatography of ethylene-methyl acrylate copolymers and
physical mixtures of the two homopolymers has been studied by Smith (633). He
describes an apparatus utilizing a thermal cracker device in detail. A typical
chromatogram of a copolymer pyrolyzate is shown in Fig. 67.

Smith (633) also applied his pyrolysis technique to methyl methacrylate-ethyl-
ene copolymer. Polymethyl methacrylate degrades to 98% monomer. At the ethylene-acrylate junctions in copolymer, however, other materials may be formed.
It followed that the amount of methyl methacrylate produced on pyrolysis was
an inverse function of the number of ethylene-methyl methacrylate junctions.

Barrall et al (595) have described a pyrolysis-gas chromatographic procedure
for the analysis of polyethylene-ethyl acrylate and polyethylene-vinyl acetate
copolymers and physical mixtures thereof. They used a specially constructed
pyrolysis chamber as described by Porter et al (596). Less than 30 seconds is

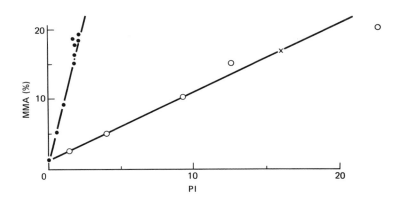

Fig. 66   Pyrolysis and vapour-phase chromatography of eth-
          ylene-methyl methacrylate copolymers
          Abscissa: PI=pyrolysis index= area under the meth-
          yl methacrylate peak multiplied by the mole-fract-
          ion of ethylene in the copolymers and divided by
          the area under the C_8-peak.
          Ordinate: MMA=methyl methacrylate (% by weight).
          -o-o-   : Random copolymers.
          -•-•-   : Homopolymer mixtures.
             x   : Graft copolymer.

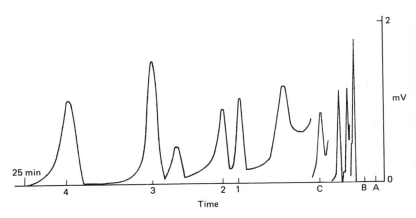

Fig. 67   Typical chromatogram of thermal pyrolyzates from
          ethylene-methyl acrylate copolymer.
          1. Methyl acetate.   A. Pyrolyzer on.
          2. Methanol.   B. Pyrolyzer off.
          3. Methyl acrylate.   C. To purge position.
          4. Methyl methacrylate.

required for the sample chamber to assume block temperature. This system has

the advantages of speed of sample introduction, controlled pyrolysis temperature, and complete exclusion of air from the pyrolysis chamber. The pyrolysis
chromatogram of poly(ethylene-vinyl acetate) contains two principal peaks. The
first is methane and the second acetic acid. Variations from 350°C to 490°C
in pyrolysis temperature produced no change in the area of the acetic acid
peak but did cause an area variation in the methane peak. The pyrolysis chromatogram of poly(ethylene-ethyl acrylate) (Fig. 68) at 475°C shows one principal peak due to ethanol. No variation in peak areas was noted in the temperature range 300°C to 480°C. Table 38 shows the analysis of 0.05 gram samples
of poly(ethylene-ethyl acrylate) (PEEA) and poly(ethylene-vinyl acetate (PEVA)
obtained at a pyrolysis temperature of 475°C.

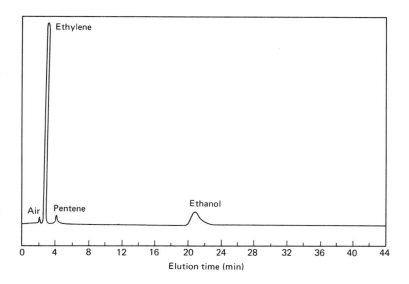

Fig. 68   Pyrolysis chromatogram of poly(ethylene-ethyl ac-
          rylate). 475°C helium carrier, 30 cc. per minute
          Carbowax column.

Table 38   Pyrolysis Results on Physical Mixtures of Poly
           (ethylene-Ethyl Acrylate) and Poly(ethylene-vin-
           yl acetate).

Mixture	Acetic acid, wt%		Ethylene, wt%		Oxygen, wt%	
	Found	Calcd.	Found	Calcd[a]	Found	Calcd.
50% PEEA-1 and						
50% PEVA-2	9.10	9.05	2.65	2.62	7.88	8.25
33.3% PEEA-2 and						
66.6% PEVA-3	12.15	12.33	0.75	0.70	7.33	7.49

a Calculated from results for acetic acid and ethylene content for individual
samples on weight per cent basis.

Bambaugh et al (597,598) utilized pyrolysis gas chromatography (599) and differential thermal analysis to elucidate comonomer distribution in ethylene-acrylate copolymers. The major pyrolyzate was shown to be inversely related to the

number of ethylene-acrylate junctions. The DTA first-order transition was shown to be related to the polyethylene chain length. Under the conditions of pyrolysis, the degradation profiles of block copolymers (i.e. copolymers with the fewest acrylate-ethylene junctions and which therefore contain long sequences of both acrylate and ethylene units) were similar to those of equivalent concentration mixtures of their respective homopolymers. These findings, particularly the thermal pyrolysis data may appear to be in conflict with other published work in the field (599-603). For example, Strasburger and Brauer (599) reported that they were able to distinguish copolymers from equivalent concentration mixtures of homopolymers, while Barrall, Porter and Johnson (600) reported no difference between random copolymers and random copolymer mixtures.

In an attempt to resolve this situation Bambaugh and Clampitt (597) carried out further investigations of ethylene-acrylate copolymers using the same techniques as Bambaugh (598). In this work they considered the thermal degradation products obtained at various temperatures of ethylene-acrylate and ethylene-vinyl acetate copolymers. The data measurements are primarily concerned with the crystalization of the ethylene chains between the branch points. Indeed, a linear relationship between DTA crystallinity and branching is indicated.

Thermal Methods

Bambaugh and Clampitt (597) applied differential thermal analysis to crystallinity measurements of random and block ethylene-methyl acrylate copolymers.

ETHYLENE-METHYL ACRYLATE COPOLYMER

The pyrolyzate produced as described by Bambaugh and Clampitt (597) was examined first with a 2M dinonyl phthalate column. Table 39 shows a typical composition as calculated from chromatographic peak areas. Methyl methacrylate and methyl acrylate are produced in nearly equal yield suggesting that chain scission is one mode of decomposition.

Table 39   Composition of Pyrolyzates Below C_{10}

Peak No.	Area %
1 Light gases,* CO, CO_2	17
2 C_3	18
3 C_4	13
4 C_5 + methanol	8
5 C_6 + methyl acetate	16
6 Methyl acrylate	3
7 C_7	9
8 Methyl methacrylate	3.6
9 C_8	6
10 Higher ester	2
11 C_{10}	4
	100 **

* Consisting of C_2H_6, C_2H_4
** The back flushed heavy ends contained higher hydrocarbons and a homologous series of esters tentatively identified as α- alkyl acrylate esters.

ETHYLENE-ETHYL ACRYLATE COPOLYMER

Barrall et al (595), in the studies with ethylene-ethyl acrylate copolymer, reported the production principally of ethylene with a trace of ethanol and suggested cleavage primarily of the alkyl linkage with the resulting formation of the acid on the residual polymer chain.Bambaugh and Clampitt (597), though cracking only 3 mg. of sample, showed that increased cleavage occurred at the other side of the oxygen linkage yielding considerably more ethanol, which was analogous to their results with ethylene-methyl acrylate copolymer. Barrall's conditions are advantageous for composition analysis; those of Bambaugh and Clampitt (597) are preferred for configurational analysis since they fragmented the carbon skeletal chain.

ETHYLENE-METHYL METHACRYLATE COPOLYMER

Thermal Methods

Many workers have established that at relatively low cracking temperatures methyl methacrylate unzips to less than 99% monomer. Bambaugh and Clampitt (597) demonstrated that at their slightly higher cracking temperature (600°C. top temperature) they could effect a decrease in methyl methacrylate monomer yield when the methyl methacrylate unit was separated by ethylene units. This occurred when the cracking temperature was high enough to crack the polyethylene chain as well as to unzip the methyl methacrylate portion of the chain.

Smith (633) applied differential thermal analysis to the examination of ethylene methyl acrylate copolymers. Differential thermal analysis substantiated the thermal pyrolysis data. Thermograms of the block copolymers showed first order transitions comparable to equivalent concentration mixtures of the respective homopolymers; material determined by thermal pyrolysis to be random copolymer showed only a second order transition.

ETHYLENE-VINYL ACETATE COPOLYMER

Chemical

Vinyl acetate in ethylene copolymers has been determined by elimination of acetic acid in molten toluene sulfonic acid at 160°C followed by acidimetric titration (608).

Wojeck and Sporysz (609) have followed the decomposition of hydrogen peroxide during vinyl acetate polymerization by polargraphy.

Matsumoto et al (610) showed that the hydroxy groups in partially saponified ethylene-vinyl acetate copolymers had a block distribution, whereas hydroxy groups in partially acetylated ethylene-vinyl alcohol copolymers had a random distribution. Many hydroxy groups in ethylene-vinyl alcohol polymer were hydrogen bonded to neighbouring carbonyl groups, whereas the hydroxy groups in ethylene-vinyl acetate copolymer existed mainly as free hydroxy groups.

Majer and Sodomka (611) determined the vinyl acetate content in ethylene-vinyl acetate copolymers by hydrolysis in toluene with alcoholic potassium hydroxide and back titration of the excess potassium hydroxide with 0.1 N hydrochloric acid. The authors also used an infrared method based on the absorbances of the infrared absorption bands at 610 and 720 cm^{-1}.

Munteanu (612) determined vinyl acetate in random and graft copolymers with ethylene by saponification in 1 N ethanolic potassium hydroxide at 80°C for 3h.

Leukroth (613) found that the best method for determination of bound vinyl
acetate in ethylene-vinyl acetate copolymers was saponification with p-toluene
sulphonic acid and titration of the acetic acid.

Infrared Spectroscopy

Munteanu et al (614) determined the degree of grafting of vinyl acetate onto
polyethylene and the content of vinyl acetate in linear ethylene-vinyl acetate copolymers from the absorptivities of the peaks at 3455 and 3420 cm^{-1}.

Siryuk (615) determined the absolute content of vinyl acetate blocks in ethylene-vinyl acetate copolymers using absorptivities at 1743 and 1245 cm^{-1}. The
absorptivity at 1378 cm^{-1} did not give satisfactory results for the determination of the degree of branching.

The infrared spectra of ethylene-vinyl acetate random copolymers, graft copolymers, and blends of polyethylene and polyvinyl acetate exhibit characteristic bands of acetoxy groups (3455, 1739, 1241, 1022, 947, 794 cm^{-1}) and of
CH, CH_2, CH_3 groups independently of the mode of bonding of vinyl acetate in
the polymer (616).

NMR Spectroscopy

Noise-decoupled pulsed Fourier transformed ^{13}C NMR spectroscopy has been used
by Ibrahim et al (617) to study tacticity in poly(vinyl acetate) and monomer
sequence distribution in ethylene-vinyl acetate copolymers.

Broad-line NMR spectroscopy was employed by Sobottka et al (618) to show that
methyl group rotation in ethylene-vinyl acetate copolymers occurred at temperatures below -196°C.

Keller (619) determined the monomer sequence distribution in ethylene-vinyl
acetate polymers from the methane and methylene carbon regions of ^{13}C Fourier
transformed NMR spectra using proton broad-line coupling. A peak assigned to
the acetate methyl protons was split into a triplet and the peak assigned to
the methine proton was split into a quintet and a broad peak in the high resolution proton NMR spectra of ethylene-vinyl acetate copolymers measured in
the presence of a shift reagent. The split peaks were assigned tentatively to
the acetate methyl and methine protons of the central vinyl acetate unit in
triad sequences, respectively (620).

Pyrolysis-Gas Chromatography

Bambaugh and Clampitt (597) showed that vinyl acetate-ethylene copolymer, when
pyrolyzed at temperatures as low as 300°C, yields acetic acid and methane. The
pyrolyzate produced at 600°C also includes a series of hydrocarbons, as is
shown in Fig. 69. The comparative ease with which the side chain is removed
may present an obstacle to structural analysis even at higher temperatures.
Nevertheless, the basic premise holds that lower cracking temperatures are
desirable for compositional analysis.

X-ray Diffraction

The application of x-ray diffraction to the examination of ethylene-vinyl
acetate copolymers has been studied by Forrster and Brand (621), Mitra and
Katti (622) and Munteanu (623).

Miscellaneous

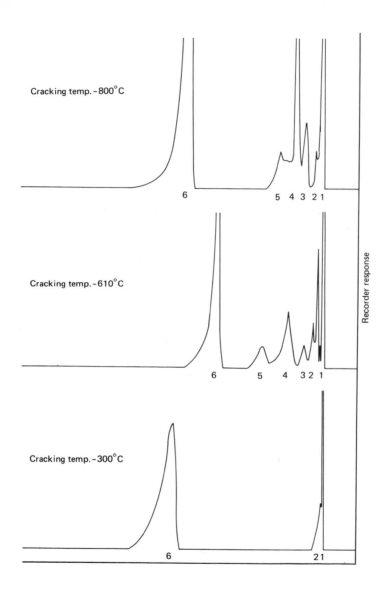

Fig. 69   Chromatogram showing the effect of cracking temp-
          erature on pyrolyzate from ethylene-vinyl acetate
          copolymer. Components: (1,3,4,5)hydrocarbons;(2)
          vinyl acetate; (6) acetic acid.

Mlejnek et al (624,625) determined vinyl acetate in ethylene-vinyl acetate
copolymers thermogravimetrically when pyrolyzed at 770ºC for 15 s or 550ºC for
10 min.

Ryasnyaskaya (626) described ebulliometric apparatus for determining number average molecular weight of ethylene-vinyl acetate copolymers.

Detailed studies of ethylene-vinyl acetate copolymers including ultracentrifugation and fractionation for molecular weight distribution have been carried out by Belyaev et al (627).

German and Heikens (628) have studied the kinetics of the polymerization of ethylene- and vinyl acetate.

Braun and Guillet (629) have used inverse gas chromatography to determine the crystallinity of ethylene-vinyl acetate copolymers.

Additives

A method has been described (630) for the determination of secondary antioxidants such as dilauryl thiodipropionate in ethylene vinyl acetate. The method involves hydrolysis to lauryl alcohol, followed by extraction of the alcohol with chloroform-ethanol-hexane then thin-layer or gas chromatography.

ETHYLENE-VINYL CYCLOHEXANE COPOLYMERS

Infrared Spectroscopy

Infrared spectroscopy has been applied (631) to the determination of the relative amounts of comonomers in ethylene-vinylcyclohexane copolymers. Since methylenes ($-CH_2$) in a six-membered ring absorb at a different wavelength (1450 cm^{-1}) than methylenes in a straight chain (1472 cm^{-1}) infrared analysis is particularly adaptable to determining the relative amounts of comonomers in the polymer. Thus, the ratio of intensity for the 1450 cm^{-1} and 1472 cm^{-1} infrared bands shows the relative amounts of ethylene and vinylcyclohexane in the copolymer chain. This method was applicable to copolymers containing from 10% to 70% incorporated vinylcyclohexane and was able to distinguish between block, random and alternating copolymers.

ETHYLENE ACROLEIN COPOLYMERS

Tanaka et al (635) have studied the electron spin resonance spectroscopy of these polymers.

ETHYLENE-VINYL CHLORIDE COPOLYMERS

Wu (636) used 220 MHz NMR data from ethylene-vinyl chloride copolymer in concluding that formation of racemic diads of vinyl chloride units is slightly favoured over meso diads.

Keller and Muegge (637) calculated chemical shifts for ethylene-vinyl chloride copolymers, ethylene-vinylidene chloride copolymers, and chlorinated polyethylene from measurements of hydrogen broad band-decoupled ^{13}C NMR spectra.

ETHYLENE-VINYL ALCOHOL COPOLYMERS

Broadline spectra of ethylene-vinyl alcohol copolymers at -196 - $+25°$ have been interpreted (638). Sequence distributions of ethylene-vinyl acetate copolymers were obtained from ^{13}C NMR spectra (639, 640). Hoffmann and Keller reviewed the use of model compounds in ^{13}C NMR for assigning triad sequences for both methine and methylene groups.

ETHYLENE-CARBON MONOXIDE COPOLYMERS

Wu and Ovenall (641) utilized 220 MHz proton and 22.6 MHz ^{13}C NMR to characterize the comonomer sequencing of ethylene-carbon monoxide copolymers.

A structure assignment has been made of a polyalcohol synthesized from 1:1 ethylene-carbon monoxide alternating copolymer reduced by sodium borohydride from results by infrared spectroscopy and x-ray diffraction. The 1,2-glycol content was obtained from the periodate reaction (642).

Braun and Guillet (643) have used inverse gas chromatography to determine the crystallinity of ethylene-carbon monoxide copolymers.

POLYETHYLENE OXYBENZOATE

Ishibashi (644) has measured CH_2 rocking vibrations in this polymer.

CHLORINATED POLYETHYLENE

Chemical

The zinc content of chlorinated polymers has been determined by fusing the sample with sodium peroxide followed by a standard complexometric titration of the resulting solution (645). The chlorine content has been determined by potentiometric titration after decomposition with sodium biphenyl (646).

Quenum et al (647) studied the infrared spectra of chlorinated polyethylene over the 19-73% chlorine range. At 73% chlorine content, chlorinated polyethylene and chlorinated PVC has nearly the same spectrum; however, they differed significantly from that of polyvinylidene chloride.

Infrared Spectroscopy

Infrared and laser-Raman spectral analysis of chlorinated polyethylene have indicated that chlorination occurred in the amorphous regions of the polyethylene (648).

Oswald and Kubu (649) used infrared analysis to characterize chlorinated polyethylene samples. Bands in the 1400-1475 cm^{-1} region (CH_2 bending) were used to determine the proportions of methylene groups centred in -$CH_2CH_2CH_2$-, CH_2-CH_2CHCl-, and $CHClCH_2CHCl$- triads. By considering the chlorinated polymers to be terpolymers of ethylene, vinyl chloride, and 1,2-dichloroethylene, glass transition temperatures of the copolymers can be predicted from the infrared results with considerable accuracy. No evidence for the presence of -CCl_2- units in the copolymers was obtained, in agreement with the results of Nambu (650) and of Fuchs and Louis (651) who studied chlorinated poly(vinyl chloride) samples.

NMR Spectroscopy

Keller and Muegge (652, 653) have determined the microstructure of chlorinated polyethylene based on high resolution proton and ^{13}C NMR studies. Keller (654) proposed introducing corrections, relating to the interactions of vicinal chlorine atoms within the chain, through the increment system proposed for calculating the chemical shift of ^{13}C NMR of chlorine-containing polymers. The deviations of ^{13}C chemical shifts to higher fields caused by the steric hinderance of vicinal carbons were 7 ppm for chlorinated polyethylene, 12 ppm for chlorinated PVC, and 7-15 (depending on the possible arrangement of chlorine) for trichloroethylene-vinyl chloride copolymers (655).

Abu Isa and Myers (656, 657) have investigated the degradation characteristics of chlorinated high density polyethylene (24-45% chlorine) utilizing a variety of techniques including NMR, infrared spectroscopy, differential scanning calorimetry and x-ray diffraction. The peak at 1.25 ppm in the NMR spectrum for the γ_+ CH_2 sequence is well enough resolved to allow an accurate evaluation of its integral, and this is the most important integral for the chlorine sequence distribution determination. For any one sample the mole fraction of γ_+ CH_2 can be calculated from these spectra by dividing the area of the γ_+CH_2 peak occurring in a chemical shift region of 0.76-1.36 ppm, with respect to hexamethyl disilazone, by the total area of unsubstituted methylenes with absorption between 0.76 and 2.76 ppm. The experimental mole fractions of γ_+CH_2 thus determined correlate well with % chlorine in the polymer (656,657).

Abu Isa and Myers (657) showed that the molecular weight of the parent polyethylene, the degree of chlorination and the degree of residual crystallinity of the samples had no bearing on the type of substitution.

CHLOROSULPHONATED POLYETHYLENE

Chemical

Braeme (658) has investigated the chlorine distribution in fractions of chlorosulphonated polyethylene.

NMR Spectroscopy

Braeme (658) carried out NMR investigations to study the chlorine distribution of chlorosulphonated polyethylene containing between 18 and 35% chlorine. They developed an NMR method to determine the chlorine distribution by measuring the relative amounts of α, β, and γ(or greater) methylenes and compared the NMR measurements with the statistical predictions made for substitution polymers (659). The comparison showed that polyethylenes chlorosulphonated by a solution reaction with gaseous chlorine and sulphur dioxide show a random chlorine distribution.

Figure 70 shows the NMR spectra obtained at 100 MHz and 220 MHz for 0 to 20% p-dichlorobenzene solutions at 150°C of chlorosulphonated polyethylene sample. This sample exhibits a pattern in the proton resonance spectrum that is typical of spectra obtained from chlorosulphonated polyethylenes whose chlorine contents are in the range from approximately 10 wt-% to approximately 40 wt-%. Based on a comparison of the two spectra in Fig. 70, it is apparent that the spectrum obtained at 220 MHz is clearly superior to that obtained at 100 MHz. Not only are the different lines better resolved at 220 MHz but also the patterns obtained for the different lines clearly show multiplet splittings.

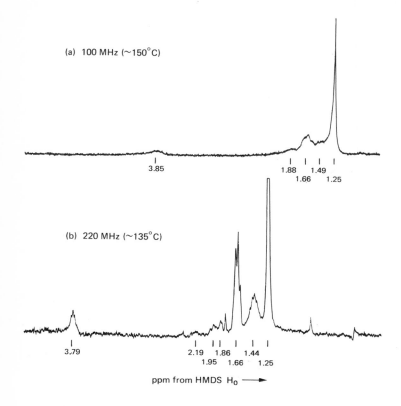

(a) 100 MHz (~150°C)

3.85

1.88   1.49
    1.66   1.25

(b) 220 MHz (~135°C)

3.79

2.19   1.86   1.44
    1.95   1.66   1.25

ppm from HMDS H$_0$ ⟶

Fig. 70   NMR spectra of chlorosulphonated polyethylene containing 25 wt% chlorine (p-dichlorobenzene solution).

CHAPTER 2

POLYPROPYLENE

Chemical methods

Oguro (660) extracted polypropylene with hydrochloric acid to remove antimony which was subsequently determined by atomic absorption spectrophotometry.

Low temperature ashing was used by Narasaki and Umezawa (661) for controlled decomposition of polypropylene prior to the determination of metals.

Citovicky et al (662) developed an iodometric method for the determination of hydroperoxides in powdered polypropylene.

Infrared Spectroscopy

Slovokhototova et al (663) have applied infrared spectroscopy to a study of structural changes in polypropylene in vacuo with fast electrons from an electron accelerator tube(200 kv accelerating field) and with gamma radiation from 60 Co. They found that the infrared spectrum of irradiated polypropylene contains absorption bands in the 1645, 890 and 735-740 cm^{-1} regions. The first two bands correspond to $RR'C=CH_2$ vinylidene groups and the band in the 735-740 cm^{-1} region to propyl branches, $R-CH_2CH_2CH_3$ (664) (Fig. 71). When polypropylene is degraded thermally these groups are formed by disproportionation between free radicals formed by rupture of the polymer backbone (665). Under the action of ionizing radiations the polymer backbone ruptures, with formation of two molecules, with vinylidene and propyl end groups, at a temperature as low as that of liquid nitrogen, because the corresponding bands are found in the infrared spectrum of polypropylene irradiated at -196° and measured at -130°. When polypropylene is irradiated with dosages greater than 350 Mrad a band appears at 910 cm^{-1}, corresponding to vinyl groups, $R-CH=CH_2$, i.e. degradation of polypropylene can also involve simultaneous rupture of two C-C bonds in the main and side chains. The strength of the vinyl-group band in the spectrum of irradiated polypropylene is lower than that of the vinylidene group although the extinction coefficients of these bands are approximately the same (664).

In the spectrum of amorphous polypropylene (Fig.72) irradiated with a dosage of 4000 Mrad at -196° and measured at -130°,in addition to the band at 1645cm^{-1}, a weaker band appears, with a maximum near 1665cm^{-1}, possibly due to internal double bonds.

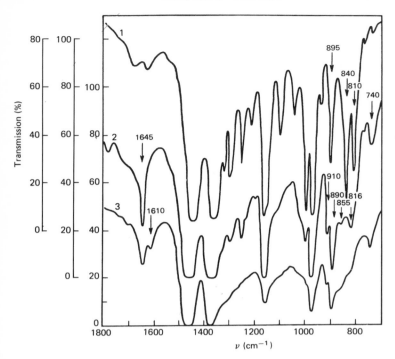

Fig. 71   Infrared spectra of isotactic polypropylene: 1-
           original (d=200μ); 2-irradiated by fast electrons
           at 25°, dosage 500 Mrad (d=300μ); 3-dosage 400Mrad
           (d=200μ).

when the spectrum of this specimen is recorded after it is heated to +25° this
maximum disappears, leaving only a shoulder on the strong band at 1645 cm^{-1}.
The extinction coeffient, ε, (664) of the band at 1665 cm^{-1} is less than that
of the 1645 cm^{-1} band by a factor of 6.7. Supplementary evidence of the formation of internal double bonds in irradiated polypropylene is provided by the
presence in the spectrum of bands in the 815-855 cm^{-1} region. In this region
lie bands due to deformation vibration of CH at double bonds in

$$\begin{array}{c} R \\ \diagdown \\ C = CHR \\ CH_3\diagup \end{array}$$

groups, existing in various conformations (666). The appearance of bands at
815 and 855 cm^{-1} in the spectrum of irradiated polypropylene can be regarded
as an indication of the formation of internal double bonds in the polymer.

A study of the ESR spectra of irradiated polypropylene (667) has shown that
the alkyl radicals formed during irradiation at -196°

$$\sim CH_2 - \overset{\bullet}{C} - CH_2 - CH \sim$$
$$\underset{CH_3}{|} \underset{CH_3}{|}$$

undergo transition to alkyl radicals when the specimen is heated

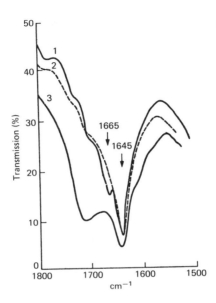

Fig. 72   Infrared spectra of amorphous polypropylene (d= 300μ); 1-irradiated by fast electrons with a dosage of 4000 Mrad at -196° (spectrum recorded at -130°); 2- the same specimen after heating (spectrum recorded at 25°); 3-the same specimen 2 weeks after irradiation (spectrum recorded at 25°).

$$\sim CH - CH = C - CH - C \sim$$
$$\quad\;\; | \qquad\qquad | \qquad\quad\; |$$
$$\quad\; CH_3 \qquad\;\; CH_3 \qquad CH_3$$

i.e. on heating the radical centres migrate to internal double bonds with the formation of stable, allyl radicals. Irradiation at room temperature leads immediately to the formation of allyl radicals.

It is very probable that the decrease in intensity of the internal double bond valency vibration band at 1665 cm^{-1} and the broadening of its maximum after a specimen irradiated at a low temperature is heated to room temperature, is associated with the formation of allyl radicals because interaction of the π-electrons of the double bond with the unpaired electron of the allyl radical must have a marked effect on the vibration of the double bond. Conjugation of two double bonds gives rise to a band of lower frequency. It is possible that conjugation with an unpaired electron also lowers the frequency of the double-bond vibration. Comparison of the intensities of the terminal double-bond bands at 890 and 910 cm^{-1} with the band at 1645 cm^{-1} in the spectrum of irradiated isotactic polypropylene shows that the intensity of absorption in the 1645 cm^{-1} region does not correlate with the intensity of absorption in the 890 and 910

cm^{-1} regions. Thus according to the known extinction coefficients for these bands (664) the ratio of their optical densities chould be

$$D_{890}/D_{1645} = 3.7, \quad D_{910}/D_{1645} = 3.2 \quad \text{and} \quad (D_{890} + D_{910})/D_{1645} = 3.5$$

In the latter case D_{1645} is the sum of the optical densities of the vinylidene and vinyl absorption bands in this region. An optical density ratio for these bands of approximately this value was found (1.75 to 3.3) for the products of thermal degradation of polypropylene.

It is seen that only in the case of amorphous polypropylene irradiated with γ-radiation from ^{60}Co is the ratio $(D_{890} + D_{910})D_{1645}$ close to the value calculated from the extinction coefficients of these bands. In the spectra of irradiated isotactic polypropylene however, the intensity of the 1645 cm^{-1} band is greater than would be expected if only vibration of terminal double bonds contributes to absorption in this region. This increase in absorption in the 1645 cm^{-1} region can be related to absorption by the internal double bond in the allyl radical, the vibrational frequency of which is lowered by conjugation of the π-electrons of the double bond with the unpaired electron of the radical. In amorphous polypropylene irradiated at room temperature the alkyl radicals can recombine rapidly, therefore there is obviously little formation of allyl radicals. This explains the fact that the ratio of the optical densities of the terminal double bond bands in the 900 and 1645 cm^{-1} regions is close to the calculated value.

The occurrence of conjugated double bonds in irradiated polypropylene is indicated by the following facts: (1) in the spectra of isotactic polypropylene irradiated with dosages of 2000 to 4000 Mrad at room temperature there is a band at 1610 cm^{-1}, which is the region in which polyene bands occur, whereas this band is absent from the spectrum of isotactic polypropylene irradiated with the same dosages at $-196°$; (2) in the electronic spectra of these polypropylene specimens the boundary of continuous absorption is shifted to a region of longer wavelength in comparison with the spectra of polypropylene irradiated with the same dosages at $-196°$.

It has been shown from ESR spectra (667) that when specimens of isotactic polypropylene are heated above $80°$ they contain polyenic free radicals

$$\sim \overset{\cdot}{C}H - (C = CH)_n - CH \sim$$
$$\qquad\quad | \qquad\qquad\quad |$$
$$\qquad\quad CH_3 \qquad\qquad CH_3$$

and this also indicates the possibility of migration of double bonds along the polymer chain.

The appearance in the spectra of irradiated polypropylene specimens, after storage in air, of strong bands in the region of 1710 cm^{-1}, corresponding to carbonyl groups, must be explained by reaction of oxygen with the long-lived allyl radicals, with formation of peroxide radicals

$$\qquad\qquad\qquad CH_3 \qquad CH_3$$
$$\qquad\qquad\qquad | \qquad\qquad |$$
$$\sim CH_2 - C - CH = C \sim$$
$$\qquad\qquad\quad |$$
$$\qquad\qquad\quad OO\cdot$$

which form carbonyl groups by decomposition.

The intensity of the 1710 cm^{-1} band (and consequently the degree of oxidation) increases sharply with time of storage of specimens in air. Irradiated amorphous specimens oxidize to a considerably smaller extent than isotactic polypropylene specimens. The degree of oxidation of specimens irradiated at -196° increases more rapidly than when the specimens are irradiated at 25°. All these facts indicate that the life time of the allyl radicals is longer in crystalline polypropylene, and that the concentration of these radicals is higher in specimens irradiated at a low temperature. The free radicals are destroyed only after heat treatment of the specimens in an inert atmosphere at 150°. After this heat treatment the intensity of the 1710 cm^{-1} band ceases to increase on storage, i.e. no further oxidation occurs.

Adams (668) has compared on a qualitative basis the non-volatile oxidation products obtained by photo and thermal oxidation of poylpropylene. He used infrared spectroscopy and chemical reactions. The major functional group obtained by photo decompositon is ester followed by vinyl alkene, then acid. In comparison, thermally oxidized polypropylene contains relatively more aldehyde, ketone, and γ-lactone and much less ester and vinyl alkene. Photodegraded polyethylene contains mostly vinyl alkene followed by carboxylic acid. Gel-permeation chromatography determined the decrease in polypropylene molecular weights with exposure time. Adams' determined that there is one functional group formed per chain scission; in thermal oxidation there are two groups formed per scission.

Adams (668) makes the following comments regarding the infrared spectrum of oxidized polypropylene.

Hydroxyl region. The hydroxyl absorptions in the infrared for polypropylene has a broad band centred at 3450 cm^{-1} (associated alcohols) with a definite shoulder at 3610 cm^{-1} (unassociated alcohols). At a similar extent of degradation, thermally oxidized polyolefins show hydroxyl bands of roughly half the absorbance values of the photooxidized polyolefins. Thus, thermal oxidation produces about half as many hydroxyl groups as photooxidation in polyolefins.

A portion of the polypropylene hydroxyl absorption could be due to hydroperoxides. If so, then an exposed sheet, with the volatiles removed, heated in a nitrogen atmosphere for two days at 140°C should show a decrease in the hydroxyl infrared band and an increase in the carbonyl band due to the decomposition of hydroperoxides under such treatment.

The infrared spectrum of the photodegraded polypropylene sheet subjected to this thermal treatment showed a 20% decrease in the hydroxyl band. However, the broad carbonyl band at 1740 cm^{-1} did not increase but showed a 5% decrease. The small γ-lactone (1780 cm^{-1}) and vinyl alkene (1645 cm^{-1}) bands did show a slight increase, however. Thus, these results are due not to hydroperoxide decomposition but to some carboxylic acids converting to γ-lactones and some terminal alcohols dehydrating to vinyl alkenes at the high temperature. While hydroperoxides are undoubtedly an intermediate in the photooxidation process, they decompose too rapidly under ultraviolet light to build up any significant concentration.

Carbonyl region. The polypropylene carbonyl band after 335 hr exposure is broad with few discernible features except for the vinyl alkene band at 1645 cm^{-1}. The broadness of the carbonyl band indicates a large variety of functional groups and makes accurate quantitative analysis difficult. The large vinyl alkene at 1645 cm^{-1} stands out clearly, and distinct carboxylic acid (1715 cm^{-1}) and γ-lactone (1790 cm^{-1}) spikes can be readily identified.

After the volatile products are removed by the vacuum oven, the carbonyl band for polypropylene decreases. Isopropanol extraction removes about 40% of the polypropylene carbonyl. The carbonyl band is then narrow and appears to centre at the ester absorption at 1740 cm^{-1}.

Treatment with base converts lactones, esters and acids to carboxylates (1580 cm^{-1}), leaving only a small band at 1720 cm^{-1}, which is due to aldehyde and ketone.

Upon reacidification of the polypropylene, some of the original esters at 1740 cm^{-1} do not reform but become carboxylic acids and γ-lactones. Curiously, the vinyl alkene band becomes less intense with each step and broader, shifting down to 1640-1600 cm^{-1}. The vinyl groups may be isomerized into internal alkenes or become conjugated during the various treatments, although no such change occurs with either the polyethylene vinyl alkene or with the process-degraded polypropylene vinyl alkene.

Wool and Statton (669) developed a new technique to study molecular mechanics of oriented polypropylene during creep and stress relaxation based on use of the stress-sensitive 975 cm^{-1} band and orientation-sensitive 899 cm^{-1} band. The far infrared spectrum of isotactic polypropylene was obtained from 400 to 10 cm^{-1} and several band assignments were made (670). Isotacticity of polypropylene has been measured from infrared spectra and pyrolysis-gas chromatography following calibration from standard mixtures of isotactic and atactic polypropylene. The infrared spectrum of oxidized polypropylene indicated small amounts of OOH groups plus larger concentrations of stable cyclic peroxides or epoxides in the polypropylene chain (670).

The Raman spectra of isotactic polypropylene has at 5-523°K showed bands characteristic of the unit cell (673). Relations between crystallinity and chain orientation have been observed in the Raman spectra of oriented polypropylene (674, 675).

Chalmers (676) has discussed the laser Raman spectrum of helical syndiotactic polypropylene.

Willis and Cudby (677) described the use of low-frequency spectroscopy in the far infrared region for studying the crystallinity of polypropylene.

Chen (678) determined the isotacticity of polypropylene from the ratios of the optical densities of peaks at 1170, 998, 973, and 841 cm^{-1} to that of the peak at 1460 cm^{-1} (internal standard).

Maier and Brettschneiderova (679) used the 975/1000 cm^{-1} infrared double band to determine the tacticity of polypropylene.

Painter et al (680) isolated infrared bands characteristic of isotactic polypropylene in the preferred helical conformation and in the irregular conformation of the amorphous phase using the absorbance subtraction technique.

Infrared bands in the region of CH_2 bending and rocking vibrations in the infrared spectra of isotactic poly-α-olefins have been found to be sensitive to changes in the state of order (681).

Blais et al (682) investigated surface changes during the photooxidation of polypropylene using infrared spectroscopy and elution microscopy.

To determine the isotactic content of polypropylene Peraldo (822) carried out a normal vibrational analysis. He considered the primary unit as an isolated threefold helix. From this work and a number of subsequent publications (823-826) it was suggested that the absorptions at 8.57μ (1167 cm^1), 10.02μ (997 cm^{-1}) and 11.9μ (841 cm^{-1}) were indicative of the helical conformation of the isotactic form. Measurements of the isotactic contents of a series of polypropylene fractions based on these three bands were made (826) and compared with results from Flory's melting point theory (827). Melting points were determined as the point of disappearance of the birefringence on highly annealed samples. All three bands give qualitative agreement with the melting point data, however, only the method based on the 8.57μ band gives quantitative agreement. Therefore, the method based on the 8.57μ band appears to give a good measure of the isotactic content in polypropylene, at least in the 60-100% range.

Nuclear Magnetic Resonance Spectroscopy

Randell (687, 703, 704) has discussed the NMR spectroscopy of polypropylene in detail. With the discovery of crystalline polypropylene in the early 1950's, polymer stereochemical configuration was established as a property fundamental to formulating both polymer physical characteristics and mechanical behaviour. Although molecular asymmetry was well understood, polymer asymmetry presented a new type of problem. Both a description and measurement of polymer asymmetry were essential for an understanding of the polymer structure.

Technically, each methine carbon in a poly(1-olefin) is asymmetric; however, this symmetry cannot be observed because two of the attached groups are essentially equivalent for long chains. Thus a specific polymer unit configuration can be converted into its opposite configuration by simple end-to-end rotation and subsequent translation. It is possible, however, to specify relative configurational differences and Natta introduced the terms isotactic to describe adjacent units with the same configurations and syndiotactic to describe adjacent units with opposite configurations (683). Although originally used to describe dyad configurations, isotactic now describes a polymer sequence of any number of like configurations and syndiotactic describes any number of alternating configurations. Dyad configurations are called meso if they are alike and racemic if they are unalike (684). Thus from a configurational standpoint, a poly(1-olefin) can be viewed as a copolymer of meso and racemic dyads.

The measurement of polymer configuration was difficult and sometimes speculative until the early 1960's when it was shown that proton NMR could be used, in several instances, to define clearly polymer stereochemical configuration. In the case of polypropylene, configurational information appeared available but was not unambiguously accessible because severe overlap complicated the identification of resonances from the mm, mr, and rr triads (685). Several papers appeared on the subject of polypropylene tacticity but none totally resolved the problem (686). With the advent of C-13 NMR in the early 1970's, the measurement of polymer stereochemical configuration became routine and reasonably unambiguous.

The advantages of C-13 NMR in measurements of polymer stereochemical configuration arise primarily from a useful chemical shift range which is approximately 20 times that of proton NMR. The structural sensitivity is enhanced through an existence of well separated resonances for different types of carbon atoms. Overlap is generally not a limiting problem. The low natural abundance (~1%) of C-13 nuclei is another favourably contributing factor. Spin-spin interactions among C-13 nuclei can be safely neglected and proton inter-

actions can be eliminated entirely through heteronuclear decoupling. Thus each
resonance in a C-13 NMR spectrum represents the carbon chemical shift of a
particular polymer moiety. In this respect, C-13 NMR resembles mass spectrometry because each signal represents some fragment of the whole polymer molecule. Finally, carbon chemical shifts are well behaved from an analytical viewpoint because each can be dissected, in a strictly additive manner, into contributions from neighbouring carbon atoms and constituents. This additive behaviour led to the Grant and Paul rules (688), which have been carefully applied in polymer analyses, for predicting alkane carbon chemical shifts.

The advantages so clearly evident when applying C-13 NMR to polymer configurational analyses are not devoid of difficulties. The sensitivity of C-13 NMR
to subtle changes in molecular structure creates a wealth of chemical shift-
structural information which must be "sorted out". Extensive assignments are
required because the chemical shifts relate to sequences from three to seven
units in length. Model compounds, which are often used in C-13 analyses, must
be very close structurally to the polymer moiety reproduced. For this reason,
appropriate model compounds are difficult to obtain. A model compound found
useful in polypropylene configurational assignments was a heptamethylhepta-
decane where the relative configurations were known (689). To be completely
accurate, the model compounds should reproduce the conformational as well as
the configurational polymer structure. Thus reference polymers such as predominantly isotactic and syndiotactic polymers form the best model systems.
Even when available, only two assignments are obtained from these particular
polymers. Pure reference polymers can be used to generate other assignments
(686).

To obtain good quantitative C-13 NMR data, one must understand the dynamic
characteristics of the polymer under study. Fourier transform techniques combined with signal averaging are normally used to obtain C-13 NMR spectra. Equilibrium conditions must be established during signal averaging to ensure that
the experimental conditions have not led to distorted spectral information.
The nuclear Overhauser effect (NOE), which arises from H-1, C-13 heteronuclear
decoupling during data acquisition, must also be considered.

Energy transfer, occurring between the H-1 and C-13 nuclear energy levels during spin decoupling, can lead to enhancements of the C-13 resonances by factors between 1 and 3. Thus the spectral relative intensities will only reflect
the polymer's moiety concentrations if the NOE's are equal or else taken into
consideration. Experience has shown that polymer NOE's are generally maximal,
and consequently equal, because of a polymer's restricted mobility)690, 691).
To be sure, one should examine the polymer NOE's through either gated decoupling or paramagnetic quenching and thereby avoid any misinterpretation of the
spectral intensity data.

The C-13 configurational sensitivity falls within a range from triad to pentad for most vinyl polymers. In non-crystalline polypropylenes, three distinct
regions corresponding to methylene (~46 ppm), methine (~28 ppm) and methyl
(~20 ppm) carbons are observed in the C-13 NMR spectrum. (The chemical shifts
are reported with respect to an internal tetramethylsilane (TMS) standard)
The C-13 spectrum of a 1.2.4. trichlorbenzene solution at $125^{o}C$ of a typical
amorphous polypropylene is shown in Fig. 73. Although a configurational sensitivity is shown by all three spectral regions, the methyl region exhibits by
far the greatest sensitivity and is consequently of the most value. At least
ten resonances, assigned to the unique pentad sequences, are observed in order,
mmmm, mmmr, rmmr, mmrr, mmrm, rmrr, mrmr, rrrr, rrrm and mrrm, from low to
high field (692, 693, 694). These assignments will be discussed in more detail later.

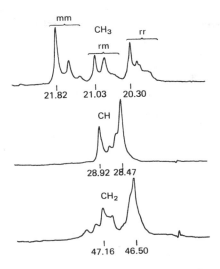

Fig. 73 Methyl, methine and methylene regions of the C-
13 NMR spectrum of a non-crystalline polypropylene (30).

The vinyl polymer studied most thoroughly with respect to configuration has
been polypropylene (695-702). The C-13 NMR spectrum of a crystalline polypropylene shown in Fig. 74 contains only three lines which can be identified as
methylene, methine and methyl from low to high field by off-resonance decoupling. An amorphous polypropylene exhibits a C-13 spectrum which contains not
only these three lines but additional resonances in each of the methyl, methine and methylene regions as shown previously in Fig. 73. The crystalline polypropylene must, therefore, be characterized by a single type of configurational
structure. In this case, the crystalline polypropylene structure is predominantly isotactic, thus the three lines in Fig. 74 must result from some particular length of meso sequences. This sequence length information is not available from the spectrum of the crystalline polymer but can be determined from
a corresponding spectrum of the amorphous polymer. To do so, one must examine
the structural symmetry of each carbon atom to the various possible monomer
sequences. Randall (687, 703, 704) carried out a detailed study of the polypropylene methyl group in triad and pentad configurational environments.

Woodward (709) has utilized NMR spectroscopy in his study of relaxation phenomena in polypropylene.

Stehling (710) has studied stereochemical configurations of polypropylenes
by high resolution nuclear magnetic resonance.

Brosio and coworkers (711) reported on ^{13}C NMR spectra for measurements of
tacticity, terminal conformation and configuration of polypropylene.

High resolution NMR spectra of isotactic and syndiotactic polypropylene have
been used by Cavalli (712) to provide conformational information.

Mitani (713) found that the diad and triad content of isotactic polypropylene,

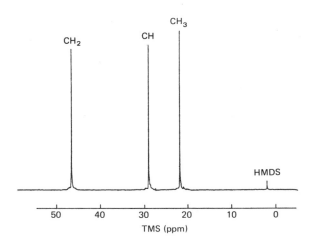

Fig. 74   C-13 NMR spectrum at 25.2 MHz of crystalline polypropylene.

syndiotactic polypropylene, and atactic polypropylene, as determined from 100
-MHz NMR spectra, were in agreement with the values determined from the 220-
MHz NMR and ^{13}C NMR spectra.

Inoue et al (714), Zambelli et al (715) and Randall (716) have shown that [13]C
NMR is an informative technique for measuring stereochemical sequence distributions in polypropylene. These workers reported chemical shift sensitivities
to configurational tetrad, pentad and hexad placements for this polymer.

Randall (717) has described work on the application of quantitative ^{13}C NMR to
measurements of average sequence length of like stereochemical additions in
polypropylene. He describes sequence lengths of stereochemical additions in
vinyl polymers in terms of the number average lengths of like configurational
placements. Under these circumstances, a pure syndiotactic polymer has a number
average sequence length of 1.0; a polymer with 50:50 meso, racemic additions
has a number average sequence length of 2.0 and polymers with more meso than
racemic additions have number average sequence lengths greater than 2. Amorphous and crystalline polypropylenes were examined using ^{13}C NMR as examples of
the applicability of the average sequence length method. The results appear
to be accurate for amorphous and semi-crstalline polymers but limitations are
present when this method is applied to highly stereoregular vinyl polymers
containing predominantly isotactic sequences.

Randall (718) has measured the ^{13}C NMR spin lattice relaxation times of isotactic and syndiotactic sequences in amorphous polypropylene. Spin-lattice
relaxation times for methyl, methylene, and methine carbons in an amorphous
polypropylene were measured as a function of temperature from 46 to 138°C. The
carbons from isotactic sequences characteristically exhibited the longest spin
relaxation times of those observed. The spin relaxation times differences increased with temperature with the largest difference occurring for methine
carbons where a 32% difference was observed. Randall determined activation
energies for the motional processes affecting spin relaxation times for isotactic and syndiotactic sequences. Essentially no dependence upon configurat-

ion was noted.

The sequence length of stereochemical additions in amorphous and semi-crystalline polypropylene were accurately measured using ^{13}C NMR (719). The method
has some limitations for addition polymers having predominantly isotactic sequences.

Asakura and coworkers (720) observed several new peaks in the ^{13}C NMR spectra
of polypropylene prepared with VCl_4-Et_2AlCl catalyst and assigned the peaks
to isolated head-to-head or tail-to-tail units.

Chlorine distributions in chlorinated polypropylene have been obtained from
100- and 220 MHz data on solutions and suspensions in tetrachloroethane.(721)

Reilly (722) has investigated an NMR method for determining the syndiotactic
content of polypropylene. The results are not entirely consistent with the
syndiotactic crystallinity as determined by alternate methods such as the density method. A possible explanation for the lack of consistency is that the
NMR method does not require as long a block in the chain in order for the syndiotactic placement of methyl groups to be detected. He attempted to determine
the syndiotactic content of some experimental polypropylenes. NMR spectra of
samples dissolved in o-dichlorobenzene were obtained at $170^{\circ}C$ at 100 Mc/sec.
Spectra of syndiotactic and isotactic samples gave the following NMR parameters:

	Syndiotactic	Isotactic
δCH_3	0.835 ppm	0.895 ppm
δCH_2	1.075 ppm	1.895 and 1.310 ppm
δCH	1.570 ppm	1.615 ppm
J_{CH_3-CH}	6.0 cps	6.0 cps
J_{CH-CH_2}	6.0 cps	6.0 cps
J_{H-H} (geminal)	Indeterminate	-14.0 cps

The reference for the chemical shift measurements was hexamethyldisiloxane.
The methylene hydrogens in the isotactic material are nonequivalent - as expected on geometrical grounds. Spectra calculated with the above parameters
agreed reasonably well with the observed spectra.

Woodbrey and Tremenozzi (731) have applied high resolution proton magnetic
resonance spectroscopy to an examination of very highly isotactic, very highly syndiotactic, and stereoblock polypropylenes in o-dichlorobenzene solutions.
They discuss the effects of stereoregulation on proton shieldings and some of
the complexities of the methylene proton resonances and determine tactic placement contents for several polymers by a method based on the methylene proton
resonances. Tactic pair contents were determined for two stereoblock fractions
by a method based on the methyl proton resonances. Their results revealed much
higher stereoblock characters than those determined for the same fractions
from melting data. All of the PMR results were in very good accord with the
results obtained on several polymers by infrared, x-ray diffraction, and differential thermal analysis.

Electron Spin Resonance Spectroscopy

Loy (723) has reported on the application of electron spin resonance to a study
of the radicals produced in the 60 Co irradiation of normal and 1,1 d_2 polypropylene. He concludes that 80% of the radicals are $-CH_2-CH(CH_2)-$ and 20%
are $-CH_2-C(CH_3)-CH_2$ in both atactic and isotactic polymer. In the isotactic,
but not in the atactic material, the primary radical disappears much faster

than the secondary one when heated. The primary spectrum of gamma-irradiated isotactic polypropylene as shown in Fig. 75 is seen to consist principally of four lines. Assuming that this is the result of equal interaction with three protons, three possible radical structures are;

$$- CH_2 - \underset{\underset{CH_2}{|}}{CH} - CH_2 -, \quad - CH - \underset{\underset{CH_3}{|}}{CH} - CH -, \quad and \quad - \underset{\underset{CH_3}{|}}{CH} - \overset{\overset{CH_2}{||}}{CH} - CH_2$$

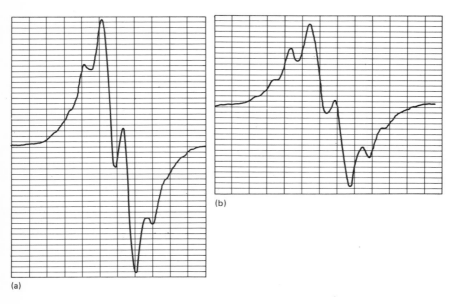

(a)

(b)

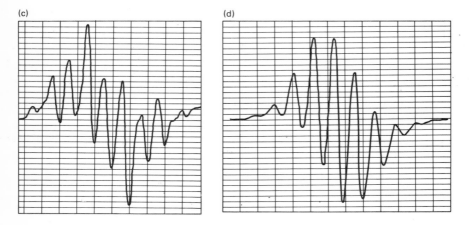

(c)

(d)

Fig. 75   Isotactic polypropylene:(b)heat treated 15 min. at 78°C,(c) heat treated 15 min. at 41°C.

The first two possibilities result from the loss of a hydrogen atom, either
in the primary event or abstraction by free hydrogen atoms produced in the
primary event. The third is the scission product from the reaction suggested
by Wall (724).

$$
\begin{array}{cccc}
CH_3 & CH_2^{\cdot} & CH_3 & CH_2 \\
| & | & | & || \\
- CH - CH_2 - CH - & = & - CH - CH_2^{\cdot} + CH -
\end{array}
$$

The secondary spectrum after 15 min. heat treatment at $41^{\circ}C$ as shown in Fig.
75 consists of nine lines. These could be the result of equal interaction of
the unpaired electron with the protons of the

$$
CH_3 - \overset{\cdot}{C} - CH_2 - \\
| \\
CH_3
$$

radical. This radical presumably results from the rearrangement of the

$$
CH_3 \\
| \\
- CH - CH_2 \cdot
$$

radical considered above.

Buch (725), on the other hand considers the primary spectrum to be a combination of all possible radical species that can be obtained by the removal of
hydrogen atoms. The origin of the nine-line spectrum is thought to be the

$$
- CH_2 - \overset{\cdot}{C} - CH_2 - \\
| \\
CH_3
$$

radical. The nine-line spectrum is explained by assuming that the methyl group
is not freely rotating at $77^{\circ}K$ but frozen in such a position that two of its
hydrogens are equivalent to the methylene hydrogens and the interaction with
the third hydrogen is twice as great. The plausibility of this argument is
shown by a rigorous mathematical treatment of the problem. Buch also states
that the secondary spectrum is present only in the isotactic material.

Kusomoto et al (726) carried out an ESR study of radical sites in crystalline
texture of irradiated polypropylene by means of nitric acid etching.

Forrestal and Hodgson (727) carried out electron spin resonance studies of
irradiated polypropylene.

The electron spin resonance of radiation oxidized polypropylene has been studied by Ohnlishi et al (728).

Stehling and Knox (729) defined the stereochemical structure of polypropylene
from the PMR spectra of normal, deuterated and epimerized polypropylene.

Ooi et al (730) attributed both the 17- and 9-line ESR spectrum of γ-irradiated
isotactic polypropylene and the 6-line spectrum of atactic polypropylene to
the $\cdot CH_2 \overset{\cdot}{C} MeCH_2$ - radical.

Pyrolysis-Gas Chromatography

The studies of Wall and Straus (732) indicate that pyrolysis of polypropylene proceeds principally by random cleavage of the polymer chain. Other workers have applied pyrolysis and polymer degradation studies to study of polyolefin degradation (733-735).

Brauer (736) has applied pyrolysis-gas chromatography to the elucidation of the structure of atactic and isotactic polypropylenes.

Voigt (737) used a platinum filament type flash pyrolyzer with a filament temperature of 550°C to pyrolyze polypropylene. The decomposition products from up to 2 mg polymer were then led directly on to the gas chromatographic column. Figure 76 shows the pyrogram obtained from Hostalen PPN polypropylene and identifies some of the pyrolysis products. The (unidentified) peak 18 in Fig. 76 is suitable for qualitatively identifying polypropylene in mixtures with other polymers such as polyethylene, polybutene-1 and poly-4-methylpentene-1.

In the case of polypropylenes, no straight-chain decomposition products with a length greater than C_5 were observed, a fact which is in agreement with the prediction readily deduced from the head-tail structure of the polymers. The larger decomposition products all contain methyl branches, and the multiple-branched chains the expected 2,4-branching.

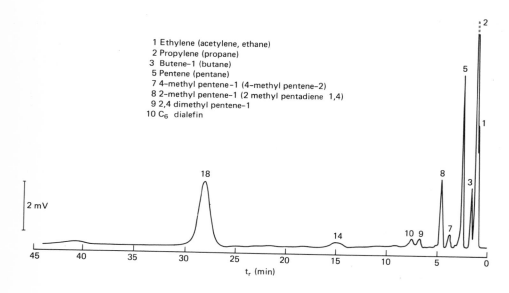

1 Ethylene (acetylene, ethane)
2 Propylene (propane)
3 Butene-1 (butane)
5 Pentene (pentane)
7 4-methyl pentene-1 (4-methyl pentene-2)
8 2-methyl pentene-1 (2 methyl pentadiene 1,4)
9 2,4 dimethyl pentene-1
10 C_6 dialefin

Fig. 76   Pyrolysis-gas chromatography of polypropylene
(Hostalen PPN).

Van Schooten and Evenhuis (738, 739) found identical pyrolysis patterns for various polyproylene samples. The surface areas of the main peaks are shown in Fig. 77. Depolymerization is much more important for polypropylene, as shown by the large propane peak. Two other very large peaks in the polypropylene pyrogram can be interpreted as originating from intramolecular hydrogen transfer with the fifth carbon atom of the secondary radical:
I, n pentane + II, 2.4 dimethyl heptane.

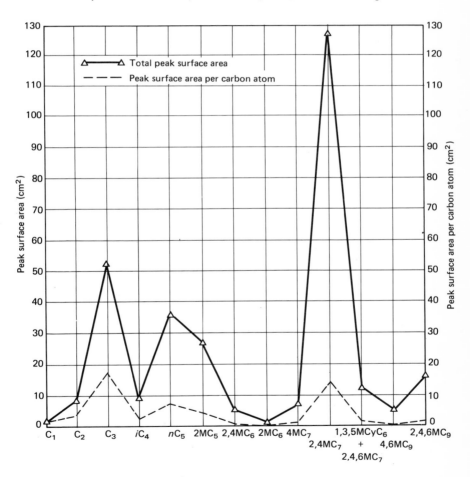

The primary radical gives in this way:

I, isobutane + II, 2.4 dimethyl heptane + III CH_3

Fig. 77. Pyrolysis-gas chromatography of polypropylene.
Surface areas of the main peaks.

The isobutane and especially the methane peaks are rather small, (Fig. 77), indicating that this reaction is not very frequent for the primary radical. For this radical hydrogen transfer with the sixth carbon atom might be more important as the 2 methyl pentane peak could be explained in this way. Another possibility is that this peak has to be ascribed to intramolecular hydrogen transfer reactions, (i.e. formation of 2 methyl pentadiene). In Table 40 are listed the products expected from intramolecular hydrogen transfer during the pyrolysis of polypropylene. These can be compared with the products found in the pyrogram in Fig.77.

Table 40 Products Expected from Intramolecular Hydrogen Transfer during Pyrolysis of Polypropylene.(Main Peaks found in Pyrogram in Fig. 77 are underlined)

	Number of the Hydrogen Donating Carbon Atom				
	5	6	7	8	9
Secondary radical	nC_5	CH_4	$4MC_7$	CH_4	$4.6MC_9$
~C-C-C-C-C-C-C	$\underline{2.4MC_7}$	$2MC_5$	$2.4.6MC_9$	$2.4MC_7$	$2.4.6.8MC_{11}$
| | | |		$\underline{4MC_7}$*		$4.6MC_9$*	
C C C C		$4.6MC_9$		$4.6.8MC_{11}$	
Primary radical	iC_4	$2MC_5$	CH_4	$2.4MC_7$	CH_4
~C-C-C-C-C-C-C-	$2MC_5$*	$\underline{2.4.6MC_7}$	$2.4MC_5$	$2.4.6.8MC_9$	$2.4.6MC_7$
| | |	$\underline{2.4MC_7}$		$2.4MC_7$*		$2.4.6MC_9$*
C C C	$\underline{CH_4}$		$2.4.6MC_9$		$2.4.6.8MC_{11}$

* Formed by hydrogen exchange with a methyl group.

Pyrolysis-gas chromatography has been used to carry out micro-structural studies on isotacticity in polypropylene (740, 741).

Thermal Degradation

Moiseev and Neiman (742) have reviewed the thermal degradation by radical mechanisms of polyolefins and other linear polymers and shown that in the thermal degradation of such polymers in a vacuum the main body of the products volatile at the temperature of degradation arises in intramolecular transfer of the chain.

Moiseev and Neiman (742) conclude that there is quite definite correspondence between the amounts of the volatile products of the degradation of polypropylene expected and those found.

The pyrolysis product distribution of polypropylene is large; and the monomer yield, 2% is low (744). The higher decomposition temperature of polypropylene, as opposed to polyisobutylene, $380^{\circ}C$ and $340^{\circ}C$, respectively (745) further indicates a large difference in chain stability towards thermal cleavage brought about by a lack of a quaternary carbon atom in polypropylene.

Schwenken et Zuccarello (746) and Donald et al (747) have applied differential thermal analysis to the characterization of polypropylene and crystallization studies on polypropylene.

The nature of products obtained in the thermal oxidation of polypropylene has been studied by Beachall and Beck (748) and thermal decomposition has been studied by Seeger and Gritter (749) and Tsuchiya and Sumi (750).

Kamida and Yamaguchi (751) and other workers (752, 753) have investigated multiple meeting peaks in isotactic polypropylene using differential thermal analysis and differential scanning calorimetry.

Bosch (754) has used differential thermal analysis to identify polypropylene.

Three general types of polyolefin degradation are: sustained heat or long-term aging below the melting point, exposure to ultraviolet light, and high-temperature processing in the melt. Oxidative degradation caused by the first two cases has been investigated on unstabilized polyolefins by use of a variety of methods (755).

Of the few studies of high temperature processing in the melt that have been made, Rugg et al (756) and Meyer (757) have obtained the infrared spectra of unstabilized polyethylene sheets after milling above its melting point. Carlsson et al (758) obtained the infrared spectrum of polyethylene film after oxidation in the melt at 225°C. In all three cases, ketone was claimed to be the principal product of oxidation. Losses in intrinsic viscosities have been noted after oxidation in the melt, (755, 758) and decreases in weight-average molecular weights M_w after extrusions, (759, 760) but no detailed analyses of the changes in molecular weight have been made.

Adams and Goodrich (761) have studied the products of degradation of unstabilized polypropylene processed in the melt and the molecular weight changes that occur. They used a Brabender torque rheometer to simulate polymer processing conditions. The type and amount of carbonyl functional groups formed were determined by infrared analysis and chemical reactions of the functional groups by treatment with base and acid. Gel-permeation chromatography was used to determine the molecular weights of the oxidized sample in order that the number of chain scissions per functional group could be calculated.

X-ray Diffraction

X-ray diffraction measurements can be used for routine estimation of degree of crystallinity of polypropylene films (762). Changes in orientation and crystalline state of stretched films have been studied by x-ray'diffraction, optical, and calorimetric methods (763).

Neutron Scattering

Ballard et al (764) have studied the relationship between morphology and chain conformation of molten and crystalline isotactic polypropylene by small angle neutron scattering using samples containing small amounts of polypropylene-d_4 in a protonated polypropylene matrix.

Fractionation

Several workers (765-772) have applied fractionation techniques to the measurement of the molecular weight distribution of polypropylene.

Crystalline polypropylene is mainly characterized by three factors: tacticity, molecular weight, and its distribution. Molecular weight distribution is a very ambiguous factor, in spite of the amount of attention it has received, because the observed distribution curves depend on the determination method and no definite method has been established.

The molecular weight distribution of crystalline polypropylene is generally determined by column fractionation or gel-permeation chromatography. To obtain the molecular weight distribution curve by column fractionation takes many hours and the polymer must be protected from thermal degradation. Nevertheless, the column fractionation method is still widely used. Gel-permeation chromatography has been rapidly extended over the field of polymer characterization and it becomes possible to obtain the distribution curve very easily by this method.

Ogawa et al (776) have compared column fractionation and gel-permeation chromatography for determining the molecular weight distribution of polypropylene. They calculated statistical parameters such as average molecular weight, standard deviation, skewness, and kurtosis for each distribution curve, and also the number-average and weight-average molecular weights were determined by osmometry and light scattering to compare with these from distribution curves.

The results of their investigations on the comparison of the determination methods of molecular weight distribution for crystalline polypropylene are summarized as follows.
(1) The molecular weight distribution curve obtained from column fractionation was narrower than that from gel-permeation chromatography, and the D value from gel-permeation chromatography was more close to that from the absolute methods (osmometry and light scattering).
(2) It was confirmed that the distribution curve from column fractionation became similar to that from gel-permeation chromatography on correcting the overlapping of the distribution of fractionated polymers.
(3) The broadening effect in gel-permeation chromatography and thermal degradation during fractionations were both found to be of little importance.
(4) It was assumed by Crouzet (775) that a broader distribution curve had been obtained from gel-permeation chromatography due to a broadening effect as compared with that from column fractionation and that the distribution curve from column fractionation was more accurate than that from gel-permeation chromatography. However, Ogawa et al (776) consider that the distribution curve obtained from gel-permeation is more accurate and reliable than that from column fractionation.

Nakajima and Fujiwara (228,229) carried out fractionation of polypropylene in a vapour-jacketed Soxhlet-type apparatus. These workers prepared plots of the integral weight fraction against the limiting viscosity number, the density, and the optical density ratio (D995 cm^{-1}/D974 cm^{-1}, Luongo (232)) of the fractions respectively. A characteristic feature of the curve is that the limiting viscosity number of the fractions is almost constant up to about a 0.2 integral weight fraction. This means that the fractionation in the low molecular weight range occurred according to the crystallinity or tacticity of the fractions, and that the higher fractions were fractionated according to molecular weight and have almost the same crystallinity or tacticity.

Natta et al (233, 234, 235) from x-ray data obtained the following relationship between the degree of crstallization (X_c) and the density of propylene:

$$X_c = 11.68 - (9.966/d)$$

Vaughan (777) fractionated polypropylene using preparative gel-permeation chromatography. He discussed the effect of concentration and sample size on polydispersity. Wims (778) extracted polypropylene with tetrahydrofuran for gel-permeation chromatography and with methylene chloride for liquid chromatographic determination of antioxidants.

Lovric (779) fractionated polypropylene by coacervate extraction and column elution. Fractionation data from polypropylene were examined to compare bulk polymer, fibres, and slightly degraded material (780). Ogawa and Hoshino (781) compared fractionation of isotactic polypropylene by using temperature and solvent gradient methods. Comparison of results agreed fairly well on fractionation of polyethylene-polypropylene blends between hypothetical calculations and experimental data by solvent gradient method (782).

Ogawa, Tanaka and Inaba (783) studied the effects of the polymer deposition conditions in column fractionation of isotactic polypropylene and showed the particle size to be an important factor to obtain good fractionation.

Molecular Weight

Tanaka et al (791) have used the high temperature turbidimetric titration procedure originally described by Morey and Tamblyn (792) for determining the molecular weight distribution of cellulose esters. This method has been applied to the measurement of the molecular weight distribution of polypropylene. They found that the type of molecular weight distribution of these polymers is a log-normal distribution function in a range of I(M), (cumulative weight %), between 5-90%. The effect of heterogeneity in the molecular weight distribution of polypropylene on the viscosity-molecular weight equation was examined experimentally; the results agreed with those calculated from theory.

Light scattering has been applied successfully by Oth (784), Harris and Miller (785), Scholten (786), Bischoff (787), Kyogoku and Kimura (788), Goobermann (789), Tanaka et al (790) and others in experiments at room temperature. Tanaka et al (791) determined the molecular weight distribution of polypropylenes using an automatic turbidimetric titration apparatus with which measurments at high temperatures can be carried out; the effect of heterogeneity in molecular weight distribution of polypropylene on the viscosity-molecular weight equation was investigated.

Several workers (793, 794, 795) have reported on the molecular weight distribution of polypropylene.

Specific and reduced viscosities of polypropylene have been used to calculate the intrinsic viscosity by a one-point method using the Huggins- Martin, and Solomon-Ciuta equations. The (η) values were used for the calculation of Mn of polypropylene (796).

Narrow molecular weight distribution standards of polypropylene are available from National Physics Laboratories of Teddington, England (797).

Electron Probe Microanalysis

Crompton (798) has applied electron probe microanalysis in an attempt to elucidate a phenomenon known as lensing in polypropylene film. When the film in question was drawn down to between 1 and 2 thou thickness the imperfection became apparent (Fig.78). Electron probe microanalysis was used as a direct check on whether or not a higher concentration of chlorine is associated with a lens than is present in the surrounding lens-free polymer. The polypropylene film was first coated with a thin layer of copper, to keep the sample cool by conduction during electron bombardment. Of the lenses examined by EPM, a number were seen to have a speck at the centre. Figure 79 shows a back scattered

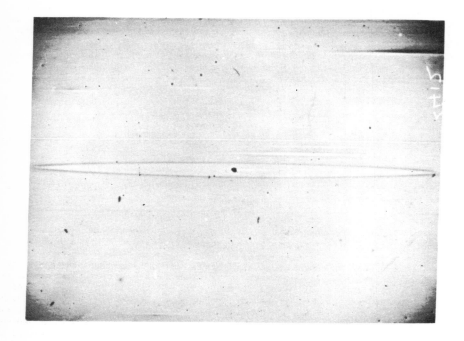

Fig. 78 Lens in polypropylene film. Magnification x 18.

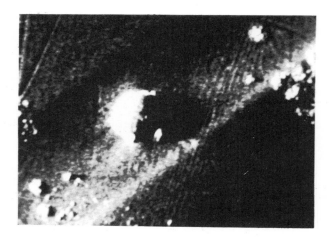

Fig. 79 Back scattered electron microanalysis of a speck
within a lens in polypropylene.

electron image of such a speck; part of the lens is also seen. The specks
examined were shown by EPM to contain the elements sodium and chlorine. Figures 80 a, b and c show respectively the image in back scattered electrons of
a speck within a lens, the emitted chlorine radiation, and the emitted sodium
radiation. The chlorine and sodium distributions in Fig. 80 b and c show as
a ring round the circumference of the speck.

a) **ELECTRON**
   IMAGE IN BACK
   SCATTERED
   ELECTRONS

b) **CHLORINE**
   DISTRIBUTION

c) **SODIUM**
   DISTRIBUTION

Fig. 80   Electron probe microanalysis of a speck within a
lens in polypropylene.

The x-rays emitted by the sample in EPM can be detected and counted by a proportional counter. In Fig. 81 are shown three pulse height distribution curves.
Curve 81a was obtained with the electron probe positioned on the sodium and
chlorine containing ring shown in Fig. 80b and c. Curve 81b was obtained with
the electron probe displaced from the ring shown in Fig. 80 b and c, i.e. on
a non-lensed part of the polypropylene film. The curve in 81 was obtained with
the electron probe positioned on a crystal of pure sodium chloride. Comparison of Fig.81a and b demonstrates that chlorine and sodium only occur in high
concentrations in the speck occurring in a lens, and that these elements are
not generally distributed throughout the polymer. Comparison of Fig. 81 a and
c shows that the material which surrounds the central speck within the lens
is not very different from pure sodium chloride.

Miscellaneous

Samuels (799) has discussed use of small angle light scattering birefringence,
and refractive index measurements to provide information on internal crystallization processes in polymers such as isotactic polypropylene.

Radtsig (800) studied the structure and conformation of free radicals formed
during the radiolysis and mechanical degradation of isotactic polypropylene
and poly-1-butene.

Van Sickle (801) has studied the oxidation of polypropylene, polyethylene and
copolymers.

Russell (802) has investigated the determination of the relative isotactic
content of polypropylene. He concluded that a measure of this parameter may
be obtained by successive extractions with diethyl and n-heptane or with n-
heptane alone under controlled conditions. Although the values obtained cannot
be regarded as absolute, they are reasonably reproducible and could serve to
characterize polypropylenes. These measurements are useful in the correlation of the isotactic content of polypropylene with its mechanical properties
(803, 804).

Several methods have been used to estimate the isotactic content of polyprop-

Analysis of Plastics

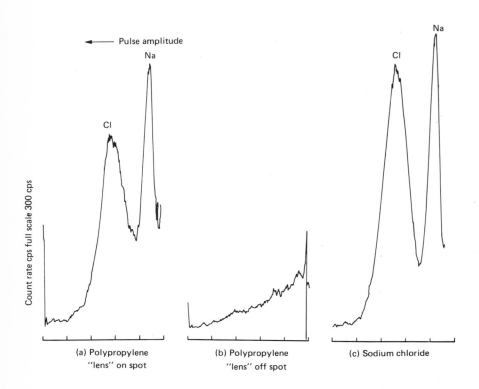

Fig. 81 Pulse height distribution curves obtained in electron probe microanalysis of polypropylene lenses.

ylene, such as x-ray studies (805), density determinations (806), infrared
spectroscopy (809) and solvent extraction procedures (806-808). None of these
has as yet been universally accepted as an absolute standard, and agreement
between two different methods is not always good. The density and x-ray methods
depend upon the completeness of crystallization of the isotactic material in
the sample, and as a result the values are difficult to reproduce, even with
careful annealing.

One of the most commonly used estimates of steric regularity is the amount of
material unextracted by hot heptane. This has been referred to as an isotacticity index by Natta and his coworkers (803) who studied the solubility of
polypropylenes in various solvents. They found that diethyl ether removed
lower molecular weight polymer which showed no crystallinity by the x-ray method. Subsequent extraction with n-pentane removed material having about the
same intrinsic viscosity as the ether-soluble material, but which was about
30% crystalline. In going from C_5 to C_8 saturated hydrocarbon solvents, they
found progressive increases in both the x-ray crystallinities and intrinsic
viscosities. The normal octane-extractable material was nearly as crystalline
as the residue, but its intrinsic viscosity was about half that of the residue.
They reasoned that these intermediate fractions were essentially block poly-

mers of the d and l forms, caused by occasional inversion of configuration as the chain grew, giving chain configuration such as: ddddlllllddddddlllll, etc. They term such polymers "stereoblock". Thus, the ether-soluble fractions are generally considered atactic and the heptane-soluble to be stereoblock polymer.

Natta and his coworkers describe a procedure involving use of a relatively large amount of polymer and extraction for up to 40 hr with acetone, then repeating with ether and heptane at the boiling point of the solvent under a nitrogen atmosphere to avoid the possibility of oxidation. These same authors also reported that one fraction tended to solubilize the less soluble material. Most authors, however, give only general descriptions of the extraction procedure used. In ordinary Soxhlet-type extractions, both the ether- and heptane-extractable values may vary as much as 5% for polypropylenes of medium isotacticity.

Braun and Guillet (810) have used inverse gas chromatography to determine the crystallinity of polypropylene.

Moisture in polypropylene has been reported to enhance the x-ray absorption coefficient but appears to have no influence on the crystal lattice (811).

Water

Traces of water has been estimated in polypropylene by a procedure involving heating the polymer under nitrogen, followed by sweeping the released water vapour into a Karl Fischer titration cell (812).

Volatiles

Low boiling hydrocarbons have been determined in polypropylene using a procedure in which the sample is heated under nitrogen and the volatiles swept into a gas chromatograph (812).

Additives

Antioxidants in Polypropylene (see also Polyethylene)

Gas Chromatography

Ionox 330 (trimethyl 2,4,6 tri(3,5 di-t-butyl 4-hydroxybenzyl) benzene)	Solvent extraction - glc. (813,814)
antioxidants	Solvent extraction - glc (815)
2,6, di-t-butyl-p-cresol laurylthiodipropionate	p-xylene extraction - glc (816)

4,4^1 methylene bis(2,6 di-t-butylphenol)
2,2^1 methylene bis(6-C^1-methylcyclohexyl)-p-cresol)
4,4^1 thio bis(6-t-butyl-m-cresol)
4,4^1 butylidene bis(6-t-butyl-m-cresol)
Ionox 330 (135 trimethyl-2,4,6,tris
(3,5, di-t-butyl 4 hydroxybenzyl) benzene
tris (2,6 di-t-butyl-4-hydroxyphenyl)phosphate

octadecyl 3,5 di-t-butyl 4-hydroxyhydrocinnamate
2-hydroxy 4-(octyloxy)benzophenone

2-hydroxy-4-methoxybenzophenone
4-dodecyloxy)-2-hydroxybenzophenone
2-(2 4 benzotriazol- 2 yl)-p-cresol
2,4 di-t-butyl-6-(5-chloro-2 4 benzotriazol-2yl)phenol

Thin-layer Chromatography

dilauryl thio
dipropionate

Hydrolysis to lauryl alcohol,
chloroform-ethanol-alcohol-
hexane extraction.
Thin-layer chromatography then
glc.   (817)

4,4[1] butylidene antioxidants
(2-t-butyl-5-methyl)phenol
4,4[1] thio bis(6-t-butyl-m-cresol)
pentoacryltritol tetra bis 3,5,di-t-butyl
4-hydroxyhydrocinnamate

n-heptane extraction.
Thin-layer chromatography for
quantitative analysis
                            (818)

2,2[1] methylene bis(4 methyl-6-t-butyl phenol)

octadecyl (3,5 di-t-butyl-4-hydroxyphenyl)acelate
2,6 di-t-butyl-p-cresol

Miscellaneous Methods

alkyl phenol and bis phenol
antioxidants

Volatile preconcentration -
mass spectrometry   (819)

diorgano sulphides and tert phosphites
e.g. distearylthio dipropionate
     dilaurylthiodipropionate
     triphenylphosphite
     triethylphosphite
     didecylphosphite

Heptane extraction.
Oxidation with m-chloro-peroxy-
benzoic acid.
Unreacted oxidant estimated
iodimetrically.   (820)

Stabilizers in Polypropylene

Stabilizers for polypropylene have been analyzed by spectrophotometry. For
example, 2,4-ditert-butyl-6-(5-chlorobenzo-triazol-2-yl)phenol ("Tinuvin" 327)
was extracted from the polymer with chloroform-ethyl alcohol (19:1) and determined at 315 or 355 nm. Where necessary, the extract was fractionated by
thin-layer chromatography on silica gel with chloroform as solvent (821).

CHAPTER 3

HIGHER ALKENE AND COPOLYMERS

Infrared Spectroscopy

Various workers (828-830) have discussed the existence of polymorphic forms
of polyisobutylene. Three modifications have been described to date as modifications I, II and III. Crystallographic studies have shown that modifications I and II are hexagonal and tetragonal crystal structures respectively.
Modification III has been tentatively assigned an orthorhombic structure.

The different modifications are obtained by either crystallizing from the melt
or casting a film from solution (828, 833). Crystallographic (834) and differential thermal analysis (833, 834) studies have shown that the unstable
modification II, obtained by moulding a sample from the melt, gradually trans-
forms into the stable modification I. Modification II is reported to assume
a four-fold helical conformation and modification I a three-fold helical conformation (831, 832). The transformation, however, can be accelerated by sub-
jecting the film to stretching or pressure.

The third form, modification III, is obtained by casting a film of the polymer
from solution. This form, although dependent on solution temperature and solvent (833) appears stable even after 3 weeks.

Luongo and Salovey (835) have shown that these three crystalline forms of the
polymer produce three significantly different infrared spectra from 2000-400
cm^{-1}, which are shown in Fig. 82.

The top curve is of the unstable modification II immediately after the hot
moulding of a film. The middle trace is that of the same film after 3 weeks
at room temperature and is the spectrum of the stable modification I. The
spectral changes associated with the transformation of modification II to
modification I begin approximately 1/2 hr after moulding of the film. The
bottom curve is the spectrum of a cast film (from carbon tetrachloride) and
represents the stable modification III.

Luongo and Salovey (835) comment on the differences in the profiles and intensities of the bands in the three spectra between 1350 and 500 cm^{-1}.

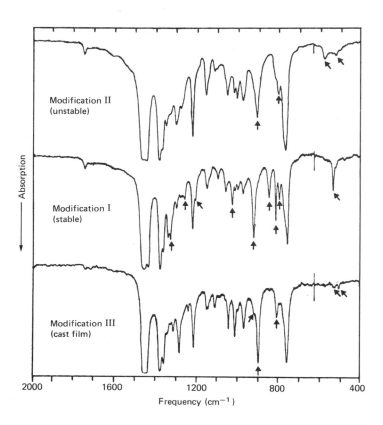

Fig. 82   Infrared spectra of polybutene-1 polymorphs.

Higgins and Turner (836) studied the radiolysis of polyisobutene by means of infrared and ultraviolet spectroscopy.

Toman et al (837) carried out spectrophotometric investigation of the mechanism of polymerization of isobutylene with vanadium tetrachloride and visible light.

Nuclear Magnetic Resonance Spectroscopy

Barrall et al (838) have applied nuclear magnetic resonance and pyrolysis-gas chromatography (839, 840) at 430 to 600°C, to an investigation of the composition analysis of polybutenes prepared by the ionic polymerization of various mixtures of propylene, 1-butene, cis- and trans-2-butene, and isobutylene. Isobutylene content ranged from 10% to 100%. The nature of the polymerization caused the average molecular weight of the copolymers to depend directly on the isobutylene content of the reaction mixture.

Figure 83 illustrates the pyrolysis-gas chromatogram obtained for relatively pure polyisobutylene. In general, polyisobutylene pyrolyzes in the temperature

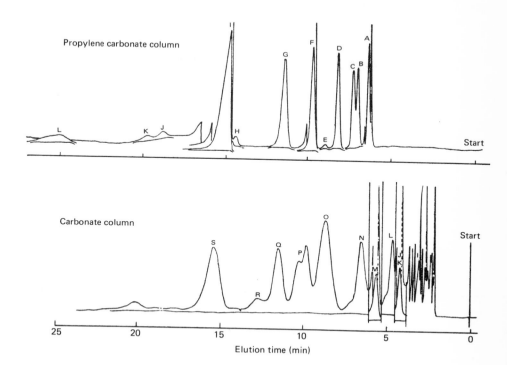

Fig. 83 Pyrolysis/gas chromatograms of polyisobutylene
A. A sample equivalent to 0.0198 g of polymer was
pyrolyzed at 485° in a 35 cc/min helium carrier
gas stream. The Carbowax column record was recorded at four times the base sensitivity of the
propylene carbonate column. The chromatographic
peaks are identified as follows: A=Air and methane. B=Ethane. C=Ethylene. D=Propane. E=Butane.
F=Propane. G=Isobutane. H=Butene-1. I=Isobutyl-
ene. J=trans-2-Butene. K=cis-2-Butene. L=2-Methyl
-1-Butene. M=2,2,4-trimethyl-2-pentene. N,O=Dimer
olefins. P,Q,R=Trimer olefins. S=Tetramer olefin.

range 430° to 600° to yield, in varying ratios, methane, ethane, propane, butane, ethylene, propylene, neopentane, 1-butene, isobutylene, trans- and cis-
2-butene, 2-methyl-1-butene and trans- and cis-2-pentenes. In addition to
these lighter products, the dimer of isobutylene, 2,4,4-trimethyl-2-pentene,
is produced. Isobutylene and propylene make up over 30% of the pyrolysis products. Isobutylene trimer and tetramer materials are also found. In the pyrolysis temperature range 400° to 500°, 80-86% of the polymer sample is pyrolyzed
to materials in the carbon range C_1 to C_{16}.

Barrall et al (838) applied pyrolysis-gas chromatography and NMR spectroscopy
to polyisobutylenes made from various proportions of isobutylene, propylene
and butene-1. They concluded that polyisobutylene decomposes primarily by a
random cleavage of the polymer chain. Due to the instability of the quaternary

carbon atom in the chain, cleavage at these sites is favoured. The high thermal
stability of the isobutylene fragment over that of possible higher cleavage
products favours the apparent high monomer yield.

Warren et al (841) carried out an NMR identification of C_{11} to C_{40} branched
hydrocarbons derived from the decomposition of polyisobutylene.

Mauzac and co-workers (842) determined the overall isotactic content of crude
polybutene-1 by ^{13}C NMR.

Suzuki (843) reported 220-MHz NMR results from poly-1-butene (ether-soluble
and ether-insoluble fractions). Comparison of NMR spectra of poly(4-methyl-1-
pentene) and poly(3-methyl-1-butene) indicated that meso diad chemical shifts
were dependent on side chain constituents.

Proton Magnetic Resonance Spectroscopy

Proton NMR signals for olefinic end groups, $-CH_2C(CH_3)=CH_2$, $CH=C(Me)_2$, and
$-CH_2C-(=CH_2)CH_2-$ were observed in high molecular weight polyisobutylenes by
Manaff et al (844).

Electron Spin Resonance Spectroscopy

Lou (845) has reported the ESR spectrum of ^{60}Co irradiated polybutylene and
polyisobutylene recorded at various temperatures between -55 and -190°C.

Pyrolysis Gas Chromatography

Voigt (846) used a platinum filament type flash pyrolyzer with a filament temperature of 550°C to pyrolyze polybutene-1. The decomposition products from up
to 2 mg polymer were then lead directly onto the gas chromatographic column.
Di-n-decylphthalate on Kieselguhr was used as the separation column. Figure 84
shows the pyrogram obtained from polybutene-1 and identifies some of the pyrolysis products.

Whereas, in the case of polyethylene and polypropylene, the splitting up of
the main chain is virtually the only result of thermal reaction, the stripping
off of side chains, in polybutene-1 is clearly indicated by the nature of the
pyrolysis products.

In the case of polybutene-1 a mechanism involving stripping off of side chains
is involved in the formation of ethylene;

$$\sim CH_2 - CH \sim \qquad\qquad \sim CH_2 - CH \sim$$
$$\qquad\quad |\qquad\qquad\qquad\qquad\qquad\qquad | $$
$$\qquad\quad CH_2 \qquad\qquad\quad + \qquad\quad CH_2$$
$$\qquad\quad |\qquad\qquad\qquad\qquad\qquad\qquad ||$$
$$\qquad\quad CH_2' \qquad\qquad\qquad\qquad\qquad CH_2$$

In actual fact, a considerable yield of ethylene is obtained from polybutene-1,
whereas the production of propylene is much smaller because this can only emanate from the principal chain. The complicated structure of polybutene-1 as
compared with those of polyethylene and polypropylene, gives rise to the formation of a markedly larger number of structurally isomeric degradation products.
Theoretically with a maximum molecular size of C_8, 37 different hydrocarbon
decomposition products should be obtainable from polybutene-1, and, in fact,
Voigt (846) was able to identify, or indicate the probable existence of 33
compounds for polybutene-1.

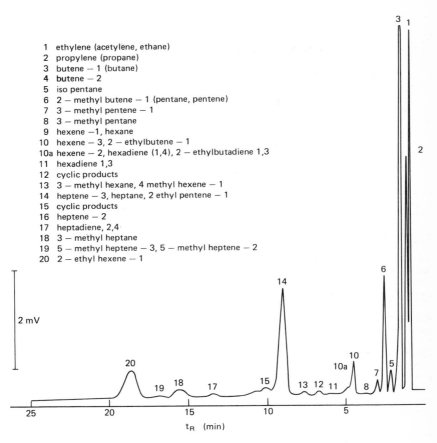

1 ethylene (acetylene, ethane)
2 propylene (propane)
3 butene — 1 (butane)
4 butene — 2
5 iso pentane
6 2 — methyl butene — 1 (pentane, pentene)
7 3 — methyl pentene — 1
8 3 — methyl pentane
9 hexene —1, hexane
10 hexene — 3, 2 — ethylbutene — 1
10a hexene — 2, hexadiene (1,4), 2 — ethylbutadiene 1,3
11 hexadiene 1,3
12 cyclic products
13 3 — methyl hexane, 4 methyl hexene — 1
14 heptene — 3, heptane, 2 ethyl pentene — 1
15 cyclic products
16 heptene — 2
17 heptadiene, 2,4
18 3 — methyl heptane
19 5 — methyl heptene — 3, 5 — methyl heptene — 2
20 2 — ethyl hexene — 1

Fig. 84 Pyrolysis-gas chromatography of polybutene-1

Peaks 6 and 14 (2 methylbutene-1 and heptene-3) (Fig.84) are suitable for quantitatively identifying polybutene-1. The presence of these peaks, by themselves,
is not sufficient to identify polybutene-1, because peaks could also emanate
from other polymers. Much more decisive is the height of the peak as compared
with that of another peak. If, for example, the presence of polybutene-1 was to
be established in polyethylene, the ratio of the height of the polyethylene
peaks is determined as the ratio heptene-1 to hexene-1. In the presence of
polybutene-1 this ratio is increased by the polybutene-1 peak heptene-3 as the
polybutene-1 content increases until a maximum end value of heptene-3/hexene-
3 is obtained for the polybutene-1 peaks when only pure polybutene-1 is present. Comparative measurements on known mixtures are indispensible for this
method. The studies of Wall and Straus (847) indicate that pyrolysis of polyisobutylene proceeds principally by random cleavage of the polymer chain. However, the exceptionally high yield of monomer (20%) from polyisobutylene reported in their studies suggests a non-statistical distribution of pyrolysis
products. Barrall et al (848) investigated the structure and composition of
the homopolymers and copolymers of isobutylene. The workers state that each

polymer and copolymer exhibits a specific pyrolysis temperature for maximum isobutylene yield.

Thermal Analysis

Geacintov et al (849) have applied differential thermal analysis, dilatometry X-ray diffraction and photomicrography to a study of crystalline form transitions in polybutene-1.

Holden (850) and Clampitt and Hughes (851) have applied differential thermal analysis to a study, respectively, of the effect of annealing on the low melting modifications of isotactic polybutene-1 and of the three polymorphic forms of polybutene-1.

McNeill (852) has discussed the application of the technique of thermal volatilization analysis to polyisobutene, butyl rubber and chlorobutyl rubber.

Radiochemical Methods

McNeill (854) has described a radiochemical method involving the use of ^{36}Cl for the measurement of higher levels of rubber unsaturation 0.5-2 mole % in butyl rubbers. Pepper and Reilly (853) have measured terminal unsaturation in polystyrene, but their hydrogenation technique was limited to oligomers which have a relatively high mole per cent unsaturation. The same restriction would apply to the use of conventional techniques (856, 857) which have been successfully employed for butyl rubbers having unsaturations in the range 0.5-5.0 mole %. Polyisobutylene prepared by a cationic mechanism contain unsaturated end groups at very low concentrations and in order to facilitate the determination of such low levels of unsaturation, McGuchan and McNeill (855) developed the radiochemical method of McNeill (854) involving reaction of polyisobutene with radiochlorine (^{36}Cl) which is applicable to the determination of end group unsaturation of the order of 0.01 to 0.1 mole % in butyl rubbers. If the specific activity of the radiochlorine in the gas phase is known, the weight of chlorine in the polymer can be found by counting. The mole per cent unsaturation (U_m) of the polymer was calculated from the weight per cent unsaturation (U_w) by assuming tentatively that one mole of chlorine enters the polymer per double bond. The experimental unsaturations can be compared to the unsaturations in polymers having one double bond per molecule to give the unsaturation of the unknown sample in terms of double bonds per molecule. The values of unsaturation exhibited by the unfractionated polymers are given in Table 41. The results of chlorination for the fractionated polyisobutenes are given in Table 42 and illustrated graphically in Fig. 85. The calculated unsaturations are tabulated in Table 43.

Table 41   Unsaturations of Unfractionated Polyisobutenes.

Polymer No.	Molecular weight ($\overline{M}_n$)	10^5 $\overline{M}_n$	U_w	U_m	U'_m	Double bonds per molecule
1	45,700	2.19	0.090	0.070	0.125	0.56
2	49,100	2.04	0.083	0.065	0.118	0.55
3	40,500	2.47	0.095	0.075	0.142	0.53
4	100,000	1.00	0.049	0.040	0.058	0.69

The chlorination data for the fractionated polyisobutenes clearly indicate that the mole per cent unsaturation is inversely proportional to the molecular weight. This is illustrated in Fig. 86. The points for the unfractionated samples are also shown in the figure. The two sets of results are in good

Table 42   Chlorination of Fractionated Polyisobutenes.

Polyiso-butene fraction	Cl in reaction mixture, wt.-%	Count rate of solution, counts/min. per 10 ml.	Wt. polymer in solution, mg. per 10 ml	Wt. Cl in polymer, mg.	Cl in polymer, wt-%

Specific Activity of Cl = 814 counts/min. per mg. Cl in CCl$_4$ solution.

F1	3.52	18.4	164	0.0226	0.0138
	2.90	17.9	168	0.0220	0.0132
	2.49	19.1	168	0.0234	0.0139
	2.12	17.7	164	0.0219	0.0132

Specific Activity of Cl = 406 counts/min. per mg. Cl in CCl$_4$ solution.

F2	3.25	17.7	165	0.0439	0.0264
	2.93	21.4	168	0.0527	0.0312
	2.55	20.1	172	0.0495	0.0289
	1.99	19.4	168	0.0478	0.0285
F3	3.80	25.5	169	0.0628	0.0372
	2.88	26.1	166	0.0643	0.0388
	2.36	16.4	95	0.0406	0.0426
	2.15	28.1	169	0.0692	0.0410
F4	3.67	36.2	166	0.0894	0.0538
	2.84	34.6	165	0.0854	0.0517
	2.47	36.6	166	0.0902	0.0546
	2.18	38.1	168	0.0940	0.0560
F5	4.01	90.8	167	0.224	0.134
	3.40	90.9	168	0.224	0.133
	3.10	88.0	166	0.217	0.131
	2.57	90.1	169	0.222	0.131
	2.18	76.2	166	0.188	0.113

Table 43   Unsaturation of Fractionated Polyisobutenes.

Polymer Fraction	Molecular weight ($\bar{M}_n$)	$\frac{10^5}{\bar{M}_n}$	U_w	U_m	U'_m	Double bonds per molecule
F1	376,000	0.266	0.012	0.009	0.015	0.60
F2	190,000	0.526	0.026	0.020	0.030	0.67
F3	125,000	0.800	0.036	0.028	0.046	0.61
F4	77,000	1.299	0.052	0.041	0.075	0.55
F5	32,000	3.120	0.128	0.101	0.181	0.56

agreement. It can be concluded from Fig. 86 that the unsaturation is terminal; this quantitative proof is consistent with spectroscopic evidence (859, 860) which indicates the predominance of methylenic unsaturation in oligomers and low molecular polymers of isobutene prepared by other catalysts. This result is as expected for a linear polyisobutene chain with regular head-to-tail

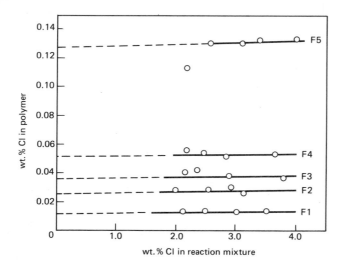

Fig. 85. Reaction of radiochlorine with polyisobutene
fractions.

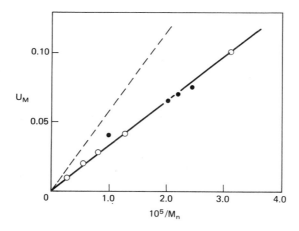

Fig. 86   Relationship between unsaturation of polyisobutenes and molecular weight: (•)unfractionated polymers; (o)fractionated polymers; (--)behaviour
expected for polymers having one terminal double
bond per chain.

linkages. The experimental line in Fig. 86 corresponds to approximately 0.6
double bonds per chain. Kinetic investigations (861) of this system show that
transfer reactions are unimportant and that the main chain-breaking process

is a bimolecular collision involving the cation and the anionic catalyst residue. It may be concluded, therefore, that the termination step in this polymerization involves the two reactions shown in the following equation:

$$\sim CH_2 - \underset{\underset{CH_3}{|}}{\overset{\overset{CH_3}{|}}{C^+}} + SnCl_4OH^- \quad \rightarrow \quad \sim CH_2 - \underset{}{\overset{\overset{CH_3}{|}}{C}} = CH_2$$

$$\sim CH_2 - \underset{\underset{CH_3}{|}}{\overset{\overset{CH_3}{|}}{C}} - OH$$

Hydroxyl end groups have been observed in oligomeric isobutenes prepared by
a titanium tetrachloride-water catalyst (860, 862) and in oligomers prepared
by a boron fluoride-water catalyst (859). The ratio of saturated to unsaturated end groups would be expected to vary with the conditions of polymerization, but the double bonds can be analyzed quantitatively by this method.

In further work McGuchan and McNeill (855) investigated the chlorination of
double in bonds in further detail by examining pure model compounds. They
concluded that, in fact, the reaction leads to monochloro compounds not di-
chloro compounds as assumed above. With this assumption, they found that high
molecular weight polyisobutene consumes a mean of 1.19 chlorine atoms per
chain. This is probably an indication that side reactions such as addition of
chlorine are occurring to a small extent. Since the true unsaturation is close
to one double bond per chain, it follows that the main chain-breaking reaction
during the cationic polymerization of isobutene is proton transfer.

Thin-layer Chromatography

Thin-layer chromatography has been used by Geymer (863) to separate mixtures
of homopolymers, e.g. polyisobutylene and polybutadiene, and to distinguish
such mixtures from copolymers. It has been found that thin-layer chromatography
can also separate polymers on the basis of molecular weight, i.e. the lower
molecular weight polymers migrate more rapidly. Best results are obtained
with polar solvents using Kieselguhr as the adsorbent.

X-ray Diffraction

Lederer et al (864) carried out small angle x-ray scattering measurements on
dilute solutions of polyisobutene using Kratsky cone collimation.

Tanaka et al (865) have also determined the crystal structure of polyisobutylene.

Fractionation

Kenyen et al (866) have described methods for the large scale fractionation
of polybutene-1, 500 g samples can be fractionated in 20 hours.

Molecular Weight

Cantow et al (867) have described an apparatus for the thermoelectric determination of molecular weights and have applied it to the examination of polyisobutenes of narrow and defined molecular weight distributions. With this impr-

oved instrument, temperature differences of down to approximately $5 \times 10^{-5}°C$ can be recorded. This allows determination of molecular weights up to 40,000 for the case of coiled macromolecules in a poor solvent. The accuracy of measurements, for all molecular weights was established.

<u>Miscellaneous</u>

Studies of molecular motions in polyisobutylene were reported by Wall (868).

POLY 3 METHYL-BUTENE

Infrared Spectroscopy

Elliot and Kennedy (869) measured the infrared spectra of the three crystalline forms of cationically synthesized poly-3-methyl-1-butene. The α, β, and γcrystalline phases of cationically synthesized poly-3-methyl-1-butene have been characterized by infrared spectroscopy (870).

The approximate degree of isotacticity of crystalline poly(3-methyl-1-butene) was determined (871) from the ratio of the absorbance at 778 cm^{-1} to 1180 cm^{-1}.

Nuclear Magnetic Resonance Spectroscopy

McGuchan and McNeill (872) have conducted proton magnetic resonance and thermal volatilization studies of chlorinated polyisobutenes.

POLY(4-METHYL-1-PENTENE)

Infrared Spectroscopy

The infrared spectra of isotactic poly(1-pentene), poly(4-methyl-1-pentene), and atactic poly(4-methyl-1-pentene) has been reported by Gabbay and Stivala (873).

Nuclear Magnetic Resonance Spectroscopy

Wanless (874) investigated poly 4-methyl-1-pentene and poly 1-pentene i.e.polymers and copolymers which had low zip lengths and low monomer yields on pyrolysis at 375°C. He correlated mass spectral data with NMR spectra taken on the same depolymerization products. Figure 87 gives a calibration curve for copolymer content of a 4-methyl-1-pentene 1 pentene copolymer system based on NMR measurements. The results which were obtained by these procedures showed good consistency between the mass spectrometric and NMR data.

Woodward (875) has utilized NMR spectroscopy in his study of relaxation phenomena in poly(4-methyl pentene-1).

Pyrolysis-Gas Chromatography

Voigt (876) used a platinum filament type flash pyrolyzer with a filament temperature of 550°C to pyrolyze poly-4-methyl-pentene-1. The decomposition products from up to 2 mg polymer were lead directly onto the gas chromatographic column. Figure 88 shows the pyrogram obtained from this polymer and identifies some of the pyrolysis products. Peak 7 (4-methyl pentene-1) is suitable for quantatively identifying poly-4-methyl pentene-1. The presence of this peak, by itself, is not sufficient to identify this polymer, because this peak can also

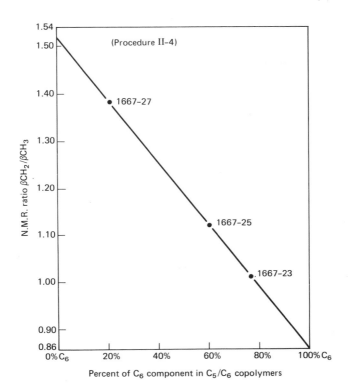

Fig. 87   Nuclear magnetic resonance calibration curve for
copolymers of 1-pentene and 4-methyl-1-pentene.

emanate from other polymers. Much more decisive is the height of this peak as
compared with that of another peak.

Whereas, in the case of polyethylene and polypropylene, the splitting up of
the main chain is virtually the only result of thermal reaction, the stripping
off of side-chains in poly-4-methyl pentene-1 is clearly indicated by the nature of the pyrolysis products obtained with this polymer, the predominant products are isobutene and propylene, substances which can only be attributed to
the stripping off of side chains together with a small proportion of ethylene.

POLY(-4 METHYL-1-HEXANE)

Analysis of the 300-MHz proton NMR spectra of poly(4-methyl-1-hexane) by Kennedy and Johnston (878) has shown that the molecular structure was not that of
a completely isomerized 1,4 structure. These findings are in contrast to those
for poly(4-methyl-1-pentene) which can be modified by changing the polymerization temperature.

POLY-1-OCTADECENE

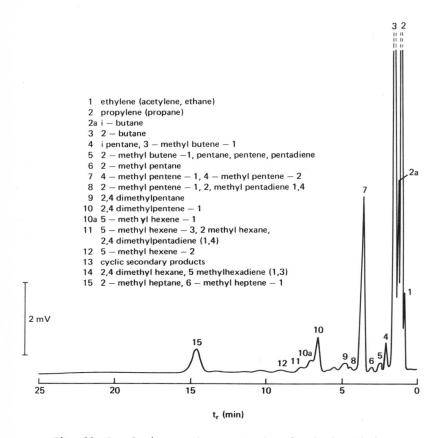

1   ethylene (acetylene, ethane)
2   propylene (propane)
2a i — butane
3   2 — butane
4   i pentane, 3 — methyl butene — 1
5   2 — methyl butene —1, pentane, pentene, pentadiene
6   2 — methyl pentane
7   4 — methyl pentene — 1, 4 — methyl pentene — 2
8   2 — methyl pentene — 1, 2, methyl pentadiene 1,4
9   2,4 dimethylpentane
10 2,4 dimethylpentene — 1
10a 5 — methyl hexene — 1
11 5 — methyl hexene — 3, 2 methyl hexane,
    2,4 dimethylpentadiene (1,4)
12 5 — methyl hexene — 2
13 cyclic secondary products
14 2,4 dimethyl hexane, 5 methylhexadiene (1,3)
15 2 — methyl heptane, 6 — methyl heptene — 1

Fig. 88   Pyrolysis-gas chromatography of poly-4-methyl
          pentene-1.

Influence of side chain on stereoregularity, conformation, and crystallinity
of isotactic poly-α-olefins such as poly-1-octadecene was determined from infrared and Raman data. These workers (879) described an infrared cell for measurements at -120 to 83°C.

POLY-1-DOCOSENE

Raman and infrared spectra of melt crystallized and solution crystallized isotactic poly-1-docosene showed that the second endothermic peak (328 K) in the
differential thermal analysis curve of melt crystallized polymer was due to
a change of the polyethylene-like side chain order (880).

POLY 1-C_{1-20} ALKYLETHYLENES

Modric and coworkers (881) observed a distinct change in the wavenumber and the intensity of the absorbance bands between 600 and 500 cm^{-1} with increasing side chain length of poly ($1-C_{1-20}$ alkylethylenes).

COPOLYMERS OF HIGHER ALKENES

BUTENE-PROPYLENE COPOLYMERS

Nuclear Magnetic Resonance Spectroscopy

The propylene content of 1-butene-propylene copolymers has been determined from the area of the $C_3H_6-C_4H_8$ diad in the ^{13}C NMR spectra (889).

The composition and diad, triad, and tetrad sequences of isotactic 1-butene-propylene copolymers have been determined by ^{13}C NMR (890). The tetrad was the longest sequence detected at 25.2 MHz.

Pyrolysis-Gas Chromatography

Van Schooten and Evenhuis (891, 892) applied their pyrolysis-hydrogenation-gas chromatographic technique to propylene-butene-1 copolymers prepared by the titanium trichloride-triethylaluminium catalyst route. In this method, the sample, about 0.4 mg, is pyrolyzed at a temperature of about 500°C on a nichrome spiral in a stream of hydrogen which serves as the carrier gas. The pyrolysis products are then lead over a hydrogenation catalyst and all products are thus converted into saturated hydrocarbons which are separated by a gas chromatographic column with a flame ionization detector. These workers found peaks stemming from propylene-butene-1 (p-b) or b-p sequences, from p-p-b or b-p-p sequences and from p-b-b or b-b-p sequences. Characteristic peaks for the alternating copolymer should include, for example, n-C_6, 2C_6, 3methyl-C_6, 3,5methyl C_9, 2methyl 4ethyl C_7, 3,5,7methyl C_9, 2,6methyl 4ethyl C_7 peaks; for a -(p-p-b)$_n$ - polymer characteristic peaks would be the 2,4methyl C_8, 3, 5methyl C_8 and 2,4,6methyl C_8 peaks; and for a -(b-b-p)$_n$ - polymer the 2methyl 4ethyl C_8 and 3methyl 5ethyl C_8 peaks. The areas of the main peaks of the pyrograms are given in Fig. 89 while peak areas are given as a function of butene content in Fig. 90. In Fig. 90 are plotted the areas of several peaks which stem from a p-b or b-p sequence - the n-C_6, 2methyl C_6 and the 3,5methyl C_9 peaks - and one peak (2,4methyl C_8) which stems from a p-p-b or b-p-p sequence. The former peaks are a measure of the degree of alternation and should be at their maximum height at about 50% methyl butene, the latter peak should be at its maximum height at higher propylene contents. The shapes of the curves in Fig. 90 is roughly in agreement with these expectations.

Gas Chromatography

Myers and Lord (893) have applied gas chromatography and infrared and NMR spectroscopy to the analysis of propylene-butene-1 codimerization products produced by a Ziegler reaction and showed that the major products are 2-methyl pentene-1, 2-methyl hexene-1, 2-ethyl pentene-1 and 2-ethyl hexene-1.

3-METHYL -1-BUTENE-PROPYLENE COPOLYMERS

Pyrolysis-Gas Chromatography

Eggertson et al (894) have discussed an apparatus for the pyrolysis of ethylene-olefin copolymers at 550°C. To simplify interpretation of the patterns,

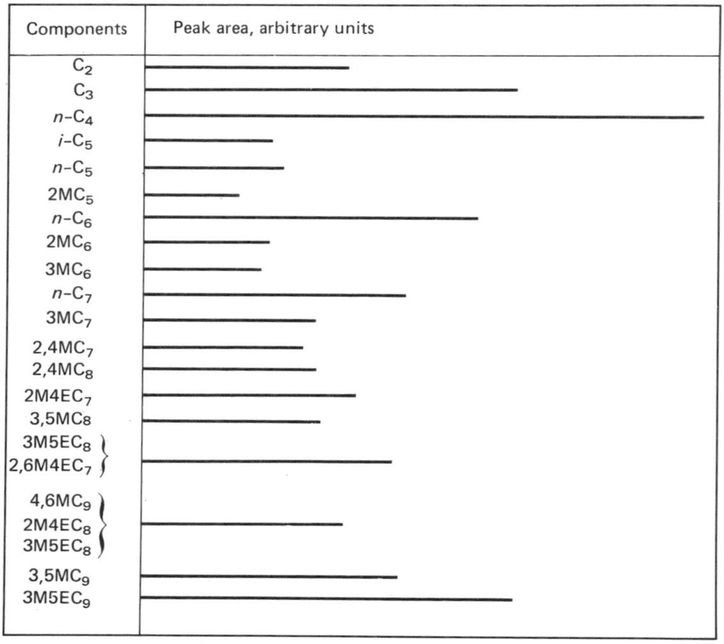

Components	Peak area, arbitrary units
C_2	
C_3	
n-C_4	
i-C_5	
n-C_5	
$2MC_5$	
n-C_6	
$2MC_6$	
$3MC_6$	
n-C_7	
$3MC_7$	
$2,4MC_7$	
$2,4MC_8$	
$2M4EC_7$	
$3,5MC_8$	
$3M5EC_8$ } $2,6M4EC_7$ }	
$4,6MC_9$ } $2M4EC_8$ } $3M5EC_8$ }	
$3,5MC_9$ $3M5EC_9$	

Fig. 89   Areas of the main peaks in the pyrogram of a prop-
ylene-butene-1 copolymer containing 50% M of but-
ene-1.

the pyrolysis products were hydrogenated by use of hydrogen carrier gas and a separating column containing platinum as hydrogenation catalyst. A number of polymers were successfully analyzed with this equipment, for example, 3-methyl-1-butene - propylene copolymers. (Table 44)

Table 44   Specific Peaks found in 3-Methyl-1-Butene.

	Pyrolysis Products Composition				
		Homopolymer			Random Copolymer
	3-methyl-1-butene	Propylene			
Peak position[a]	4.7 9.3 12.5 (i-C_5)	5.0 8.4 8.8 10.7 (n-C_5)			9.9[c]
%m[b]	13.5 4.1 2.5	8.5 17.6 1.7 1.8			0.6

a   The peak designations are according to their positions in the n-paraffin time scale (n-C_5 = 5.0, n-C_6 = 6.0, etc.)
b   Percentages are based on C_{13} and lighter pyrolysis products obtained from the homopolymers.
c   This peak was found in an experimental random copolymer containing about 10% 3-methyl-1-butene in polypropylene, but was not observed with the homopolymers or with experimental block copolymers.

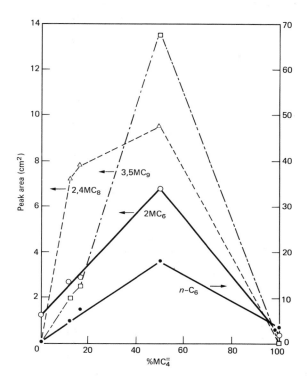

Fig. 90 Propylene-butene-1 copolymers. Peak areas versus
         butene-1 content.

4-METHYL -1-PENTENE-VINYLCYCLOHEXANE COPOLYMERS

Kinetic and infrared data showed that a random copolymer was formed from 4-
methyl-1-pentene and vinylcyclohexane (895).

1-HEXENE-4-METHYL-1-PENTENE COPOLYMERS

A method for determining the block structure in 1-hexene-4-methyl-1-pentene
polymers from infrared spectral data has been developed (896).

CHAPTER 4

POLYSTYRENE, POLY α-METHYL/STYRENE AND COPOLYMERS

Chemical Methods

Urbanski (898) discussed the analysis of polystyrene and its copolymers.

Kalinina and Doros (899) have reviewed chemical methods for qualitative and quantitative analyses of polystyrene.

Low temperature ashing has been used by Narasaki and Umezawa (900) for controlled decomposition of polystyrene prior to the determination of metals.

Nitrile groups incorporated in polystyrene by initiation or copolymerization have been detected and estimated by dye-partition techniques after reduction with lithium aluminium hydride in tetrahydrofuran (901).

Kato (902) has followed the regeneration of carbonyl groups from 2,4-dinitro-phenylhydrazones formed on the surface of irradiated polystyrene films by absorption measurements at 378 nm.

Pepper and Reilly (903) have measured terminal unsaturation in polystyrene, but their hydrogenation technique was limited to oligomers which have a relatively high mole per cent unsaturation.

Ghosh et al (904) have carried out end-group analysis of persulphate initiated polystyrene using a dye partition and a dye interaction technique. Sulphate and hydroxyl end-groups are generally found to be incorporated in the polymer to an average total of 1.5 to 2.5 end-groups per polymer chain.

Sretnagel and Palit (905) applied dye partition end-group analysis procedures to the examination of sulphate end-groups in persulphate initiated polystyrene.

Banthia and coworkers (906) determined sulphate sulphonate and isothioronium salt end-groups in polystyrene by the dye-partition technique. Polymer polarity did not affect the results of the end-group determination.

Crompton (2794) has described a method for the determination of traces of

bromine in polystyrene in amounts down to 100 pm bromine equivalent. In this
method a known weight of polymer is mixed intimately with pure sodium peroxide
and sucrose in a micro Parr bomb which is then ignited. The sodium bromate
produced is converted to sodium bromide by the addition of hydrazine sulphate.
The combustion mixture is dissolved in water and acidified with nitric acid.
The bromine content of this solution is determined by potentiometric titration
with standard silver nitrate solution. The organic bromo content of the polymer can then be calculated from the determined bromine content.

Infrared Spectroscopy

Staneson (907) has investigated in detail the infrared spectrum of polymers
containing aromatic rings such as styrene and terphthalic acid and isophthalic
acid.

Lindley (908) applied specular reflectance infrared techniques to the examination of cast polystyrene film.

Sloan et al (909) described a specular reflectance system for the infrared
analysis of micro-sized samples in the 10 to 100 microgram range of polystyrene and other materials. They compared the advantages and disadvantages of
this technique with those of other micro infrared techniques. Samples are
mounted on small metal mirrors which reflect the light beam back through the
sample. A transmission spectrum of the sample is thereby obtained but the effective path length is twice that of the actual thickness. A given absorption
band therefore has twice the absorbance obtained by conventional transmission
measurements. Major limitations include a "stray light" artifact, polymorphism
effects, and difficulties in achieving sample uniformity.

Raman Spectroscopy

Yoshino and Shinomiya (910) have measured the Raman spectra of carbon tetrachloride solutions of polystyrene. The intensity and the depolarization ratio
of a Raman line measured on a turbid solution of the polymer were confirmed
experimentally. (Fig. 91)

Jasse and Monnerie (911) used infrared and Raman spectroscopy in the region
below 420 cm^{-1} to characterize 2,4-diphenylpentane and 2,4,5-triphenylheptane
as model compounds for styrene polymers.

Spells et al (912) described Raman spectrum at 5-560 cm^{-1} of amorphous and
partially crystalline isotactic polystyrene. Infrared absorption data for the
fluoro, chloro, and bromo substituted polystyrene derivatives are given over
the same frequency range.

Nuclear Magnetic Resonance Spectroscopy

Woodward (913) has utilized NMR spectroscopy in his study of relaxation phenomena in polystyrene.

Recent studies have shown that ^{13}C NMR is a useful tool for detecting and measuring the distributions of stereochemical configurations that can occur in
polystyrene (915, 916).

Randall (914) has made ^{13}C NMR configurational assignments for an amorphous
polystyrene sample examined at 25.2 MHz and 120°C (Fig.92) The assignments are
based strictly on a one-parameter Beroullian fit that was in satisfactory agreement with the nine observed methylene relative intensities. The methylene reg-

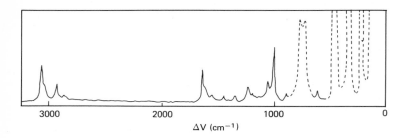

Fig. 91   Raman spectrum of polystyrene in carbon tetra-
           chloride solution.

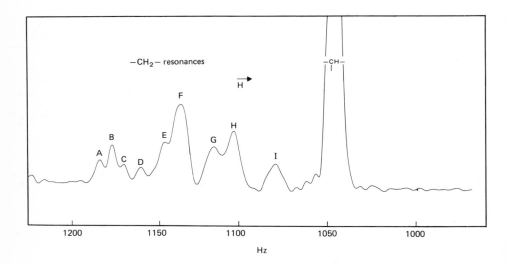

Fig. 92   The methylene and methine region of a ^{13}C NMR
           spectrum of an amorphous polystyrene at 25.2 MHz
           and 120°C. The MHz values are relative to an in-
           ternal tetramethylsilane (TMS) standard.

ions of the ^{13}C NMR spectra of a polystyrene were examined before and after
hydrogenation of the side chain phenyl substituents. Randall concluded that
ring current effects have influenced the ^{13}C methylene chemical shifts substan-
tially and are limited largely to contributions from adjacent phenyl substituents. In addition, aromatic substituent parameters are reported that can be
used in conjunction with the Grant and Paul parameters for calculating chemical shifts in aromatic hydrocarbons and polymers. Free-radical and n-butyl-
lithium-prepared polystyrenes had essentially atactic structures with meso
additions favoured over racemic additions by approximately 55/45. An ^{1}H noise
decoupled ^{13}C NMR spectrum is reproduced in Fig. 92. Nine methylene resonances
are present that have been labelled A through I. A high field strong methine
resonance is also present but shows no apparent configurational splitting.

The spectrum was obtained in 1,2,4-trichlorobenzene at 120°C where resolution of the nine methylene resonances became possible. The observation of nine resonances is in itself interesting since a ^{13}C NMR sensitivity to just tetrad sequences would have produced six resonances while a complete hexad sensitivity would have produced twenty resonances.

Inoue et al (917) has shown that ^{13}C NMR is an informative technique for measuring stereochemical sequence distributions in polystyrene. They reported chemical shift sensitivities to configurational tetrad, pentad and hexad placements for this polymer.

Randall (918) has also described work on the application of quantitative ^{13}C NMR to measurements of average sequence length of like stereochemical additions in amorphous polystyrene. He describes sequence lengths in vinyl polymers in terms of the number average lengths of like configurational placements. Under these circumstances, a pure syndiotactic polymer has a number average sequence length of 1.0, a polymer with 50:50 meso, racemic additions has a number average sequence length of 2.0 and polymers with more meso than racemic additions has a number average sequence length of greater than 2.0. The results appear to be accurate for amorphous and semi-crystalline polymers but limitations are present when this method is applied to highly stereo-regular vinyl polymers containing predominantly isotactic sequences.

Borsa and Lanzi (919) have investigated the NMR absorption in polystyrene-styrene systems. Stereoregularity of polystyrene determined by ^{13}C NMR when analyzed in terms of pentads agreed with calculated peak areas and obeyed Bernoullian statistics (1077). Evans et al (1078) using ^{13}C NMR on methyl substituted polystyrene, confirmed infrared data indicating that metalation of polystyrene with n-butyllithium-tetramethylethylenediamine leads to meta- and para-substitution on the benzene ring.

Based on a comparison of the ^{13}C NMR spectra of polystyrene and of deuterated polystyrene, Ebdon and Huckerby (1080) have assigned the high field peak at 40.5 ppm to the aliphatic methine (CH) group.

Morita and coworkers (1081) and Morita and Shen (1082) used broadline NMR to study plasma polymerized polystyrene. Two types of NMR line shapes were observed: a broad component attributed to cross-linked polystyrene and a narrow component attributed to low molecular weight species.

Jasse and coworkers (1083) compared the ^{13}C NMR chemical shifts of various structural sequences of model compounds with the chemical shifts of polystyrene and concluded that methyl, methylene, and carbon atoms attached to the backbone chain are sensitive to tacticity.

Electron Spin Resonance Spectroscopy

Florin et al (1084) have studied the electron spin resonance spectra of gamma irradiated polystyrene. Electron spin resonance has been applied to a study of azodiisobutyronitrile-initiated oxidation of polystyrene in benzene solution (1085). This technique has also been used to study peroxy radicals formed on surface treatment of polystyrene with fluorine (1086) and characterization of meta-labelled polystyrene (920).

Pyrolysis-Gas Chromatography

Lehmann and Brauer (921) investigated the pyrolysis-gas chromatography of polystyrene under helium at temperatures ranging from 400 to 1100°C utilizing for

Analysis of Plastics

the pyrolysis a silica boat surrounded by a platinum heating coil (922). Separations were carried out on a column of Apiezon (30%) on Chlorowax 70 and
dinonyl phthalate on firebrick operated at 100-140°C.

Typical chromatograms obtained on pyrolyzing polystyrene are shown in Fig. 93.
At 425°C only styrene monomer is eluted from the column; degradation at 825°C
produces a number of identifiable products. Reactions leading to these products
are even more important at a temperature of 1025°C. Results of the analyses
at various pyrolysis temperatures are shown in Table 45.

Table 45 Composition, Weight Per Cent, of Pyrolyzates of
          Polystyrene.

Product	Pyrolysis Temperature, °C							
	425	525	625	725	825	1025	1025	1125
Carbon dioxide	present	-	-	-	-	Trace	Trace	
Ethylene	present	-	-	Trace	4.1	6.9	6.9	6.8
Acetylene	present	-	-	Trace	4.1	6.9	6.7	6.8
Benzene	-	-	-	Trace	8.1	13.4	12.0	13.0
Toluene	Trace	Trace	Trace	0.9	2.5	5.6	2.7	5.8
Ethylbenzene							3.7	
Styrene	64.3	67.5	74.4	83.9	73.7	62.8	63.8	64.3
Material retained in column	35.7*	32.5	25.6	14.4	7.5	4.3	4.3	3.3

* Mainly styrene dimer (923, 924).

Comparison of the chromatographic results with mass spectrometric analyses
of polystyrene degraded in a helium atmosphere by Madorsky and Straus (925)
indicates that consistently larger quantities of monomer are obtained by the
chromatographic procedure. The high flow rate of the carrier gas rapidly sweeps
the primary reaction products from the hot zone into the chromatographic column.
On the other hand, the pyrolysis chamber used by Madorsky and Straus (925) is
quite large, and the primary products are kept at the elevated temperature
for a considerable time period, during which they can undergo secondary de-
compositions.

Sindelfingen (926) has discussed the pyrolysis-gas chromatography of polystyrene using the open boat combustion technique.

Several other workers (927-929) have discussed the gas chromatography pyrolysis of polystyrene and its copolymers using a packed column with a liquid
coating on a solid substrate.

Esposito (930) has described the semi-quantitative determination of polystyrene by means of pyrolysis-gas chromatography using an internal polymeric standard. A silicone grease column was used, programmed from 60°C to 225°C at 5.6°
C/min.

Feuerberg and Weigel (931) pyrolyze polystyrene samples in a glass tube at
620°C and then chromatograph the condensed pyrolyzate at 150°C on a 2 metres
long column, using di-n-decyl phthalate as the stationary phase.

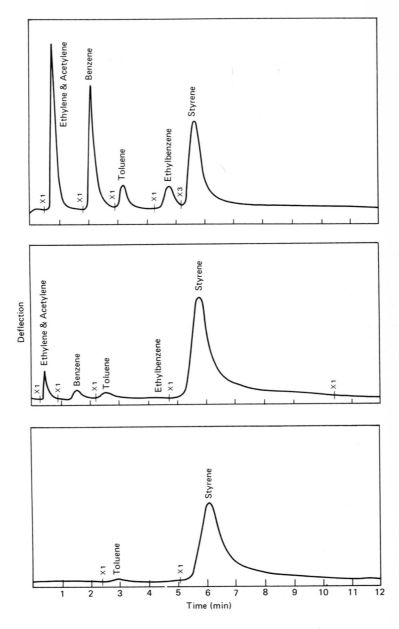

Fig. 93 Chromatograms of pyrolysis products of polystyrene.
Column, Apiezon L; col. temp., 140°C; flow rate,
60 ml/min; pyrolysis temp; °C, top, 1025, middle,
825, bottom, 425. (Attenuation scale indicated
by numbers in figure)

Brauer (932) has applied pyrolysis-gas chromatography to the elucidation of the structure of polystyrene.

Klein and Widdecke (933) have discussed the use of pyrolysis-gas chromatography for characterizing polystyrene. Retention indexes were determined for benzene, ethanol, ethyl acetate, nitromethane, and pyridine as polar test moleculaes.

Cabasso (934) applied pyrolysis-mass spectrometry to an examination of various polystyrenes.

Oxidative Degradation

Grassie and Weir (935) described an apparatus for the measurement of the up-take of small amounts of oxygen by polystyrene with a high degree of precision. The design of the apparatus is such that photoinitiated oxidation and pure photolysis can be studied. They describe methods for the preparation of poly-styrene films for photooxidation and vacuum photolysis investigations. Grassie and Weir (936) also investigated the application of ultraviolet and infrared spectroscopy to the assessment of polystyrene films after vacuum photolysis in the presence of 253.7 nm radiation using the apparatus mentioned above. During irradiation there is a general increase in absorption in the region 230-350 nm. Rates of increase are relatively much greater in the 240 and 290-300 nm regions, however, as shown in Fig. 94. Absorption in the 240 nm region is characteristic of compounds having a carbon-carbon double bond in conjunction with a benzene ring. Styrene, for example, has an absorption band at 244 nm.

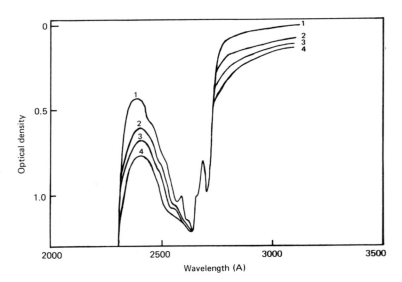

Fig. 94   Effect of irradiation on the ultraviolet spectrum
          of polystyrene: (1)unirradiated; (2) 60 min; (3)
          120 min; (4) 180 min.

Schole et al (937) have applied their oxidative degradation technique to the

study of polystyrene. In this technique the polystyrene sample is mixed with
a support in a precolumn which is mounted at the inlet to a gas chromatographic column. Figure 95 shows a typical gas chromatogram obtained under these
conditions. The oxidative changes in polymers which occur during polymerization (as opposed to after polymerization) have been investigated only to a
limited extent (938-940).

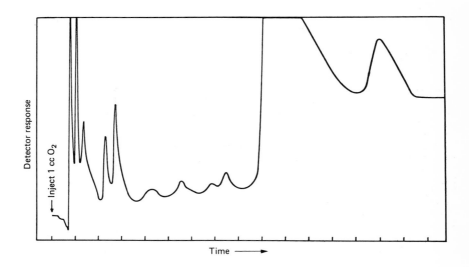

Fig. 95   Gas chromatogram of oxidation products for poly-
          styrene. 15% polystyrene on 35- to 80-mesh Chrom-
          osorb P. Column length, 1 foot. Precolumn temper-
          ature, 240°C.

Shaw and Marshall (941) have carried out an infrared spectroscopic examination of emulsion polymerized polystyrene which has been oxidized during polymerization. Previous electrokinetic and radiotracer studies carried out by
Shaw and others (942-944) indicated that the emulsifiers could be removed
from the polymer by extensive dialysis to leave particles with surface carboxyl groups of pk value 4.60. Dye interaction (with Thodamine 6GBN) and potentiometric titration studies also provided evidence for the presence of surface carboxyl groups. The latter were found to be an integral part of the
polymer chains and were presumably formed by oxidation during polymerization.
Infrared spectroscopy was employed to obtain more direct evidence for the
presence of carboxyl end groups on emulsion polymers.

Shaw and Marshall (941) found that irrespective of the nature of the initiat-
or-emulsifier combination employed, all of the polymer spectra revealed bands
at 1705 and 1770 cm^{-1}. The band at 1705 cm^{-1} was assigned in part to the carbonyl stretching mode of dimeric carboxylic acid, formed by oxidation, in the
polystyrene chains. Absorption at 1770 cm^{-1}, which was very weak, was tentatively attributed to the carbonyl stretching mode of the monomeric form of
this acid. The structure of the acid end group was not established, but the
results obtained suggest that it was possibly a phenylacetic acid residue or
a residue with a similar structure. Figure 96 and 97. compare the infrared
spectra of standard (unoxidized) and of oxidized emulsion polymerized poly-

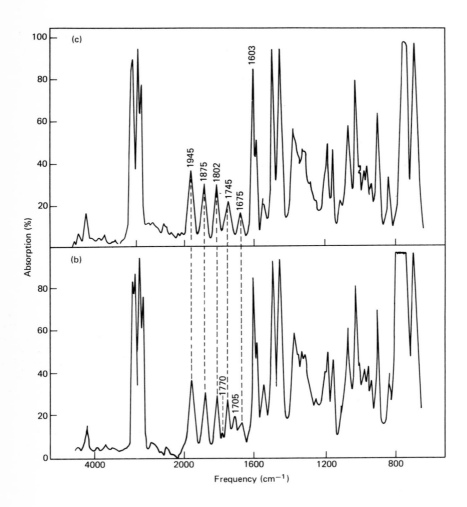

Fig. 96   Infrared spectra of (a) standard polystyrene sample and (b) latex Bl in the range 800-4000 cm^{-1}.

styrene in the region 800-4000 cm^{-1}. The spectrum of latex Bl, shown in Fig.
96b contains weak bands at 1770 and 1705 cm^{-1}.

The expanded and compensated spectra of two latex samples Bl and B2, in the
region 1500-1800 cm^{-1}, are compared in Fig. 97a and 97b. The film thickness
of the samples was so chosen that the intensities of the characteristic polystyrene bands would be of equal magnitude in both spectra. It can be seen that
the third compensation C_3, removed the absorption at 1802, 1745 and 1675 cm^{-1}
in each spectrum. Absorption at 1770 and 1705 cm^{-1} was unaffected in each case.
In addition, the compensation technique resolved very weak bands at 1630 and
1608 cm^{-1}. The former is probably attributable to the C=C stretching mode of

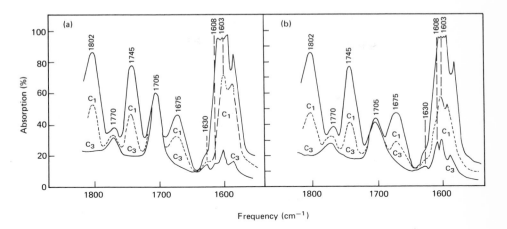

Fig. 97   Comparison of the infrared spectra of (a) latex
B1 and (b) latex B2 in the range 1600-1800 cm^{-1}.

residual styrene in the emulsion polymer.

Although the band at 1705 cm^{-1} is common to both spectra, the intensity of
this band in the spectrum of latex B1 (Fig. 97a) is significantly greater than
it is in that of latex B2 (Fig. 97b). This difference would be expected if the
polymer contained carboxylic acid groups, since in the case of B2, owing to
the presence of alkali, some of these groups would be ionized and consequently
a reduced acid carbonyl absorption at 1705 cm^{-1} would be anticipated. In the
spectra of carboxylic acids the carbonyl stretching mode at about 1700 cm^{-1}
disappears on salt formation and is replaced by the asymmetric (1610-1550 cm^{-1})
and symmetric (1420-1330 cm^{-1}) modes of the carboxylate ion.

The fact the 1705 cm^{-1} band in the spectrum of B2 was not removed suggests
that either all of the acid groups did not ionize, or that some other carbon-
yl-containing structures such as ketones or aldehydes, or both, contributed
to this absorption.

Thermal Degradation

Various workers have investigated the cause of multiple melting peaks, observed by differential analysis and differential scanning calorimetry, in polystyrene and showed that multiple melting originated from melting followed by
recrystallization or reorganization during the dynamic heating process; for
instance, in the polymer systems of isotactic polystyrene (945-947).

Taylor (948) has evaluated column thermal diffusion as a means of characterizing polystyrene. He elucidated the mechanism of polymer fractionation in a
Clusius-Dickel thermal diffusion column by a systematic study of the effects
of important experimental factors on the separations attainable combined with
an analytical description of column operation and theories of polymer thermal
diffusion.

Wunderlich and Bodily (949) have applied the technique of dynamic differential thermal analysis to a study of the glass transition interval of polystyrene.

Shibazaki and Kamebe (950) analyzed the thermal degradation products of poly-styrene utilizing gas chromatography. The monomer yield was observed in a vac-uum and in the air at the same temperature. Benzene and acetophenone and sty-rene were detected in the pyrolysis products.

McNeill (951) has discussed the application of the technique of thermal volat-ilization analysis to polystyrene.

The combination of a low power carbon dioxide laser with time-of-flight mass spectrometry permitted Coloff and Vandenburgh (952) to interpret the molecular processes occurring during thermal degradation of polystyrene.

The fragmentation mechanism and the composition of polystyrene oligomers has been determined by Beckewitz and Heusinger (953).

The kinetic parameters in thermal degradation of polystyrene have been measured by thermal volatilization analysis, and the most suitable method for computat-ion of results established (954).

Wall (955) has discussed the mechanisms of pyrolysis of polystyrene.

Mitera et al (956) found that polystyrene started to decompose in air at about 290°C with the main products being carbon dioxide, benzene, alkylbenzenes, sty-rene, phenol, indane, styrene oxide, acetophenol, vinylbenzaldehyde, dimer, diphenylpropane, and diphenylpropene.

A Setchkin ignition apparatus has been used by Schmit for the mass spectromet-ric analysis of toxic gaseous products from the thermal degradation polystyr-ene (957).

X-ray Diffraction

Nielson et al (958) used small-angle x-ray and electron spin resonance to study fracture processes in polystyrene and acrylonitrile-butadiene-styrene resins.

A numerical desmearing technique has been described by Lovell and Windle (959) to improve wide-angle x-ray diffraction patterns of aligned atactic polystyrene and quenched isotactic polystyrene.

Jasse and Monnerie (960) used the vibrational spectra of polystyrene model com-pounds to demonstrate the effects of alkyl chain conformation on the wave num-ber of the 16b and 10b out of plane vibrations of the benzene ring. The 16b mode was influenced only by the local conformation of the alkyl chain whereas the 10b mode is sensitive to the length of the alkyl chain conformation.

Hamada et al (961) calculated the molecular weight, radius of gyration, second virial coefficient, mass per unit length, hydrodynamic length, radius of gyra-tion of cross-section for polystyrene in methyl ethyl ketone from measurements of small-angle x-ray scattering.

Fractionation and Molecular Weight

Moore (962) has investigated the application of gel permeation chromatography to the measurement of the molecular weight distribution of high polymers. He used as a column packing polystyrene gels crosslinked in the presence of di-luents and made into fine bead form suitable for column packing. A series of narrow molecular weight range polystyrene fractions was eluted through such columns with aromatic and chlorinated solvents. Effluent concentrations were detected snd recorded by a continuous differential refractometer.

Gruber and Elias (963) have discussed apparatus suitable for the cloud point titration to determine the critical non-solvent fraction ($\mathscr{V}$crit) of polystyrene. They emphasize that to obtain optimum results one should use good solvents, non-solvents of moderate precipitation power, and small volumes of the cuvettes. Observing these working conditions, a standard error of ± 0.3% can be obtained for $\mathscr{V}$ crit. To work within this limit or error, the absolute temperature should deviate no more than ± 0.15°C.

Moore and Hendrickson (964) applied gel permeation chromatography to the determination of the molecular weight distribution of polystyrene.

Ambler et al (965) have given a detailed theoretical treatment of the use of gel permeation chromatography as a basis for the determination of branching in polymers. Randomly branched polystyrenes are discussed in particular. As Ambler et al, point out along with chemical composition, molecular weight, molecular weight distribution, and type and amount of gel, branching is considered one of the fundamental parameters needed to characterize polymers fully (966-970).

Ambler et al (965) extended the technique for determining branching in polymers by using a combination of gel permeation chromatography and intrinsic viscosity data and developed suitable equations.

Peaker and Robb (971) fractionated polystyrene dissolved in benzene by slowly freezing the solutions with dry ice and alcohol. A more detailed treatment of this method is given by Loconti and Cahill (972). The polymer in the first frozen-out portions was of higher molecular weight than later. Ruskin and Parravano (973) were able to fractionate polymer dissolved in cyclohexane by both the zone-melting and freezing techniques.

Bryson et al (974) investigated the fractionation of polystyrene by slow freezing of dilute benzene solutions of the polymer using ice water mixtures instead of dry ice-alcohol. Bryson et al (974) concluded that no fractionation of the polymer occurs according to molecular weight from benzene solutions.

The gel permeation chromatographic elution behaviour of polystyrene homopolymers was studied by Kranz and coworkers (975) and a calibration curve prepared for different solvents.

Tung and Runyon (976) carried out a molecular weight distribution determination and sedimentation velocity analysis of standard polystyrenes using gel permeation chromatography.

Zhdanov et al (977) reported on the influence of the structure of macroporous glasses on separations achieved in gel permeation chromatography using polystyrene as standard polymer.

Belenki et al (978) carried out a micro-fractionation of mixtures of linear and branched polystyrenes by means of gel permeation chromatography followed by thin-layer chromatography of the fractions. This permitted the quantitative determination of the percentage of linear and branched components, the molecular weight of linear polystyrenes, and the hydrodynamic radii of linear and branched components (978).

Otocka (979) has proposed a high speed gel permeation chromatographic method
for determining the molecular weight distribution of polystyrene.

Uglea (980) used gel permeation chromatography in quantitative analysis of
polystyrene and oligomers covering the molecular weight range 200-4500. Poly-
disperse and monodisperse polystyrene samples were fractionated with cyclohexane as elution solvent (981).

Guener et al (982) used gel permeation chromatography to determine the molecular weight of isotactic polystyrene.

Turbidimetric titration has been used by Gooberman (983) for the determination of the molecular weight distribution of polystyrene. (Urwin et al 984 and
Stearne and Urwin 985). In this procedure a solution of polystyrene in but-
anone is titrated with isopropanol. In their very detailed investigation
Urwin et al (984) studied the effect of factors such as size and shape of
the suspended particles, correction of turbidities as a function of concentration and polymer solubility as a function of the concentration and molecular
weight. They also devised a correction procedure for the loss of precipitate
during the titration and devised a method of location of the point of precipitation in relation to the weight average molecular weight.

Yamada et al (986) applied elastoosmometry to the determination of the number
average molecular weights of polystyrene.

Hall (987) has reviewed the main disadvantages of the turbidimetric method,
as applied to polystyrene. Beattie (988) overcame many of these disadvantages.
He discusses only the determination of solubility distributions. It is of
interest to point out that the method is capable of giving molecular weight
distributions provided that a relationship between solubility and molecular
weight is known.

Beattie (989) developed an absolute turbidimetric titration method for determining solubility distribution (which are closely related to molecular weight
distribution) of polystyrene. In this method polymer is precipitated from its
solution in methylethyl ketone by addition of a nonsolvent (isopropanol) of
the same refractive index as that of the solvent. He showed by the use of
light-scattering theory that under these conditions the concentration of polymer which is precipitated can be calculated from the maximum turbidity on
an absolute basis. He also discusses the effect of particle size and particle
size distribution. Beattie (989) concluded that with polystyrene, under the
specified conditions, the reproducibility of turbidimetric precipitation
curves is very good and that the method is accurate.

Beattie, (990) has also studied the application of turbidimetric titration
to the determination of the solubility distribution of polystyrenes. In this
method, polymer is precipitated from its solution by addition of a nonsolvent
of the same refractive index as that of the solvent. The precipitating particles of swollen polymer are allowed to agglomerate until the turbidity reach-
es a maximum. He showed by the use of light-scattering theory that under these
conditions the concentration of polymer which is precipitated can be calculated from the maximum turbidity on an absolute basis. He discusses the effects
of particle size and particle size distribution.

Klenin and Shchegolev (991) have demonstrated a turbidimetric titration method
to determine the molecular weight of polystyrene.

Rolfson and Coll (992) have described an automatic osmometer to determine number average molecular weights of styrene polymers. An operator need only pour in a sample, then isolate it in the cell using the inlet and outlet valves. After a period of about 6 to 9 minutes the osmotic pressure is recorded.

Figure 98 illustrates the type of pressure head-time plots obtained by these workers, for commercial samples of polyalphamethylstyrene and polystyrene (Dow Styron 666).

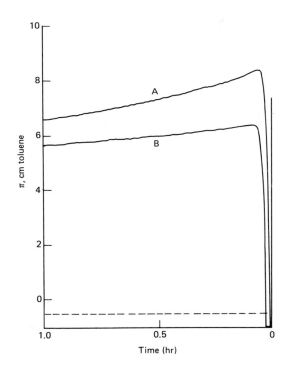

Fig. 98 Low-molecular weight poly-a-methylstyrenes. Variation of osmotic pressure with time.
A:PAMS 1, 0.0913 g/100 ml.
B:PAMS 2, 0.104 g/100 ml.
Pressures in cm toluene.

Table 46 illustrates the agreement obtained between the Rolfson and Coll (992), automatic osmometer method and manual osmometer and light-scattering measurements.

Kallistov (993) described a rapid light-scattering method for $\overline{M}_w$ and virial coefficients of polystyrene solutions in methylethyl ketone or toluene.

Kamata and Nakahara (994) proposed a rapid method for determining the molecular weight of polystyrene by light-scattering involving a single concentration approximation.

Table 46 Molecular Weights of Polystyrene Samples.

Polymer	Wt.-av. mol. wt. x 10^{-3}			Remarks referring to automatic osmometry
	Automatic	Manual	Light scattering	
1	2.87	2.2	3.4	Solute permeation linear extrapolation
2	4.30	3.7	5.0	Solute permeation linear extrapolation
3	23.3	20 (23.5)	25	19 experimental points linear extrapolation
4	58.4	57	64	
5	186	(196)	(213)	
6	77.5	77.6	90	
7	207	210	237	
8	50		(295)	Unprecipitated
9	98.6	(104)	(295)	Twice precipitated

Accurate molecular weights of polystyrene have been determined by light-scattering measurements in a mixed solvent of benzene-cyclohexane and benzene-isopropanol which was a θ solvent (995).

Blair (996) has discussed sedimentation velocity determinations of molecular weight distributions of polystyrene.

Goodrich and Cantow (997) have carried out a numerical analysis and kinetic interpretation of molecular weight distribution data for polystyrene.

Harrington and Pecoraro (998) have described a method for the determination of the precise molecular weight distribution of mono-disperse polystyrene of molecular weight up to 4 x 10^6 without appreciable "tail effect", using a semi-automatic solvent extraction procedure.

Morton et al (999) have reported on the determination of molecular weights of polystyrene produced by sodium napthalene initiation.

Brewer and McCormick (1000) have discussed molecular weight determination in narrow molecular weight distribution polystyrene.

The National Bureau of Standards offers standard polymers such as polystyrene for instrument and method calibration (1001).

Miscellaneous

Spitzbergen and Beachell (1002) carried out a light-scattering study of the oxidative degradation of polystyrene.

Rayleigh light-scattering line width measurements has provided information on local concentration fluctuations in polystyrene/solvent systems (1003, 1004).

Experimental optical anisotropies of atactic and isotactic polystyrene have been obtained (1005) and interpreted (1006) based on depolarized Rayleigh scattering.

Rayleigh scattering from mono-disperse polystyrene has been discussed by
Yoshimura and coworkers (1007).

Lange (1008) has studied the effects of solvent and wavelength of light on
microgel evaluations of light-scattering measurements for polystyrene.

Hansen and Hvidt (1009) have reported on osmotic pressure measurements on
polystyrene in toluene solution covering wide ranges of concentration and
degree of polymerization of the polystyrene.

An empirical method has been proposed for estimating osmotic second viral coefficients, using polystyrene as reference polymer (1010).

Cotton and coworkers (1011) used neutron diffraction to determine chain conformation of polystyrene in dilute solutions (10^{-3} - 10^{-1} g/cm^3), corresponding to the concentration range in which coils begin to overlap.

Cotton et al (1012) compared light- and neutron-scattering data with osmotic
pressure measurement on polystyrene solutions up to a concentration of 0.1 g/
cm^3. Results showed two different ranges with the transition occurring at the
concentration in which the average distance between chains is about equivalent to the radius of gyration.

Billmeyer and Siebert (1013) have extended the summative fractionation method
to narrow distribution polymers having $\overline{M}_w/\overline{M}_n$ < 1.12; a fractionation parameter
was calculated. For anionic polystyrene of M_w = 97,000 the estimated polydispersity by the method was 1.02.

Clark and Dilks (1014) have carried electron scan chemical analysis studies
on shake-up phenomena in subsituted polystyrenes.

In studies of photodegradation of polystyrene films, George (1015) used low
temperature polarized phosphorescence spectroscopy to detect emission from
phenyl alkyl ketone end groups.

The kinetics of isothermal crystallization of polystyrene in dilute solutions
have been determined by a photoresistance method for measuring turbidity; relation between turbidity and resistance was established by examination of polystyrene (1016).

Venediktov et al (1017) described ultrasonic apparatus for determining viscoelasticity of polystyrene.

Cloud point curves have been reported for polystyrenes of low polydispersity
in cyclohexane solution near the critical point (1018, 1019).

Wide aperture spectronephelometry has been used to estimate sizes of randomly
dispersed spherical particles, e.g. mono-disperse polystyrene latex, involving a beam of parallel light rays and wide aperture objective (1020).

The random coil configuration and the radius of gyration of polystyrene employing hydrogen and deuterium-tagged polymers, has been determined by Schelten
and coworkers (1021).

The depolarization thermocurrent method shows promise as a useful tool for
examination of transitions and relaxations in polystyrene (1022).

Thermoluminescence studies have been reported for polystrene (1023).

Beevers (1024) reviewed refractometry for polymer characterization with emphasis on use of molar refractivity as defined by the Lorenze-Lorentz equation. Critical angle refractometry was discussed in studies of polystyrene films.

Looyenga (1025) discussed the relation between refractive index and density of polystyrene.

Elias and Gruber (1026) have investigated the applicability of cloud point titration to the determination of the constitution of polystyrene.

Kämmerer et al (1027) have described an ultraviolet spectrophotometric method for the determination of end-groups in polystyrenes with a molecular weight up to one million.

Monomers and Volatiles

Adcock (1028) discusses the determination of styrene monomer in polystyrene packaging film. Pfabs and Noffz (1029) describe a gas chromatographic method for the analysis of polystyrene for styrene, ethyl benzene and cumene or xylene. Ragelis and Gajan (1030) and Shapras and Claver (1031) also describe gas chromatographic methods for determining volatiles in polystyrene. Barbul et al (1032) describe the analysis of styrene monomer for its major impurities. For a complete separation of the compounds present, three gas chromatographic columns are used.

Determination of expanding agents in polystyrene. Two methods, described below, are suitable for the determination of down to 500 ppm expanding agents in polystyrene. The first solution method which determines η- and isopentane can also be adapted with minor modifications to determine other expanding agents such as isobutane, 2-methylbutene-1, 2,2 dimethylbutane, 2,3 dimethylbutane, 2-methylpentane, 3, methylpentane, cyclopentane, methylcyclopentane and 2,2,4 trimethylpentane. The second method involves heating the polymer, without solvent, in a heated copper block and using the carrier gas sweeping the released volatiles from the expanded polymer into the gas chromatograph for quantitative analysis.

Method 1. Solution procedure. A known weight of the polystyrene is made up to standard volume with a 10% solution of cyclohexane dissolved in pure propylene oxide. A calibration blend comprising 1 ml of each expanding agent and 1 ml cyclohexane internal standard accurately made up to 10 ml by weight is also prepared. A portion of this solution (15 µl) is injected into a gas chromatograph operated under pressure (Fig. 99). The chromatogram is developed until the cyclohexane peak emerges.

From the calibration chromatogram obtain the response factors F(i) and F(n) for isopentane and n-pentane respectively:

$$F(i) = \frac{\text{iso-pentane wt.}}{\text{cyclohexane wt.}} \times \frac{\text{cyclohexane pk. ht.}}{\text{iso-pentane pk. ht.}}$$

$$F(n) = \frac{\text{n-pentane wt.}}{\text{cyclohexane wt.}} \times \frac{\text{cyclohexane pk. ht.}}{\text{n-pentane pk. ht.}}$$

From the analysis chromatogram, calculate the % weight in the polymer of iso-pentane and n-pentane: (Fig. 100) isopentane (% w/w) = $\dfrac{\text{isopentane pk. ht.}}{\text{cyclohexane pk. ht.}}$ F(i) x P x $\dfrac{10}{\text{sample wt.}}$

n-pentane (% w/w) = $\dfrac{\text{n-pentane pk. ht.}}{\text{cyclohexane pk. ht.}}$ F(n) x P x $\dfrac{10}{\text{sample wt.}}$

where P = % w/v concentration of cyclohexane in propylene oxide used as a solvent.

Gas chromatographic conditions: 8' x 3/16 ID of 20% w/w di-isodecyl phthalate on acid washed Celite (40-6) mesh) preceeded with 6" x 3/16 ID acid washed Celite to trap polymeric material. Injection port, 130°C; Column, 50°C; Detector, 50°C (katharometer).

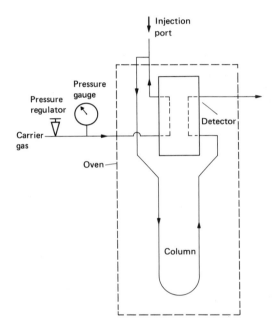

Fig. 99   Gas chromatographic apparatus for the determination of volatiles in polystyrene.

Method 2. Solventless procedure. The apparatus is illustrated diagrammatically in Fig. 102. The gas chromatograph carrier gas is fed to the instrument via a gas sampling valve, coupled as shown. The sampling valve is attached to a glass ignition tube, alsoas shown, and the latter is heated by means of a removable hot copper block.

Operating conditions.
          Instrument: Katharometer instrument or any
                                         suitable alternative.
          Column: Copper tube (9 ft x 3/16 in.
                                         O.D. x 1/8 in. I.D.) packed

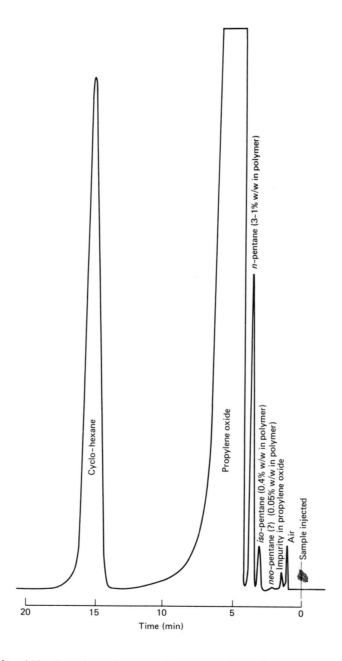

Fig. 100   Gas chromatogram of pentanes in polystyrene.

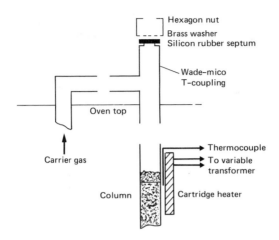

Fig. 101   Heated injection port. Gas chromatographic det-
           ermination of volatiles in polystyrene.

	with 10% w/w 50/50 Bentone 34/ di-isodecyl phthalate or 100- 120 Celite.
Carrier gas:	Helium, 15 psig., 50 ml/min.
Temperatures:	Injection 25°C
	Column 25°C } Ambient
	Detector 25°C
Copper block:	240°C
Bridge current:	250 mA. Negative polarity.

Procedure. Lower the heater block and turn the sample valve to the "by pass"
position. Into a new rimless ignition tube, weigh a sample (2.5-50 mg) of the
polymer and connect it to the ignition tube to the apparatus. Turn the sample
valve to the "analysis" position for 1 minute to remove air. Turn the sample
valve to the "by pass" position and raise the heater block (at 240°C) to sur-
round the ignition tube and heat for 5 minutes. Turn the sample valve to the
"analysis" position and allow the chromatogram to develop, Fig. 102.

Determination of Aromatic Volatiles in Polystyrene

Scope. The method describes a gas chromatographic procedure for the determination of aromatic hydrocarbons in polystyrene. Benzene, toluene, ethyl benzene,
n-propyl benzene, cumene, iso-/tert-butyl benzenes, o-xylene, m-p-xylenes,m-/
p-ethyl toluenes, styrene/sec-butyl benzene and α-methyl styrene are all measurable and the lower limit of detection for any one of these is 0.001% by
weight.

Procedure.
Preparation of calibration blend. Into a 10 ml volumetric flask weigh in turn
1.1 ml n-undecane, 1.0 ml methyl styrene, 0.9 ml styrene, 0.5 ml iso- or tert-
butyl benzene, 0.5 ml meta- or para-ethyl toluene, 0.5 ml n-propyl benzene,

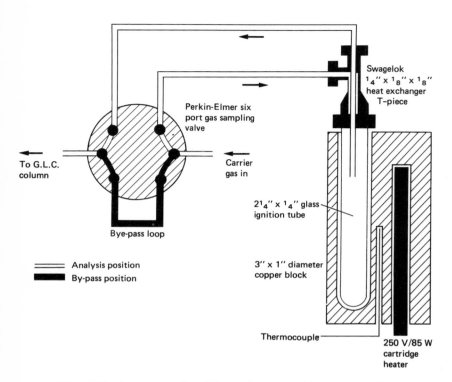

Fig. 102 Apparatus for liberating expanding agents from
expandable and expanded polystyrenes.

0.5 ml ortho-xylene, 0.8 ml cumene, 0.2 ml meta- or para-xylene, 0.4 ml ethyl
benzene, 0.2 ml toluene and 0.1 ml benzene. Keep the flask sealed during each
weighing. Dilute to 10 ml with propylene oxide and then further, dilute 0.2
ml of the solution to 25 ml with propylene oxide.

Calibration. Chromatograph 5 micro-litres of the calibration blend, attenuating suitably. Measure the peak height of each aromatic hydrocarbon.

Preparation of polymer solvent. Into a 100ml volumetric flask weigh 0.1 ml
n-undecane keeping the flask sealed during weighing. Dilute to 100 ml with
propylene oxide and re-seal.

Preparation of sample for analysis. Into a 25 ml stoppered measuring cylinder
accurately weigh about 1 g sample and add exactly 10 ml of the prepared solvent from a pipette. Seal with a rubber serum cap and shake to dissolve, using a laboratory shaking machine.

Analysis. Chromatograph 5 micro-litres of the solution obtained in the previous
section, attenuating the peaks suitably. Measure the peak heights of the aromatic hydrocarbons (Fig. 103).

Calculation. From the calibration chromatogram calculate the response factor

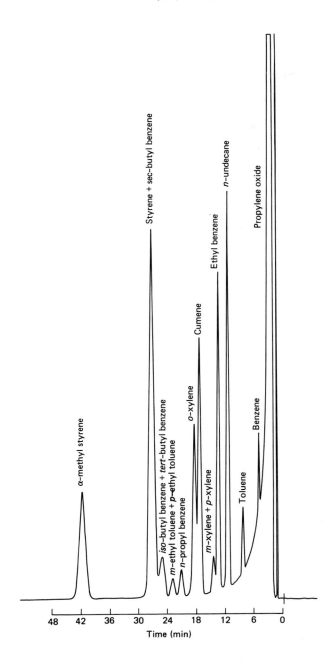

Fig. 103   Gas chromatogram of blend of aromatic hydrocarbons
           in propylene oxide.

(F) for each aromatic hydrocarbon as follows:-

$$F = \frac{wt.of\ hydrocarbon}{wt.of\ n-undecane} \quad x \quad \frac{pk.\ ht.\ of\ n-undecane}{pk.\ ht.\ of\ hydrocarbon}$$

From the analysis chromatogram calculate the concentration of each aromatic hydrocarbon as follows:-

$$\%\ wt/wt\ hydrocarbon = \frac{pk.ht.\ of\ hydrocarbon}{pk.ht.\ of\ n-undecane} \quad x \quad \frac{F\ x\ P\ x\ 10}{wt.\ of\ sample}$$

where P = % wt/vol concentration of n-undecane in the polymer solvent.

Notes 1. Gas chromatograph operating conditions. These conditions apply to an F & M Model 1609 gas chromatograph but may be readily modified to suit any alternative equivalent apparatus.
Column. Copper tube (15 ft x 3/16 in i.d.) packed with 10% wt/wt Carbowax 15-20 M on 60-72 BS mesh acid washed Celite.
Gas flows. Helium, 30 psig, rotameter = 10.0 (100 ml/min).
          Hydrogen, 12 psig, rotameter = 10.0 (75 ml/min).
          Air, 7 psig, rotameter = 10.0 (650 ml/min).
Temperatures. Injection, 155°C   Column, 80°C
              Detector, 125°C   Flame, 200°C
Recorder. Honeywell-Brown, 1 mV full scale deflection, 1 second response, 10 in/hr chart speed.

Styrene and α- methyl styrene have been determined simultaneously by polarographic reduction of their pseudonitrosites (1033).

Maeda and coworkers (1034) have determined traces of styrene in the presence of durene by dual-wavelength spectrophotometry of tetrahydrofuran solutions at 261.8 and 280,7 nm.

A fluorescence method has been developed by Vasilev et al (1035) to determine residual monomer and low molecular weight fraction in polystyrene using a choroform solution of rhodamine as fluorescent reagent.

Morita and Nawata (1036) have determined the degree of cross-linking and residual monomer and oligomer content of polystyrene films formed by glow discharge polymerization of styrene.

Markelov and Semerenko (1037) have studied the kinetics of the migration of styrene monomer from polystyrene using gas chromatography.

Residual monomer in polystyrene have been detected by an ultraviolet method in tests to determine suitability of the plastic as food containers (1038).

Corson et al (1039) chromatographed styrene dimer mixtures using a polyethylene glycol column at 206°C and an injection port temperature of 300°C.

Klesper and Hartmann (1040) have described an apparatus for chromatography with supercritical dense gases comprising a pressure cascade of three consecutive levels of decreasing pressure. Separations of styrene oligomers were carried out an analytical and preparative scales.

Nowak and Klemm (1041) determined pentane expanding agents in heat expandable polystyrene using gas chromatography. They used a by-pass loop and then swept the liberated expanding agent onto a di-iso-decyl phthalate column with the carrier gas.

Further methods for the determination of monomers and volatiles in polystyrene
are reviewed below.

Monomers and Volatiles in Polystyrene		Reference
Styrene o-xylene	head space analysis, glc	1042
Styrene	head space analysis, glc	1043 1042
Styrene	solution, glc	1044-1049
Styrene ethyl benzene	solution in o-dichlorobenzene or methylene di- chloride, glc	1050
Styrene	solution in DMF, glc	1051
Styrene o-methyl styrene benzene toluene ethyl benzene xylene cumene propyl benzenes ethyl toluenes butyl benzenes	solution in propylene oxide, glc	1052 1053
Styrene	solution in THF, glc	1054 1055
alkanes	heating sample, glc	1056
Styrene	chloroform extraction, methanol precipitation of polymer, ultraviolet spectroscopy	1052

Additives

Methods are reviewed below for the determination in polystyrene of antioxidants and optical brighteners and of organic peroxide residues.

Antioxidants and Optical Brighteners in Polystyrene		
Antioxidants Phenolic acid amine type	solvent extraction, thin-layer chromatography for quantitative analysis	1058

Ionol (2,6 di-t-butyl -p-cresol)	methylene dichloride solution, glc	1059
Optical brightener Tinuvin P (hydroxybenzyl- benzotriazole	methylene dichloride solution, glc	1059
Ionol, Tinuvin P	methylene dichloride solution, glc	1057

Organic Peroxide Residues in Polystyrene Gas Chromatography

t-butylhydroperoxide t-pentylhydroperoxide t-butylperoxyisobutyrate di-t-butylperoxide di-t-pentylhydroperoxide	solvent extraction, glc	1060
di-t-butylhydroperoxide	thermal decomposition of benzene solution, glc estimation of acetone and ethane produced	1061

Thin-layer Chromatography

Dicumylperoxide	acetone extraction, thin-layer separation and iodometric estimation	1062

Polarography

benzoyl peroxide succinic acid peroxide lauroyl peroxide acetyl peroxide peroxy acetic acid methyl ethyl ketone peroxide phenyl cyclohexane peroxide di-t-butyl perphthalate p-t-butyl perbenzoate	toluene extraction, polarography	1063

POLYSTYRENE SULPHONATE RESINS

Chemical

The water contents of anion resins in the hydroxy-form have been determined by oven drying and by Karl Fischer reagent titration; results between the two methods were in good agreement (1064).

Potentiometric titrations of cation exchange resins has been discussed by Simonov and coworkers (1065).

Kurenkov and coworkers (1066) determined sodium ρ-styrene-sulphonate polarographically either alone or in combination with acrylamide after conversion to
the corresponding pseudonitrosite by treatment with sodium nitrite in acetic
acid. the acrylamide content of the mixture was determined by direct polarography.

Infrared Spectroscopy

An infrared method for the determination of the degree of sulphonation of cation-exchange resins and for determining the chelate complexes formed from carboxylic acids and metals has been discussed by Hlavay et al (1067).

Nuclear Magnetic Resonance Spectroscopy

Gordon (1068, 1069) has shown that aqueous suspensions of sulphonated polystyrene resins in the hydrogen form give particularly well defined proton magnetic
resonance spectra, and that the observed shift of the internal water peak may
be used to calculate the molality of the internal solution.

De Villiers and Parrish (1070) showed that the proton magnetic resonance spectrum of an ion-exchange resin is often characteristic of that type of resin.
They showed that the internal molality of a strongly acidic or strongly basic
polystyrene resin can be determined from its NMR spectrum. These workers found
a straight line relationship between water molalities obtained by conventional
methods and chemical shifts. They also showed that the water content of a sulphonated polystyrene resin can also be obtained from the NMR spectra.

Pyrolysis-Mass Spectrometry

Fingerprint spectra of ion-exchange resins obtained by pyrolysis mass spectrometry has given information on the type of network and functional groups present in polystyrene sulphonates (1071).

Fractionation

An aqueous system has been used to characterize sodium (polystyrene sulphonate)
(1072). Interaction of the porous alumina packing with polyacrylamide was observed.

GPC results obtained for sulphonated polystyrene agreed with those of viscosity
measurements (1073).

POLYHALOGENATED STYRENES

Chlorinated polystyrene has been pyrolyzed to produce characteristic chlorosty-
renes and dichlorostyrenes which provided information on substitution of the
aromatic nucleus in the original polystyrene (1074).

Durchschlag et al (1075) have applied the high resolution small angle x-ray
technique to poly α bromostyrene solutions in benzene. Ayrey established a
relationship between $\overline{Mn}$ and viscosity of unfractionated low conversion poly
(p-chlorostyrene and poly(o-methoxystyrene) (1076).

STYRENE COPOLYMERS

STYRENE-BUTADIENE COPOLYMERS

Chemical Methods

Crompton and Reid (1087) have described procedures for the separation of high
impact polystyrene into the free rubber plus rubber grafted polystyrene plus
copolymerized rubber and a gel fraction and for estimating total unsaturation
in the two separated fractions. To separate a sample into gel and soluble fractions it was first dissolved in toluene. Only gel remains undissolved. Methanol is then added, which precipitates the polystyrene-rubber graft, ungrafted
rubber, and polystyrene. Any styrene monomer, soap, or lubricant remain in the
liquid phase, which is separated from the solids and rejected.

The addition of toluene to the solids dissolved all the polymeric material
with the exception of the gel. The toluene solubles are separated from the
solid gel by centrifuging and made up to a standard volume with toluene. The
gel is then dried in vacuo and weighed.

Both the gel and toluene soluble fractions are reserved for determination of
unsaturation. The iodine monochloride method developed by Cheyney and Kelley
(1088) was used for measuring unsaturation. To determine unsaturation in styrene-butadiene rubbers with good accuracy using the iodine monochloride procedure it was found necessary to contact the sample with chloroform for 15 hours
before reaction with iodine monochloride. In Fig. 104 a plot is made of sample
size against the determined iodine value; it may be seen that, even with a
30-hr reaction period, a constant iodine value (ca. 320) is obtained only when
the sample size is 0.05 g or less, i.e. a fivefold excess of iodine monochloride reagent.

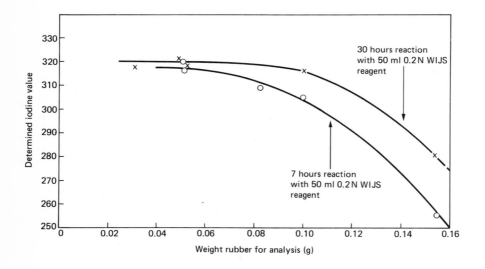

Fig. 104   Influence of excess iodine monochloride reagent
           and reaction time on the determination of the
           iodine value of rubber.

The solid gel, separated from a high impact polystyrene by solvent extraction
procedures, is completely insoluble in chloroform and in the iodine monochlor-

ide reagent solution. A contact time with chloroform of 90 hr with a 75 hr
reaction period with reagent is required.

Crompton and Reid (1087) used these procedures to study the distribution of
rubber added in several laboratory preparations of high impact polystyrene
containing 6 wt.-% of a styrene-butadiene rubber and 94% styrene, i.e. theoretical 4.1% butadiene. The results in Table 47 show the way in which the added
unsaturation of 4.1% butadiene distributes between the gel and soluble fractions. The butadiene content of the separated gel remains fairly constant, in
the 20-25% region, regardless of the quantity of gel present in the sample.
As the gel content increases, therefore, so more of the rubber becomes incorporated into the gel and less remains as free rubber or soluble graft. The recovered unsaturation lies mainly in the 90-95% region, indicating that loss
of unsaturation due to grafting or crosslinking reactions occurs only to the
extent of some 5-10%.

Infrared Spectroscopy

The analysis and characterization of high impact polystyrene by chemical and
spectral methods has been reviewed by Hobbs et al (1089).

Littke et al (1090) applied a computer analysis to the complex infrared peaks
at 760 cm^{-1} in the spectra of butadiene-styrene and acrylonitrile-styrene copolymers.

Sefton and Merrill (1091) discussed the infrared spectra of butadiene-styrene
block copolymers hydroxylated with peracetic acid.

Schmolke et al (1092) used infrared spectroscopy to determine the types and
quantities of rubber components in rubber-modified styrene polymers.

Sequence distribution of styrene-1-chloro 1,3-butadiene copolymers has been
established from infrared, ultraviolet and nuclear resonance spectroscopy and
chemical analysis (1093).

Albert (1094) has compared determinations of butadiene in high impact polystyrene by an infrared method and by the iodine monochloride method described by
Crompton (1087). The infrared method is based on a characteristic absorbance
in the infrared spectrum associated with the transconfiguration in polybutadiene and is calibrated with standards containing a standard polybutadiene.
Since different grades of high impact polystyrene may contain elastomers with
different trans contents, calibration curves based on the standard rubber are
not always suitable for analyzing these products. The results obtained by
the two methods for several high impact grades are compared in Table 48. The
rubber content of high impact polystyrene sample 1 determined by titration is
lower than the value obtained by the infrared method. This is expected on in-
terpolymerized polymers because of crosslinking which reduces the unsaturation
of the rubber. The other polymers (except sample 3), appear to contain Diene
55 type rubber since reasonable agreement was obtained between the iodine monochloride and infrared methods. High impact polystyrene 3, however, must contain a polybutadiene of high cis content to explain the low (1.2%w) amount of
rubber found by the infrared method compared to the 9.0% found by the titration method.

Raman Spectroscopy

Raman spectra have been used for quantitative analysis of styrene-butadiene-
methyl methacrylate graft copolymer,C=C stretching bands were observed for the

Table 47   Distribution of Butadiene between Soluble and Gel Fractions Obtained from Polystyrenes
           Containing Different Amounts of Gel.

Gel content of sample, wt.-%	Butadiene content of isolated gel, wt.-%	Soluble graft butadiene content A (calculated on original sample) wt.-%	Gel butadiene content B (calculated on original sample), wt.-%	Total butadiene content (A + B) (calculated on original sample), wt.-%	Amount of original rubber unsaturation in the sample, $C = \dfrac{(A + B)}{4.1} \times 100\%$
0	–	3.5	–	3.5	85
4.7	19.5	2.8	0.9	3.7	90
5.6	16.2	2.9	0.9	3.8	93
8.9	23.3	1.5	2.1	3.6	88
11.8	20.0	1.5	2.4	3.9	95

Table 48   Rubber Content of High Impact Polystyrenes (based on PBD).

Sample	Polybutadiene, %w (Iodine Monochloride method)	Polybutadiene, %w (IR method)
Standard: 6.0% w Diene 55	6.2	-
Standard: 12.0% w Diene 55	12.2	-
Standard: 15.0% w Diene 55	14.8	-
High Impact polystyrene 1	8.6	9.7
" " " 2	5.6	5.8
" " " 3	9.0	1.2
" " " 4	11.2	11.4
" " " 5	5.8	5.9

three configurations of polybutadiene (1095).

Nuclear Magnetic Resonance Spectroscopy

Spectra have been published on ^{13}C NMR on styrene-butadiene copolymers (1096).

Gronski et al (1097) showed the segmental motion in butadiene-styrene copolymers to be influenced by sequence distribution as determined by ^{13}C spin lattice relaxation times. The relaxation times for 1,4-diene units at boundaries with styrene or 1,2 units are different from those in block sequences; the relaxation of the o-phenyl carbon atoms is slower than p-phenyl relaxations.

Randall (1099) described a general method for the characterization of terpolymers using ^{13}C NMR. The method as applied to hydrogenated butadiene-styrene copolymers shows how the number average sequence length can be obtained from diad, triad and higher order monomer distributions.

Conti and coworkers (1100) concluded that the off-resonance ^{13}C NMR data on butadiene-styrene copolymers were consistent with that for atactic polystyrene.

Hydrocarbon mineral oil can be determined in impact polystyrene by Soxhlet extraction of the polymer with pentane followed by nuclear magnetic resonance (NMR) analysis of the extract to determine the amount of oligomers present (1098).

Pyrolysis-Gas Chromatography

Sindelfinger (1102) has discussed the pyrolysis-gas chromatography of styrene-butadiene copolymers using the open boat combustion technique.

Vien Virus (1103) has applied pyrolysis-gas chromatography to the examination of styrene-butadiene and acrylonitrile-butadiene-styrene polymers at temperatures up to 700°C and to the radioactively labelled products.. Residence times in the pyrolyzer of 3 to 30 seconds were used.

Brauer (1104) has applied pyrolysis-gas chromatography to the elucidation of the structure of styrene-butadiene copolymers.

Dimbat and Shaw (1105) have investigated the application of pyrolysis-gas chromatography of ABA type block copolymers where A is polystyrene or poly- α-methyl-styrene and B is polybutadiene or polyisoprene. In the case of α-methyl styrene/butadiene/α-methylstyrene block copolymers, better than 99% of the

pyrolysis products of α-methylstyrene appear in a single peak that is completely separated from the polybutadiene peaks. One of the peaks from polybutadiene is sensitive to the 1,4 content and can be used to estimate the 1,4
and 1,2 content of the polybutadiene segment. Cis- and trans-polybutadiene
produce the same pyrolysis pattern so the cis content cannot be determined.

Various other workers have discussed the application of pyrolysis-gas chromatography to examination of styrene-butadiene copolymers (1106-1111).

Tutorskii and coworkers (1108) employed data based on the pyrolysis-gas chromatographic analysis of mixtures of polystyrene and polybutadiene to determine
the distribution of polystyrene chains in butadiene-styrene block copolymers.

Bound styrene in butadiene-styrene copolymers has been determined using pyrolysis-gas chromatography with a pyrolysis temperature of 610°C (1109).

Haeusler and coworkers (1110) determined the microstructure of styrene-butadiene copolymers by pyrolysis-gas chromatography.

Thermal Methods

McNeill (1112) has discussed the application of the technique of thermal volatilization analysis to styrene-butadiene copolymers. The incorporation of
large amounts of butadiene into the polystyrene chain does not result in lower thermal stability, but it was clear that less volatile degradation products
are obtained in the case of the copolymer.

X-Ray Diffraction

X-ray diffraction has been used in studies on styrene butadiene blends with
polypropylene (1113, 1114) and determination of the structural parameters of
styrene-butadiene block copolymers in heptane using small-angle x-ray scattering (1115).

Thin-layer Chromatography

Random, tapered, diblock, and triblock butadiene-styrene copolymers have been
separated by thin-layer chromatography (1116).

Thin-layer chromatography has also been used to fractionate styrene-butadiene
copolymers (1117).

Fractionation

Cragg and coworkers (1119, 1120) and Morton (1118) have examined the relative
amounts of crosslinking in styrene-butadiene produced under various conditions,
and concluded that the Huggins' constant (1121) increases with the extent of
crosslinking.

Blackford and Robertson (1122, 1123) made a quantitative determination of the
degree of crosslinking and the distribution of the crosslinks in a sample of
styrene-butadiene rubber.

To determine the number of crosslinks per molecule it is necessary to know
the ratio of the intrinsic viscosity of the molecule to that of a linear molecule of the same molecular weight. Blackford and Robertson (1122, 1123) proposed a relationship between the intrinsic viscosity and the sedimentation coefficient to obtain this ratio. This method can be used to estimate the degree

of crosslinking in fractions of a sample of styrene-butadiene rubber. By using the weights of the fractions together with the degree of crosslinking in
each fraction, the amount of crosslinking in the unfractionated sample could
be calculated.

Gel permeation chromatographic separations have been made on styrene-butadiene
block copolymers using viscosity and osmometry for molecular weight (1124).
The effect of molecular weight distribution on cloud point curves has been
studied on the system polystyrene-polybutadiene-tetralin (1125).

The distribution of chemical compositions of butadiene-styrene block copolymer molecules in a gel permeation chromatographic elute has been derived from
the increase of turbidity with volume at different points of the turbidimetric
titration curve. By plotting such a distribution perpendicularly over the
curve representing constant elution volume at the molecular weight-composit-
ion plane, a three-dimensional surface was obtained which characterized completely the molecular and compositional distribution of the polymer (1126).

Gel permeation chromatography has been used to determine the chemical composition of butadiene-styrene copolymer and butadiene-α-methylstyrene copolymer with considerations of the refractive index and extinction coefficient,
etc. The results agreed well with those obtained from infrared and NMR spectroscopy (1127).

Ambler et al (1128) have applied the gel permeation technique to obtaining
branching results on randomly branched styrene-butadiene copolymers and star-
branched polybutadiene. These analyses in styrene-butadiene rubbers and polybutadiene rubbers have revealed large differences in branching between rubbers
polymerized in different ways. They calculated the functionalities of several
star-branched solution-polymerized styrene-butadiene rubbers and compared them
to their expected structures.

The dielectric constant detector (Model 410, Applied Automation Ltd) was introduced commercially by Benningfield (1129). Although it was initially designed as a detector for process liquid chromatography, a laboratory version
also is available, which is suitable for polymer fractionation studies. Benningfield and Mowery (1130) have described this detector in detail. This is a
bulk property detector that complements, rather than competes with, the refractive index detector. The dielectric constant detector offers specific
advantages for certain analyses, while the refractive index detector is pref-
erable in other applications. Bode et al (1131) used this technique for the
examination of styrene-butadiene polymers.

The dielectric constant detector has been primarily used as a detector for
size-exclusion chromatography of polymers (1132, 1133). The dielectric constant detector has several advantages over the refractive index detector for
some size-exclusion chromatography analyses. For example, the refractive index detector loses sensitivity faster as a function of temperature than does
the dielectric constant detector. Therefore, at elevated temperatures the
dielectric constant detector may be more sensitive than the refractive index
detector, even when the opposite is true at room temperature. This is caused
by the reduced output of the light source and the reduced sensitivity of the
photodiodes commonly used in refractive index detectors. As shown in Table
49, monomer ratios also greatly affect the response of the refractive index
detector in some copolymers. In the case of styrene-butadiene copolymers, the
refractive index response variation due to changes in the monomer ratio is
approximately an order of magnitude higher than the variation observed with
a dielectric constant detector. This effect can be significant for size-ex-

Table 49 Detector Response.

	RI	ΔRI*	ε	Δε*
Polybutadiene	1.52	0.11	2.3	5.3
Polystyrene	1.60	0.19	2.6	5.0
%		42%		5.6%

* Versus tetrahydrofuran; RI = 1.41, ε = 7.6.

clusion chromatographic analyses of copolymers with unknown or variable monomer ratios.

Figure 105 shows size exclusion chromatographs of a polystyrene calibration
standard. the composite chromatogram of Fig. 106 shows the size exclusion
chromatographic separations of four different styrene-butadiene copolymers.

Figure 107 is a composite chromatogram showing the results obtained with a
refractive index detector and a dielectric constant detector in series. Some
differences in the responses are evident. They both clearly show the bimodal
molecular weight distribution of the copolymer, but the additive that elutes
after the process solvent is barely seen by the refractive index detector.
Note also the difference in response to the process solvent of the two detectors.

Molecular Weight

Kratochvil (1134) has reviewed the potentials of the light scattering method
and of some related aspects for the investigation of dilute polymer solutions,
multicomponent with respect to both solvents and polymers. He also suggested
the use of acetophenone-cyclohexane-ethanol mixed solvent for the $\overline{M}w$ determination of butadiene-styrene polymers by light scattering (1135). A computer
program in FORTRAN IV language has been written for a numerical variant of
the graphical Zimm method for interpretation of light scattering experimental
data (1136).

Miscellaneous

Blanchford and Robertson (1137) have applied viscometry to a study of radiation induced branching in styrene-butadiene copolymers.

Bradford and Vanzo (1138, 1139) have studied the ordered structures of styrene-butadiene block copolymers in the solid state.

Roth (1140) determined the degree of crosslinking of dibutyl maleate-cross-
linked butadiene-styrene copolymers by determining the temperature dependence
of the dielectric loss factor and by a dye take-up procedure.

Sedimentation equilibrium in a density gradient has given accurate composition analysis of butadiene-styrene copolymers (1140, 1141).

Monomers

Styrene monomer in polyester laminates, styrene-butadiene rubber, and acrylic
polymer dispersions can be determined polarographically after conversion to
$PhCH(NO)CH_2NO_2$ and $PhC(:NOH)CH_2NO_2$ by reaction with sodium nitrate in acetic

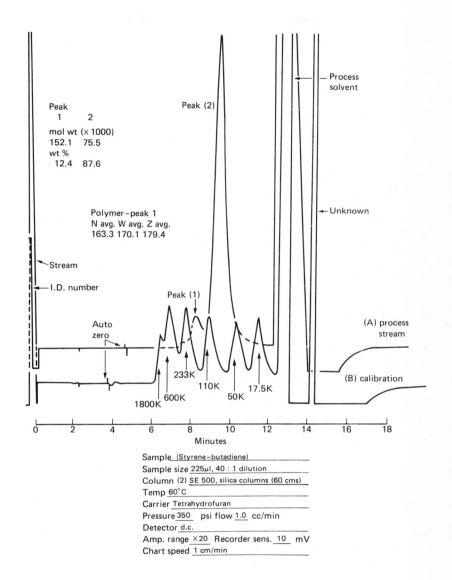

Fig. 105   SEC separations of a styrene-butadiene copolymer
           and polystyrene reference standards.

acid (1142).

Styrene and butadiene monomers have been determined in these polymers by procedures involving head-space analysis-gas chromatography and gas chromatography of a dimethyl formamide solution of the polymer (1143).

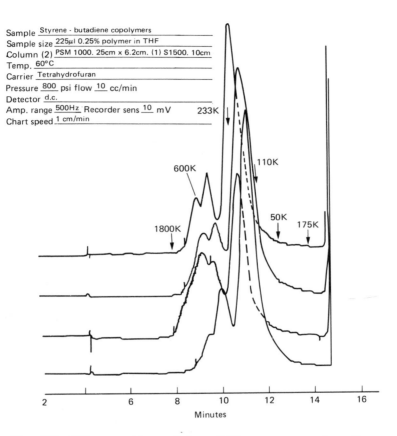

Sample _Styrene - butadiene copolymers_
Sample size _225µl 0.25% polymer in THF_
Column (2) _PSM 1000. 25cm x 6.2cm. (1) S1500. 10cm_
Temp. _60°C_
Carrier _Tetrahydrofuran_
Pressure _800_ psi flow _10_ cc/min
Detector _d.c._
Amp. range _500Hz_ Recorder sens _10_ mV   233K
Chart speed _1 cm/min_

Fig. 106   SEC separations of four different styrene-butadi-
           ene copolymers.

Additives

Methods for the determination of antioxidants in styrene-butadiene copolymers
are reviewed below;

Antioxidants in Styrene-butadiene copolymers

	Spectrophotometric Methods		Ref.
Polygard (tris-nonylated phenyl)phosphite	methanol or ethanol extract- ion coupling with diazotized-p-nitro- aniline-spectrophoto- metric finish.	low levels	1144
Polygard (tris-nonylated	solvent extraction-		1145

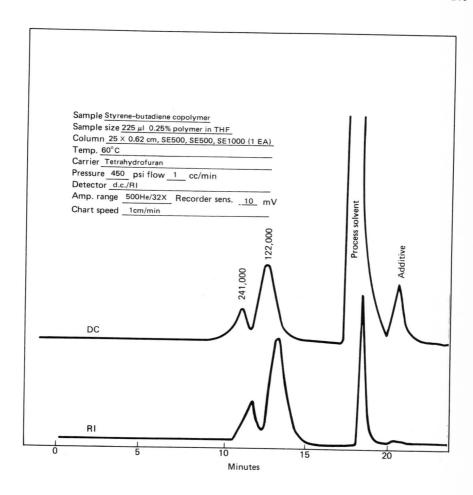

Sample Styrene–butadiene copolymer
Sample size 225 μl 0.25% polymer in THF
Column 25 X 0.62 cm, SE500, SE500, SE1000 (1 EA)
Temp. 60°C
Carrier Tetrahydrofuran
Pressure 450 psi flow 1 cc/min
Detector d.c./RI
Amp. range 500He/32X Recorder sens. 10 mV
Chart speed 1cm/min

Fig. 107 Comparison of RI and DC detectors for the SEC
separation of a styrene-butadiene copolymer.

phenyl) phosphite Agerite Supealite Nevastain A	hydrolysis to nonyl phenol-diazotization and spectrophotometric evaluation.	
Polygard (tris-nonylated phenyl)phosphite	solvent extraction- addition of sodium hydroxide and measurement of bathochromic shift spectrophotometrically.	1146
Various types	solvent extractions-	1147

Phenyl β napthylamine   colormetric tests.
Wingstays (styrenated phenols)
Polygard

Thin-layer Chromatography

amine acid   solvent extraction-thin-   1148
phenolic types   layer chromatography for
                                    quantitative analysis.

Mass Spectrometry

N-phenyl-β napthylamine   volatile preconcentration-   1149
                                    mass spectrometry.

6 dodecyl-2,2,4-tri methyl-   hydrolysis to lauryl alcohol- 1150
1,2 dihydro-quinaline   chloroform-ethanol-hexane
trisnonyl phenyl phosphate   extraction and thin-layer
dilaurylthio   chromatography then glc.
dipropionate

alkylated cresols   ethanol extraction-   1151
(2,6-di-t-butyl-p-cresol)   glc.
amine type
(N-phenyl-2 napthylamene)
Subst'd p-phenylene
        diamine type
        Santoflex 13
sec heptyl phenyl-p-phenylene diamine

Liquid Chromatography

Additives and low molecular   solvent extraction -   1152
weight compounds   column chromatography.

STYRENE-ACRYLONITRILE COPOLYMERS

Chemical Methods

Acrylonitrile-styrene copolymer has been separated from acrylonitrile-butadiene-styrene graft copolymer by extraction with ethyl acetate (1153). the ungrafted rubber was extracted from the insoluble portion with a 1:1.5 ethyl-
acetate-hexachlorobutadiene mixture.

Infrared Spectroscopy

Scheddel (1154) has reported an infrared method for the compositional analysis
of styrene-acrylonitrile copolymers. In this method relative absorbance between a nitrile $\acute{v}$(CN) mode at 4.4μ and a phenyl v(CC) mode at 6.2μ is used.

Takeuchi et al (1155) have described a near infrared method using combination
and overtone bands for carrying out the same analyses. Near infrared spectra
of four random copolymers and homopolymers are shown in Fig. 108. The bands
which occur in the region 1.6-2.2μ result from overtones and combination tones
which occur, respectively, in the regions 1.6-1.8μ and 1.9-2.2μ.

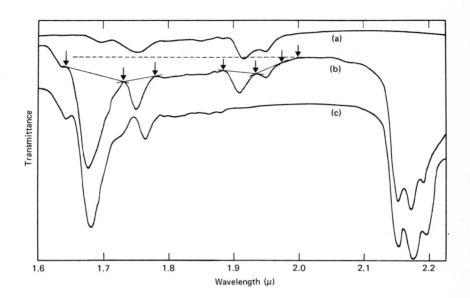

Fig. 108   Near-infrared spectra: (a)polyacrylonitrile;(b)
          styrene-acrylonitrile copolymer (AH content 25.7
          wt-%); (c)polystyrene.

The band near 1.68µ is assigned as an overtone of phenyl v(CH) mode near 3000
cm^{-1} (3000 x 2 = 6000 cm^{-1} = 1.67µ), and its absorbance is directly proportional to the styrene content of the copolymer and the film thickness. the band
near 1.75µ is assigned as an overtone of aliphatic v(CH) mode near 2900 cm^{-1}
(2900 x 2 =5800 cm^{-1} = 1.724µ) and both acrylonitrile and styrene absorb at
this wavelength. bands at 1.910 and 1.952µ are combination tones of acrylonitrile and are assigned as, respectively, v(CN) + v_{asym}(CH$_3$)(2237 + 2940 =
5177 cm^{-1} = 1.932µ) and v(CN)+ v_{asym} (CH$_3$)(2237 + 2870 = 5107 cm^{-1} = 1.958µ).
For calculation of the absorbance of these four characteristic bands, two
different baseline methods were used, as shown in Fig. 108. The method using
the extrapolated baseline from 2µ introduces less deviation than the other
baseline method. The absorbance ratio $A_{1.675}/A_{1.910}$ proved to be the best one
for analytical measurements.

Sequence distribution of acrylonitrile in acrylonitrile-styrene copolymers
has been determined from measurements of the C$\equiv$N stretching mode with shift
to lower frequency with decreasing acrylonitrile content (1156).

Iyer and Padhye (1157) used the C$\equiv$N stretching mode for quantitative determination of acrylonitrile in butadiene-acrylonitrile copolymers.

Comparison of attenuated total reflectance with transmission spectra on solid
polymer indicated that acrylonitrile concentration on the surface was the same
as in the bulk. Kimmer and Scholke (1158) used infrared and NMR in analyses
of acrylonitrile-styrene, acrylonitrile-butadiene and methacrylonitrile-but-
adiene copolymers.

Infrared spectroscopy has been used to study styrene oligomers containing
acrylonitrile active ends (1159). The addition of acrylonitrile to oligostyrene
alkylmetal salt solutions resulted in attachment of acrylonitrile across the
double bond of the oligostyrene and by formation of the carbanion C-HCN.

The sequence distribution of styrene in acrylonitrile styrene copolymers has
been calculated by Oi and Moriguchi (1160) from the ratio of the C-H out-of-
plane bending vibration band at 760 cm^{-1} to the out-of-plane ring deformation
band at 700 cm^{-1}.

Mikhailov and coworkers (1161) showed that the equation for determining triad
distribution in styrene-acrylonitrile copolymers from the intensities of bands
at 1585 and 1603 cm^{-1} is invalid.

The intensity of the absorption peak at 1200 to 1250 cm^{-1} was used to obtain
information concerning the length of the $CH_2C(CH_3)R$ sequences in copolymers
of 2-substituted propene and styrene or acrylonitrile. The absorption band
reached its full intensity at a length of approximately 5 monomer units. With
decreasing sequence length the band intensity decreased and the position of
the band shifted to higher wave numbers (1162).

Nuclear Magnetic Resonance Spectroscopy

Sandner et al (1163) calculated the diad and triad distribution in acrylonitrile-styrene copolymers from the resonance of the quaternary carbon of the
phenyl groups in the styrene units, the CN group of the acrylonitrile units,
and the CH_2 groups in the main chain from ^{13}C NMR spectra.

Carbon-13 NMR spectrum of alternating copolymers of α-methyl styrene and methacrylonitrile were shown to be dominantly syndiotactic in contrast to PMR results (1164).

Pyrolysis-Gas Chromatography

Various workers (1165-1169) have discussed the pyrolysis-gas chromatography
of styrene-acrylonitrile block copolymers and homopolymer blends. According
to Voight (1165) it was possible to identify styrene-acrylonitrile copolymer
by this method. Lebel (1166) reported that this method could successfully be
used to distinguish a homopolymer blend from the random styrene-acrylonitrile
copolymer.

Shibasaki and Kambe (1167, 1168) determined that on the basis of styrene and
acrylonitrile yields it was possible to get data on the structural composition
of the copolymer. These authors succeeded in identifying, in addition to styrene and acrylonitrile, benzene, α-methylstyrene, and ethylbenzene by this
method. The latter two components were found in trace amounts only.

Sindelfingen (1170) has discussed the pyrolysis-gas chromatography of styrene-
acrylonitrile copolymers using the open-boat combustion technique.

Shibasaki and Kambe (1171, 1172) used gas chromatography to determine the monomer yields obtained on pyrolysis of styrene-acrylonitrile and styrene-methyl-
methacrylate copolymers. Their results were interpreted in terms of "boundary-
effect" parameters β_A and β_B for the monomers A and B.

Shibasaki and Kambe (1173) investigated the high temperature pyrolysis of
styrene-acrylonitrile copolymers using a preheated quartz tube. The degradation products were swept directly into the column of the gas chromatograph.

Copolymer composition and temperature had no appreciable influence on monomer yield.

Vukovic and Gnjatovic (1174) and Deur-Siflar et al (1175) investigated the pyrolysis of styrene-acrylonitrile copolymer at 645°C and identified eleven components in the pyrolyzate, the most important of them being styrene, acrylonitrile, and propionitrile. By examination of the pyrolyzate composition during pyrolysis of the styrene-acrylonitrile copolymer of different compositions, it was established that the propionitrile yield was definitely decreased when the acrylonitrile concentration in copolymer was about 60 mole-%. Further, from the propionitrile yield, they could distinguish random styrene-acrylonitrile copolymer from the styrene-acrylonitrile homopolymer blend, and on the basis of propionitrile yield some information on the molecular structure of the copolymer could be obtained. The styrene yield depends linearly on the composition. This permitted determination of copolymer composition on the basis of the styrene yield. The following substances were identified on the pyrogram obtained for styrene-acrylonitrile copolymers: acetonitrile, acrylonitrile, propionitrile, benzene, allylcyanide, toluene, ethylbenzene, styrene, α-methylstyrene, diethylbenzene, and ethylvinylbenzene. Figure 109 shows the yields of various components obtained on pyrolysis of random copolymers containing between 7% and 64% acrylonitrile.

Thermal Methods

Chaigneau (1176) found that the main gas phase pyrolysis products from poly-acrylonitrile-styrene copolymer and acrylonitrile-butadiene-styrene copolymer are hydrocyanic acid, hydrogen and methane.

Grassie and Bain (1177, 1178) have studied the thermal degradation of styrene-acrylonitrile copolymers.

Thin-layer Chromatography

Gloeckner and Kahle (1179) separated acrylonitrile-styrene copolymers into fractions differing in acrylonitrile content by temperature gradient thin-layer chromatography on silica gel using a toluene-acetone mixture as the eluent.

Styrene-acrylonitrile copolymers have been fractionated by Gloeckner and Kahle (1180) by thin-layer chromatography on silica gel D using a benzene or toluene solvent to which acetone was gradually added.

Fractionation

Solvent-nonsolvent systems such as ethylene carbonate-ethylene cyanohydrine and methyl ethyl ketone-cyclohexane have been used for the fractionation of acrylonitrile-styrene random copolymers (1181).

Miscellaneous

Polydispersity and molecular weight studies have been reported for styrene-acrylonitrile copolymers polymerized by peroxide (1182) and azo (1183) initiation.

Viscosity measurements have been reported (1184) on styrene-acrylonitrile copolymers.

Refractive index increment measurements have been used to provide structural

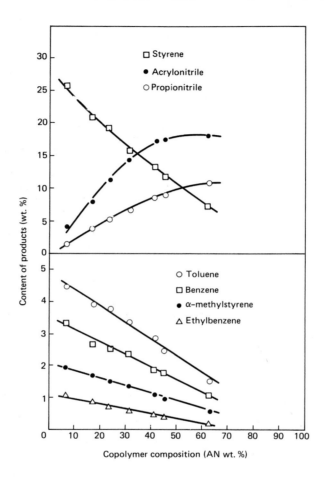

Fig. 109 Content of various components of the pyrolyzate
          of random styrene-acrylonitrile copolymer of
          different compositions calculated on the sample
          weight. Pyrolysis temperature 645°C, in nitrogen
          stream.

information on styrene-acrylonitrile-methyl methacrylate terpolymers (1185).

Monomers

Roy (1186) has described a titrimetric method for the determination of residual
monomers in styrene-acrylonitrile copolymers.

Various workers have described methods for the determination of monomers in
these copolymers. Acrylonitrile has been determined in the polar extract by
reaction with dodecyl mercaptan (1026) also by polarography (1270) and by gas
chromatography (1268) of a dimethyl formamide solution. The polarographic meth-

od (1270) also determines methacrylonitrile. Stryene monomer has been determined by the polarographic (1187) and the gas chromatographic (1188) methods.

ACRYLONITRILE BUTADIENE-STYRENE COPOLYMERS

Nuclear Magnetic Resonance Spectroscopy

A method has been described by Turner et al (1189) for determining the amount
of ungrafted rubber in ABS polymers of the G type, namely a mixture of polybutadiene grafted with styrene and acrylonitrile and styrene-acrylonitrile.
First the styrene-acrylonitrile resin is removed by extraction with ethyl acetate. The composition of the styrene-acrylonitrile is determined by NMR analysis. The ethyl acetate insoluble material which consists of graft polymer
and ungrafted rubber is then extracted with a mixture of hexachlorobutadiene
and ethyl acetate. The ungrafted rubber dissolves in the extraction solvent
mixture while the graft polymer remains insoluble. Filtration, concentration
of the filtrate by vacuum distillation and NMR analysis of the concentrated
filtrate using an internal standard technique provided the amount and composition of the ungrafted· rubber. The data also permitted the amount and composition of the graft polymer to be calculated.

Since the ethyl acetate insoluble portion was not completely soluble in any
solvent, NMR could not be used to determine the overall composition of this
fraction. Therefore, infrared and nitrogen analysis were used to determine
the composition of this fraction. Nitrogen analysis provided the amount of acrylonitrile while infrared provided the ratio of butadiene to styrene. Infrared spectra clearly showed the vinyl content of the butadiene portion to be
about 20% which is typical for emulsion polybutadienes. No good measure of the
cis microstructure could be obtained because of the presence of styrene. The
trans band (10.35μ) was sharp and easily measured. The ratio of butadiene to
styrene was determined by assuming the trans content of the butadiene portion
was 60%. This is about the average value of trans in emulsion polybutadienes.
The ratio of styrene to acrylonitrile in the ethyl acetate insoluble portion
is similar to the ratio of the monomers in the styrene-acrylonitrile portion
of the ABS polymers. The similarity of the styrene to acrylonitrile ratios of
the graft and styrene-acrylonitrile portions was expected and shows that the
infrared analysis of the ethyl acetate insoluble was satisfactory.

Analysis showed that about 60 to 80% of the butadiene in the entire sample is
present as ungrafted rubber. using the compositional analysis of the ethyl acetate insoluble fraction (graft and ungrafted rubber) and the amount of ungrafted rubber extracted, one can calculate the composition of the graft. Figure
110 shows a typical NMR spectrum of the material extracted from the ethyl acetate insoluble portion with the hexachlorobutadiene-ethyl acetate mixture. No
aromatic protons of styrene or acrylonitrile protons are seen in the NMR spectra. The vinyl content of the polybutadiene is about 20%. Figure 111 shows a
typical NMR spectrum of the material extracted from the ethyl acetate insoluble portion with hexachlorobutadiene-ethyl acetate mixture for some of the
samples examined. The aromatic protons of styrene are seen around 7 ppm. The
styrene content was 10%.

Turner et al (1189) were unable to detect any significant amount of nitrogen
in these rubber fractions which indicate that there is no acrylonitrile present. Infrared analyses also did not detect any acrylonitrile. Therefore, the
styrene appears to be due to the presence of styrene-butadiene rubber rather
than polybutadiene compared to the other possibility that the styrene is due
to the presence of graft.

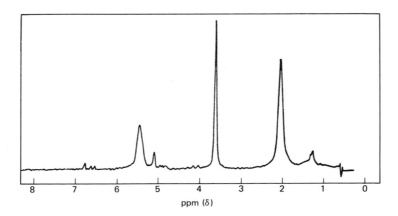

Fig. 110   Typical NMR spectrum of extracted ungrafted rubber
           from Samples 1 through 6.
           Signal at 5.4 ppm - olefinic protons of 1,4-poly-
           butadiene.
           Signal at 5.1 ppm - olefinic protons of 1,2-poly-
           butadiene.
           Signal at 3.5 ppm - dioxane (internal standard).
           Signal at 2.0 ppm - aliphatic protons of 1,4-
           polybutadiene.
           Signal at 1.2 ppm - methylene protons of the soap.
           Signals between 6 and 7 ppm -impurities in hexa-
           chlorobutadiene.

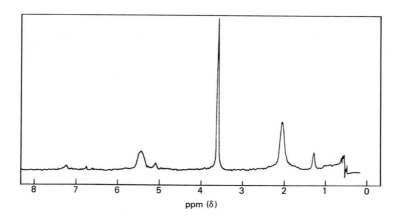

Fig. 111   Typical NMR spectrum of extracted ungrafted rubber
           from Samples 7 and 8.
           Signal at 7.2 ppm - styrene aromatic protons.
           Other signals were identified in Fig. 110.

Lindsay et al (1190, 1191) prepared 300 MHz NMR spectra of highly alternating and conventional butadiene-acrylonitrile copolymers. In the alternating copolymers, the olefinic proton resonance of acrylonitrile was observed as two patterns of equal intensity assigned to protons in the meso and racemic acrylonitrile-butadiene-acrylonitrile triads. The resonance of acrylonitrile methine and methylene protons was observed as four patterns of approximately equal intensity. The methine proton resonance of the free radical-initiated acrylonitrile (A)-butadiene (B) polymer was observed in three areas and was used to measure the relative amounts of AAA, AAB, and BAB triads.

Pyrolysis-Gas Chromatography

Sindelfingen (1192) has discussed the pyrolysis-gas chromatography of acrylonitrile-butadiene-styrene terpolymers using the open-boat combustion technique.

Araki (1191) discussed a method for the determination of butadiene in acrylonitrile-butadiene-styrene resins by quantitative pyrolysis-gas chromatography.

Voigt (1193, 1194) applied pyrolysis-gas chromatography to the examination of ABS terpolymers.

Fractionation

The gel permeation chromatographic elution behaviour of ABS and acrylonitrile homopolymers, has been studied by Kranz and coworkers (1196) and a calibration curve prepared for different solvents.

Crompton (1197) has investigated methods, described below, for the analysis of acrylonitrile-butadiene-styrene terpolymers. This included the development of methods determining solvent-insoluble polymer (gel), solvent-soluble polymer and total non-polymer additives.

Methods were also developed for determining bound acrylonitrile, butadiene and styrene both in the original polymers and in the separated insoluble polymer (gel) and soluble polymer fractions. All these methods are also applicable to the intermediate styrene-acrylonitrile-butadiene graft samples and the blends of this graft with low molecular weight styrene-acrylonitrile polymer which are used in the manufacture of ABS.

To remove non-polymer additives the polymer is dissolved (or dispersed) in chloroform. This solution is slowly poured into an excess of stirred methanol to reprecipitate the polymer and leave the soluble non-polymer additives in a clear chloroform-methanol phase which can be separated from the polymer by filtration. The polymer is then washed with methanol and vacuum dried.

Crompton (1197) examined the effect of centrifuging in various organic solvents on samples of additive-free ABS polymer. The polymer (5 g) was contacted with 100 ml of solvent and left for several hours to dissolve, then centrifuged at up to 2,000 g for two to three hours then clear soluble polymer phase was separated off.. The solvent phase was poured into an excess of methanol and the precipitated polymer filtered off, vacuum dried and weighed.

It is evident from the results obtained in these experiments (Table 50) that of the various solvents examined, chlorobenzene and bromobenzene dissolve the highest proportion of an ABS terpolymer.

It is seen in Table 50 that a very poor separation is obtained of insolubles from soluble polymer when a 5% methyl ethyl ketone solution of nibs polymer

Table 50   Insoluble/Soluble Polymer Distribution in Various Centrifuging Solvents.

Sample	Centrifuging solvent	% Insolubles	% Solubles	Comments on separation after first centrifuging
natural nibs	Trichlorethylene	92.6	7.4	Good separation of soluble and insoluble fractions.
natural powder	Trichlorethylene	90.1	10.6	Good separation of soluble and insoluble fractions.
natural nibs	Butyl acetate	69.2	27.3	Insolubles settle out. Upper phase rather hazy.
natural nibs	Chlorobenzene	56.4, 55.0 (Av. 55.7)	44.3	Insolubles float on solvent. Lower solvent phase clear.
natural nibs	Bromobenzene	51.6, 51.5 (Av. 51.6)	48.4	Insolubles float on solvent. Lower solvent phase clear.
natural powder	Bromobenzene	52.6, 53.4 (Av. 53.0)	47.0	Insolubles float on solvent. Lower phase clear.
natural nibs	Methyl ethyl ketone	-	-	Very poor separation of insoluble polymer phases after centrifuging.

was centrifuged at speeds of up to 2,000 g. At 25,000 g excellent separations resulted.

Crompton (1197) explored the possibility of using methyl ethyl ketone as a separation solvent on the centrifuge (M.S.E. super medium) with high speed attachment. In these experiments the additive-free ABS polymer (1 g) was contacted with methyl ethyl ketone (20 ml) and either refluxed for six hours or left to disperse overnight at room temperature. The solutions were then centrifuged at 16,000 to 25,000 g for 30 minutes to provide an almost clear upper phase, i.e. a good separation of soluble and insoluble polymer. The clear phase was pipetted off and the insolubles recentrifuged with a further 20 ml of methyl ethyl ketone. The two methyl ethyl ketone extracts were combined, dried in vacuo and weighed. The insoluble fraction was washed from the centrifuge tube with methyl ethyl ketone and dried in vacuo and weighed. The results obtained in these experiments (Table 51) indicate that for natural nibs appreciably lower insoluble contents are obtained in methyl ethyl ketone than are obtained using either chorobenzene or bromobenzene.

Table 51   Comparison of % Insolubles and % Solubles obtained from ABS Nibs in MEK and in Halogenated Solvents.

Centrifuging solvent	Centrifugal force, g	% Insolubles	% Solubles
MEK (refluxed before centrifuging)	16,000-20,000	30.1	69.7
MEK (not refluxed before centrifuging)	16,000-20,000	39.3	66.7
Chlorobenzene	1,000-2,000	55.7	45.3
Bromobenzene	1,000-2,000	51.6	48.4

Crompton (1197) also investigated methods for the determination of bound acrylonitrile, styrene and butadiene units in ABS terpolymers and in the soluble and insoluble fractions obtained from these polymers by the solvent separation procedures. In Table 52 are reported complete gel and soluble fraction analyses carried out on various ABS polymers. In both the whole ABS polymer and in its soluble and insoluble fractions the sum of the determined constituents usually adds up to 100 ± 2%.

Miscellaneous

Ambler (1128) has investigated molecular weight and glass transitions in ABS terpolymers.

Monomers

Streichen (1199) has determined residual butadiene and styrene in polymers with an analytical sensitivity of 0.05 to 5 ppm by analysis of the equilibrated headspace over polymer solutions and determined acrylonitrile, α-methyl styrene, and styrene monomers by headspace analysis over heated solid polymer samples.

Methods have been described for determining styrene and acrylonitrile (dimethyl formamide extraction-polarography, 1200) and styrene, acrylonitrile and butadiene monomers (solution in dimethyl formamide-gas chromatography, 1188).

Table 52   Determination of Monomer Units in Competitors' A.B.S. Terpolymers.

Determined	Cycolac T 1000 natural nibs	Cycolac H 1000 natural nibs	Kralastic MH nibs
	Analysis of additive free polymer		
Butadiene % a	20.5	28.6	19.1
Acrylonitrile % b	23.8	20.9	20.9
Styrene % c	54.0 (Total = 98.3%)	48.5 (Total = 98.0)	
	Analysis of insoluble fraction		
Insolubles in polymer %	56.0*	63.5*	11.0*
Butadiene %	30.8	38.2	49.9
Acrylonitrile %	22.1	19.2	10.7
Styrene %	48.0 (Total = 100.9%)	–	–
		48.0 (Total = 99.8%)	
	Analysis of soluble fraction		
Solubles in polymer %	44.0*	48.5**	–
Butadiene %	10.3	7.0	13.9
Acrylonitrile %	27.6	28.7	21.5
Styrene	62.0 (Total = 99.9%)	64.0 (Total = 99.7%)	61.0 (Total = 96.4%)

*   Determined using chlorobenzene
** Determined using bromobenzene
a   iodine monochloride method
b   kjeldahl method
c   infrared method

Additives

A method has been described for the determination of dilauryl thiopropionate secondary antioxidant in ABS. The procedure (1201) involves hydrolysis to alcohol, extraction with chloroform-ethanol-hexane and thin-layer chromatography and gas chromatography.

          STYRENE COPOLYMERS

          STYRENE (α- METHYLSTYRENE) METHACRYLATE

Infrared Spectroscopy

Yanagisawa (1202) investigated the absorption of a series of styrene copolymers in the 1000-1300 cm^{-1} and 500-650 cm^{-1} regions and concluded that strong phenyl absorption (coupled with skeletal vibrations) occurs at 543 cm^{-1} only when the average styrene sequence length is greater than three. Copolymers having shorter average styrene sequences absorb at higher frequencies. These results suggest that the 543 cm^{-1} band might provide a measure of SSS structure (S=styrene) and that bands at higher frequencies might be found for other styrene-centres triads. Similarly, the absorption observed at 1068 cm^{-1} in polystyrene shifts to about 1075 cm^{-1} in copolymers having low styrene contents. Shifts of bands observed at 1191 and 1148 cm^{-1} in poly(methyl methacrylate) to higher and lower frequencies, respectively, in styrene-methyl methacrylate copolymers were also noted. The magnitude of the shifts depends on the average sequence length of the methyl methacrylate units in the copolymers. Similar results were obtained with styrene-ethyl methacrylate copolymers.

Nuclear Magnetic Resonance Spectroscopy

Nishioka and coworkers (1203, 1204) and also Bovey (1205) measured the NMR spectra of styrene-methyl methacrylate and p-xylene-methyl methacrylate copolymers in chloroform and carbon tetrachloride solution. They observed three types of methoxy resonance and concluded that this resonance might be used to characterize sequence distribution and possibly also tacticity in the copolymers. Similar results (1206) have been obtained for styrene-methyl methacrylate copolymers in thionyl chloride solution. Bovey (1205) correlated the methoxy resonance patterns of styrene (S)-methyl methacrylate (M) copolymers (chloroform solution) with calculated distributions of M centred triads by considering the resonance of methoxy protons in SMM and SMS triads to be distributed among several methoxy resonance areas, rather than being associated with a given area. This was attributed to triad configurational effects. According to further interpretation (1207) the highest field, methoxy resonance is due to SMS triads in which both styrene units have the same configuration (coisotactic) as the central methyl methacrylate unit. The central methoxy resonance area is due to methoxy protons of methyl methacrylate units centred in SMM and SMS traids, where one of the styrene units has the same configuration as the central methyl methacrylate unit. The upfield shift of the resonance of such methoxy protons is probably due to shielding by diamagnetic regions of neighbouring phenyl units. The resonance of the remaining methoxy protons (MMM and SMM or SMS where none of the styrene units has the same configuration as the central methyl methacrylate unit) occurs in the lowest-field resonance area.

On the basis of this interpretation, which permits the methoxy resonance patterns to be accounted for quantitatively, the probability that adjacent styrene-methyl methacrylate units have the same configuration (coisotactic) in their copolymers is 0.48. The probability that adjacent styrene-methyl acrylate units have the same configuration in their copolymers was estimated to be much higher (0.80).

Ito and Yamashita (1207) and also Overberger and Yamamoto (1208) have studied the NMR spectra of styrene-methyl methacrylate copolymers prepared by anionic polymerization. They conclude that there are very few styrene-methyl methacrylate linkages in the copolymers (or polyblends).

Harwood and Ritchey (1209) obtained spectra of the type shown in Fig. 112 for styrene-methyl methacrylate copolymers in deuterobenzene solution. As in the other studies, methoxy resonance was observed in three general areas. For copolymers containing less than 50 mole percent styrene, the methoxy patterns observed were correlated with calculated distributions of M centred pentads; the fraction of methoxy resonance observed in the lowest-field peak is equal to the fraction of M units centred in MMMMM, MMMMS, SMMMS, and SMMSM pentads, for example, and the fraction of methoxy resonance observed in the highest-field peak to the fraction of M units centred in MSMSM pentads. Copolymers containing higher amounts of styrene have appreciable quantities of M units centred in SSMSS and SSMSM pentads and these units seem to be responsible for resonance in more than one general area. This phenomenon was attributed to pentad configurational effects.

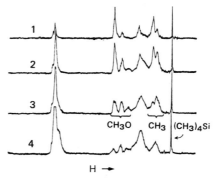

Fig. 112   NMR spectra of styrene-methyl methacrylate copolymers. Styrene contents, Spectrum 1, 16.8; 2,
33.3; 3, 50.2; 4, 72.6 mole%.

α-methyl resonance was observed for these copolymers in two general areas, the higher-field area being about the intensity expected for α-methyl protons centred in SMM and SMS triads.

Yabumoto et al (1210) carried out an NMR study of styrene-methyl methacrylate copolymer.

Wang and coworkers (1211) studied the 300 MHz proton NMR spectra of methacrylic acid-styrene copolymers and found that the spectra contained sufficiently well resolved methine proton resonances that could be used directly for sequence distribution characterization. The monomer sequence distribution in styrene-sulphur dioxide copolymers was determined from ^{13}C NMR spectra. The spectra showed that 1:1 alternating sequences were not formed above approximately 40°C. Triad monomer sequences were resolved and assigned from the multiple resonances observed for the methine, methylene, and quaternary carbons.

Mori et al (1212) calculated the sequence distribution and stereoregularity
in radical and anionic initiated methyl acrylate-methyl methacrylate copolymers from proton NMR spectra (1212).

The microstructure of styrene-methyl methacrylate copolymers has been studied
by Brockrath and Harwood (1213) by Stroganov and coworkers (1214) and by Blouin
et al (1215) using PMR and by Kato et al (1216) and by Katritzky (1217) using
^{13}C NMR. Diad, triad, and pentad configurational distributions were evaluated
from the NMR spectra. Katritzky et al (1217) found that the aromatic C-1 in
styrene and the methyl carbon in methyl methacrylate units of the copolymer
were sensitive to tacticity in monomer sequence triads.

Yakato and Hirabayashi (1218) determined the coisotacticities of some alternating styrene-acrylic copolymers by NMR spectroscopy.

The comonomer composition of random and alternating copolymers of styrene with
ethyl methacrylate, butyl methacrylate, or octyl methacrylate has been calculated from proton NMR spectra (1219).

Pyrolysis-Gas Chromatography

Macleod (1220) showed that pyrolysis-gas chromatography of a known polymer,
containing typical acrylic monomers, methyl methacrylate, styrene, butyl methacrylate, butyl acrylate, hydroxyl propyl methacrylate, and methacrylic acid,
gave monomer yields quantitatively related to the original concentrations,
allowing calculation of the original composition.

Hummel and Düssel (1221) applied pyrolysis-gas chromatography to α-methyl styrene-methyl acrylate and α-methyl styrene-acrylonitrile copolymers.

Tsuge and Takeucki (1222) designed a furnace pyrolyzer stated to yield highly
specific quantitation and reproducible pyrograms for any form of polymer sample. The concept was applied to the characterization of diad sequence distribution in acrylonitrile-m-chlorostyrene copolymer, acrylonitrile-p-chlorosty-
rene copolymer, chlorostyrene-styrene copolymer, methyl acrylate-styrene copolymer, and acrylonitrile-styrene copolymer.

Milina and Pankova (1223) determined the acrylonitrile in copolymers with
styrene and methyl methacrylate with an error of approximately 2.2% by pyrolysis-gas chromatography.

Combined methyl methacrylate in methyl methacrylate-butadiene-styrene copolymers has been determined by pyrolysis-gas chromatography (1224).

Evans et al (1225) have carried out a compositional determination of styrene-
methacrylate copolymers by pyrolysis-gas chromatography, proton nuclear magnetic resonance spectrometry, and carbon analysis.

Evans and coworkers (1226) analyzed methyl methacrylate-styrene copolymers
and butyl methacrylate-styrene copolymers by pyrolysis-gas chromatography.

A novel method of determination of reactivity ratios in binary and ternary
copolymerizations was based on the use of gas chromatography to determine
unreacted monomers. The method was tested with polymerization of styrene and
methyl methacrylate (1227).

Thin-layer Chromatography

The compositional heterogeneity of styrene-methyl methacrylate block copolymers has been determined by thin-layer chromatography and compared with the
results obtained from cross-fractionation and a computer simulation method
(1228). The compositional distributions agreed in all three methods and were
found to be unexpectedly broad.

A combination of gel permeation chromatography followed by thin-layer chromatography of the fractions followed by pyrolysis-gas chromatography of the
thin-layer fractions has been used to study the polydispersity of block copolymers of styrene and methyl methacrylate; pyrolysis-gas chromatography
yielded quantitative data on the polymer composition of fractions previously
separated by gel permeation chromatography on the basis of size and by thin-
layer chromatography on the basis of composition (1229).

Fractionation

Methacrylic acid-styrene copolymers have been fractionated according to molecular weight and chemical composition by precipitation with methanol from
benzene solution (1230).

Methyl methacrylate-styrene copolymers have been fractionated according to
molecular weight by batch-wise and continuous column elution and by coacerv-
ation extraction techniques using different solvent-nonsolvent combinations
(1231).

Dautzenberg (1232) examined the effects of molecular weight, chain structure
and conditions for sample preparation on light scattering on moderately concentrated solutions of polystyrene and styrene methyl methacrylate copolymer.

Miscellaneous

The polarization of the phosphorescence of styrene-vinylbenzophenone copolymers and of the fluorescence of methyl methacrylate-styrene co-olymers and
methyl acrylate-styrene copolymers has been discussed by David et al (1233)
in terms of copolymer composition, low molecular weight compounds, and chain
conformation.

Luminescence studies have been reported for styrene-acrylic copolymers; relative and absolute intensities of emission spectra varied with temperature
and copolymer composition (1234).

Rafikov et al (1235) have carried out studies of styrene-methacrylic acid copolymer involving viscosity, diffusion, and light scattering experiments.

The composition heterogeneities of random methyl methacrylate-styrene copolymer (1236) and acrylonitrile-α-methylstyrene have been determined from light
scattering and osmotic pressure measurements in various solvents. The apparent molecular weight of the copolymers in all solvents agreed with each other
(1237).

Srinivasan and Santappa (1238) applied a combination of light scattering, viscosity and osmometry to butyl methacrylate-styrene copolymers.

Small-angle neutron scattering and light scattering studies of styrene-methyl
methacrylate diblock copolymers indicated that the poly(methyl methacrylate)
block forms an interior core from which the polystyrene block radiates out-
wardly as an expanded chain (1239).

Marked solvent effects have been observed by Gallo and Russo (1240) on the hypochromism of styrene-methyl methacrylate copolymer measured at 269.5nm. Intensity was inversely proportional to dielectric constant of the solvent.

Shima (1241) determined the molar polarizations and average dipole moments of a series of styrene-methyl methacrylate copolymers. the effective dipole moment of methyl methacrylate units in the copolymers increases as the percentage of styrene increases; the limiting value is close to that observed for simple esters. The results were correlated by a least squares procedure with calculated sequence distributions. Although good correlations are obtained in these studies, this approach to sequence distribution measurement has limited sensitivity. The molar polarizability of methyl methacrylate units centred in SMS triads in styrene-methyl methacrylate copolymers is only about 20% higher than that of units centred in MMM triads. The difference is even less for methyl acrylate units in styrene-methyl acrylate copolymers.

Polarographic methods have been used for differential determination of sodium and potassium in styrene-methacrylate copolymers (1242).

Monomers

Holtmann and Souren (1243) and Yamaoka et al (1244) have described methods for the determination of unreacted styrene, diallyl phthalate, and methyl methacrylate in polyester resin moldings.

Acosta and Satre (1245) have reviewed the theoretical and practical aspects of determining monomeric compositions of polymers, using a methyl methacrylate-styrene copolymer as an example, by radioactive tracer techniques.

Tweet and Miller (1246) determined monomers in ethyl acrylate-styrene copolymer emulsions by distilling off the volatile components in the presence of toluene and analyzing the distillate by gas chromatography.

Residual monomers in some polymer mixtures have been determined (3% relative) by a mercurimetric procedure; advantage was taken of differing reaction rates of unsaturated compounds and unsaturated end-groups with methanolic mercuric acetate for monitoring polymerization of styrene, methacrylic acid, and copolymers (1247, 1248).

STYRENE-POLYETHYLENE TEREPHTHALATE COPOLYMERS

Sakurada et al (1249) investigated the chemical structure of copolymers from polyethylene terephthalate fibres prepared by the radiation induced graft copolymerization of styrene. The graft copolymers were isolated from the products by extraction and characterized by hydrolysis in 35% hydrochloric acid dioxan (4:30) at 95°C for 20 hours and osmometry. Among the swelling agents employed, methanol was most effective for increasing the extent of grafting onto polyethylene terephthalate using both irradiation and preirradiation methods of grafting.The molecular weight of polystyrene formed in the substrate matrix was higher by one million if no chain-transfer agent was added to the monomer solution. The isolated graft copolymer carried only one branch per copolymer molecule in both cases. Of great interest is the particularly low extent of grafting in the case of polyethylene terephthalate-styrene. This should be attributed to the low sensitivity of polyethylene terephthalate to radiation.

Graft copolymers of styrene on poly(ethylene terephthalate) and nylon have been characterized by hydrolysis and osmometry (1250).

Hori et al (1251) used thin-layer chromatography to estimate the purity of
polyethylene terephthalate-styrene copolymers. The graft copolymers were produced by radiation grafting to produce polyethylene terephthalate-styrene.
Prior to the chromatographic development the homopolymers were removed as rigorously as possible by extract-on or selective precipitation. The characterization study proved that all the graft copolymers have one branch per molecule
on the average, the number-average molecular weight of the branches being 1
x 10^5

Polyethylene phthalate-styrene give clear chromatograms with a purity of 90%
or greater and some small contamination with homopolymer.

STYRENE-NYLON COPOLYMERS

Sakurada et al (1252) investigated the chemical structure of copolymers from
nylon fibres prepared by the radiation induced graft copolymerization of styrene. They used hydrolysis in 35% hydrochloric acid - dioxan (4:30) at $95^{\circ}C$
for 20 hours and osmometry to examine the chemical structure of these copolymers.

Hori et al (1251) used thin-layer chromatography to estimate the purity of
nylon-styrene copolymers. The graft copolymers were produced by radiation
grafting to produce nylon-styrene. Prior to the chromatographic development
the homopolymers were removed as rigorously as possible by extraction or selective precipitation. The charcterization study proved that all the graft copolymers have one branch per moleculae on the average, the number-average
molecular weight of the branches being 1 x 10^5.

STYRENE-MALEIC ANHYDRIDE COPOLYMERS

Ang and Harwood (1253) noted that phenyl absorption at 700 cm^{-1} in styrene
maleic anhydride copolymers is independent of sequence distribution ($P_{SSS} \neq
P_{MSS} = P_{MSM}$) and hence useful for determining copolymer compositions. Phenyl
absorption occurring at 760 cm^{-1} is not independent of sequence distribution,
however; the relative intensities of the 700 and 760 cm^{-1} bands (E_{760}/E_{700})
decrease as the styrene content of the copolymers decreases.

A random distribution of styrene monomer units in styrene-maleic anhydride
copolymers has been indicated by the NMR chemical shifts of four triads 1:3,
1:2, and 1:1 maleic anhydride-styrene copolymers (1254).

The molecular weight of styrene-maleic anhydride with maleic anhydride content
from 5-50 mol % has been determined and found to be comparable to values obtained from the hydrodynamic volume calibration (1255).

Goel determined K, α values in the Mark-Houwink equation of ethyl acrylate-
maleic anhydride-styrene polymers in 1,4-dioxan (1256).

STYRENE-ITACONIC ACID COPOLYMERS

High frequency titration has been applied by Douglas et al (1257) to the analysis of itaconic acid-styrene and maleic acid-styrene copolymers and ethyl
esters of itaconic anhydride-styrene copolymers. The method can also be used
to detect traces of acidic impurities in polymers and in the identification
of mixtures of similar acidic copolymers. Titration indicates that the acid

segments in the copolymers of itaconic acid-styrene, maleic acid-styrene, and the hompolymer polyitaconic acid act as dibasic acids. The method has a sensitivity that permits identification and approximate resolution of two carboxylate species in the same polymer.

High frequency titration gives a precise location of the inflection points related to the polymer carboxyl groups, and is a sensitive method for the determination of the freedom of the copolymer samples from monobasic acid impurities (comonomer acids), since mixtures of copolymer acids with monobasic and dibasic acids show definite inflection points that can be related to the individual carboxylate species present.

A titration curve (Fig. 113) is shown for a monomethyl ester of an itaconic acid-styrene copolymer.

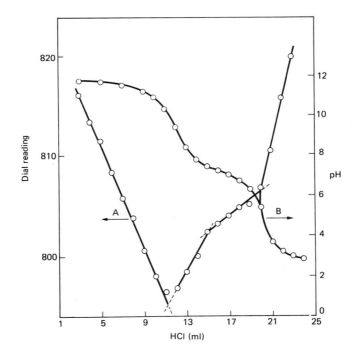

Fig. 113   High frequency (A) and potentiometric (B) displacement titration of the monosodium salts of the monomethyl esters of poly(itaconic acid-co-styrene). Titration of 0.2345 g. 57:43 anhydride-styrene copolymer + MeOH (heat) + excess NaOH with 0.1286 N HCl.

Hen (1259) has determined the concentration of acid bound at the surface of itaconic acid-styrene copolymer latexes and free acid in the water phase by conductometric titration. The concentration of acid bound within the latex particles was determined from the difference between the acid charge during polymerization and the conductometric titration data.

STYRENE-MALEIC ACID COPOLYMERS

Garrett and Guile (1260) have described a potentiometric titration technique
using standard hydrochloric acid for the estimation of carboxyl groups in the
sodium salts of maleic acid-styrene copolymers. Difficulties were encountered
by these workers and later by Bamford and Barb (1261) in locating with precision the inflection points in the potentiometric curves. These difficulties
were partially overcome by using more dilute solutions (0.2 g/180 ml).

STYRENE-ACRYLATE COPOLYMERS

Chemical Methods

Functional groups can be incorporated in polymers by copolymerization with
monomers bearing the specific groups. Estimation of the specific groups thus
introduced in the copolymers provides an easy and simple determination of copolymer composition and monomer reactivity ratio. This has been done for carboxyl-bearing monomers (1263) by using a dye-interaction method of estimation
of carboxyl groups in copolymers.

Saha et al (1262) and Palit (1264) have developed a dye-partition method for
the determination of halogen atoms in copolymers of styrene, methyl methacrylate, methyl acrylate, or vinyl acetate with a chlorine-bearing monomer such
as allyl chloride and tetrachloroethylene. The quaternized copolymers were
quaternized with pyridine then precipitated with petroleum ether or alcohol
and further purified by repeated precipitation from their benzene solutions
with a mixture of alcohol and petroleum ether as the nonsolvent. The finally
precipitated polymers were then washed with petroleum ether and dried in air.
The test for quaternary halide groups in polymers was carried out with a reagent consisting of disulphine blue dissolved in 0.01 M hydrochloric acid and
the colour evaluated spectrophotometrically at 630 nm. Saha et al (1262) and
Palit (1264) found that there may be some uncertainty in the quantitative
aspects of this method. Nevertheless, the experimental r_1 values for styrene
allyl chloride (1267) and styrene-tetrachloroethylene (1268, 1269) copolymers
are in close agreement with the literature values, while those for MMA-allyl
chloride (1270) and vinyl acetate-allyl chloride (1271) are higher than the
values given in the literature.

Quaternization of polyvinylpyridine with butyl bromide has been extensively
studied by Fuoss (1265, 1266, 1272) and it has been shown that the extent of
quaternization in this system is not always complete and varies from about
70 to 100%.

Monomers

A method involving solution in a solvent followed by gas chromatography has
been described (1273) for the determination of ethyl acrylate and styrene
monomers in this copolymer.

STYRENE-MALEATE COPOLYMERS

The pyrolysis-gas chromatography of styrene-butyl maleate copolymer has been
studied by Cobler and Samsel (1274). These workers showed that in this pyrol-

ysis copolymer, chain-scission competes with the non-chain scission reaction. Butyl alcohol, which is split off the maleate ester, is the main component. This may not be a non-chain scission reaction in the classical sense since it is possible that butyl maleate monomer is formed first and then cleaved. Although styrene is also formed, the yield is considerably lower than would be expected from the total amount present.

STYRENE-DIVINYL BENZENE COPOLYMERS

Infrared Spectroscopy

Parrish (1275) related the capacity of sulphonated divinylbenzene-styrene copolymers to the amount of styrene produced on pyrolysis and to the absorbance of the infrared peak at 1008 cm^{-1}.

The presence of 1,2,4-substituted C_6H_6 rings in carboxylated styrene-divinylbenzene cation-exchange resins has been confirmed by infrared analysis (1276).

Miscellaneous

The approximate amount of cross-linking in ion-exchange resins such as styrene-divinylbenzene copolymers has been estimated from the nature and relative amounts of the pyrolysis products (1277, 1278).

Blasius and Haeusler (1279) has used pyrolysis-gas chromatography to estimate the degree of cross-linking, type of functional group, and position of substitution in styrene-divinylbenzene copolymer ion-exchange resins.

Oehme et al (1280) studied the structure of styrene-divinylbenzene copolymers substituted with functional groups (e.g. -CH$_2$CN, -CH$_2$OEt, -CH$_2$Bu) in the para position by a combination of pyrolysis-infrared, pyrolysis-gas chromatography-infrared and pyrolysis-gas chromatography-mass spectrometry.

Benoit et al (1281) have used small angle neutron scattering to study the structure of networks formed by anionic block copolymerization of styrene with divinylbenzene.

Gel permeation chromatography combined with viscometry has been used to analyze branching in styrene-divinylbenzene copolymer (1282).

Monomers

A method has been described (1283) involving pyrolysis and mass spectrometry for the determination of low levels of divinylbenzene monomer in styrene-divinylbenzene copolymers.

STYRENE-ALKENE COPOLYMERS

Kobayashi et al determined the NMR spectrum of propylene-styrene copolymers (1284).

Eda et al (1285) applied electron spin resonance spectroscopy to studies on graft copolymerization of gaseous styrene onto preirradiated polypropylene.

Forette and Rozek (1286) analyzed styrene-isobutylene copolymers using infrared, nuclear magnetic resonance spectroscopy and carbon and hydrogen determinations.

Yamashita and coworkers (1287) used high-resolution NMR to show that styrene-isobutene and α-methylvinyl methyl ether-methyl vinyl ether copolymers had a random structure.

The structure of vinylcyclohexane-styrene copolymer has been determined by infrared spectroscopy (1288). Experimental data showed that activity of the following monomers to copolymerization was in the order: propylene > 4-methyl-1-pentene styrene > 3-methyl-1-butene > vinylcyclohexane.

COPOLYMERS OF STYRENE AND HALOGENATED MONOMERS

Cabasso et al (1289) have examined styrene-vinylidene chloride copolymers using pyrolysis-mass spectrometry.

The sequence distribution of dyads in several vinyl-type copolymers has been investigated by pyrolysis-gas chromatography, including styrene-chlorostyrene copolymers (1290).

Soluble chloromethylated polystyrene and copolymers with vinylidene chloride or poly(phenyl oxides) brominated in the side chains and ring have been characterized by NMR (1291). The halogenated polymers were reacted with alkyl phosphites to form pendent phosphonate groups and studied by NMR, thermogravimetry, and mass spectrometry.

Turbidimetric titration has been used to define systems capable of fractionating chlorinated PVC-styrene graft copolymer. Chlorobenzene-methyl alcohol was preferred (1292).

Regel and coworkers (1293) used ^{19}F NMR and PMR to study the syniotactic and heterotactic triads in poly(p-fluoro-α-methylstyrene) copolymers. The triad distribution determined by ^{19}F NMR agreed with the distributions determined from PMR of the α-methyl protons even though the fluorine nucleus was relatively far removed from the main chain.

Work and Trehu (1294) developed a general expression for the average dipole moments of polar-polar and polar-nonpolar copolymers in terms of sequence distribution, the dipole moments of equivalent isolated units present, and the average cosines of the angles between neighbouring dipoles in the copolymers. The dielectric polarization of p-chlorostyrene copolymers was then correlated via this expression with calculated sequence distributions, and the average cosines of the angles between various dipoles in the copolymers were estimated. Excellent agreement is obtained between theory and experiment when only nearest-neighbour dipolar interactions are treated.

STYRENE-1-ACRYLOXYAMINOANTHRAQUINONE COPOLYMER

Molecular weight distribution studies have been reported for the structurally coloured copolymer styrene-1-acryloxyaminoanthraquinone (1295).

STYRENE-GLYCIDYL-P-ISOPROPENYL ETHER COPOLYMERS

Aliev (1296) determined the composition of styrene-glycidyl p-isopropenylphenyl ether copolymers using the optical density ratios of the infrared absorption bands at 1250 and 1490 cm^{-1}.

STYRENE AMYLOSE COPOLYMERS

Patel and Patel (1297) used chemical, instrumental and physical methods for characterization of styrene-amylose graft copolymer.

STYRENE METHACROLEIN COPOLYMERS

Quantitative infrared in the 4000-450 cm^{-1} region has been used to characterize free radical styrene-methacrolein copolymers (1298). Of particular value were the carbonyl band at 1722 cm^{-1} for free aldehyde groups, styrene band at 540 cm^{-1} related to shortening of sequence length of styrene units, and styrene band at 760 cm^{-1} related with distribution of MSM triads and copolymer structure.

STYRENE-VINYL ALCOHOL COPOLYMERS

Typical chromatograms of polyvinyl acetate-styrene copolymers graft copolymers are demonstrated in Fig. 114. The thin-layer chromatography was always carried out not only for the graft copolymer sample, but for mixtures of known quantities of sample and homopolymer to evaluate quantitatively the amount of the homopolymer contaminating the starting graft copolymer sample.

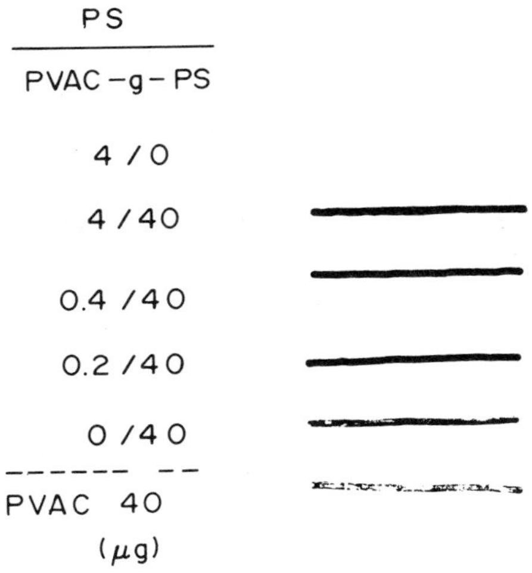

Fig. 114 Thin-layer chromatograms of the polystyrene homo-
          polymer, the polyvinyl acetate homopolymer, the
          isolated polyvinyl acetate-g-polystyrene, and
          mixtures of polystyrene and polyvinyl acetate-
          g-polystyrene with different mixing ratios.
          Developer: chloroform.

Few efforts have been directed to the isolation of a pure graft copolymer from a grafting product (1299-1306). Density-gradient ultracentrifugation (1307) has been used but is a very complicated technique.

It has been reported that thin-layer chromatography makes it possible to char-

acterize polymers with respect to the differences in composition (1308-1313)
monomer arrangement (1314), steric isomerism (1315-1319) and molecular weight
(1320-1324).

Inagaki et al (1315) observed that a properly chosen solvent does not develop
block copolymers on a chromatographic plate, whereas the corresponding homopolymers and statistical and alternating copolymers have a high R_f value (rate
of flow). Thus, thin-layer chromatography seems to be a suitable technique for
distinct estimation of the purity of polyvinyl alcohol-styrene graft copolymers.
Hori et al (1251) applied thin-layer chromatography to acetylated graft copolymers and the purity was evaluated by comparing the chromatogram of the graft
copolymer sample with that of graft copolymer-homopolymer mixtures of given
mixing ratios.

Pure graft copolymers have been effectively isolated from their reaction product by a column adsorption chromatography on silica gel (1325).

STYRENE-VINYL ACETATE

Ebdon et al (1326) have studied the effects of overall composition and monomer sequence distribution on the infrared carbonyl stretching frequency of sty-
rene-vinyl acetate copolymers.

STYRENE-UNSATURATED ALDEHYDE COPOLYMERS

Svec et al (1327) have described a method for determining carbonyl groups in
styrene- $\alpha\beta$-unsaturated aldehyde copolymers.

Miscellaneous

Solvent systems for light scattering and also suitable for optical and thermo-
dynamic studies have been proposed for styrene copolymers (1328).

An on-line infrared detector has been used to determine compositional distribution of styrene copolymers (1329, 1330).

POLY-α-METHYL STYRENE

Ultraviolet Spectroscopy

Bogomolova et al (1331) employed ultraviolet spectroscopy to follow interactions of styrene, α-methyl styrene, and polystyrene in complexes with stannic
chloride.

Nuclear Magnetic Resonance Spectroscopy

Woodward (1332) has utilized NMR spectroscopy in his study of relaxation phenomena in poly-α-methyl styrene.

The NMR spectrum of poly-α-methyl styrene has been discussed by Elgert et al.
(1333).

Mass NMR infrared, and kinetic data were obtained by Richards and Wilhams (1334)
to determine the structure of the tetramer of α-methylstyrene. The various proton resonances in the NMR spectra of poly(p-isopropyl-α-methylstyrene) have
been assigned by Leonard and Malhotra (1335) to isotactic, heterotactic, and

syndiotactic triads.

Tacticity information has been published on ^{13}C NMR of poly(α-methylstyrene)
and polybutadienes (1337).

Pyrolysis-Gas Chromatography

Cobler and Samsel (1338) showed by pyrolysis-gas chromatography that when poly-
α-methylstyrene is pyrolyzed at 300°C under combustion boat conditions a small
amount of α-methylstyrene is produced together with the small amounts of styrene. As the temperature is increased to 410°C and then to 495°C the amount
of α-methylstyrene produced increases. At these latter temperatures however,
significant quantities of styrene, toluene, benzene, and other volatile products
are formed.

Thermal Methods

McNeill (1339) has discussed the application of the technique of thermal volatilization analysis to poly-α-methylstyrene.

Thin-layer Chromatography

Thin-layer chromatography has been used by Geymer (1340) to separate mixtures
of poly-α-methylstyrene with molecular weights of 14,000, 86,000 and 730,000.
Methylethyl ketone was found to be the best migration solvent for this separation. Figure 115 shows what can be achieved using sequential migration. A microlitre of 2% decahydronapthalate solutions of three polymers were applied to
this plate and eluted with ethyl acetate. After drying the plate in a stream of

Fig. 115   Chromatograph of three poly(α-methylstyrene frac-
           tions and a mixture of the three.
           Starting point is at the left. Sequential migra-
           tions were used. The molecular weights of the
           fractions are 14 thousand, 86 thousand and 730
           thousand. Adsorbent: Kieselguhr. Migration Sol-
           vent. 7cm with 10% chloroform, 90% ethyl acetate.
           10 cm with ethyl acetate. 12 cm with acetone.
           Polymer initially dissolved in decalin, 2% for fractions
           fractions, 6% for mixture.

nitrogen, a further migration was accomplished using acetone. It can be seen
in Fig. 115 that the lowest molecular weight fraction has the highest R_f of
the three.

Fractionation

Attainable resolution in gel permeation chromatography using polystyrene gel particle packing of about 5μ diameter and mixtures of two monodisperse poly-methylstyrenes has been investigated by Kato and coworkers (1341).

Molecular Weight

Bradley (1342) applied light scattering methods to the estimation of the Z-average molecular weight of poly-α-methylstyrene. He describes a light scatter-ing method for the estimation of the Z-average molecular weight which is applic-able to random coil polymers in the molecular weight range between 0.2×10^6 and 2×10^6 and for which a suitable solvent is available. The solvent may be either a theta solvent or a good solvent for which the relation between intrin-sic viscosity and molecular weight is known.

Klenin et al (1343) have used spectrophotometric titration in studies of the molecular weight distribution of polymethylstyrene obtained under various polymerization conditions.

Chaula and Huang (1344) have carried out molecular weight distribution studies in the radiation induced polymerization of liquid α-methylstyrene.

Miscellaneous

Neutron scattering has provided fundamental torional frequency data for heter-otactic, syndiotactic and head-to-head isomers of poly-α-methylstyrene (1345).

Monomers and Low Polymers and Oligomers

Yamamoto et al (1346) reported mass spectra for α-methylstyrene dimer, trimer, and other oligomers prepared in living systems.

Gankina and coworkers (1347) have discussed the feasibility of separation of oligomers according to molecular weight using thin-layer chromatography based on the difference in absorption activity of the terminal and central units in the macromolecule. They provide illustrations for the separation of oligomer-ic polystyrene and poly(α-methylstyrene) without terminal functional groups.

POLY-α-METHYLSTYRENE COPOLYMERS

Zizin et al (1348) determined the composition of styrene, α-methylstyrene, and butadiene-styrene-α-methylstyrene block copolymers by pyrolysis-gas chrom-atography using mixtures of the corresponding homopolymers for standardization.

Gel permeation chromatography has been used to evaluate copolymers, styrene or α-methylstyrene and butadiene (1349).

α-METHYLSTYRENE-BUTADIENE COPOLYMERS

Elgert and coworkers (1350, 1351) reported structural information on α-methyl-styrene-butadiene copolymers from 220-MHz NMR data.

α-METHYLSTYRENE-STYRENE COPOLYMERS

The codimerization of styrene with α-methylstyrene has been studied by Zwierzak and Pines (1352) who gave retention data for 15 dimers produced in the reaction.

The composition of styrene-α-methylstyrene copolymers has been calculated from the ratio of the extinction coefficients of the absorption bands (1353).

Baras and Juveland (1354) have carried out an NMR study of α-methylstyrene-p-methyl-α-methylstyrene copolymers.

α-METHYLSTYRENE-METHYLENE COPOLYMERS

Shirakawa, Yamazaki and Kambara (1355) used ionic polymerization techniques to prepare α-methylstyrene-methylene block copolymers having controlled sequence lengths. Many bands in the infrared spectra

$$\{ \ -(\alpha\text{-methylstyrene})_m \ - \ (CH_2)_x \ - \ \}_n$$

(600-1300 cm^{-1}) of these copolymers were noted to increase or decrease in intensity with increasing length of the α-methylstyrene sequences. The spectra of - methylstyrene blocks in the copolymers are essentially identical to those of poly(α-methylstyrene) samples of comparable size. These workers concluded that planar C-H deformation of monosubstituted benzene rings increase in intensity as the number of α-methylstyrene units in a sequence increase. The opposite seems to hold for out of plane deformations.

α-METHYLSTYRENE POLYISOPROPENYL CYCLOHEXANE COPOLYMERS

Gritten et al (1356) have used pyrolysis-gas chromatography and pyrolysis-mass spectrometry to study the structure of this copolymer.

CHAPTER 5

POLYVINYLCHLORIDE, POLYVINYLIDENE CHLORIDE AND OTHER CHLORINE CONTAINING POLYMERS

POLYVINYLCHLORIDE

Chemical Methods

Labile chlorine has been determined in PVC by ultraviolet spectral analysis of the phenolysis reaction product (1358).

Mitterberger and Gross (1359) determined chlorine in·PVC by fusion with sodium peroxide and silver nitrate titration of the chloride produced.

Tanaka and Morikawa (1360) described a semimicro technique for the determination of total chlorine in PVC using a semimicro method based on the Schöniger flask and Fajans method.

Low temperature ashing was used by Narasaki and Umezawa (1361) for controlled decomposition of poly(vinylchloride).

Getmanenko and Perepletchikoya (1362) determined hydroperoxides in benzene-methanol-tetrahydrofuran solutions of PVC by oxidation of Fe^{2+} of Mohr's salt to Fe^{3+} and formation of the o-phenanthroline coloured complex.

Petiaud (1363) observed that very low molecular weight PVC contained 1-chloro-2-pentene groups which can be hydrochlorinated to form 1,2-dichloroethylene. The polymer also contained 2 to 3 double bonds per 1000 mer units.

Girgis-Takla (1364) have discussed the determination of lead in PVC.

Purdon and Mate (1365, 1366) have described a method for the determination of the gel content of poly(vinylchloride) based on the dissolution of the polymer in a suitable solvent followed by the separation of the solution containing the soluble fraction of the polymer, from the insoluble gel which are separated by centrifugation and weighed off. This method is inaccurate for polymers with low gel contents.

Rogozinski and Kramer (1367) were interested in determining the gel contents of poly(vinylchloride) and vinyl chloride-propylene copolymers of low gel content and therefore sought a more direct method than that of Purdon and Mate (1365, 1366). The method developed is based on the repeated centrifugation of the gel with fresh portions of solvent in order to wash the gel completely free of soluble polymer. Three such extractions proved sufficient. This leaves a solvent-swollen gel, and the weight of the gel can then be determined after removal of the solvent. Rogozinski and Kramer washed the solvent out of the gel with isopropanol which is miscible with the solvent but is not a solvent for the polymer. this causes the swollen gel to collapse and re-form the normal non-swollen polymer particle, which is readily filtered from the liquid and weighed. This technique gives better results than the Purdon-Mate (1365, 1366) technique for polymers with low gel contents. For polymers with high gel contents the Purdon-Mate method appears to be more accurate.

Ultraviolet Spectroscopy

Braun (1368) has analyzed the visible ultraviolet spectrum of a mixture of polymers formed by degradation of poly(vinylchloride). These groups are believed to be responsible for the discolouration of poly(vinylchloride) upon thermal degradation.

Marks et al (1369) considered in a qualitative way, how the absorption spectrum of degraded poly(vinylchloride) is formed by overlap of the spectrum of a mixture of polyenes.

Daniels and Rees (1370) described a more sophisticated computer based spectrum matching procedure which enables the concentration of the individual polymers to be determined from the broad absorption spectrum of degraded poly(vinylchloride).

In Table 53 are shown ultraviolet-visible spectral details for a range of polyenes containing between 3 and 14 conjugated double bonds. These workers determined polyene concentrations in degraded PVC by ultraviolet spectra. They obtained extinction coefficients on known polyenes, observing that dienes usually have only one absorption maximum; trienes, three; higher polyenes, four.

Table 53   Ultraviolet/Visible Spectral Details (λ and ε) for Polyenes H-$(CH=CH)_n$-H where n varies from 3 to 14.

n	λ_1, nm	λ_2, nm	λ_3, nm	λ_4, nm	λ_5, nm
3		240	248	257(42.7)	268
4		267	278	290(78.6)	304
5	279	290	303	317(121)	334
6	300	313	328	344(138)	364
7	316	332	350	368(177.1)	390
8	332	349	367	386(210.0)	410
9	345	363	380	404(242.8)	430
10	358	376	397	420(275.6)	447
11	370	390	412	435(308.4)	461
12	382	402	422	448(341.6)	476
13	392	412	433	462(374.0)	490
14	402	424	447	474(406.8)	505

a_ε values x 10^{-3} in parentheses.

Infrared Spectroscopy

Krimm et al (1371) carried out infrared spectra and assignments for PVC and
its deuterated analogues. Krimm and Enomoto (1372) and Enomoto et al (1373)
carried out infrared studies of chain conformations in PVC.

Germar (1374) has published an infrared study of the microstructure of chlorinated poly(vinylchloride).

In the region of 600-700 cm^{-1}, the infrared spectrum of poly(vinylchloride)
shows bands assigned to C-Cl stretching vibrations. The intensities of these
bands depend on the conformational and configurational structure and on the
crystallinity of the sample, and they have been used for tacticity determinations (1375-1380).

Schneider et al (1381) showed that the absorption around 690 cm^{-1} was proportional to the number of isotactic diads, and in the region of 600-640 cm^{-1} to
syndiotactic diads. Based on this finding they proposed a method for determining the tacticity of amorphous samples of PVC. As some samples cannot be easily
transformed into an amorphous state Schneider et al (1382) devised an infrared
method of tacticity determination which is independent of the crystallinity of
the samples. From the temperature dependence of infrared spectra of poly(vinylchloride) samples prepared by different methods, the intensity of the band at
690 cm^{-1} (proportional to the number of isotactic diads in the sample), as well
as that of the tacticity-independent C-H stretching band, was found to be independent of the crystallinity of the sample. These lines were applied for the
tacticity determination in poly(vinylchloride), measured in the form of potassium bromide pellets. The numerical tacticity value was obtained from the known
values of absorbance coefficients of S_{CH} and S_{HH} type C-Cl stretching bands in
solution, and from the shape of the spectrum.

Park (1383) has discussed the determination of branching in poly(vinylchloride).

Enomoto et al (1384) carried out spectroscopic studies on poly(vinylchloride)
and its deuterated derivatives.

Baker et al (1385) improved on the method for quantitative estimation of chain
branching in PVC, based on use of the CH_3 deformation band at 1378 cm^{-1}. These
investigators determined the extinction coefficient by use of carefully calibrated polyethylene containing known numbers of C_1, C_2, and C_4 side chains.
Measurements were made on the melt at 150°C to eliminate crystallinity effects.
Millan and Dela Pena (1386) reviewed infrared and NMR methods for identification of stereoisomers in PVC.

Stanciu (1387) used infrared analysis to show that the macromolecular orientation of poly(vinylchloride) was affected by thermal and solvent treatment.

The degree of branching of PVC was determined by Bezadea et al (1388).

Raman Spectroscopy

Laser-Raman spectroscopy has been reviewed by Brandmueller and Schroetter (1389)
including studies on poly(vinylchloride).

The bands at 1124 and 1511 cm^{-1} in the Raman spectra of PVC, after thermal degradation were shown by Gerrard and Maddams (1390) to be due to a resonance
Raman effect from conjugated polyene sequences.

Maddams (1391) used Raman spectroscopy to characterize polyene sequences in degraded PVC samples and made assignments of the C-Cl stretching modes.

Nuclear magnetic resonance Spectroscopy

High-resolution nuclear magnetic resonance spectroscopy has proved to be an
effective tool for studying microstructure of chlorine-containing polymers
(1392-1404). Even though much of the work reported deals with studies on the
stereochemical configuration of poly(vinylchloride) (1392-1396), a number of
workers have been involved with studies on sequence distributions of chlorine-
containing copolymers (1397-1404). Three of these studies (1397-1399) involved
the microstructure investigation of vinylidene chloride-vinyl chloride copolymers. The NMR results reported from these studies showed that (1) even at high
concentrations of vinylidene chloride some vinyl chloride sequences were observed and (2) two types of copolymers were found. One of the two types contained sequences of head-to-tail and head-to-head structure of vinylidene chloride and sequences of vinyl chloride and vinylidene chloride. the other type
contained vinyl chloride sequences in addition to those mentioned in the first
type.

Besides the NMR studies reported on vinylidene chloride-vinyl chloride copolymers sequences distribution measurements have been reported on other kinds of
chlorine-containing copolymers. Among them were vinylidene chloride-isobutylene
copolymers (1401), vinylidene chloride-vinyl acetate copolymers (1402, 1403)
and vinyl chloride-ethylene copolymers (1400, 1404). For the first two kinds
of copolymers, the sequence distribution measurements which were made in terms
of diad or triad sequences showed that (1) for vinylidene chloride-isobutylene
copolymers the NMR determination was as accurate as the chemical analysis and
(2) for vinylidene chloride-vinyl acetate copolymer, a single NMR measurement
could be used to specify the monomer sequence distribution. It was also found
for the latter system that as the polymerization proceeds, the monomer sequence
distribution becomes broader.

Schaefer (1400) reported that vinyl chloride-ethylene copolymers prepared by
free-radical polymerizations show a monomer distribution that is random or
zeroth-order Markovian. However, Wilkes et al (1404) showed that for vinyl
chloride-ethylene copolymers prepared in bulk the copolymerization is not random and is first-order Markovian.

^{13}C NMR has been shown to be a useful tool for detecting and measuring the
distribution of stereochemical configurations that can occur in poly(vinylchloride) (1405).

Carman (1406) has shown that ^{13}C NMR is an informative technique for measuring stereochemical sequence distributions in poly(vinylchloride). He reported
chemical shift sensitivities to configurational tetrad, pentad and hexad placements for this polymer. The NMR spectrum of PVC is shown in Fig. 116. The
methine resonances in PVC occur as rr, mr and mm from low to high field. As
shown in Fig. 116, the methylene and methine carbons of poly(vinylchloride)
show a sensitivity toward configuration and complete, internally consistent
assignments have been given by Carman (1406).

Abe et al (1407) have investigated the NMR spectra of model compounds of poly
(vinylchloride) in the hope that these investigations may offer useful information for the analysis of vinyl polymer spectra. They studied the NMR spectra
of three stereoisomers of 2,4,6-trichloroheptane as model compounds of poly
(vinylchloride). Spectra were observed at 60 and 100 Mc/sec., both at room
temperature and at high temperatures and spin-decoupling experiments were performed. The difference of the chemical shifts of the two meso methylene protons

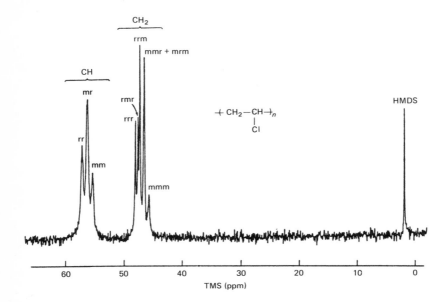

Fig. 116   C-13 NMR spectrum at 25.2 MHz of PVC.

at 60 Mc/sec. was found to be ca. 7 cps. for the isotactic three unit model while it was ca. 16 cps. for the isotactic two-unit model or heterotactic three -unit model. The spectra of poly(vinylchloride) can be interpreted reasonably on the basis of this result. Observed values of vicinal coupling constants of model compounds were interpreted as the weighted means of those for several conformations, and the stable conformations of the models were determined.

Satoh (1408) determined the undecoupled and decoupled high resolution nuclear magnetic resonance study of poly(vinylchloride).

Tho and Taieb (1409) studied the NMR spectrum of PVC - tetrahydrofuran and PVC n-butyraldehyde complexes.

Chemical shifts of PVC and model compounds such as meso- and racemic-2,4-dichloropentane have been measured from NMR spectra (1410).

Abe and Nishioka (1411) measured diad tacticities in PVC; triad and pentad, in poly(vinylacetate). Unsaturated end-groups in PVC have been estimated from NMR spectra, using reference model compounds such as 1-chloropentene-2, 1,1-dichloropentene-2, and 1,2-dichloropentene-2 (1413).

The differential enthalpy and entropy changes for bulk PVC have been estimated from triad tacticity data obtained from the pentad measurements (1414).

High resolution NMR spectra of PVC were used by Cavalli (1415) to provide conformational information.

Repko and coworkers (1416) used wide band NMR to determine the crystallinity of PVC and poly(vinylidene chloride). the crystallinity for PVC agreed well with that determined by x-ray analysis. The crystallinity of poly(vinylidene chloride)calculated from NMR data was not in agreement with that determined by x-ray analysis.

Bovey and coworkers (1417) demonstrated that the branch structure of PVC was -CH$_2$CHClC(CH$_2$Cl)HCH$_2$CHCl- by ^{13}C NMR after reduction of the polymer with lithium aluminium deuteride.

Fourier transform proton NMR studies of PVC indicated that the unsaturated end-groups contained allylic chlorine atoms. The presence of the unsaturated end-groups would partially explain the PVC degradation (1418).

Caraculacu and Bezdadea (1419) have used Fourier Transform Proton NMR spectroscopy to determine unsaturated structures in PVC.

Electron Spin Resonance Spectroscopy

Some workers have concluded that under low extent of thermal dehydrochlorination, poly(vinylchloride) did not show any evidence of the presence of radicals. Accordingly, evidence cited for a radical mechanism in this polymer based on electron spin resonance (ESR) measurements of a recorded singlet signal had little justification. The investigation of Liebman et al (1420) of carefully purified PVC and systems related to PVC gives evidence which contradicts the above conclusion.

Liebman et al (1420) utilized electron spin resonance spectroscopy and thermogravimetric studies in the investigation of the thermal decomposition of poly (vinylchloride) and chlorinated poly(vinylchloride). These workers examined powdered samples of poly(vinylchloride), a chlorinated PVC series with varying chloride content, and a reference sample of poly(vinylchloride) by ESR and thermogravimetric analysis. the temperature of the initially detected ESR signal, its generation and decay rates under certain conditions, and the corresponding temperature for initial weight loss and maximum dehydrochlorination rate from derivative thermogravimetric data were recorded for these samples. In addition, apparent activation energies and frequency factors were calculated and dependence of the former on conversion were described. Their data on PVC and chlorinated PVC systems demonstrated the existance of macroradicals in the early stage of thermal decomposition under inert and oxidative atmospheres.

Ohnishi et al (1421) carried out an ESR study of radiation oxidation of PVC. Ouchi (1422) and Suzuki (1423) measured the ESR spectrum of heat treated PVC.

Takikai et al (1424) carried out ESR studies of reaction intermediates in γ-irradiated poly(vinylchloride).

Pyrolysis-Gas Chromatography

Poly(vinylchloride) upon pyrolysis in a combustion furnace undergoes non-chain scission reactions (1425). Non-chain scission reactions are common to short chain esters or halogenated molecules which have a hydrogen atom attached to the carbon atom beta to the substituted group. These molecules dissociate into an olefinic structure and an acid. the main pyrolysis products of poly(vinylchloride) are volatile products, including hydrochloric acid and benzene (Fig. 117). The benzene is formed as the result of a cyclization of the unsaturated chain ends.

Feuerberg and Weigel (1426) have carried out pyrolysis-gas chromatography of PVC. They pyrolyzed polymer samples (500 mg) in a glass tube at 620°C and then

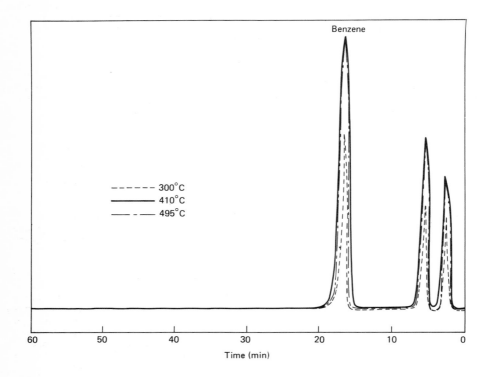

Fig. 117   Gas chromatogram of pyrolyzate of poly(vinylchlor-
           ide).

chromatographed the condensed pyrolyzate at 150°C on a 2 metres long column,
using di-n-decyl phthalate as the stationary phase.

The mechanism of the thermal dehydrochlorination of poly(vinylchloride) has
been the subject of many studies in the general area of polymer degradation.
A number of the important contributions in this area have been reviewed by
Levy (1427, 1428).

Stromberg (1429) has studied the thermal degradation of poly(vinylchloride)
in vacuum by first stripping the resin of hydrogen chloride at temperatures
up to 350°C for 30 min. The resulting residue was then heated at 400°C for
30 min. and the volatile hydrocarbon products resulting from this second heating were identified via mass spectroscopy. In this particular study, Stromberg
identified approximately 25 hydrocarbon products.

Ohtani and Ihikawa (1430) pyrolyzed PVC at 425°C in a nitrogen atmosphere and
studied the resulting degradation products by infrared and ultraviolet spectroscopy. Their findings showed that: (1) the pyrolysis products (other than hydrogen chloride) consisted of aliphatic and aromatic hydrocarbons, and (2)
the types of pyrolysis products from PVC were a function of the stereoregularity (tacticity) in the ungraded polymer. Additional qualitative or quantitative
measurements were not made.

Noffz et al (1431) degraded PVC in a high frequency pyrolyzer and separated the degradation products by gas chromatography. Their findings revealed that the hydrocarbon degradation products consisted of only aromatic compounds; a volatile aliphatic hydrocarbon fraction was not identified.

Coleman and Thomas (1432) have studied the combustion of PVC by thermally degrading the polymer in an excess of oxygen. Under the experimental conditions employed, the hydrocarbons formed as a result of polymer degradation were completely oxidized to carbon oxides. In addition to hydrogen chloride, carbon monoxide and carbon dioxide, a trace amount (5 ppm) of carbonyl chloride was also identified (1433).

Boettner et al (1434, 1435) found that when PVC homopolymer and compounds were heated in air from ambient to 600°C, approximately 95% of the PVC was reacted to hydrogen chloride, carbon monoxide and carbon dioxide (no phosgene above 0.1 ppm). the 5% hydrocarbon fraction consisted of 18 aliphatic and aromatic hydrocarbons.

Analogous degradation experiments of PVC plastisols (PVC plus a phthalate ester) revealed that the hydrocarbon fraction of the pyrolyzate increased due to products from thermal breakdown of the plasticizer. Characteristic degradation products derivable from the plasticizer alone were not formed.

O'Mara (1436, 1438) has described a pyrolysis-gas chromatographic-mass spectrometric technique for analyzing the pyrolysis products from PVC in helium atmosphere. This technique has provided qualitative and semi-quantitative analysis of the pyrolysis products from these materials. PVC resin yields a series of aliphatic and aromatic hydrocarbons when pyrolyzed at 600°C; the amount of aromatic products is greater than the amount of aliphatic products. Benzene is the major organic degradation product. A typical PVC plastisol (PVC/o-dioctyl phthalate (100/60)) yields, upon pyrolysis, products that are characteristic of both the PVC matrix and the phthalate plasticizer. The pyrolysis products from the plasticizer dilute those from the PVC portion of the plastisol and are, in turn, the major degradation products. There are no degradation products resulting from an interaction of the PVC with the plastisol. The pyrograms resulting from pyrolysis of the various plastisols of PVC can be used for purposes of "fingerprinting". Identification of the major peaks in a typical plastisol pyrogram provides information leading to a precise identification of the plasticizer.

O'Mara (1436, 1438) carried out thermolysis of PVC resin(Geon 103, Cl=57.4%) by two general techniques. the first method involved heating the resin in the heated (325°C) inlet of a mass spectrometer in order to obtain a mass spectrum of the total pyrolyzate.

The second, more detailed method consisted of degrading the resin in a pyrolysis-gas chromatograph interfaced with a mass spectrometer through a molecule enricher (1437).

Samples of PVC resin and plastisols (10-20 mg) were pyrolyzed at 600°C in a helium carrier gas flow. Since a stoichiometric amount of hydrogen chloride is released (58.3%) from PVC when heated at 600°C, over half of the degradation products, by weight, is hydrogen chloride.

A typical pyrogram of a PVC resin obtained by this method using an SE32 column is shown in Fig. 118. Component identifications are summarized in Table 54. The components obtained using the Poropak QS column are asteristed in Table 54. The major components resulting from the pyrolysis of PVC are hydrogen

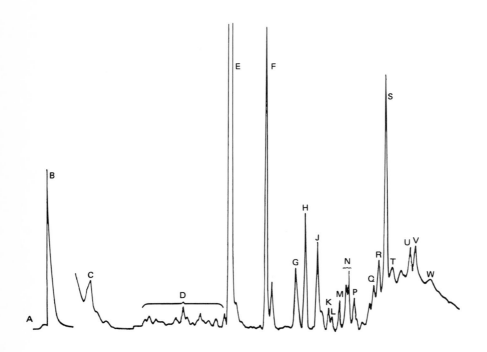

Fig. 118   Pyrogram of PVC resin from 20ft x 3/16 in. 10%
           SE32 on 80/100 CRW chromatographic column. Lett-
           ered peaks refer to identifications in Table 54.

chloride, benzene, toluene and naphthalene. In addition to these major products, an homologous series of aliphatic and olefinic hydrocarbons ranging
from C_1 to C_4 are formed.

O'Mara (1436, 1438) obtained a linear correlation between weight PVC pyrolyzed and weight of hydrogen chloride obtained by gas chromatography. (Fig. 120)
In Fig. 119 is shown a pyrogram obtained for various PVC-phthalate plastisols
and resins from which it can be seen that the technique is very useful for
identifying such materials by "fingerprinting".

Frye et al (1440) in his studies on the stabilization of PVC by organotin
esters has shown that the ester moiety enters the polymer backbone during
thermal (175°C) stabilization with the formation of organotin chlorides. Frye
and Horst (1441) had shown that barium, cadmium, and zinc 2-ethylhexanoates
undergo a similar interaction with PVC during stabilization. O'Mara (1439)
applied his technique to an investigation on the absorption of hydrogen chloride by inorganic fillers, ($CaCO_3$, CaO, $Al(OH)_3$, Na_2CO_3, Al_2O_3, $LiOH$) in PVC
compounds during pyrolysis and combustion processes. The inorganic fillers
were dispersed in a PVC-dioctyl phthalate plastisol and fused at 150°C. Hydrogen chloride evolution from these matrices during pyrolysis at 600°C and during combustion was measured gas chromatographically.

O'Mara analyzed a number of PVC compounds of known composition. A summary of

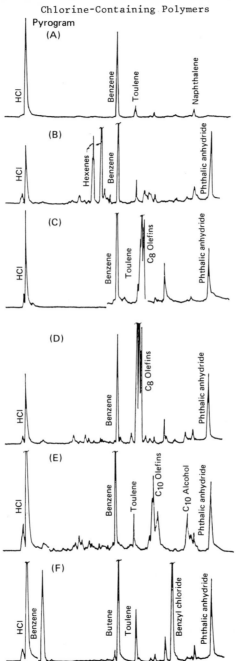

Fig. 119   Pyrograms of (A)PVC resin;(B)PVC/o-dihexyl phthal-
           ate;(C)PVC/o-di-2-ethylhexyl phthalate;(D)PVC/o-
           dicapryl phthalate;(E)PVC/o-diisodecyl phthalate;
           (F)PVC/o-butyl benzyl phthalate.

Table 54   Identification of Components in Pyrogram (Fig.
            118) of PVC.

Peak Identification   Components

A	CH_4 * CO, * CO_2 *, C_2H_4, * C_2H_6 *
B	HCl, C_3H_6, * C_3H_8 *
C	Butane[a], butene[a], butadiene[a], diacetylene[a]
D	C_5 and C_6 aliphatic and olefinic hydrocarbons
E	Benzene
F	Toluene
G	Chlorobenzene
H	Xylene
J	Allylbenzene
K	C_9H_{12}
L	C_9H_{12}
M	Indane
N	Indene, ethyltoluene
P	Methylindane
R	Methylindenes
S	Naphthalene
T	Dimethylindane
U	Methylnaphthalene
V	Methylnaphthalene, acenaphthalene
W	Dimethylnaphthalene

* Separated and identified on an 8ft. Poropak QS.
HCl ~ 58.3%, ash 3-4%.

the compounds studied, the percent hydrogen chloride found, and the theoretical
hydrogen chloride content is contained in Table 55. This analysis has been extended to chlorinated polyethylenes and chlorinated poly(vinylchloride) and in
general show good agreement between expected and determined values.

The thermal decomposition of poly(vinylchloride) has usually been represented
as involving the dehydrochlorination of the polymer backbone. On the other
hand, Stromberg (1442) assumed the release of chlorine molecules in his kinetics of the decomposition of poly(vinylchloride). The isolation of small amounts
of hydrogen and chlorine was reported by Tsuchiya (1443) on pyrolysis of poly
(vinylchloride) in an inert gas. Ohta (1444) suggested that a part of the hydrogen chloride released from poly(vinylchloride) may recombine with the double
bonds along the chain introduced by the dehydrochlorination. After pyrolysis
of poly(vinylchloride) many kinds of hydrocarbons consisting of aliphatics and
aromatics in addition to hydrogen chloride were detected by Stromberg and other
workers (1445-1448).

The nature of these hydrocarbons has been further investigated by Iida et al
(1449). These workers pyrolyzed poly(vinylchloride), preheated poly(vinylchloride) and poly(vinylchloride) containing various metal oxides. The pyrolysis
products were identified by using gas chromatography and included aromatic
hydrocarbons such as benzene, toluene, ethylbenzene, or various aryl chlorides.
Different proportions of volatile products were obtained from the various pyrolyses of poly(vinylchloride). Therefore, it was believed by these workers, that
the pyrolysis products did not depend on the original structure of poly(vinylchloride) but on the final structure of polymer releasing their small molecules.
In particular, aryl chlorides may be derived from the recombination of double

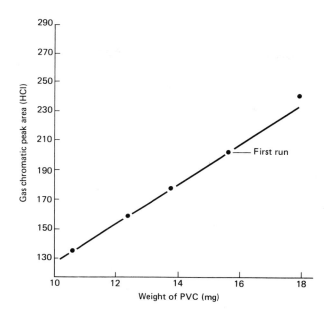

Fig. 120   Relationship of hydrogen chloride chromatographic
           peak area and amount of PVC pyrolyzed at 600°C.

Table 55   Pyrolytic Analysis of HCl from PVC Compounds,
           Polyethylene and Chlorinated Poly(vinylchloride).

Sample type	%HCl (found)	%HCl (theoretical)
PVC plastisol	30.1	29.1
"	37.0	37.3
"	44.7	44.5
PVC-vinyl acetate copolymer	53.5	52.9
"	46.3	46.3
PVC compound	38.5	38.6
PVC compound	47.5	47.7
chlorinated polyethylene	34.9	34.3
"	40.7	41.1
chlorinated poly(vinylchloride) *	63.9	69.2
"	68.5	71.5

* Impure material, theoretical recovery not expected.

bonds with chlorine as hydrogen chloride or chlorine assuming that no secondary
reactions of the volatile decomposition products take place. Poly(vinylchloride)
samples with or without 10 wt.% and 50 wt-% TiO_2, SnO_2, ZnO, and Al_2O_3 were
pyrolyzed at 200-800°C. Gas chromatograms of pure poly(vinylchloride) and poly
(vinylchloride) with 10 wt-% SnO_2 pyrolyzed at 700°C are shown in Fig. 121
and 122 and illustrate the effect of active metal oxide on the course of decomposition of poly(vinylchloride). At 200°C no hydrocarbons were detected in

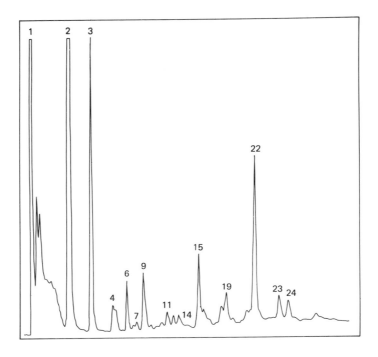

Fig.121   Pyrogram obtained at 700°C of untreated poly(vin-
          ylchloride).

pyrolysis products in any of the samples. There are 11 peaks at 300°C, 20 peaks
at 400°C, 26 peaks at 500°C and more at the higher temperatures in the pyrograms of pure poly(vinylchloride). The percentage ratio γ of each peak height
to the sum of all peak heights is shown in Table 56. No variation of peak height ratio γ was observed with pyrolysis periods of 15 sec. and 20 min., respectively. In pyrolyzing PVC, an elevated temperature favours formation of aliphatic hydrocarbons, while aromatics are more easily released than aliphatics
at lower temperatures.

It has been recognized since the work of Cotman (1450) that by studying the
polyolefin obtained from the reduction of PVC with lithium aluminium hydride,
valuable structural information can be gained concerning the starting molecule.
This reduction reaction has been investigated and refined (1451) to the point
where conditions have been established so that the chlorine may be efficiently
removed from the polymer without degradation. The reduced polymer is similar
to high density polyethylene in almost all respects (1452). Thus, studies that
have been applied to polyethylene may also be applied to reduced PVC. Indeed
better qualitative agreement has resulted when γ-ray radiolysis followed by
identification of the gaseous hydrocarbons by mass spectrometry used in conjunction with infrared measurements is applied to reduced PVC than when this
technique is applied to conventional polyethylene. It was concluded from the
large yields of methane relative to butane and ethane that the predominant
side chains along the PVC backbone are mainly one carbon long. It has been
demonstrated by ^{13}C NMR that most of the short branches in PVC are pendent

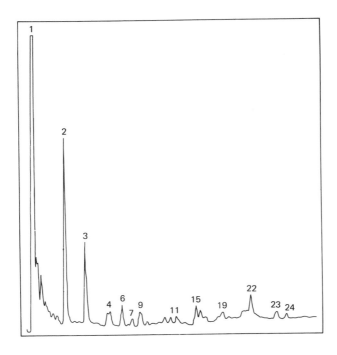

Fig. 122   Pyrogram obtained at 700°C of poly(vinylchloride)
-SnO$_2$ system (10 wt-% SnO$_2$).

chloromethyl groups (1453). This information was obtained from PVC samples reduced with lithium aluminium hydride and lithium aluminium deuteride respectively.

Ahlstrom et al (1454) applied the techniques of pyrolysis-gas chromatography
and pyrolysis-hydrogenation-gas chromatography to the determination of short-
chain branches in PVC and reduced PVC. Their attempts to determine the short-
chain branches in PVC by pyrolysis-gas chromatography were complicated by an
inability to separate all of the parameters affecting the degradation of the
polymer. Not only does degree of branching change the pyrolysis pattern, but
so do tacticity (1455) and crosslinking (1456).

In order to eliminate some of the above parameters and improve on the, then
existing methods of polymer characterization these workers examined the pyrolysis of polyethylene and studied several PVCs and lithium aluminium hydride-
reduced PVCs differing in the amount of branch content, as obtained by infrared
spectroscopy and ^{13}C NMR, but not in tacticity.

For the pyrolysis of PVC, a ribbon probe was used. The polymer was dissolved
in tetrahydrofuran and a few microlitres of the solution was deposited on the
platinum ribbon. the tetrahydrofuran was allowed to evaporate at a low temperature, and the remaining polymer film was pyrolyzed. On-line hydrogenation of
the pyrolysis products was accomplished by using hydrogen as the carrier gas
with 1% palladium on Chromosorb-P catalyst inserted in the injection port liner.

Table 56   Y Values of Pyrolysis Products

#	Peak Component	Untreated		Preheated		10% TiO₂ (Y)		10% SnO₂		10% ZnO		10% Al₂O₃	
		500°C	700°C	500°C	700°C	500°C	700°C	500°C	700°C	500°C	700°C	500°C	700°C
1	Aliphatic hydrocarbon	3.60	31.87	5.07	33.26	9.35	42.39	32.70	64.46	24.26	60.27	7.76	37.25
2	Benzene	70.23	36.74	71.68	38.35	61.65	31.87	34.57	18.49	41.16	15.20	61.62	32.04
3	Toluene	4.98	9.87	5.40	8.14	6.98	10.57	7.24	6.79	6.03	6.67	7.67	10.46
4	Ethylbenzene	0.76	0.81	0.80	0.50	1.09	0.91	2.04	0.44	1.94	1.12	1.20	0.96
5	A	0.48	0.70	0.00	0.51	0.87	0.97	1.38	0.92	0.00	1.40	0.85	0.86
6	o-Xylene	1.12	1.48	1.33	1.33	1.93	1.78	2.46	1.16	1.46	1.13	1.75	1.57
7	Monochlorobenzene	0.37	0.14	0.30	0.00	0.42	0.12	0.57	0.69	0.69	0.18	0.47	0.00
8	B	0.22	0.27	0.32	0.13	0.42	0.34	1.04	0.30	1.11	0.80	0.47	0.33
9	Styrene	1.63	1.85	1.52	2.08	2.06	1.32	2.06	1.45	0.97	0.75	2.14	2.02
10	C	0.08	0.14	0.12	0.12	0.16	0.19	0.37	0.24	0.42	0.48	0.15	0.19
11	Vinyltoluene	0.47	0.50	0.39	0.55	0.76	0.42	0.49	0.60	3.60	0.36	0.70	0.58
12	D	0.64	0.40	0.67	0.11	0.90	0.38	0.73	0.23	2.15	1.20	0.96	0.35
13	E	0.59	0.33	0.57	0.14	0.71	0.31	0.79	0.09	0.28	0.68	0.83	0.38
14	p-Dichlorobenzene	0.09	0.06	0.08	0.00	0.20	0.06	0.45	0.14	1.94	0.13	0.16	0.04
15	o-Dichlorobenzene												
	Indene	1.58	2.46	1.40	3.19	2.15	1.68	0.70	1.25	2.91	1.70	2.15	2.56
16	F	0.40	0.29	0.53	0.00	0.74	0.71	1.10	0.30	0.97	1.25	0.75	0.16
17	1,3,5-Trichlorobenzene	0.25	0.11	0.39	0.09	0.33	0.19	0.59	0.05	1.11	0.37	0.31	0.00
18	G	0.52	0.39	0.38	0.45	0.85	0.22	0.25	0.23	1.39	0.48	0.82	0.44
19	1,2,4-Trichlorobenzene	1.35	0.66	1.26	0.70	1.33	0.55	0.36	0.60	0.83	0.75	0.31	0.76
20	H	0.13	0.06	0.20	0.00	0.25	0.07	0.18	0.13	0.44	0.50	0.21	0.13
21	I	0.70	0.27	0.26	0.28	0.55	0.32	2.50	0.15	1.18	0.36	0.78	0.36
22	Naphthalene	6.22	7.28	5.39	7.06	3.82	3.46	4.79	4.26	3.19	2.04	4.51	5.75
23	α-Methylnaphthalene	0.94	1.40	0.99	1.16	0.97	0.77	1.11	0.86	1.94	1.22	1.06	1.22
24	β-Methylnaphthalene	0.73	1.14	0.66	1.03	0.63	0.62	0.78	0.88	0.00	0.59	0.78	1.01
25	J	0.18	0.19	0.14	0.09	0.15	0.00	0.00	0.00	0.00	0.00	0.18	0.13
26	K	0.50	0.48	0.45	0.50	0.40	0.00	0.00	0.00	0.00	0.22	0.45	0.47
27	L	0.00	0.00	0.08	0.13	0.00	0.00	0.00	0.00	0.00	0.34	0.00	0.00

A-L unidentified materials.

The pattern of the pyrograms of all of the reduced PVCs were similar on the SCOT column (C_9-C_{18} hydrocarbons). However, quantitative differences in the total branch content were detected. On this column, maximum triplet formation occurred at C_{14} for low density polyethylene and for reduced PVC, and at C_{15} for high density polyethylene. The occurrence of the peak maxima at C_{14} for reduced PVC indicates that the total branch content is higher than that of high density polyethylene. But aside from the C_{14}, C_{15} peak maxima difference, the pyrolysis pattern for even the most highly branched PVCs resembles high density polyethylene more than low density polyethylene. These data indicate that the type of short-chain branch in PVC is qualitatively more like that of high density polyethylene, but that the sequence length between branch sites is shorter in low density polyethylene and PVC. Since low density polyethylene contains a large amount of ethyl and butyl branches and PVC and high density polyethylene contain mainly methyl and some ethyl branches, this qualitative resemblance would be expected. A relative measure of the total amount of short-chain branches for these polymers can be obtained by calculating the percent of branched products formed (Table 57).

Table 57   Relative Total Branch Content of High Density
           Polyethylene, LAH reduced PVC and Low Density
           Polyethylene

Sample	Branched products,%
high density polyethylene	12.0
reduced PVC[a]	19.0
low density polyethylene	26.0

a average value

More information about the specific type of short-chain branch in PVC can be found from an examination of the C_1-C_{11} hydrocarbons (Fig. 124). Here quantitative differences between the reduced PVCs becomes more apparent. The most obvious differences occur in the amounts of iso-C_7 and iso-C_8 products formed which indicate differences in the total branch content. As the amount of short-chain branching in the reduced PVCs increases, there is a decrease in the amount of isoalkanes formed (Table 58). The data in Table 58 show small but distinguishable differences in the short-chain branch content of the reduced PVCs.

Ahlstrom et al (1454) also studied the pyrolysis-hydrogenation-gas chromatography of lithium aluminium hydride reduced PVCs which were structurally very similar to high density polyethylenes. The fact that polyethylene is a good model for the study of reduced PVC can be seen from a comparison of the pyrolysis patterns of reduced PVC with high and low density polyethylene,(Fig. 123 and 124). As with polyethylene, on-line hydrogenation of the pyrolysis products greatly simplifies the pyrogram (compare Fig. 124 and 125), thus simplifying the interpretation of the pyrogram. An examination of the overall pyrolysis pattern in the C_1-C_{11} region, coupled with a determination of the 3-methyl-/2-methylalkane ratios for these polymers, indicates that the short-chain branches in PVC are mainly methyl with some ethyl groups. No large increase in the relative amount of the n-C_4 peak was observed for the PVC samples as was the case with low density polyethylene. However, the possibility of the presence of minor amounts of propyl, butyl, or longer chain branches cannot be completely discounted from this data, since some differences in the iso-C_8, iso-C_9 and iso-C_{10} products were also observed. The most linear reduced PVCs, as determined by pyrolysis-gas chromatography (Pevikon R-341 and Nordforsk E-80), have slightly greater 3-methyl/2-methylalkane ratios than does high density polyethylene (0.96 g/cm^3). This ratio generally increased for the more highly bran-

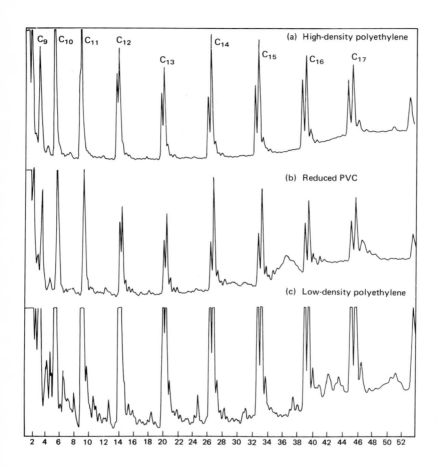

Fig. 123　Pyrolysis of polyethylenes and reduced PVC: (a)
high density polyethylene; (b) reduced PVC; (c)
low density polyethylene. SCOT column; fragments
C_9-C_{18}.

ched polymers.

Unlike polyethylene, the effect of polymer microstructure on the hydrocarbon
pyrolysis products obtained from PVC has not been well defined in the literature. O'Mara (1457), Noffz (1458), and recently Chang and Salovey (1459) have
identified the pyrolysis products by gas-chromatographic/mass spectrometric
techniques. The major products formed are the C_1-C_5 alkenes and alkanes, benzene, toluene, and naphthalene. Other aromatic and substituted aromatic compounds make up the minor products (Table 59). The order of branching, determined from pyrolysis of the corresponding lithium aluminium hydride-reduced
PVCs and the number of methyl groups/1000 carbon atoms determined from the
infrared measurements, is included in Table 59 for comparison. As the amount
of branching along the polymer backbone increases, fragmentation increases

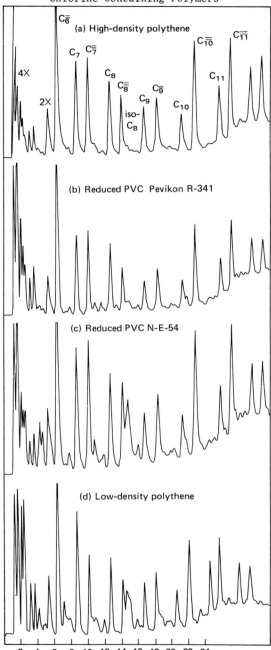

Fig. 124 Pyrolysis of polyethylenes and reduced PVC;(a) high-density polyethylene; (b) reduced PVC Pevi kon R-341; (c) reduced PVC N-E-54; (d) low density polyethylene. Durapak column fragments C_1-C_{11}.

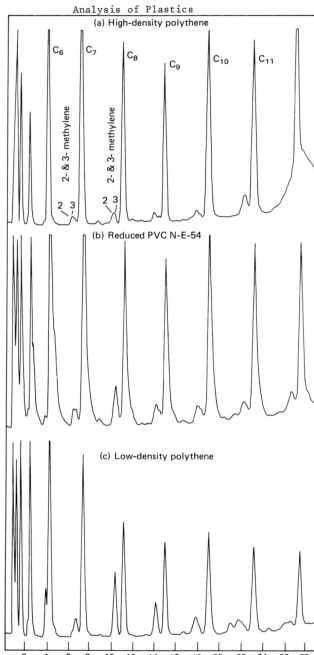

Fig. 125 Pyrolysis hydrogenation of polyethylenes and reduced PVC;)a)high density polyethylene;(b)reduced
PVC N-E-54;(c)low density polyethylene. Durapak
column; fragments C_1-C_{11}.

Table 58   Short-Chain Branch Content in LAH-Reduced PVC.

Sample	iso-$C_8^=$/n-$C_8^=$	$C_{10}^=$, %
high density polyethylene	0.1	8.6
Pevikon R-341	0.52	7.7
Nordforsk E-80	0.71	7.2
Nordforsk S-80	1.20	7.0
Nordforsk S-54	1.59	6.8
Nordforsk E-54	1.63	6.5
Ravinil R100/650	1.63	6.5
Shin-Etsu TK1000	1.87	6.3
low density polyethylene	1.83[a]	5.2

a This ratio does not include C_4 branch content.

Table 59   PVC Pyrolysis Products.

Sample	C_1-C_5 hydrocarbons wt-%	Benzene/toluene ratio	Benzene wt-%	Naphthalene wt-%	CH_3/1000 CH_2[b]	Relative branch order from reduced PVC a
Nordforsk E-80	14.1	11.2	59.9	5.7	4.1	2
Nordforsk S-80	14.0	11.1	59.3	5.6	4.1	3
Pevikon R-341	14.8	10.7	57.9	5.6	5.4	1
Nordforsk S-54	14.9	10.5	57.7	6.0	7.8	4
Nordforsk E-54	16.9	9.7	56.2	4.7	8.0	5
Shin-Etsu TK1000	16.0	9.2	55.1	3.9	5.6	6

a Relative to 1 = least branched by PGC
b Infrared

and results in the formation of increased amounts of C_1-C_5 hydrocarbons. A further consequence of reduced sequence length is to decrease the amounts of benzene and naphthalene formed. The specific effect of branch type on the pyrolysis products has not been determined; however, if it is assumed that the formation of toluene is at least in part an indication of methyl branching, then the benzene/toluene ratio should serve as an indicator of the amount of methyl branching in PVC. It can be seen from the data in Table 59 that if the above criteria are used as an indication of the degree of branching, then the two pyrolysis methods and the infrared method agree within the limits of the experimental methods.

Thermal Methods

McNeill (1460) has discussed the application of the technique of thermal volatilization analysis to poly(vinylchloride).

Danforth and Takenchi (1461) and Bataille and Van (1462) and Gupta and St Pierre (1463, 1464) have studied the mechanism of the thermal degradation of PVC.

Guyot et al (1465) have studied the thermal degradation of PVC in the presence of a second polymer. Chany and Salovey (1466) have studied the pyrolysis of PVC.

Gedemer (1467) used derivative thermo-gravimetric analysis to estimate the degree of thermal degradation of PVC.

Bosche (1468) used differential thermal analysis to identify PVC.

Iida et al (1469 have pointed out that recombination of chlorine atoms released
from the polymer chain with the double bonds in the polyene chain takes place
during the thermal decomposition of poly(vinylchloride), since chlorobenzenes,
ethylbenzene, and o-xylene were detected in the pyrolysis products from this
polymer.

During the thermal degradation of poly(vinylchloride) in nitrogen, discolouration is observed even at low conversions. The colour has been ascribed to the
formation of polyene sequences long enough to absorb in the visible region,
i.e. a length exceeding about five units. As the discolouration is noticeable
at very low conversions, relatively long polyene sequences must be formed at
the early stages of the reaction. There is general agreement that the average
polyene sequence length decreases with increasing degradation temperature (1470,
1471). Estimates of the average polyene sequence length, however, differ considerably (1470-1476). Several procedures for determination of the average
polyene sequence lengths have been suggested (1472-1479). The use of ultraviolet-visible spectroscopy makes it possible to study the distribution of the
polyene lengths between 3 to 14. Spectra of PVC degraded in nitrogen show 10
to 11 not very well resolved peaks, which have been ascribed to certain polyene sequence lengths (1472, 1475, 1479-1486). There is, however, some contro-
versy as to the exact wavelength where the different polyenes absorb. Braun
and coworkers have derived a method to measure the relative concentration of
different polyenes, showing the distribution of the polyene sequences between
4 and 12 (1471, 1473, 1478). In another extensive study, Tüdös and coworkers
(1475, 1476, 1487-1489) treated the kinetics of the polyene sequence formation
by a probabilistic approach. From this they were able also to derive the trun-
cated polyene sequence length distribution. They found a very low average polyene sequence length (1471, 1472) for degradation at about 200°C and observed
that the polyene sequence lengths were distributed geometrically. the results
were obtained for a plasticized system.

Abbas and Lawrence (1490) have outlined the general features of the dehydrochlorination kinetics and compared the resulting polyene sequence distribution with the experimental results from ultraviolet-visible measurements. They
proposed a kinetic model for the formation of polyene sequences during the
thermal degradation of 190°C of poly(vinylchloride) in nitrogen. the model includes a propagation step of the "zipper" kind and termination by crosslinking.

Abney et al (1491) have studied the thermal degradation of PVC.

Ahlstrom and Fultz (1492) used pyrolysis-gas chromatography and derivative
thermogravimetric analysis in their thermal degradation studies of PVC.

X-ray Diffraction

Small-angle scattering studies have been made on PVC plasticized with dioctyl
phthalate (1493).

Small-angle x-ray scattering of PVC has been interpreted in terms of a two-
phase system containing ordered and disordered regions (1494).

Gouinlock (1495) described methodology and apparatus for the determination of
degrees of order in highly crystalline PVC by wide-angle x-ray diffraction.

Annealing of highly syndiotactic PVC at a series of increasing temperatures, above 180°C, caused the crystallinity to increase steadily as indicated by x-ray diffraction measurements. However, the infrared spectra did not change indicating that C-Cl bands cannot be used as a measure of crystallinity (1496).

Gaidarova et al (1497) used small-angle x-ray diffraction to determine the packing density of the super molecular structure of PVC.

Radioactivity

Buriana et al (1498) determined the allyl chlorine content of PVC by measurement of the ^{36}Cl radioactivity of the PVC after exchange with $SO^{36}Cl_2$ in tetrachloroethane at 60°C.

Schroeder and Byrdy (1499) converted PVC to polyethylene by treatment with lithium aluminium hydride and determined the short-chain branching by gas chromatography of the radiolytic reaction products.

Fractionation

Jisova et al (1500) have discussed the fractionation and solution properties of high molecular weight PVC.

Lin et al (1501) used gel permeation chromatography to study PVC degraded by heat.

Ambler and Mate (1502) assessed osmotic pressure measurements for $\overline{M}_n$ for various molecular weight ranges on PVC. When $\overline{M}_n$ was greater than 10,000 osmotic, M_n and gel permeation chromatography results agreed. Gel permeation chromatography combined with thin-layer chromatography has been used for characterization of aqueous gel networks (1503) and for quantitative molecular weight determination on linear and branched components in polystyrene (1504).

Molecular Weight

Daley (1505) has applied gel permeation chromatography to the determination of the molecular weight distribution of PVC.

Freeman and Manning (1506) discussed the molecular weight of PVC.

Viscosity-MN relationships have been established for PVC in cyclohexanone solution (1507) and in cyclohexanone-ethylene glycol (1508).

Cantow et al (1509) used turbidimetric titration to determine the molecular weight of PVC.

Janca and Kalinsky (1510) have reviewed the determination of the molecular weight distribution of PVC and its copolymers by gel permeation chromatography, especially the problems of calibration and effect of aggregates.

Nakoa and Kuramoto (1511) obtained molecular weight and molecular weight distribution data for PVC by gel permeation chromatography.

Miscellaneous

Kahn et al (1512) determined conjugated polyenes in solid PVC by selective photooxidation.

Time-of-flight mass spectrometric techniques have been employed to analyze volatile products from mechanical degradation of PVC (1513).

Reviews on PVC have discussed morphology of suspension PVC (1514), PVC characterization and properties of suspensions (1515).

Composition analysis of products and toxicity studies have been reviewed for PVC (1516).

Venediktov et al (1517) have described ultrasonic apparatus for determining viscoelasticity of PVC.

Volatiles

Liao (1518) determined polynuclear aromatic hydrocarbons in PVC smoke particles by high pressure liquid chromatography and gas chromatography-mass spectrometry.

A method involving head space analysis-gas chromatography has been described (1519, 1520) for the determination of monomers and volatiles in PVC.

Monomers

Headspace methods for the determination of residual monomers and volatiles in polymers has been reported by Steichen (1521) and by Shanks (1522). Steichen determined residual vinylchloride, butadiene, acrylonitrile, styrene, and 2-ethylhexyl acrylate in polymers with an analytical sensitivity of 0.05 to 5 ppm by analysis of the equilibrated headspace over polymer solutions. Shanks (1522) determined acrylonitrile, α-methyl styrene, and styrene monomers by headspace analysis over heated solid polymer samples.

Baba (1523) determined residual vinylchloride monomer from 0.02 to 0.1 ppm in PVC by gas chromatographic-mass spectrometric monitoring.

A gas chromatographic-mass spectrometric method was developed by Gilbert et al (1524) for the determination of ppm quantities of vinylchloride in plasticized and unplasticized PVC.

Water

A procedure involving devolatilization of the polymer followed by gas chromatography has been applied to the determination of low concentrations of water in PVC (2795).

Additives

Antioxidants. Procedures involving solvent extraction followed by thin-layer chromatography have been described for the determination of phenolic and amine types of antioxidants (1525) and of optical whiteners in PVC. A procedure involving solution in dimethyl formamide followed by anodic voltammetry has been described (1526) for the determination of Ionol and 4,4'isopropylidene diphenol in PVC.

Gel permeation chromatography and thin-layer chromatography have been used for analysis of antioxidant and stabilizer additives in polyolefin and PVC food wrappings (1527).

Stabilizers. The determination of various types of stabilizers in PVC is reviewed in Tables 60 to 62.

Table 60   Organotin Type Stabilizers in Poly(vinylchloride).

Thin-layer chromatography

Dibutyltin-dilaurate	solvent extraction,	
Dioctyltin-dilaurate	thin-layer chromatography	
Butyltintrichloride	for identification	
Dimethyltindichloride	of organotin compounds	(1528)
Diphenyltin dichloride		
Hexabutylditin		
Tributyltin laurate		
Dibutyltin bis-(2-ethyl-		
hexylthioglycollate)		
dibutyltin bis(2-ethylhexyl)	solvent extraction,	
thioglycollate	thin-layer chromatography	(1528)
di-n-octyltin maleate	solvent extraction,	
	thin-layer chromatography	(1529)
stabilizers	various methods	(1530-1535)
dibenzyltin bis(isooctyl-	solvent extraction,	
mercaptoacetate)	hydrolysis to glycol,	
dibenzyltin bis(butyl-	thin-layer chromatography of	
mercaptoacetate)	3,5 dinitrobenzoates	(1536)
dibenzyltin bis(cyclohexylmer-		
captoacetate)		

Titration methods

Stabilizers	titration of carboxylate	
	end-groups with sodium	
	methoxide in pyridine	(1537, 1538)
Stabilizers	determination of sulphur by	
	potentiometric titration with	
	potassium periodate in glacial	
	acetic acid	(1539)
dialkyltin thio compounds	potentiometric titration with	
	silver nitrate in aqueous iso-	
	propanol. Also, identification	
	by infrared spectroscopy	(1537)

Miscellaneous methods

Stabilizers	heptane-acetic acid extraction	
	various instrumental finishes	(1540)
dialkyltin types	identification by infrared	
dialkylthioglycollates	spectroscopy and chemical methods	
dialkyltin laurylmercaptides		(1537)
e.g. dibutyltin oxide, dioctyl-		
tin oxide, dibutyltin maleate		
dioctyltin laurate, dioctyltin		
thioglycollate		

Table 61   Non-tin Type Stabilizers in Poly(vinylchloride).

Thin-layer chromatography

3 amino crotonic ester of 2,2' thiodiethanol	mascerate with water, acetic acid-ethanol or heptane, then thin-layer chromatography	(1541)
Phenylamine 2 phenylindal dicyandiamide aminocrotonic esters	solvent extraction, thin-layer chromatography	(1542-1545)
Stabilizers	solvent extraction, thin-layer chromatography	(1546)
epoxy types alkyl epoxy steanates epoxidized soyabean oil epoxidized linseed oil	solvent extraction, thin-layer chromatography	(1547)
epoxy types epoxidized soyabean oil epoxidized linseed oil isooctylepoxysteanate 2-ethylhexylepoxystallate butyl epoxy tallate butyl ester of epoxy linseed oil di(isoamyl)4,4,epoxytetrahydro-phthalate	methanol extraction, thin-layer chromatography	(1548, 1549)

Miscellaneous methods

Stabilizers	solvent extraction, glc	(1550-1552)
Stabilizers	diethyl ether extraction, various instrumental finishes	(1553)
diphenylthioureas 2-phenylindole	methanol or diethyl ether extraction, spectrophotometric finish	(1554)

Plasticizers. Mittenberger and Gross (1563) described methods for separating pigments, fillers, stabilizers and lubricants from PVC.

Vasil'yanova et al (1564) found that a 1:10 solvent mixture did not dissolve low molecular weight PVC and allowed maximum extraction of plasticizers such as dibutyl phthalate and tricresyl phosphate.

Kohman and Ciecierski (1565) determined the concentration of plasticizers in PVC from specific volume/concentration calibration curves.

The determination of plasticizers in PVC is reviewed in Table 63.

Initiators. Residual initiators in suspension PVC were estimated by polarography following extraction in benzene-methanol (1615). A sensitivity of 0.0001% was reported using 5-g samples.

Table 62   Metal Stearate Stabilizers in Poly(vinylchloride).

metal stearates	solvent extraction, thin-layer chromatography	(1555)
Ba Cd stearates Zn	solvent extraction, column chromatography	(1556)
Cd Pb stearates Zn	solvent extraction, paper chromatography	(1557)
Cd Pb stearates Zn	solvent extraction or ashing, polarography	(1558)
Cd Pb stearates Zn	solvent extraction, infrared spectroscopy	(1559)
Na K stearates Ba	solvent extraction, chemioluminescence techniques	(1560)
Cd Zn stearates and laurates Ba	ashing and elemental analysis by polarography	(1561)
Na K Ba stearates Cd Pb	ashing, flame photometry	(1562)

Table 63   Plasticizers in PVC.

Solvent extraction methods

e.g. dioctylphthalate tritolylphthalate	diethyl ether extraction, weighing	(1566,1567)
plasticizers	solvent extraction, weighing	(1568)
polypropylene adipate dioctylphthalate tritolylphthalate	diethyl ether then methanol extraction, weighing	(1569)
polypropylene sebacate polypropylene adipate	methanol extraction	(1570)

Gas chromatography

plasticizers	solvent extraction, glc	(1551, 1552, 1571, 1574)
alkyl adipate alkyl phthalate	solvent extraction, glc	(1573, 1574)

Table 63   (continued)

Benzylbutylphthalate	solvent extraction, glc	(1575)
alkyl phthalates fatty acid esters	solvent extraction, glc	(1576,1577)
dimethylphthalate dimethyl sebacate triacetin diacetin diethylphthalate	dichloromethane extraction, glc	(1578)
adipate and phthalate type	mild pyrolysis-gas chromatogr- aphy	(1579)
Plasticizers	solvent extraction, gas chromatography	(1580)
Plasticizers	hydrolysis to alcohols, gas chromatography	(1581)
dibutylphthalate diisobutylphthalate bis(2-ethylhexyl)phthalate di-n-octyl phthalate	pyrolysis-gas chromatography	(1582,1583)
alkyl phthalates alkyl adipates alkyl azealates alkyl citrates alkyl sebacates alkylglycollates alkyl phosphates	tetrahydrofuran extraction, glc (includes determination of alcohol degradation products)	(1584)
Plasticizers	pyrolysis-gas chromatography of polymer	(1585)

Thin-layer chromatography

Plasticizers	dimethoxymethane extraction, thin-layer chromatography	(1586)
phthalic acid esters sebacic acid esters phosphoric acid esters epoxy esters amino crotonic esters diphenylthio adipic acid type	solvent extraction, thin-layer chromatography for identifica- tion and determination	(1587-1606)
chlorinated paraffin waxes tricresyl phosphate dibutyl succinate di-n-octylphthalate bis(2-ethylhexyl)phthalate tributyl phosphate esters and carboxylic acids and	solvent extraction, glc	(1607)

Table 63 (continued)

dicarboxylic acids
diethyl phthalate
dimethylsebacate

phthalate type	solvent extraction, thin-layer chromatography and infrared identification (1608)

Infrared and NMR spectroscopy

Polypropylene adipate Polypropylene sebacate tritolylphosphate Bisoflex 791 Bisoflex 220 Bisoflex 795	extraction diethyl ether then carbon tetrachloride then with methanol followed by infrared spectroscopy (1609)
tricresylphosphate dibutyl succinate di-n-octylphthalate bis(2 ethylhexyl)phthalate tributylphosphate esters of carboxylic acids, dicarboxylic acids diethylphthalate dimethylsebacate	solvent extraction, infrared spectroscopy (1607)
diiso-octylphthalate	solvent extraction, NMR spectroscopy (1610)

Miscellaneous methods

Plasticizers	benzene extraction, measurement of saponification value (1611)
Plasticizers	diethyl ether extraction, various instrumental finishes(1612)
Phthalate plasticizers	extraction with boiling hydrochloric acid, polarography (1613,1614)

PVC COPOLYMERS

VINYLCHLORIDE POLYETHYLENE OXIDE COPOLYMERS

Laventy and Gardlund (1616) have discussed the characterization of these polymers.

VINYLCHLORIDE-VINYL ACETATE COPOLYMERS

Infrared spectroscopy has been applied to the determination of free and combined vinyl acetate in vinylchloride-vinyl acetate copolymers (1617). This method is based upon the quantitative measurement of the intensity of absorp-

tion bands in the near infrared spectral region arising from vinyl acetate.
A band at 1.63µ, due to the vinyl group, enables the free vinyl acetate content of the sample to be determined. A band at 2.15µ is characteristic for
the acetate group and arises from both free and combined vinyl acetate. Thus,
the free vinyl acetate content having been determined at 1.63µ, the combined
vinyl acetate content may determined by difference at 2.15µ. **Polymerized viny-
l chloride** does not influence either measurement.

VINYL ACETATE-VINYL CHLORIDE COPOLYMERS

Small amounts of carbonyl groups in PVC and vinyl chloride-vinyl acetate copolymers have been determined following formation of the 2,4-dinitrophenyl-
hydrazones by mineralization with sulphuric acid and photometrically determining nitrogen content using Nessler's reagent (1618).

Radical reactivities have been determined by ESR of copolymerization of vinyl
acetate and vinyl chloride,ethylene, vinylidene chloride, acrylonitrile, and
acrylic acid (1619).

Comonomer ratio in films of vinyl chloride-vinyl acetate copolymers have been
determined by infrared spectroscopy. Additives have been determined by gas
chromatography (1620).

Chia and Chen (1621) showed that vinyl chloride-vinyl acetate copolymers could
be pressed without decomposition between hot copper plates at 110 -120°C in
preparation for infrared analysis.

Terpolymers of vinyl chloride-vinyl acetate-vinyl fluoride gave all the infrared absorption bands of the chloride-acetate copolymer plus peaks for the
fluoride (1622) and also specific NMR signals.

Campbell (1623) determined combined vinyl acetate in vinyl acetate-vinyl chloride and vinyl acetate-ethylene copolymers by an isotope dilution derivative
method. The vinyl acetate was hydrolyzed with sodium hydroxide in methyl ethyl
ketone in the presence of a known amount of $^{14}CH_3CO_2H$. The resulting labelled
sodium acetate was converted by reaction with $o\text{-}C_6H_4(NH_2)_2$ to labelled 2-meth-
ylbenzimidazole.

Kalal et al (1624) has discussed the fractionation of vinyl chloride-vinyl
acetate copolymers by precipitation chromatography employing precipitant-sol-
vent systems such as acetone-methanol, acetone-heptane, tetrahydrofuran-water
and tetrahydrofuran.

Gel permeation chromatography has been applied to the determination of the
molecular weight distribution of vinyl acetate-vinyl chloride copolymers (1625).

Hirai and Tanaka (1626) discussed the dependence of the iodine affinity of
vinyl acetate-vinyl chloride.

VINYL CHLORIDE-BUTADIENE COPOLYMERS

An NMR technique has been developed by Wilks (1627) to determine the average
block length in copolymers. The technique was illustrated by a monomer sequence study of stereoregular butadiene-vinyl chloride copolymers.

VINYL CHLORIDE-TRICHLOROETHYLENE COPOLYMERS

Keller (1628) proposed introducing corrections, relating to the interactions of vicinal chlorine atoms within the chain, through the increment system proposed for calculating the chemical shift of ^{13}C chlorine-containing polymers. The deviations of ^{13}C chemical shifts to higher fields caused by the steric hinderance of vicinal carbons were 7-15 (depending on the possible arrangement of chlorine) for trichloroethylene-vinyl chloride copolymers (1629).

VINYL CHLORIDE-VINYLIDINE CHLORIDE COPOLYMERS

Chujo and coworkers (1630) reported the NMR spectrum of a vinyl chloride-vinylidine chloride copolymer. Separate resonance areas were observed for methylene protons centred in $-CHClCH_2CHCl-$ and $-CCl_2CH_2CCl_2-$ environments. From the relative resonance areas observed, they calculated the percentages of vinyl-vinyl, etc, linkages present in the copolymers. This spectrum has been reinterpreted (1631) in terms of head-head linkages in vinylidine chloride sequences, these linkages occurring only at sequence junctions. This latter interpretation is based on a comparison of the copolymer spectrum with that of partially chlorinated poly(2,3-dichlorobutadiene).

Germar studied methylene absorption (1390 to 1440 cm^{-1}) in vinyl chloride-vinylidine dichloride copolymers (1632). Absorption at 1405, 1420, and 1434 cm^{-1} provides a measure of methylene units in $-CCl_2-CH_2-CCl_2-$, $-CHCl-CH_2-CCl_2-$, and $CHCl-CH_2-CHCl-$ environments, respectively. The relative intensities of these bands thus provide a measure of the relative amounts of VDC-VDC, VDC-VC, and VC-VC linkages in the polymers.

Enomoto (1633) noted that the tertiary C-H deformation mode of vinyl chloride units in vinyl chloride-vinylidine dichloride copolymers shifts to lower frequencies as the proportion of vinylidine chloride in the copolymers increases. He concluded that the C-H deformation of vinyl chloride units occurs at 1247, 1235, and 1197 cm^{-1} when these units are centred in VC-VC-VC, VDC-VC-VC and VDC-VC-VDC triads. By measuring the absorption of copolymers at the three wavelengths and by using three simultaneous equations, Enomoto calculated the quantities of vinyl chloride units present in the various triadic environments in the copolymers.

The C-H deformation mode of vinyl bromide units in vinyl-bromide-vinylidine chloride copolymers is sensitive to environment (1633).

McClanahan and Previtera(1634) have investigated the high resolution NMR spectra of vinyl chloride-vinylidine chloride copolymers. Okuda (1635) has also investigated the structure of these polymers.

Revillon et al (1636) have determined the molecular weight of vinyl chloride-vinylidine chloride copolymer using gel permeation chromatography and viscometry.

VINYL CHLORIDE-SULPHUR DIOXIDE COPOLYMERS

Cais and O'Donnell (1637) followed the dehydrochlorination of sulphur dioxide-vinyl chloride copolymers. The data obtained from infrared and ultraviolet spectroscopy, proton and ^{13}C NMR, suggested that hydrogen chloride was eliminated preferentially from chloroethylene units occurring between two sulphonyl units. These workers (1638) also analyzed the ^{13}C proton NMR spectra of sulphur dioxide-vinyl chloride copolymers in terms of comonomer sequences and configurational placements.

VINYL CHLORIDE-ISOBUTENE COPOLYMERS

Brück and Hummel provided detailed information on infrared characterization of a series of vinyl chloride-isobutene copolymers polymerized by γ-radiation at $23^{\circ}C$ (1639) using high resolution spectra in the range 5000-400 cm^{-1} from which it was shown that head-to-tail structures predominate. Also used was the separation of the superimposed CH_2, CH_3CCl bands between 1500 and 1350 cm^{-1}, 1000 and 850 cm^{-1}, 730 and 600 cm^{-1} with determinations of overall composition, concentration of diads, and mean sequence lengths (1641).

VINYL CHLORIDE-PROPYLENE COPOLYMERS

Abarino et al (1642) used infrared spectroscopy to study the structure of irradiated vinyl chloride-propylene copolymers.

Kubik et al (1643) developed a proton NMR method for determining the ratios of vinyl chloride and propylene in vinyl chloride-propylene copolymers. Kubik and coworkers (1644) showed that the NMR spectrometry of propylene-vinyl chloride copolymers at $160^{\circ}C$ is improved when oxygen has been removed beforehand by blowing helium over the copolymer at a temperature of liquid nitrogen.

The propylene content of vinyl chloride-propylene copolymer fibers was calculated from absorbance at 1380 cm^{-1} using reference bands at 690 and 1420 cm^{-1} and sum of absorbances at 1420 and 1460 cm^{-1} (1645).

PVC-TOLUENE/BENZENE/NAPHTHALENE CONDENSATION POLYMERS

Cascaval et al (1646) have used pyrolysis in combination with gas chromatography technique to determine the structure and study the thermal degradation mechanism of the condensation polymers obtained by Friedel-Crafts reactions of poly(vinylchloride) with benzene, toluene and naphthalene.

POLYVINYLIDINE CHLORIDE

Nuclear Magnetic Resonance Spectroscopy

Hay (1647) has conducted a detailed ESR study of the thermal dehydrochlorination of poly(vinylchloride) and poly(vinylidinechloride). He concludes that most of the evidence from ESR spectroscopy suggests that the elimination of hydrogen chloride from poly(vinylchloride) does not create radicals in sufficient concentration to be detected.

^{13}C resonances of head-to-head and tail-to-tail sequences in poly(vinylidinechloride) have been assigned by Bovey and coworkers (1648).

Thermal Decomposition

Roberts and Gatzke (1649) have studied the thermal decomposition of this polymer prepared in the presence of oxygen.

Agostina and Gatzke (1650) have studied the solution decomposition of poly (vinylidinechloride), in particular the effect of solvents and evolved hydrogen chloride.

Miscellaneous

Studies of molecular motions in partially crystalline polymers have been reported by Bergmann (1651).

CHLORINATED POLY(VINYLCHLORIDE)

Nuclear Magnetic Resonance Spectroscopy

Doskočilová et al (1652) have determined the structure of chlorine atoms in the chlorinated PVC polymer chain.

Keller et al (1653) determined the structure and the number of CH_2, CHCl and CCl_2 groups in chlorinated PVC using proton and ^{13}C high resolution NMR.

Lucas and coworkers (1654) suggested that the chlorination of PVC involves a dehydrochlorination-chlorine addition mechanism. The ^{13}C NMR spectra of chlorinated PVC was used to determine the fraction of vinyl chloride units in unchlorinated sequences.

Keller (1655) proposed introducing corrections, relating to the interactions of vicinal chlorine atoms within the chain, through the increment system proposed for calculating the chemical shift of ^{13}C NMR of chlorine-containing polymers. The deviations of ^{13}C chemical shifts to higher fields caused by the steric hinderance of vicinal carbons were 12 ppm for chlorinated PVC.

Keller and coworkers (1656, 1657) interpreted the ^{13}C NMR spectra of chlorinated PVC in terms of structure and the fraction of vinyl chloride units in unchlorinated sequences. Copolymers of vinyl chloride and trichloroethylene were prepared and used as models for the NMR determination of the structures of chlorinated PVC (1658).

POLYEPICHLOROHYDRIN

Infrared Spectroscopy

Gasan-Zade et al (1659) used multiple attenuated total internal reflection infrared spectroscopy to follow the reaction of epichlorohydrin and diphenylolpropane. As the reaction proceeded, the intensity of the 1040 cm^{-1} absorption band (ether bonds) increased while the intensity of the 1520 cm^{-1} absorption band (benzene rings) remained constant.

Nuclear Magnetic Resonance Spectroscopy

The NMR spectra of cyclic tetramers obtained as byproducts in boron trifluoride initiated polymerization of epichlorohydrin showed the presence of methyl groups (1660). Formation of the methyl groups was ascribed to a hydride shift in the carbanion-ion-terminated growing chain rather than isomerization of the epoxy group to the corresponding aldehyde.

Column Chromatography

Impurities in bisphenol A have been determined by a reverse phase liquid chromatography with a gradient elution system (1661).

Polymer mixtures of polyoxyethylene, polyoxypropylene, polytetrahydrofuran, polyepichlorohydrin or oligobutadiene have been quantitatively separated by column chromatography on silica gel (1662).

Molecular Weight

Myers and Dagon (1663) determined the distribution of molecular weight and of branching in high molecular weight polymers from bisphenol A and epichloro-hydrin.

POLY-3,3 BIS(CHLOROMETHYL)OXTANE

An ESR study has been carried out on this polymer irradiated with electron beams and ultraviolet light (1664, 1665).

CHAPTER 6

POLYMETHYLMETHACRYLATES, POLYMETHACRYLATES AND COPOLYMERS

POLYMETHACRYLATES

Chemical

Determination of End-Groups

Ghosh et al (1666, 1668) have developed methods for the determination of end-groups in persulphate initiated polymethyl methacrylate. Sulphate and other anionic sulphoxy end-groups were determined by shaking a chloroform solution of the sample with aqueous methylene blue dye reagent. The blue colour is evaluated spectrophotometrically at 660 nm. The quantity of anionic sulphoxy end-group present in the polymer is obtained by comparing the experimental optical density values with a calibration curve of pure sodium lauryl sulphate obtained by following a similar procedure (1667).

Ghosh et al (1666, 1668) also described dye partition methods for distinguishing between sulphate (OSO_3^-) and sulphonate (SO_3^-) end-groups, for determining sulphate, sulphonate and carboxy end-groups and for determining hydroxyl end-groups. The method for carboxy end-groups has been checked by quantitative study of carboxyl groups in a large number of copolymers (1669) in which one of the reactant monomers is a carboxylic monomer.

Hydroxyl end-groups have been usually determined after transforming them to carboxyl groups by the phthalic anhydride-pyridine technique (1670, 1671). Sulphate end-groups are destroyed by hydrolysis during this process and are converted to carboxyl groups by the phthalic anhydride present.

Results obtained by Ghosh et al (1666, 1672) in end-group analyses of polymethyl methacrylate indicate that all the polymer samples exhibit a positive response to methylene blue reagent in the dye partition test, indicating the presence of at least some sulphate (OSO_3^-) end-groups.

Ghosh et al (1673) also studied sulphate and hydroxy end-group analysis in polymethylmethacrylate initiated by Redox persulphate systems in conjunction with other activators such as Ag^+, Fe^{++} hydrazine and hydroxylamine, aliphatic amines, alcohols, reducing acids and their salts, etc.

Ghosh et al (1674, 1675) have also described a dye partition method for the
determination of hydroxyl end-groups in poly(methyl methacrylate) samples prepared in aqueous media with the use of hydrogen peroxide as the photoinitiator.
In this method, the dried polymers were treated with chlorosulphonic acid under
suitable conditions whereby the hydroxyl end-groups present in them were transformed to sulphate end-groups. Spectrophotometric analysis of sulphate end-
groups in the treated polymers was carried out by the application of the dye-
partition technique, and thus a measure of hydroxyl and-groups in the original
polymers was obtained (average 1 hydroxy end-group per polymer chain).

Maiti and Saha (1676) have described a dye partition technique (1677, 1678)
utilizing disulphine blue for the qualitative detection and, in some cases,
the determination of amino end-groups in the free radical polymerization of
polymethyl methacrylate. They found only 0.01-0.62 amino end-groups per chain
in polymethyl methacrylate made by the amino-azo-bisbutyronitrile system,
whereas in polymer made by the titanous chloride and acidic hydroxylamine systems they found 1.10 to 1.90 amino end-groups per chain.

Arai et al (1679) have reported on end-group analysis of isolated trunk grafted
polymethyl methacrylate chains in graft copolymers of wool. In this method the
graft polymers were isolated almost completely from the wool trunk by a hydrochloric acid-digestion method, leaving a few amino acid residues on the end
of the graft polymers. Dinitrophenylation of the isolated polymer was then
carried out. The spectral features were almost the same as for dinitrophenylated amino acids of the usual type such as valine, leucine, and methionine,
with a maximum in ultraviolet light at 340-345 nm. From colorimetric analysis
of the number of dinitrophenylated amino acid end-groups and the measurement
of the average molecular weight of the isolated polymers, the number of amino
acid end-groups linked to the graft polymers was calculated to be about one
and two per polymer chain in reduced and unreduced wool, respectively, independent of the reaction system, graft-on, and molecular weight of graft polymers. From these facts, it was suggested that most of the isolated polymers
are the truly grafted polymers.

It has been suggested on the basis of kinetic considerations of graft copolymerization (1683) and infrared spectroscopical investigation for the polymer
fractions obtained by the solvent-extractive fractionation technique (1684)
that the cysteine residues on wool keratin are the main sites of graft. Thus,
the absorption spectra of dinitrophenyl-methionine might be expected to be
taken as the reference standard curve for estimation of amino acid end-groups
incorporated in the isolated poly(methyl methacrylate). There is a linear relation between optical density and molar concentration of these dinitrophenyl
-amino acids. The molar extinction coefficient is 1.0×10^{4}.

Kobayashi (1685) used neutron activation analysis to determine trace elements
in polymethyl methacrylate.

Raman Spectroscopy

Yoshino and Shinomija (1686) have measured the Raman spectra of chloroform
solutions of polymethyl methacrylate. the intensity of a Raman line measured
on a turbid solution of the polymer was related to the true intensity and the
depolarization ratio, and the relations were confirmed experimentally. Apparent
depolarization ratios measured on a polymer rod or solution are different from
the corrected values obtained by using these relations, especially for polarized lines. the intensity per methyl methacrylate unit of the polymer Raman
lines is equal to the intensity of the corresponding line of a model compound.
Yoshino and Shinomija (1686) found no effect of polymer tacticity on Raman sp-

ectra. An example of a Raman spectrum of polymethyl methacrylate in chloroform solution is shown in Fig. 126. The optimum polymer concentration was
3 to 10%.

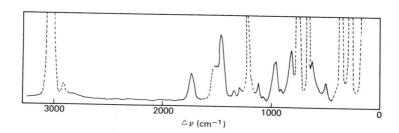

$\triangle \nu$ (cm^{-1})

Fig. 126   Raman spectra of poly(methyl methacrylate) in
chloroform solution.

Low frequency Raman spectra (less than 100 cm^{-1}) of solid amorphous polymethyl
methacrylate have shown two previously unresolved broad bands attributed to
changes in the density of states function for skeletal normal modes (1687).

NMR Spectroscopy

Woodward (1688) has utilized NMR spectroscopy in his study of relaxation phenomena in polymethyl methacrylate and polyethyl methacrylate.

The NMR spectrum of polymethyl methacrylate has been discussed by Peat and
Reynolds (1689), Girad and Mongol (1690) and Matsuzaki et al (1691).

Nagai and Nishioka (1692) studied the solvent effect on NMR spectra of polymethyl methacrylate.

Brosio and coworkers (1693) have reported on the ^{13}C NMR spectra of polymethyl
methacrylate in relation to measurements for tacticity, terminal conformation
and configuration.

Isotactic triads in dimethylformamide solutions of polymethyl methacrylate
have been measured by NMR (1694).

Amiya and coworkers (1695, 1696) obtained conformational information on polymethyl methacrylate in benzene solution with use of tris(dipivalomethanato)
europium as shift agent.

Miyamoto (1697) used NMR to follow the absorption and desorption behaviour of
tactic polymethyl methacrylate in chloroform solution on silica gel.

Cavelli (1698) used high resolution NMR spectroscopy to obtain conformational
information on polymethyl methacrylate.

Spevacek and Schneider used high resolution NMR (1699, 1700) and broadline
NMR (1701) to study the structure of the stereocomplex of isotactic and syndiotactic polymethyl methacrylate.

Suzuki and Harwood (1702) obtained NMR spectra, 300 MHz and 220 MHz for polymethyl methacrylate. The methylene proton resonances in the 300 MHz spectra

were well-resolved and defined. Tetrad, pentad, and hexad resonances were
recognizable in the 300 MHz spectra.

Schaefer and coworkers (1703) determined the individual spin-lattice nuclear
Overhauser, and rotating-frame cross-relaxation parameters for individual carbon atoms in solid polymethyl methacrylate using natural abundance, dipolar-
decoupled ^{13}C NMR.

Strasilla and Klesper (1704) showed that in syndiotactic polymethyl methacrylate only the units once removed are responsible for resolving the methoxy
resonance into three peaks.

The tacticities of poly(alkyl methacrylates) have been determined by using
the large difference in the spin-relaxation times of protons in α-methyl and
ester groups to eliminate the ester group resonance overlap with the α-methyl
signal (1705).

The spin-lattice relaxation of the α-methyl protons in isotactic polymethyl
methacrylate and syndiotactic polymethyl methacrylate have been shown to arise
mainly from dipolar interactions between the protons in the methyl group itself (1706) and the relaxation times of carbon atoms in isotactic polymers
are consistently longer than those of comparable carbon atoms in syndiotactic
polymers (1707).

Electron Spin Resonance Spectroscopy

Hajimoto et al (1708) measured the electron spin resonance spectra of polymethyl methacrylate irradiated at 77°K.

Michel et al (1709) determined the decay of the ESR signal in ultraviolet-irr-
adiated polymethyl methacrylate.

Iwasaki and Sakai (1710) have interpreted the ESR spectra of polymethacrylic
acid and polymethyl methacrylate.

Harris et al (1711) have used ESR to examine methacrylate propagating radicals
in polymerization of methacrylic acid and several methacrylates. Studies of
radical formation on mechanical fracture of polymethyl methacrylates have
been reported by Sakaguchi et al (1712). Experimental observations were reported by Bullock and coworkers (1713) on the synthesis of polymethyl methacrylate with nitroxide spin-label at the chain end.

Genkins and David (1714) identified six free radicals produced in the gamma
irradiation of polymethyl methacrylate.

Yoshioka et al (1715) carried out an ESR study of radiation-induced polymerization of methyl methacrylate on solid surfaces and Torikai and Okamoto (1716)
carried out ESR studies of the reaction intermediates in γ- irradiated polymethyl methacrylate.

Pyrolysis-Gas Chromatography

de Angelis et al (1717) have investigated the pyrolysis in an oxygen free atmosphere of methyl methacrylate at temperatures not above 550ºC. Pyrolysis
gases were identified by gas chromatography. These workers quote reproducibility data and describe their pyrolysis apparatus.

Haslam et al (1718, 1719) have described a technique for the determination of

poly(ethyl esters) in methyl methacrylate copolymers. The alkoxyl groups in
the sample are converted to their corresponding iodides which are then determined by gas chromatography on a dionyl sebacate column at 75°C.

Radell and Strutz (1720) have reported satisfactory results, both qualitative
and quantitative, for the pyrolysis-gas chromatography of poly(methyl methacrylate) at controlled pyrolysis temperatures. Guillet, Wooten and Combs
(1721) have carried out quantitative pyrolysis on various polymethacrylates.

Several other workers (1722, 1723) have described gas chromatography-pyrolysis
of polymethacrylate polymers and copolymers using a packed column with a liquid coating on a solid substrate.

Lehmann and Brouer (1724) and Brouer (1725) investigated the pyrolysis-gas
chromatography of polymethyl methacrylate at temperatures between 400° and
1100°C utilizing for the pyrolysis a silica boat surrounded by a platinum heating coil (see section on polystyrene).

Chromatograms obtained from pyrolyzing poly(methyl methacrylate) at 425° and
1025°C using a dinonyl phthalate column are shown in Fig. 127. Monomer is
formed nearly exclusively at 425°C whereas a number of compounds are detected
at 1025°C. No volatile products are retained in the column. Table 64 gives

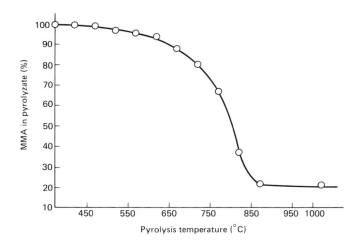

Fig. 127 Effect of degradation temperatures on amount of
methyl methacrylate monomer in pyrolyzate.

the composition of the pyrolyzates at degradation temperatures from 425° to
1025°C. Fig. 128 shows the percentage of monomer product as a function of
degradation temperature. At 425°C the product contains 99.4% monomer decreasing to about 20% at 875°C.

Stanley and Peterson (1726) used filament pyrolysis at 500°C in conjunction
with gas chromatography using a capilliary column and flame ionization detect-
or to quantitatively and qualitatively analyse methyl methacrylate, ethyl acrylate, 2-ethyl hexyl acrylate, styrene, butadiene, acrylonitrile and vinyl
acetate polymers both as homopolymers and copolymers.Optimum relative retention times for pyrolysis products of various acrylate copolymers and other poly-

Table 64   Composition, Weight Per Cent, of Pyrolysis Products of Polymethyl Methacrylate.

(Column: 30% (wt) dinonyl phthalate on ground firebrick 30-60 mesh; col. temp. 128°C).

Product	Pyrolysis temperature, ° centigrade						
	425	525	625	725	825	875	1025[a]
Gaseous compounds*	trace	3.6	4.2	18.2	60.3	73.6	76.3
Methanol	-	-	trace	trace	trace	trace	trace
Ethanol	-	-	trace	trace	0.6	0.5	0.4
Methyl acrylate	-	trace	trace	trace	trace	trace	trace
Methyl propionate	trace	trace	0.4	0.5	0.7	0.2	0.3
Methyl isobutyrate	trace	trace	0.4	0.5	0.8	2.5	3.0
Methyl methacrylate	99.4	96.2	94.7	80.6	37.4	23.2	19.9
Residue	nil	nil	nil	nil	nil	nil	nil

* methane, CO, ethane, CO_2, ethylene, acetylene.
a column temperature, 100°C.

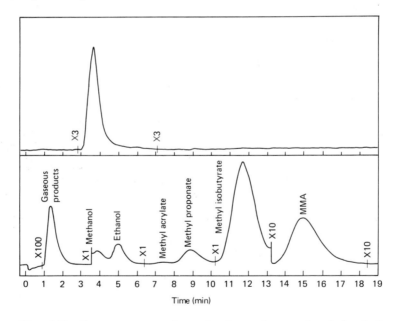

Fig. 128   Chromatograms of pyrolysis products of poly(methyl methacrylate). Column,dinonyl phthalate. Top pyrolysis temp.,425°C; col.temp.,128°C; He flow rate, 60 ml/min. Bottom pyrolysis temp., 1025°C; col. temp., 100°C; He flow rate, 20 ml/min. (Attenuation scale indicated by numbers in fig.)

mers are given in Table 65. In most cases, the peak denoted by (m) in Table 65 is the monomer. This is usually the major product and has been the peak used in the analyses. To quantitatively determine monomers Stanley and Peterson

Table 65   Relative Retention Times for Polymer Pyrolysis
           Products Relative to Methane.

(8.5 min on Column 1 and 8.4 min on Column 2)

Polymer	Relative retention times	
	Column 1	Column 2
Butadiene	1.000	1.000
	1.035 (m)	1.036 (m)
Acrylonitrile	1.224	1.238
Vinyl acetate	-	1.000
		1.131
2-Ethylhexyl Acrylate	1.000	1.393 (m)
	1.035	1.000
	1.459 (m)	1.024
Ethyl Acrylate	1.000	1.524 (m)
	1.188	1.000
	1.435 (m)	1.202
Styrene	3.094	1.500 (m)
Methyl Methacrylate	1.482 (m)	3.833
Acrylic Acid	1.000	1.571 (m)
Methacrylic Acid	1.000	1.000
Dimethyl Itaconate	1.000	1.000
		1.000

(m) denotes the peak used in analysis of the polymer.

(1726) ran known polymer along with the unknown, to give the correlation between peak height and component weight.

Cobler and Samsel (1727) have investigated the application of pyrolysis-gas chromatography to the identification of polyacrylates and polymethyl methacrylate. They used a combustion boat type of apparatus. The 20% di-2-ethylhexyl sebacate on 30 to 60 mesh firebrick column was operated at 110°C for styrene and α-methyl styrene plastics. Although monomer formation is the primary degradation reaction in the case of polyacrylates and polymethacrylates it is seen in Fig. 129 that other compounds such as alcohols and acids are usually detected. The pyrolysis of methacrylate/acrylate plastics may result in the formation of traces of hybrid monomers apparently formed by the recombination of some of the alcohol and acid fragments. For example, a small amount of methyl acrylate has occasionally been found after the pyrolysis of an ethyl acrylate-methyl methacrylate copolymer.

Hausler et al (15) have examined the light beam and Curie Point pyrolysis of polymethyl methacrylate using the pyrolysis-gas chromatographic technique.

Thermal Methods

McNeill (1728) has discussed the application of the technique of thermal volatilation analysis to the study of polymer degradation in polymethyl methacrylates. In this technique in a continuously evacuated system the volatile products are passed from a heated sample to the cold surface of a trap some distance away. A small pressure develops which varies with the rate of volatilization of the sample. If this pressure is recorded as the sample temperature is increased in a linear manner, a thermovolatilization (TVA) thermogram showing one or more peaks is obtained.

TVA thermograms for various polymethyl methacrylates are illustrated in Fig.

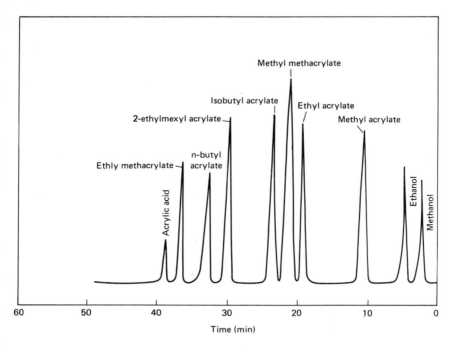

Fig. 129   Gas chromatogram of pyrolyzates of polyacrylates
           and polymethacrylates.

130. As in the case of TGA, the trace obtained is somewhat dependent on the

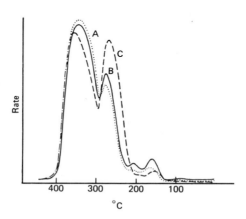

Fig. 130   Thermograms (10°C/min) for samples of poly(methyl
           methacrylate) of various molecular weights;(a)
           820,000;(b) 250,000; (c) approx. 20,000.

heating rate. With polymethyl methacrylate the two stages in the degradation
are clearly distinguished (Fig. 130). The first peak above 200°C represents
reaction initiated at unsaturated ends formed in the termination step of the
polymerization. The second, larger peak corresponds to reaction at higher temperatures, initiated by random scission of the main chain. It is apparent that
as the proportion of chain ends in the sample increases, the size of the first
peak increases also. These TVA thermograms illustrate very clearly the conclusions drawn by MacCallum (1729) in a general consideration of the mechanism
of degradation of this polymer. The peaks occurring below 200°C can be attributed to trapped solvent, precipitant etc. These show up very clearly, indicating the usefulness of TVA as a method of testing polymers for freedom from
this type of impurity.

Clark and Jellinek (1730) studied the thermal degradation of polymethyl methacrylate in a closed system.

The thermal behaviour of amorphous tactic polymethyl methacrylates and poly
(tert- butyl methacrylates) has been studied by various thermal techniques
including thermogravimetry, differential scanning calorimetry, thermomechanical
analysis, and torsional braid analysis (1731).

Differential thermal analysis has been used to identify polymethyl methacrylate (1732).

Column Chromatography

Gel permeation chromatography has been used by Meyerhoff (1733) for analysis
of polymethyl methacrylate through columns packed with polystyrene gel and
acetone solvent.

Miyamoto and coworkers (1734) separated isotactic polymethyl methacrylate from
syndiotactic polymethyl methacrylate using competitive absorption on silica
gel from chloroform.

Grubisie et al (1735) suggested the product $(\eta) \times M$, which is proportional to
the hydrodynamic volume, as a universal calibration parameter in the gel permeation chromatography of polymers. By using this method of calibration, molecular weight averages of the polymers under investigation can be calculated
from gel permeation chromatograms if the constants of the Mark-Houwink equation $(\eta) = K \times M^a$ for the polymer-solvent system are known for both the calibration standards and the polymer under study. Weiss and Cohn-Ginsberg (1736)
and Belinkii and Nefedov (1737) showed that in those cases when universal calibration is valid it is not absolutely necessary to show the relationship (η)
versus $\bar{M}$ for the polymer, but that the gel permeation chromatographic data
can be interpreted if two of the three values $\bar{M}_w$, $\bar{M}_n$, and (η) are known, or
if one of these values is known for two different samples of the polymer to
be determined. Kolinsky and Janca (1738) suggested a general method for the
evaluation of applicability of the universal calibration parameter. Gel permeation chromatograms are evaluated assuming validity of universal calibration.
The calculated values $\bar{M}_w$, $\bar{M}_n$, or (η) are correlated with those measured directly by light scattering, osmometrically, or viscometrically. The results of
correlation are used to estimate the justification of universal calibration.
Janca et al (1739) used this method in their investigations on polymethyl methacrylate and poly(2-methoxyethyl methacrylate).

The experimental conditions for gel permeation chromatography used by Janca
et al (1739) were identical to those used by Kolinsky and Janca (1738). They
were chosen so as to leave the experimental results unaffected within the lim-

its of experimental error by various influences, such as concentration effort, etc. The original calibration curve is based on polystyrene standards.

Thin-layer Chromatography

Thin-layer chromatography has been shown to be effective in separation of atactic and syndiotactic poly(methyl methacrylate) (1741).

Buter, Tan and Challa (1742) used thin-layer chromatography to distinguish mixtures of isotactic and syndiotactic poly(methyl methacrylate) from the corresponding stereoblock polymers, and were able to estimate the tacticities from the R_f values.

Mass Spectrometry

Time-of-flight mass spectrometric techniques have been employed to analyze volatile products produced during mechanical degradation of poly(methyl methacrylate) (1743).

Yasina and Pudov (1744) also used mass spectrometry to study the photolyses of poly(methyl methacrylate).

Gardella and Hercules (1745) applied static secondary ion mass spectrometry to a homologous series of polyalkyl methacrylates and this provided a direct fingerprint for members of the series.

X-ray Diffraction

X-ray diffraction and water permeation studies, made on thick and ultrathin films of poly(methyl methacrylate) and cellulose triacetate have shown that the latter had more oriented structures (1746).

The amorphous phases of poly(methyl methacrylate), has been examined by Fischer et al (1747) employing a combination of electron diffraction (determination of short-range order), light and x-ray scattering (morphology) and neutron small-angle scattering (conformation).

Patterson (1748) studied the Brillouin scattering of poly(methyl methacrylate) and polystyrene as a function of temperature. Equilibrium values of Brillouin splitting were observed at approximately $20^{\circ}C$ below the transition temperature (T).

Monomers

Kalinin et al (1749) have discussed the polargraphic determination of methyl methacrylate monomer, oxidation products and inhibitors in polymethyl methacrylate.

Waters (1750) determined methyl methacrylate monomer in poly(methyl methacrylate) using a method based on measurement of the intensity of the C=C vibration at approximately 1640 cm^{-1}.

Methyl methacrylate and methacrylic acid monomers have been determined by gas chromatography of a methylene dichloride solution of the polymer (1751).

Additives

Plasticizers have been determined in these polymers by a procedure involving pyrolysis to alkyl iodides followed by gas chromatography (1752).

Miscellaneous

Turbidimetric titration has been used by Horris and Miller (1753) for the determination of the molecular weight distribution of polymethyl methacrylate.

Gruber and Elias (1754) have discussed apparatus suitable for cloud point titration to determine the critical non-solvent fraction (γcrit) of polymethyl methacrylate. The influence and significance of possible errors are examined.

The particle size distribution of polymethyl methacrylate latices has been determined by ultra centrifugation (1755).

Tomono et al (1756) elucidated mechanisms of polymerization of methyl methacrylate by end-group analysis using radioactive isotopes. The oxidation rate of polymethyl methacrylate has been determined using an automated apparatus (1757).

Schelton et al (1758) have reported configurational information for polymethyl methacrylate in the glass state based on neutron scattering data.

Neutron scattering has provided fundamental torsional frequency data for isotactic polymethyl methacrylate (1759).

The depolarization thermocurrent method has shown promise as a useful tool for examination of transitions and relaxations in poly(methyl methacrylate) (1760).

Studies of molecular motions in partially crystalline polymers were reported by Weill (1761) on polymethyl methacrylate.

Differential scanning calorimetry has been used to determine the glass transition temperature of polymethyl methacrylate (1762).

Venediktov et al (1763) have described ultrasonic apparatus for determining viscoelasticity of polymers and demonstrated applications to polymethyl methacrylate.

The fine structure of polymethyl methacrylate has been established using high resolution light scattering methods (1764).

METHYL METHACRYLATE-METHYL ACRYLATE COPOLYMERS

Chemical Methods

Polymethyl acrylate, prepared by a radical process at 60°C, can be hydrolyzed rapidly and completely under alkaline conditions, on the other hand, the monomer units in polymethyl methacrylate prepared and treated similarly are resistant to hydrolysis (1766) although benzoate end-groups react readily (1767). This marked difference between the esters of polyacrylic and polymethacrylic acids has been studied by Baines and Bevington (1768) using tracer methods for examination of the hydrolysis of the monomer units in copolymers of methyl methacrylate with methyl acrylate and in copolymers of these monomers with styrene. Baines and Bevington (1768) hydrolyzed homopolymers of methyl methacrylate and methyl acrylate. The products were isolated and assayed. It was confirmed that only about 9% of the ester groups in polymethyl methacrylate reacted even during prolonged treatment;hydrolysis of polymethyl acrylate was complete in 0.5 hr. These workers concluded that although only about 9% of the ester groups in methyl methacrylate homopolymers are hydrolyzed by alcoholic sodium hydroxide, this proportion is increased by the introduction of comonomer units into the polymer chain.

The pyrolysis products of methyl methacrylate/methyl acrylate copolymers and styrene/acrylonitrile copolymers have been analyzed by Takeuchi and Kakugo (1769). Samples are pyrolyzed at $500^{\circ}C$ in the gas chromatograph carrier gas stream.

METHYL METHACRYLATE-METHACRYLIC ACID COPOLYMERS

Nuclear Magnetic Resonance Spectroscopy

Johnson et al (1770) have discussed the compositional analyses of methyl methacrylate-methacrylic acid copolymers by carbon-13 nuclear magnetic resonance spectroscopy. In pyridine solutions of these copolymers, the resonances arising from acid carboxyl and ester carbonyl carbons are sufficiently resolved to allow the determination of relative integrals.

The probabilities of occurrence of various compositional configurational triads to methacrylic acid-methyl methacrylate copolymers have been studied by Klesper et al (1771) using both proton and ^{13}C NMR.

Configurational pentads and compositional triads in the ^{13}C NMR of atactic methacrylic acid-methyl methacrylate copolymers have been assigned based on known chemical shifts and analysis of syndiotactic and isotactic copolymers and atactic homopolymers (1772).

Johnson et al (1773) calculated the composition of methacrylic acid-methyl methacrylate copolymers from the ^{13}C Fourier transform NMR spectra in pyridine employing the resonances arising from carboxy and ester carbonyl groups.

METHYL METHACRYLATE-BUTYLACRYLATE COPOLYMERS

Grassie and Fortune (1774) have discussed the gas chromatography-mass spectrometry of methyl methacrylate-butylacrylate copolymers.

Grassie and Torrance (1775, 1776) have discussed chain scission and reaction mechanism in the thermal degradation of methyl methacrylate-methyl acrylate copolymers.

METHYL METHACRYLATE-STYRENE COPOLYMERS

Hirai et al (1777) have carried out a ^{13}C NMR study of alternating copolymers of methyl methacrylate with styrene.

METHYL METHACRYLATE-BUTADIENE COPOLYMERS

Chemical Methods

Hill et al (1778) utilized ozoneolysis in their investigation of a butadiene-methyl methacrylate copolymer. The principal products were succinic aicd, succindialdehyde and dicarboxylic acids containing several methyl methacrylate residues. The percentage of butadiene (9.2%) recovered as succinic acid and succindialdehyde provided a measure of the 1,4-butadiene-1,4-butadiene linkages in the copolymers, and the percentage of methyl methacrylate units (> 51%) recovered as trimethyl 2-methyl-butane-1,2,4-tricarboxylate(4), n=1, provided a measure of the methyl methacrylate units in the middle of butadiene-methacrylate-butadiene triads.

Nuclear Magnetic Resonance Spectroscopy

Ebdon (1780) obtained 220-MHz NMR data on triad and pentad methyl methacrylate fractions in random and alternating methyl methacrylate-butadiene copolymers which agreed with calculations from reactivity ratios.

Suzuki et al (1781) have shown that the 220-MHz NMR spectra of alternating methyl methacrylate-butadiene and methyl methacrylate-isoprene copolymers indicated an isotactic configuration with flanking butadiene units.

Pyrolysis-Gas Chromatography

Sokolowska et al (1782) determined the chemical composition of butadiene-methyl methacrylate-styrene copolymers by pyrolysis-gas chromatography with a pyrolysis temperature of 610°C.

METHYL METHACRYLATE-ISOPROPYL STYRENE COPOLYMERS

Elias and Gruber (1783) have investigated the applicability of cloud point titration to the determination of the constitution of graft copolymers of methyl methacrylate and isopropyl-styrene.

METHYL METHACRYLATE-ACRYLONITRILE AND METHACRYLONITRILE COPOLYMERS

Infrared Spectroscopy

Fujimoto et al (1784) have carried out infrared studies of the stereoregular polymerization of methyl methacrylate and methacrylonitrile.

Nuclear Magnetic Resonance Spectroscopy

Methyl acrylate-methacrylonitrile copolymers have been characterized from methine and α-methyl proton resonance from 220-MHz NMR (1785).

The NMR spectra at 220 MHz, for copolymers of methyl methacrylate, acrylonitrile or methacrylonitrile with isoprene or chloroprene showed that a linkage was always formed between the alpha-position of the acrylic monomer and the one-position of the 1,3-diolefin (1786).

Pyrolysis-Gas Chromatography

Mihra and Pankova (1787) determined the monomer composition of acrylonitrile-methyl methacrylate copolymers by pyrolysis-gas chromatography.

Molecular Weight

The molecular weight of acrylate-acrylonitrile copolymers has been measured by a viscometric method on gel permeation chromatograph fractions (1788).

Miscellaneous

Light scattering procedures have provided structural information on acrylonitrile-methyl methacrylate-styrene terpolymer (1789).

METHYL METHACRYLATE-METHACRYLOPHENONE COPOLYMERS

Roussel and Galin (1790) have determined the sequence distribution in methacr-
ylophenone-methyl methacrylate copolymers by ^{13}C (proton) NMR.

METHYL METHACRYLATE-ISOPRENE COPOLYMERS

A pyrolysis-gas chromatography method has been developed for the quantitative
determination of the composition of multicomponent polymer systems and applied
to the analysis of polyisoprene-poly(methyl methacrylate)-polystyrene mixtures
and isoprene-methyl methacrylate-styrene block copolymers (1791).

METHYL METHACRYLATE-CHLOROPRENE

Nuclear Magnetic Resonance Spectroscopy

The sequence distribution of chloroprene-methyl methacrylate copolymers has
been studied using proton NMR (1792). The ^{13}C relaxation times at 38 and $100^{\circ}C$
were longer for isotactic polymethyl methacrylate than for syndiotactic polymethyl methacrylate indicating more segmental motion in the isotactic chain
backbone (1793).

Ebdon (1794) related the intensities of the various α-methyl signals in the
220 MHz PMR spectra of free radical polymerized chloroprene-methyl methacrylate copolymer to the relative proportions of the various methyl methacrylate
centred triads.

METHYL METHACRYLATE-GLYCIDYL METHACRYLATE COPOLYMERS

The glycidyl methacrylate content of methyl methacrylate-glycidyl methacrylate copolymers have been determined from the ratio of the infrared absorbance
at 907 and 1717 cm^{-1} (1795).

METHYL METHACRYLATE-METHYL VINYL KETONE COPOLYMERS

Fluorescence spectra were used by Amerik et al (1796) to identify interaction
of chromophore groups during photostability studies of polymers in dioxane
solutions after irradiation at 313 nm. Polymers examined included poly(methyl
vinyl ketone) homopolymer and its copolymer with methyl methacrylate and poly
(1-naphthyl methacrylate).

METHYL METHACRYLATE-VINYLIDENE CHLORIDE COPOLYMERS

Chiang et al (1797) used NMR to determine the reactivity ratio of methyl methacrylate-vinylidene chloride copolymers.

Revillon et al (1798) have determined the molecular weight of vinylidene chloride-methyl methacrylate copolymers using gel permeation chromatography and viscometry.

METHYL METHACRYLATE-VINYL CARBONATE COPOLYMERS

The structure of graft copolymers of methyl methacrylate-vinyl carbonate prepared by γ-irradiation has been established by hydrolysis and oxidative cleavage by periodate (1799).

METHYL METHACRYLATE-METHYL-α-CHLOROACRYLATE AND METHYL METHACRYLATE-TRIMETHYLSILYL METHACRYLATE COPOLYMERS

Heublein and coworkers (1800) determined the concentration of diads and mean sequence lengths in copolymers of methyl-α-chloroacrylate-methyl methacrylate and methyl methacrylate-trimethylsilyl methacrylate copolymers from ^{13}C NMR spectra.

METHYL METHACRYLATE-POLYVINYL ALCOHOL

Hori et al (1801) used thin-layer chromatography to estimate the purity of polyvinyl alcohol-methyl methacrylate copolymers. The graft copolymers were produced by radiation grafting to produce poly(vinyl alcohol)-methyl methacrylate. Prior to the chromatographic development the homopolymers were removed as rigorously as possible by extraction or selective precipitation. The hydroxyl groups were completely acetylated to obtain polyvinyl acetate-g-poly(methyl methacrylate). The characterization study proved that all the graft copolymers have one branch per molecule on the average, the number-average molecular weight of the branches being 1×10^5.

POLYMETHACRYLATES

Nuclear Magnetic Resonance Spectroscopy

Heublein and coworkers (1802) have determined the composition, tacticity, average sequence length, and degree of cross-linking of poly(allyl methacrylate) and allyl methacrylate-methyl methacrylate copolymers by proton NMR.

NMR data has provided an insight on polymerization mechanism of 2-allylphenyl methacrylate (1803).

Characterizations of poly(phenyl methacrylates) have been made from 100 MHz NMR spectra.

Electron Spin Resonance Spectroscopy

Bowden and O'Donnell (1806) have carried out an electron spin resonance study of radical reactions in irradiated octadecyl methacrylate. Sakai and Iwasaki (1807) have measured the change with temperature of the ESR spectrum of methacrylic and radicals.

Harris et (1808) have studied electron spin resonance in methacrylate prolongating radicals.

Pyrolysis-Gas Chromatography

Esposito (1809) has described the semi-quantitative determination of polymethacrylates by means of pyrolysis-gas chromatography using an internal polymeric
standard. A silicone grease column is used, programmed from 60°C to 225°C at
5.6°C/min.

The identification of polymethacrylates using pyrolysis-gas chromatography
has been described by Guillet et al (1810) who pyrolyze samples by means of
a hot wire in the injection port of the chromatograph.

Ettre and Varadi (1811, 1812) have described a method for the pyrolytic analysis of polybutymethacrylate. Qualitative and quantitative analyses are described. Product composition is shown to vary with pyrolysis temperature.

Feuerberg and Weigel (1813) pyrolyze acrylate and methacrylate polymers in a
glass tube at 620°C and then chromatograph the condensed pyrolyzate at 150°C
on a 2 metres long column, using di-n-decyl phthalate as the stationary phase.

Ettre and Varadi (1765) applied their rapid pyrolysis technique (described in
more detail in Chapter on nitrocellulose) to poly-n-butyl methacrylate. Figure
131 gives a typical gas chromatogram obtained for the pyrolysis products at
500°C whilst Table 66 tabulates the breakdown products obtained between 400
and 95°C.

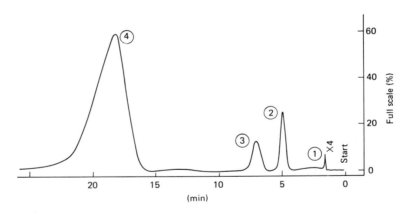

Fig. 131   Typical chromatogram obtained from pyrolysis of
           13 mg poly(n-butylmethacrylate) at 500°C.
           Peaks 1, light gases; 2, ethanol; 3, propanol;
           4, monomer n-butylmethacrylate.

Table 66 shows a very sharp change in the concentration of the breakdown products. At lower temperatures, the polymer degrades nearly completely to the
monomer n-butylmethacrylate, alcohols and gases appear only in small quantities.
With rising temperatures, the monomer content decreases slowly, while the gas
content simultaneously increases. Between 600° and 700°C, however, the monomer
concentration drops sharply from about 65% to about 6 to 7%. At the same time,
the amount of gases increases very sharply. The concentrations of alcohols
show a nearly continuous increase up to 800°C; above 900°C, however, they completely disappear. Here again, the breakdown products above 900°C are composed
only of gases.

Thermal Methods

Table 66 Composition of Breakdown Products of Poly(n-butyl methacrylate).

	Temperature °C						
	400	500	600	700	800	900	950
		weight per cent of original polymer					
Carbon dioxide	1.2	4.0	7.8	23.5	24.5	24.6	29.0
Carbon monoxide	1.8	4.3	6.9	21.5	22.5	22.2	27.0
Methane	0.8	3.0	5.8	20.0	20.5	20.4	25.0
Ethane	-	0.5	0.8	1.0	2.0	2.0	3.0
n-butyl methacrylate	.92.0	80.9	63.8	6.6	-	-	-
Methanol	-	0.5	1.0	3.0	3.5	3.6	-
Ethanol	1.8	3.0	5.0	7.0	8.0	8.0	-
Propanol	1.8	2.8	5.0	6.0	6.5	6.4	-
Methyl acetate	-	-	-	traces	1.0	1.2	-

Grassie et al (1814) have used thermal volatilization analysis to observe the photochemical degradation of polymethacrylates.

Thin-layer Chromatography

Brinkman et al (1815) have analyzed 2-hydroxyethyl methacrylate by thin-layer chromatography on silica gel with hexene-ethyl ether or hexane-methyl isobutyl ketone-octanol saturated with 25% nitric acid as the mobile phase.

Additives

Pasteur (1816) has used high performance liquid chromatography to determine Santonox and other phenolic antioxidants in tetramethylene glycol dimethacrylate.

Miscellaneous

Electron scanning chemical analysis (ESCA) has been applied by Clark and Thomas (1817) to the determination of core and valence energy levels in a series of polymethacrylates.

Wendorff et al (1818) have correlated the effects of side chain flexibility of a series of methacrylic acid polymers with thermal and x-ray diffraction properties.

POLYMETHYLACRYLATES (POLYACRYLATES)

Chemical Methods

The degree of conversion in emulsion polymerization of methylacrylate has been measured by bromide-bromate titration for double bonds (1819).

Kalinina and Doroshina (1831) reviewed chemical methods for qualitative and quantitative analyses of polyacrylates.

Coe (1832, 1833) and Kennett and Stanton (1834) discussed the analysis of polyacrylates.

The quantitative analysis of polyacrylates and polymethacrylates by the pyrolysis technique has been reported by Guillet and coworkers (1835) and a technique involving hydriodic acid hydrolysis has been described by Haslam and coworkers (1836). This latter technique is suitable for the determination of the methyl, ethyl, propyl, and butyl esters of acrylates, methacrylates, or maleates in plastics (1837). First the total alcohol content is determined using a modified Zeisel hydriodic acid hydrolysis (1838). Secondly, the various alcohols, after being converted to the corresponding alkyl iodides, are collected in a cold trap and then separated by gas chromatography.

Owing to the low volatility of the higher alkyl iodides the hydriodic acid hydrolysis technique is not suitable for the determination of alcohol groups higher than butyl alcohol. A procedure suitable for determining esters such as 2-ethylhexyl, decyl, and lauryl acrylates in copolymers consists in saponifying the esters at an elevated temperature and chromatographing the alcohols thus formed (1837, 1838).

Miller et al (1839) have determined acrylate esters in polymers by converting the alkoxy groups to alkyl iodides which are then analyzed by gas chromatography. Nelson et al (1840) have discussed the identification of acrylic resins.

Potentiometric titration provides a method of investigating changes of conformation undergone by polyelectrolytes in solution, since the environment of the dissociating groups is dependent on the conformation of the polymer chain. Wada (1841) has shown that the helix-coil transition of polyglutamic acid may be studied by this method, and this result has been confirmed by Nagasawa and Holtzer (1842). Leyte and Mandel (1843) have similarly found evidence of a transition between helix and coil conformations for poly(methacrylic acid), and Combet (1844) has reported a sharp break in the titration curve of poly(acrylic acid). Jacobson (1845) has interpreted her results on the binding of magnesium ions by poly(acrylic acid) in terms of two conformations for the polymer.

Mathieson and McLaren (1846) have presented precise potentiometric titration data for solutions of a sample of poly(acrylic acid) of high molecular weight, at constant ionic strength, which show similar features and indicate the presence of a conformational transition. The titrations were carried out at constant ionic strength, using sodium chloride, by the method of Katchalsky, Shavit and Eisenberg (1847).

Figure 132 shows the titration results for polyacrylic acid plotted as pH + $\log((1-\alpha)/\alpha)$ versus α (degree of dissociation), as points connected by full curves. The four curves at the different ionic strengths all show the same features. The first short region, labelled A in the figure, is probably due to some instability in the solution, such as aggregation preceeding precipitation. This region extends to higher values of α at the higher ionic strengths. The second region, B, represents the ionization of the first conformation of the polymer, the third, C, the conformational transition, and the fourth, D, the ionization of the second conformation. The first conformation, which exists at the lower degree of dissociation, has presumably the more tightly coiled structure, and is denoted PAA(a) (PAA=polyacrylic acid). The second conformation, stable at high degrees of dissociation, and less tightly coiled, is denoted PAA(b). The four curves of PAA(a) in Fig. 132 have been extrapolated (dashed curves) semi-empirically to zero α, and they meet there at a value of pH + $\log((1-\alpha)/\alpha)$ of 4.58, which is the value of pK'_0, the intrinsic dissociation constant of the polyacid for ionic strengths of 0.06 and 0.11 found previously (1848). This extrapolation is made with the help of plots of pH versus $\log((1-\alpha)/\alpha)$, shown in Fig. 133. These plots, though almost

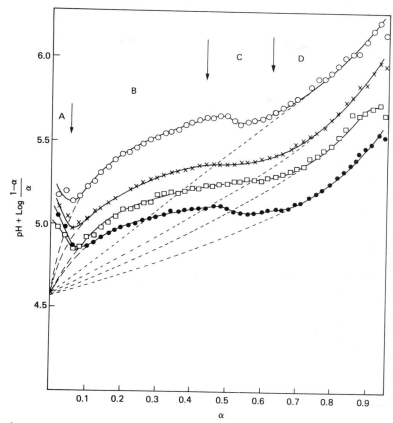

Fig. 132   Dependence of the function pH + log((1-α)/α) on
α for poly(acrylic acid) at various ionic strengths
u: (o)u=0.02; (x)u=0.065; ( )u=0.11; (+)u=0.20.

linear over the whole range, as previously reported by Mandel and Leyte (1843,
1849) are not quite so, and regions A,B,C and D can be distinguished here
also. Linear extrapolation of region B, which represents PAA(a), to higher
values of log((1-α)/α) was used to obtain the extrapolations of the PAA(a)
curves of fig. 132 to zero α.

Mattieson and McLaren (1846) used the method of Nagasawa and Holtzer (1842)
for calculating the helix content of polyglutamic acid and at different degrees of dissociation.

Infrared Spectroscopy

Brako and Wexler (1850) have described a useful technique for differentiating
the presence or absence of functional groups such as hydroxyl, carboxylic
acid or ester in polymers containing small percentage components of such
groups. Films of latices or polymers are subjected to chemical treatment which
results in marked changes in the infrared spectrum which can be associated

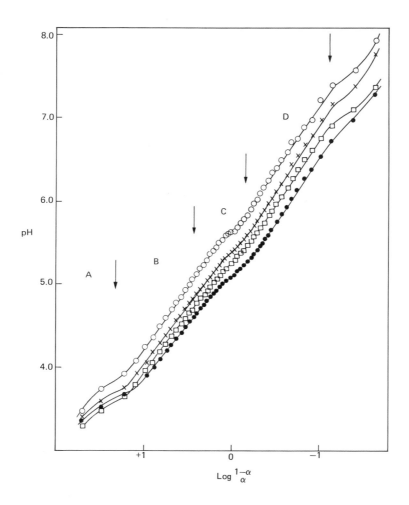

Fig. 133   pH versus $\log((1-\alpha)/\alpha)$ for poly(acrylic acid) at
           various ionic strengths u; (o)u=0.02; (x)u=0.065;
           (□)u=0.11; (●)u=0.20.

with the disappearance of a functional group or its relacement by another functional group. Infrared data may be readily interpreted negatively so that one
may definitely preclude the presence of hydroxyl, carbonyl, amine, amide, nitrile, ester, carboxylic, aromatic, methylene, tertiary butyl, and terminal vinyl groups if the corresponding group vibrations are absent in the infrared sp-
ectrogram. More difficult is the assignment of functional groups where multiple
or several alternate possibilities exist as in the mixture of a carboxylic and
keto group or in the assignment of a band to an olefinic group.

In Fig. 134 is shown the infrared spectra of a sodium polyacrylate film before
and after exposure to hydrochloric acid vapour. Exposure to acid results in

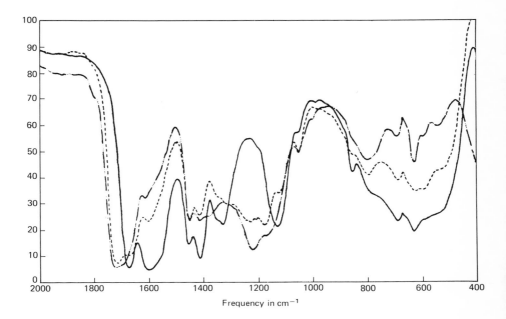

Fig. 134   Spectra of sodium polyacrylate, sodium polyacry-
            late after exposure to hydrogen chloride vapour
            and the heated film of the hydrogen chloride
            treated acrylate.
            ───────────Sodium polyacrylate film
            ───────────Sodium polyacrylate film exposed to
            HCl vapour
            ───────────Sodium polyacrylate film exposed to
            HCl vapour and then heated.

disappearance of the broad, intense band associated with carboxylate group in
polyacrylate ion at around 1600 cm^{-1}. Appearance of a broad intense absorption
at about 1720 cm^{-1} is associated with the carbonyl of the carboxylic acid group
in the polymer. Significant changes are also observed in the 1100 to 1200 cm^{-1}
region. Heating of the acidified film resulted in minor changes in the spectrum.

Figure 135 shows the changes in the infrared spectrum resulting from the exposure of acrylic acid-vinylidene chloride copolymer film to ammonia vapour.
Bands associated with the carboxylic acid carbonyl stretching frequencies at
1715 to 1740 cm^{-1} disappear on exposure to ammonia vapour. A well defined carboxylate band appears at 1570 cm^{-1}. This change is sufficient to confirm that
the copolymer contains carboxylic acid groups.

In Fig. 136 are shown spectra of polyethylacrylate films and films after hydrolysis and exposure to hydrogen chloride vapour.

Bands associated with the carbonyl stretching frequency of the ester at 1735
cm^{-1} and with the C-O bending vibrations in 1150 to 1250 cm^{-1} region disappear
after treatment of the polymer with potassium hydroxide in methanol solvent.
An intense carboxylate band is observed at 1590 cm^{-1} in the saponified residue

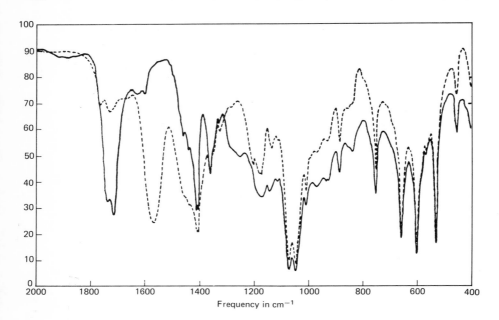

Fig. 135   Spectra of copolymer of acrylic acid and vinyl-
           idene chloride and the film after exposure to
           ammonia vapour.
           ─────────────────Acrylic acid-vinylidene chloride
           copolymer film
           --------------Acrylic acid-vinylidene chloride
           copolymer film exposed to ammonia vapour

film. Exposure of the film of saponified residue to hydrogen chloride vapour
results in a further remarkable change in the spectrum which is equivalent to
the spectrum of polyacrylic acid (compare with ------- in Fig. 134). The chemical transformations accomplished are ester → carboxylate salt → carboxylic
acid.

Nuclear Magnetic Resonance Spectrscopy

Suzuki et al (1820) obtained 300 MHz spectra on poly(methyl acrylate) and poly
(methyl α, β-dideuterioacrylate) to interpret methylene proton resonance data
in terms of tetrad configurations. Matsuzaki and coworkers (1821) studied ^{13}C
NMR spectra of poly(methyl acrylates) and poly(isopropyl acrylates) with model
compounds to interpret peaks associated with triad, tetrad, and pentad placements.

From methyl group chemical shift data, Kulkarni and Pansare (1822) developed
quantitative analyses for species obtained in production of methyl acrylate
monomer from acrylamide sulphate and methanol; average error of ± 2.0% was reported for methanol, methyl acrylate and water.

Nagata et al (1853) have discussed the ^{13}C NMR spectroscopy of acrylic monomer.

Electron Spin Resonance Spectroscopy

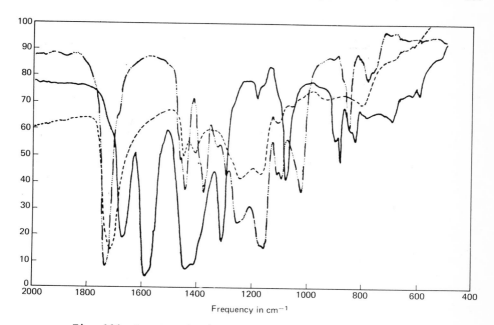

Fig. 136   Spectra showing detection of ester group in a
polymer.
——————-----——Polyethyl acrylate film
——————————Saponified residue from hydrolysis
of polyethyl acrylate
---------------Saponified residue exposed to HCl
vapour

Shioji et al (1854) have conducted an electron spin resonance study of irradiated acrylic acid and the reactions of the radicals thereby produced.

Clay and Charlesby (1851) postulated that the triplet in the elctron spin resonance spectrum of poly(acrylic acid) polymerized by gamma irradiation actually
consisted of a doublet or triplet and a singlet.

Forebioni and Chachaty (1852) have carried out NMR studies on acrylic acid.

Pyrolysis-Gas Chromatography

Several workers (1855-1858) have discussed the gas chromatography-pyrolysis of
polyacrylate polymers and copolymers using a packed column with a liquid coating on a solid substate.

Radell and Strulz (1823) pyrolyze acrylate and methyl acrylate polymers for
30 seconds at 550°C in an atmosphere of helium. Unique and reproducible pyrograms are obtained. Using samples of about 5 mg this method is applicable to
semi-quantitative determinations.

Thermal Methods

Gedemer (1859) has used thermogravimetric analysis to estimate the degree of thermal degradation of acrylic insulations.

Haken et al (1824) identified the principal high molecular weight fragments
produced in the thermal degradation of polymethylacrylate.

Molecular Weight

Pogorel'skil et al (1860) have derived equations for calculating the molecular
weight of water-soluble acrylic acid and acrylamide polymers from light scattering data.

Monomers and Solvents

Hobden (1861) has described a gas chromatographic method for the determination
of solvents in acrylic polymers. Brunn and coworkers (1862) determined the
amount of free acrylic acid in poly(acrylic acid) and of ethyl acrylate in
poly(ethyl acrylate) by ultraviolet measurements at 195 nm where the extinction coefficients of polymer and monomer differed most.

Brown (1863) has described a high-performance liquid chromatographic method
for the determination of acrylic acid monomer in natural and polluted aqueous
environments and polyacrylates.

Mel'Nikova et al (1864) have described a gas chromatographic method for the
determination of residual butyl acrylate, methyl acrylate and methacrylic
acid in isopropyl solution of acrylic polymers.

The application of gas chromatographic techniques to the determination of monomers in acrylate polymers is reviewed in Table 67.

Table 67   Monomers in Acrylic Polymers.

Gas Chromatography

Methyl acrylate   solution in methylene dichlor-
ethyl acrylate   ide, glc   (1865)
butyl acrylate and
acrylic acid

2-ethylhexyl acrylate   solution, glc   (1866)
butyl acrylate
ethyl acrylate
vinyl propionate

2-ethylhexyl acrylate   solution in propyl acetate,
vinyl acetate   cyclohexanol, glc   (1867)

butyl acrylate   solution in isopropanol, glc (1864)
methyl acrylate
methacrylic acid

Miscellaneous

Clark and Thomas (1868) have discussed the applications of electron scanning
chemical analyses to the establishment of core and valence energy levels in
polyacrylates.

Klenin (1869) has described a turbidity spectrum method for the determination of polydispersity in acrylate latexes.

Water

Devolatilization followed by gas chromatography has been used for the determination of water in acrylate polymers.

Additives

A method has been described (1870) for the determination of plasticizers in acrylate polymers. The method involves pyrolytic conversion to alkyl iodide followed by gas chromatography. Ionol antioxidant (2,6-di-t-butyl-p-cresol) has been determined (1871) in an acetone-methanol extract of acrylate polymers using anodic voltammetry.

POLYMETHACRYLIC ACID

Chemical Methods

Grant and McPhee (1825) have titrated methacrylic acid in aqueous solution with electrogenerated bromine.

Infrared Spectroscopy

A wide structureless band covering the 50-250 cm^{-1} region with a maxima at 185 cm^{-1} in the spectrum of poly(methacrylic acid) and a band with maxima at 97, 113 and 136 cm^{-1} in methacrylic acid has been assigned to the fundamental vibrations of hydrogen. Methacrylic acid has been shown to exist as cyclic dimers with three infrared active normal modes (1826).

Nuclear Magnetic Resonance Spectroscopy

Bovey (1827) has used NMR spectroscopy to study the influence of the polymerization medium on the stereochemical configuration of polymethacrylic acid.

Chacaty and Latimer (1828) have made NMR and electron spin resonance studies on methacrylic acid.

Furukawa et al (1829) have studied the proton-NMR spectra of complexes formed between methacrylic monomer and ethylaluminium dichloride in toluene and methylene chloride.

Koma et al (1830) have made NMR studies on the reaction between methacrylonitrile and trimethylaluminium.

MISCELLANEOUS POLYACRYLATES

POLY-2-ETHYLHEXYLACRYLATE

Pyrolysis-Gas Chromatography

Stanley and Peterson (1872) determined the elution times of the principal products obtained by pyrolysis at about 500°C of poly-2-ethylhexyl acrylate, polyethyl acrylate and polymethyl methacrylate; data for these and pyrolysis products of other polymers are given in Table 68.

Table 68 Elution Times for Pyrolysis Products of Acrylates
          at Column Conditions of 76°C, 27 PSIG Helium Inlet
          Pressure.

Column 300 ft capillary column of polypropylene glycol.
Flame ionization detector column 75-90°C.

Polymer	Elution times (min)
2-ethylhexyl acrylate	8.4
	8.6
	12.8
Ethyl acrylate	8.4
	10.1
	12.6
Methyl methacrylate	8.4
	13.2
Acrylonitrile	˙8.4
	10.4
Butadiene	8.4
	8.6
Vinyl acetate	8.4
	9.5
	11.7
Styrene	32.2

Monomers

Streichen (1873) has described a gas chromatographic method for the determination of traces of 2-ethylhexyl acrylate monomer in polymers.

POLY(METHYL-α-CHLOROACRYLATE)

Assignments and quantitative analyses for all tetrad peaks in the backbone
CH_2 resonance of poly(methyl-α-chloroacrylate) have been made from 300 MHz NMR
spectra (1874).

POLYETHYL ACRYLATE AND POLYBUTYL ACRYLATE

Merrill et al (1875) have measured the proton magnetic resonance spectra of
these polymers at temperatures down to 77°K.

METHYL α HYDROXY METHYL ACRYLATE AND METHYL α ACETOXY METHYL
ACRYLATE MONOMERS

Rosenthal et al (1876) have separated methyl methacrylate and the above monomers by gas chromatography on a Union LB (1800) x column at 150°C.

POLY N-ALKYL ACRYLATES

The kinetics of crystallization of polymers with long side chains such as N-
alkylacrylates and of gel formation have been investigated by small angle
light scattering and electron microscopy (1877).

POLY(p-BIPHENYL ACRYLATE)

Newman et al (1878) studied long range ordering in poly(p-biphenyl acrylate) employing x-ray diffraction techniques.

ACRYLIC ACID COPOLYMERS

ETHYLENE-ACRYLATE-TYPE

Infrared Spectroscopy

The acrylate salt in acrylate salt-ethylene ionomers has been determined by Czaja and coworkers (1879) from the ratio absorbances at 1560 cm^{-1} (asymmetric vibration of the carboxylate ion) and 1380 cm^{-1}.

Nuclear Magnetic Resonance Spectroscopy

Keller and Roth (1880) used ^{13}C NMR to determine the sequence distribution in ethylene-methyl acrylate copolymers.

Roth and coworkers (1881) compared the ^{13}C NMR chemical shifts of ethylene-methyl acrylate polymer and ethylene-ethyl acrylate polymer with the shifts calculated from electron density by the extended Nueckel molecular orbital theory.

Keller and coworkers (1882) calculated the composition of 2-chloroethyl acrylate-ethylene copolymers from the 100 MHz undecoupled proton resonance spectra. The same authors utilized high resolution NMR spectroscopy to determine the side chain content and sequence distribution of ethylene-glycidyl acrylate copolymers (1883).

Furukawa et al (1884) investigated the 220 MHz NMR spectra of acrylic monomer-2-substituted 1,3-diolefin alternating copolymers.

McGrath and Robeson (1885) used 100 MHz NMR spectroscopy to characterize ethylene-ethyl acrylate copolymers, ethylene-ethyl acrylate-carbon monoxide terpolymers, and ethylene-2-ethylhexyl acrylate-carbon monoxide terpolymers.

Pyrolysis-Gas Chromatography

Columns of Carbowax, molecular sieve and propylene carbonate were used by Barrall et al (1886) to separate the thermal degradation products of poly(ethylene-ethyl acrylate) and poly(ethylene-vinyl acetate) copolymers. The main products from polyethylene-ethyl acrylate are ethylene, pentene and ethanol; those from polyethylene vinyl acetate are methane, propane and acetic acid.

Bombaugh et al (1887) determined the comonomer distribution in ethylene-acrylate copolymers using pyrolysis-gas chromatography.

Sugimura et al (1888) found that, by pyrolysis-gas chromatography, the thermal stability of ethylene-methyl methacrylate copolymers was related to the degree of branching and the localized distribution of methyl methacrylate units in the polymer chain.

Thermal Degradation

McGaugh and Kottle (1889) studied the thermal degradation of acrylic acid-ethylene polymers.

Miscellaneous

Small- and wide-angle x-ray diffraction was used by Macknight et al (1890) to determine the nature of the distribution of ionic groups in cesium salts of ethylene-methacrylic acid and ethylene-acrylic acid copolymers.

Hull and Gilmore (1891) have described a neutron activation procedure for the determination of oxygen in poly ethylene-ethylacrylate and polyethylene-vinyl acetate copolymers and physical mixtures thereof.

ACRYLIC ACID-ACRYLAMIDE COPOLYMERS

Mukhopadhyay et al (1892) have reported reverse dye-partition technique for the estimation of acid groups in water-soluble polymers such as copolymers of acrylamide and other carboxyl-bearing monomers (e.g. acrylic acid) and the monomer reactivity ratio r_1 has been determined by measuring the carboxyl group content in those copolymers (1893, 1894). The carboxyl contents of the purified copolymers were determined by the reverse dye-partition method.

ACRYLIC ACID-VINYL ESTER AND METHYL ACRYLATE-VINYL ESTER COPOLYMERS

Chemical Methods

Aydin and Schulz (1895) have saponified copolymers of vinyl esters and esters of acrylic acid in a sealed tube with 2M sodium hydroxide. The free acids from the vinyl esters were determined by potentiometric titration or gas chromatography. The alcohols formed by the hydrolysis of the acrylate esters were determined by gas chromatography.

Garrett and Pauk (1896) have determined reactivity ratios for the copolymerization of vinyl acetate and methyl acrylate.

Pyrolysis-Mass Spectrometry

Sharp and Paterson (1897) used gas chromatography-mass spectrometry to identify small amounts of copolymerized unsaturated carboxylic acid in acrylic polymers.

ACRYLIC ACID-ETHYL ACRYLATE COPOLYMERS

Acidimetric potentiometric titrations have been employed in analyses of acrylic acid-ethyl acrylate copolymers in 4:1 isopropyl alcohol-water solution (1898) of polymeric acrylic acid, methacrylic acid, and methacrylic acid calcium salt (1899), and poly(acrylic acid) in ethyl alcohol-water solution (1900).

ACRYLIC ACID-ITACONIC ACID COPOLYMERS

Crisp et al (1901) employed a conductometric titration method to analyze polyacrylic acid and acrylic acid itaconic acid copolymers. These workers found that a pH titration gave a readily detectable end-point for the polyacrylic acid but a difficulty detectable end-point for the copolymer.

ETHYLENE-2 CHLOROACRYLATE AND ETHYLENE GLYCIDYL ACRYLATE COPOLYMERS

Keller et al (1902) have stated that the overall composition of ethylene-2-chloroacrylate copolymers and ethylene-glycidyl acrylate copolymers could be determined easily by proton NMR; however, ^{13}C NMR was a more accurate technique for sequence distribution determination.

ACRYLATE COPOLYMERS, MISCELLANEOUS

Carboxylic groups in acrylic copolymers having a three-dimensional structure have been determined by exchange reaction with calcium acetate (1903).

The National Bureau of Standards (1904) method for the pyrolytic analysis of acrylic copolymers detects as little as 0.2% polymer with a precision within 0.5%. Pyrolysis is carried out at 400°C. For qualitative analysis, samples of 2-3 mg are pyrolyzed at 350°C and the degradation products condensed prior to gas chromatographic analysis on a dinonyl phthalate column.

Viscosity measurements on solutions of branched acrylic copolymers have aided their characterization (1905) and measurement of the degree of branching (1906).

Greene (1907,1908) quantitatively determined surface carboxyl groups in acrylic and methacrylic acid modified styrene-butadiene copolymer latexes by turbidimetric titration and their distribution by activation analysis.

Haslam, Hamilton and Jeffs (1909), Haslam and Jeffs (1910), Radell and Strutz (1911), and Strassburger et al (1912) have identified acrylate and methacrylate polymers singly and in mixtures by means of pyrolysis chromatograms. The latter paper also describes a method for determining quantitatively the composition of various acrylate-methacrylate copolymers.

CHAPTER 7

POLYBUTADIENE

POLY 1,4 BUTADIENE

<u>Chemical Methods</u>

Cook et al (1913) have studied methods for the determination of metals in polymers by x-ray fluoresence spectroscopy. Although their work was limited to
polybutadiene, polyisoprene and polyester resins, the techniques would be applicable to other types of polymers, and, indeed x-ray fluoresence spectroscopy
has been applied extensively by the author to the determination of metals in
polyolefins. Cook et al (1913) studied the determination of chromium, manganese,
iron, cobalt, nickel, copper and zinc. The samples were ashed and the ash dissolved in nitric acid prior to x-ray analysis. No separation schemes are necessary and concentrations as low as 10 ppm can be determined without inter element
interference. Cook et al (1913) used a solution technique (rather than examination of the solid compressed polymer) to circumvent interelement and matrix
effects. Nitric acid was used because unlike the other mineral acids it does
not have a strong absorbing effect on the x-ray fluorescence of these metals.
In general the theoretical and found values agree within ± 10%. the determination of chromium tended to be the most erratic relative to the other metals.
Many investigators have found much higher recoveries using various ashing aids
such as sulphuric acid (1917), elemental sulphur (1914), magnesium nitrate
(1917, 1918) and benzene and xylene sulphonic acids (1915, 1916) than by dry
ashing. Possibly the reason for the good recoveries after dry ashing found by
Cook et al (1913) was due to the slowness and relatively low maximum temperature, 550°C, that was used.

Purdon and Mate (1919) have described a centrifuge/ultra filtration method for
the determination of micro gel in polybutadiene and various other gel-contain-
ing polymers such as polyisoprene, styrene-butadiene, PVC, nitrile-butadiene
and acrylonitrile.

Purdon and Mate (1919) discuss earlier approaches to this problem including
their earlier (1920) low speed centrifuging method, (1400 g). However, experience with many polymer systems has since shown that, although this technique
is satisfactory for perhaps the majority of polymer systems, there are instances where gel particles of colloidal size ("microgel") may not be separated.

308

This is particularly possible when the particles of colloidal size are dispersed in a solvent whose density approaches that of the polymer. The use of higher-speed centrifugation (e.g. 8,000-12,000 rpm or 13,000-30,000 g) for removal of microgel would be a positive approach. However, there are certain disadvantages connected with this approach, such as the removal of high molecular weight sol along with the microgel and the common problem of locating the desired transparent container which will withstand the required strains and pressures.

The alternate Purdon and Mate (1919) method, based on combined centrifugation and ultrafiltration, was developed to insure better separation of microgel than is obtained with other techniques and yet not risk removing large polymer molecules from solution. This combined centrifugation-ultrafiltration technique appears to be capable of achieving complete solution of soluble polymer and complete separation of gel from solution. It is emphasized by Purdon and Mate (1919, 1920) that in the majority of cases, gel determination can be performed satisfactorily with the centrifuge technique at 1400 g, alone. The modified technique (1919) may be required in certain cases to obtain meaningful information.

Kranz and coworkers (1921) determined the structure of polybutadiene graft polymers from acrylonitrile-butadiene-styrene employing degradation techniques utilizing Me_3CCOOH-osmium tetraoxide. The polybutadiene was completely degraded without changing the degree of polymerization of the grafted acrylonitrile-styrene polymer chains.

Microoozonolysis has been used by various workers to determine the relative amounts of 1,4 and 1,2-structures in diene polymers (1922-1925). In most cases, ozonolysis products from the polymers were converted to acids and then to esters in order to deduce structural configurations of the starting materials.

With the advent of newer analytical techniques, aldehydes and ketones have become the preferred ozonide derivatives since these can be separated and identified by gas chromatography. Lorenz and Parks (1926) converted ozonides almost quantitatively to ketones and aldehydes with triphenylphosphine. Their technique avoids formation of oxygen-containing by-products (1927) since hydroperoxides, dimeric and polymeric peroxides, and polymeric ozonides are all converted to carbonyl compounds by triphenylphosphine.

Beroza and Bierl (1928) developed a microoozonolysis technique that further decreases side-reactions and eliminates the potential danger of large-scale ozonolysis. Small samples, short reaction times, and low reaction temperatures are used. Final products are obtained with triphenylphosphine, and gas chromatography is used to separate them.

Hackathorn and Brock (1929) used microoozonolysis, to determine the head-head, and tail-tail contents of natural and synthetic polyisoprenes (1930). This method has also been used by Furukawa (1931) to determine the 1,4-content of polybutadienes. Hackathorn and Brock (1929) used the microoozonolysis technique to characterize some n-butyl-lithium-initiated polybutadienes containing both 1,4 and 1,2 structures, and to determine the amount of alternation in propylene-butadiene copolymers. They also examined a polyisoprene having nearly equal amounts of 1,4- and 3,4-structures.

Microoozonolysis followed by gas chromatography has been used to determine the amount of 1,2-structures occurring as 1,4-1,2-1,4 sequences in n-butyl-lithium-initiated polybutadienes sequences, 3-formyl-1, 6-hexanedial. The product obtained from the ozonolysis of sequences was directly proportional to the 1,2 (vinyl) content of the polymers as measured by infrared or NMR spec

troscopy. An unusual ozonolysis product, 4-octene-1,8-dial was found in the
ozonolysis products of high-1,4 polybutadienes. The product obtained from
ozonolysis of propylene-butadiene structures is 3-methyl-1,6-hexanedial. Ozonolysis of a polyisoprene containing equal amounts of 1,4 and 3,4 structures
indicated it to have a non-alternating structure with long blocks of 1,4-iso-
prene units.

Polymers having 98% cis-1,4 structure, 98% trans-1,4 structure, and a series
of polymers containing from 11% to 74% 1,2 structure were ozonized (Table 69).
The final products obtained from these polymers were succinaldehyde, 3-formyl-
1,6-hexanedial, and 4-octene-1,8-dial. Model compounds were ozonized and the
products were compared with those from the polymers (Table 70).

Table 69   Microozonolysis of Polybutadienes.

Sample	1,4 (cis + trans), %	Vinyl (1,2), %	Area (from GC) %			1,2 units occurring in 1,4-1,2-1,4 sequences
			Succin- aldehyde	3-formyl 1,6-hex- anedial	4-octene 1,8-dial	
1	98.0	2	50	1	49	0.5
2	89.1	10.9	30	10	60	5
3	89.0	11.0	43	7	50	3
4	81.0	19.0	34	14	52	6
5	76.2	23.8	36	25	39	11
6	71.8	28.2	33	27	40	11
7	69.7	30.3	48	26	26	10
8	67.7	32.3	36	26	38	9
9	64.2	35.8	38	31	31	10
10	62.8	37.2	45	27	28	8
11	50.5	49.5	26	41	33	12
12	56.0	44.0	30	39	31	11
13	26.0	74.0	33	64	3	5

Succinaldehyde is formed by ozonolysis of 1,4-1,4-polybutadiene sequences as
shown in eq.(1).

$$C - C - C = C - C - C - C = C - C \xrightarrow[\text{TPP}]{O_3 \quad O_3} O = C - C - C - C = O \quad (1)$$

TPP = triphenylphosphine

Ozonolysis of 1,4-1,2-1,4 polybutadiene sequences gives 3-formyl-1,6-hexane-
dial (eq.(2)).

$$C - C - C = C - C - \underset{\underset{\underset{C}{\overset{||}{C}}}{\overset{|}{C}}}{C} - C - C - C = C - C - C \xrightarrow[\text{TPP}]{O_3 \quad O_3} O = C - C - \underset{\underset{O}{\overset{||}{C}}}{\overset{|}{C}} - C - C - C = O \quad (2)$$

Figures 137 and 138 show chromatographic separation of the ozonolysis products
from polybutadienes having different amounts of 1,2 structure, as measured by
infrared or NMR spectroscopy. Figure 139 shows the relationship of 1,2-cont-
ent to the amount of 3-formyl-1,6-hexanedial in the ozonolysis products. This

Table 70 Model Compounds.

Compound	Product Obtained	Elution time, sec
1,5-hexadiene	Succinaldehyde, O=C-C-C-C=O	Model compound 96 Polymer product 97 (1,4-1,4 sequences)

4-vinyl-1-cyclohexene	3-formyl-1,6-hexanadial
	O=C-C-C-C-C-C=O
	     |
	     C
	    ||
	     O
		Model compound 439
		Polymer product 440
		(1,4-1,2-1,4 seq.)

Cis, cis-1,5-cyclo-octadiene	Succinaldehyde Cis-4-octene-1,8-dial, O=C-C-C-C=C-C-C-C=O	Model compound 96
		Model compound 595
		Polymer product cis
		                600
		Polymer product trans
		                585

Cyclooctene	1,8-octanedial O=C-C-C-C-C-C-C-C=O	Model compound 556

4-methyl-1-cyclohexene	3-methyl-1,6-hexanedial
	O=C-C-C-C-C-C=O
	     |
	     C
		Model compound 297
		Polymer product 298
		(BD-Prop-BD seq.)

Cyclopentane	1,5-pentanedial O=C-C-C-C=O	Model compound 143 Polymer product 143 (Polypentenamer)

linear relationship is coincidental, but may be useful as a method to determine approximate amounts of 1,2-structure if other methods are not available.

The amount of 1,4-1,2-1,4 sequences in polybutadienes can be estimated from the amounts of the different ozonolysis products (Table 69) if one considers the amount of 1,4 structure not detected. (Since the ozonolysis technique cleaves the centre of a butadiene monomer unit, one half of a 1,4 unit remains attached to each end of a block of 1,2 units after ozonolysis; these structures do not elute from the gas chromatographic column.) Using random copolymer probability theory, the maximum amounts of these undetected 1,4 structures can then be calculated.

Infrared Spectroscopy

Various workers have carried out early non-quamtitative investigations of butadiene-containing vulconizates (1965-1968).

Binder (1934) has published the polarization infrared spectra of various cumyl peroxide cured polybutadienes. Polarization spectra of two different polybutadienes containing different amounts of trans-1,4 obtained at the temperature of liquid nitrogen, are shown in Fig. 140 and of hydrogenated polybutadienes and polypiperylene (Fig. 141). The band at about 9.5μ in the Ziegler and lithium polybutadienes is due to catalyst.

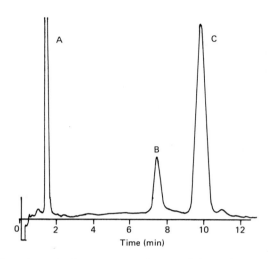

Fig. 137   Ozonolysis products from polybutadiene containing 11% vinyl structures:(A) Succinaldehyde; (B)
3-formyl-1,6-hexanedial; (C) 4-octene-1,8-dial.

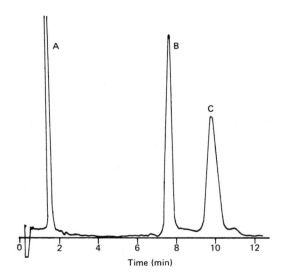

Fig. 138   Ozonolysis products from polybutadiene containing 37.2% vinyl structure: (A) Succinaldehyde;
(B) 3-formyl-1,6-hexanedial; (C) 4-octene-1,8-
dial.

Assignments of bands in polybutadiene spectra are given in Table 71.

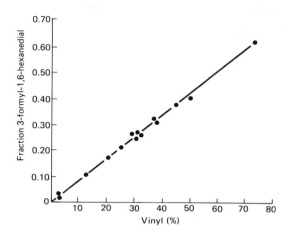

Fig. 139 Relationship of yield of 3-formyl-1,6-hexanedial
          from ozonolysis to percent vinyl structures (from
          NMR or infrared spectra) in polybutadienes.

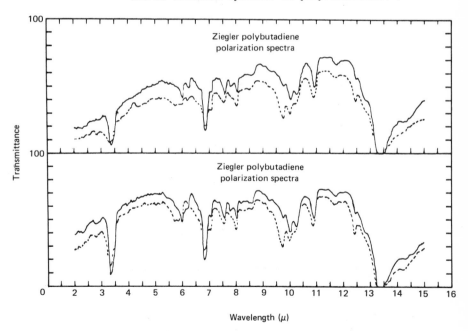

Fig. 140 Polarization spectra of high cis-1,4-polybutadienes
          at liquid nitrogen temperature.Solid curve polariz-
          ation perpendicular to direction of stretch; dotted
          curve polarization parallel.

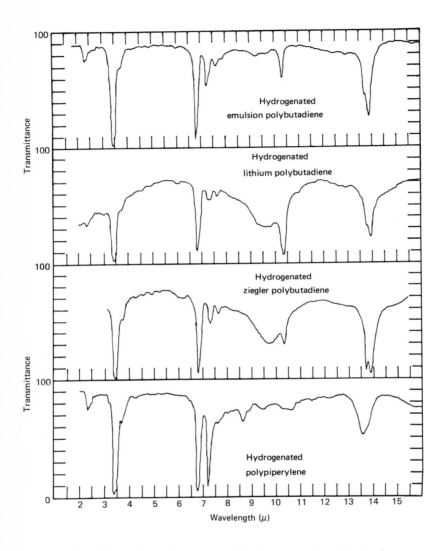

Fig. 141   Infrared spectra of hydrogenated polybutadienes.

With the preparation of very high cis-1,4 or trans-1,4 or 1,2 polybutadienes
it is now recognized, as shown by Natta (1935) that the spectrum can change
markedly depending upon the amounts of these various types of addition. However, most of the bands listed in Table 71 appear in the spectra of emulsion
polybutadienes. The assignments of those bands marked by a superscript a are
derived from olefin spectra (see Bellamy (1936) or Jones (1937)) and have been
confirmed for polybutadiene spectra by the preparation of high cis-1,4 high
trans-1,4 or high 1,2-polybutadienes. Binder (1934) has shown, by isomerization experiments, that the 1355 cm^{-1} (7.38μ) band is due to trans-CH=CH, the

Table 71   Bands in Polybutadiene Spectra.

	Band	Origin
3.25^a	3077	CH stretch of CH_2=CH
3.32^a	3012	CH stretch of cis-CH=CH
3.45^a	2900	CH stretch of CH_2
3.40	2841	CH stretch of CH_2
5.4^a	1850	Overtone of CH_2=CH
6.05^a	1660	C=C stretch of cis-CH=CH
6.10^a	1640	C-C stretch of CH_2=CH
6.8-6.9^a	1470	CH_2 def.^{c}
7.05^a	1418	CH in plane def.^{c} of CH_2=CH
7.10^a	1408	CH in plane def.^{c} of cis-CH=CH
7.38	1355	CH of trans-CH=CH;also in 1,2A, S, I^b
7.55	1325	In 1,2A, S, I^b
7.63	1311	CH of cis-CH=CH
7.65	1307	In 1,2A, S, I^b
7.75	1290	In 1,2A, S, I^b
8.1	1235	In all spectra
8.3	1205	In 1,2 I^b
8.8	1136	In 1,2 S^b
9.0	1111	In 1,2 I^b
9.25	1081	In emulsion and cis- and trans-CH=CH
9.3	1075	In 1,2 I, S^b
9.5	1053	High trans-1,4 (crystallinity)
1000	1000	High cis-1,4
10.05^a	995	CH out of plane of CH_2=CH
10.34^a	967	CH out of plane of trans-CH=CH
10.98^a	910	CH_2 out of plane of CH_2=CH
11.4	877	In 1,2 I^b
11.7	855	In 1,2 S^b
12.4	806	In 1,2 I^b
12.5	800	In some high cis-1,4
12.9	775	Crystallinity band of high trans-1,4
12.7	787	In 1,2 S^b
13.5	740	cis-CH=CH, 1,2 A, S^b
14.1	709	1,2 I^b
14.4	695	1,2 I^b
14.8	675	1,2 A, I^b
15.0	667	1,2 S^b

a These assignments established by olefin spectra.
b Here 1,2 A, S, I, means atactic, syndiotactic, and isotactic polybutadiene.
Syndiotactic and isotactic bands taken from Natta's spectra, may be in error
by up to 0.04μ.
c Def. means deformation.

1311 cm^{-1} (7.63μ) band due to cis-CH=CH as is a band at 740 cm^{-1} (13.5μ). These
experiments were done with high cis-1,4 polybutadienes and the assignments
are supported by the spectra of the hydrogenated polybutadiene spectra (Fig.
141) and also the polarization spectra (Fig. 140).

It is somewhat surprising that the intensity of the 740 cm^{-1} band has not ch-
anged more if this long-wave absorption is due only to cis-CH=CH. It is also

to be noted that the long-wave bands have not disappeared in the spectrum of
the maleic anhydride adduct of an emulsion polybutadiene and that the trans-
1,4 band at 967 cm^{-1} is still strong. A band is found at about 1310 cm^{-1} in
the spectrum of a thick film of trans polybutadiene which apparently is not
due to cis-CH=CH.

There are several features of the polarization spectrum of the high cis-1,4-
polybutadiene (Fig. 140) which are of interest and perhaps of some significance.
One of these is that, especially compared with the polarization spectrum of
Hevea, none of the bands connected with the C=C group are strongly polarized.
There is some polarization in the parallel direction, as would be expected,
and some of the difference may be due to the smaller degree of crystallinity.

Another point of interest in the polarization spectra is the shift in the positions of some bands. For example, in these polymers, the strong band is generally at 740 cm^{-1} (13.5μ) but here it is evident that the maximum must be
near 752 cm^{-1} (13.3μ). Likewise the positions of the 1311- 967- and 910 cm^{-1}
bands have shifted toward shorter wavelengths especially in the parallel polarized spectrum. The shift in the trans-CH=CH band at 967 cm^{-1} may explain the
fact that in spectra of emulsion polybutadienes the 967 cm^{-1} band has a half-
bandwidth of about 9 cm^{-1} whereas in the high trans and isomerized polybutadienes the half-bandwidth appears to be about 12 cm^{-1}. This, and other spectral evidence, raises the question as to whether there is more than one kind
(with respect to environment) of trans-CH=CH groups in some polybutadienes.

A third point of interest in the polarization spectrum of the high cis polymer
is the apparent intensity of the trans-CH=CH band at 967 cm^{-1}. From the parallel polarized spectrum, which is similar to the ordinary room temperature
spectrum, it might be concluded that the trans-CH=CH is a very weak band since
it appears as a shoulder. However, as perpendicularly polarized spectrum shows,
the trans-CH=CH is not very weak; it appears as it ordinarily does because of
adjacent bands which are evidently parallel polarized and which arise from
cis-CH=CH. This is of importance in determining the amount of trans-1,4 addition in a polybutadiene, by a base line method.

If the spectra in Fig. 142 are considered in the order, trans-1,4, 1,2 emulsion and cis-1,4 the spectrum of the cis-1,4 is surprising. By combining the
trans-1,4 and the 1,2 in about a 70/20 ratio and knowing something about correlations for cis-CH=CH, most of the bonds in the spectrum of the emulsion
polybutadiene can be assigned or explained. Such is not the case with the spectrum of the cis-1,4 polymer. While a band in the long-wave region, perhaps
740 cm^{-1}, due to cis-CH=CH is to be expected the large amount of absorption
in this region and the bands near 12 and 10μ are unexpected. Binder (1934)
discusses the origin of these additional bands.

A comparison of the spectrum of the 1,2-polybutadiene in Fig. 142, which is
atactic, with the spectra of the isotactic and syndiotactic polybutadienes
given by Natta (1935) shows many differences. The bands of interest are listed
in Table 71. The point of interest here is that some of the bands in the isotactic and syndiotactic polymer spectra probably are the same group frequency
but shifted because of differences in configuration as has been observed in
the spectra of Hevea and balata. From Table 71 it is evident that there are
many bands which are common to all three spectra and thus probably have the
same origins.

Binder (1934) concludes that in polybutadiene spectra there are bands due to
cis-CH=CH at 1311 and 740 cm^{-1} and to trans-CH=CH at 1355 cm^{-1} as well as at
967 cm^{-1}. High cis-1,4 spectra contain bands that cannot be assigned to the

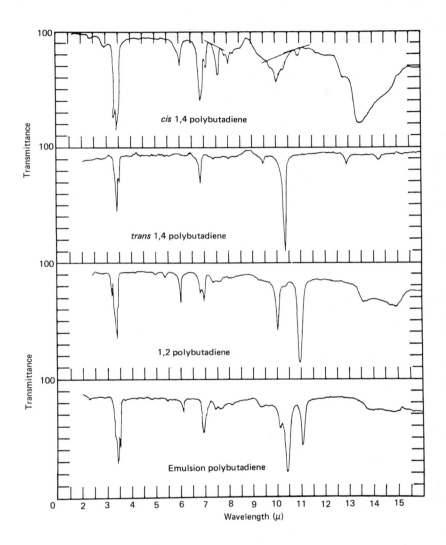

Fig. 142   Infrared spectra of various kinds of polybutadienes.

expected structures and that they may be due to additional structures, in high
cis-1,4-polybutadienes, or to changes in the configuration of the polymers or
both.

The long-wave spectrum of atactic 1,2-polybutadiene appears to be a mixture
of the bands found in the spectra of isotactic and syndiotactic polybutadienes
but shifted to shorter wavelengths.

The frequency of the hydrogen out-of-plane vibration of the cis-CH=CH group in the spectra of polybutadienes has been the subject of considerable discussion. It is important not only in connection with determining the microstructures of polybutadienes but also in regard to the actual configuration of such polymers. On the basis of the spectra of emulsion and sodium polybutadienes Hampton (1939) assigned this vibration to a band at 730 cm^{-1}, while Binder (1940) assigned it to a band at 680 cm^{-1}.

From the spectra of high cis-1,4 polybutadienes, such as described by Natta (1941) and others, Silas et al (1942) considered all the absorption in the 12.5 -15.5μ region to be due to cis-CH=CH vibrations but Morero (1943) assigned the above vibration to a band at 741 cm^{-1}. It was also observed that the position of the cis-CH=CH band apparently shifted in the spectra of polybutadienes, made with Ziegler-type catalysts, from 741 cm^{-1} toward lower frequencies as the amount of cis-1,4 addition in the polymers decreased. Despite these differences in the assignment of the frequency of the cis-CH=CH band, microstructure determinations of polymers made in emulsion systems or with sodium or potassium catalysts were very similar (1942, 1943). For polymers made with Ziegler-type catalysts the microstructure determinations were likewise similar (1942, 1943).

The shift in the maximum of the absorption from 741 cm^{-1}, in view of the uncertainty of the assignment of the frequency of the cis-CH=CH band, raised the question as to whether it was real or due to the presence of other bands in this region which remained nearly constant in intensity as the amount of cis-1,4 addition decreased. Polarization spectra by Binder (1944) showed that not only was the absorption in the 13-15μ region complex but also that, in the 10-11μ region, other bands must be present which are not apparent in the usual absorption spectrum. Various attempts to resolve these bands with prism instruments met with no success.

Binder (1945) has also shown by analytical resolutions of the spectra of isomerized cis-1,4-polybutadienes, that the band due to hydrogen out-of-plane vibrations of cis-CH=CH groups appears at 740 cm^{-1} and that the apparent shift of this band to lower frequencies, as the amount of cis-CH=CH decreases in a polybutadiene, is due to the presence of other bands in this region. He discusses and also shows that there are some bands in the 10-11μ region of polybutadiene spectra which are probably connected with vibrations of cis-CH=CH groups and that there are probably others which are not due to either cis-CH= CH or trans-CH=CH or CH$_2$=CH. He discusses possibilities regarding the origins of these bands and comments that these bands explain in large measure the appearance of polarization spectra of high-cis 1,4-polybutadienes.

Brako and Wexler(1946) have described a useful technique for testing for the presence of unsaturation in polymer films such as polybutadiene and styrene butadiene. They expose the film to bromine vapour and record its spectrum before and after exposure (Fig. 143). This results in marked changes in the infrared spectrum. Noteworthy is the almost complete disappearance of bands at 730, 910, 965 and 1640 cm^{-1} associated with unsaturation. A pronounced band possibly associated with a C-Br vibration appears at 550 cm^{-1}. Bands also appear at 785, 1145 and 1250 cm^{-1} which is due to exposure to bromine vapour. Exposure of butadiene-styrene copolymer (Fig. 144) to bromine vapour results in the disappearance of bands at 910 and 965 cm^{-1} associated with unsaturation in the butene component of the copolymer. Some alteration of the phenyl bands at 700 and 765 cm^{-1} is evident. The loss of a band at 1550 cm^{-1} and the appearance of a band at 1700 cm^{-1} are probably due to the action of acidic vapours on the carboxylate surfactant of the latex.

Fraga (1947) has developed an infrared-near infrared method of analysis of carbon tetrachloride solutions of polybutadienes suitable for the evaluation

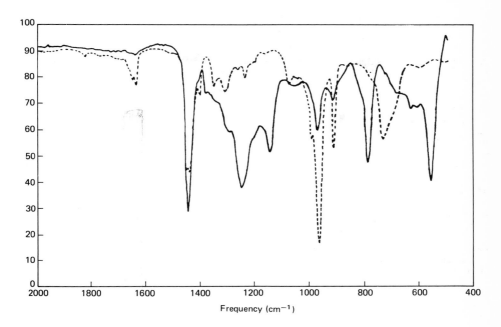

Fig. 143   Spectra of polybutadiene and of the film after
exposure to bromine.
--------- Polybutadiene
————————Polybutadiene after exposure to bromine
vapour.

of cis-1,4 (2.14μ), trans-1,4 (10.3μ) and vinyl (11.0μ) structures. Only polybutadiene is required for calibration purposes. The method is applicable to
carbon tetrachloride soluble polybutadienes containing between 0-97% cis-1,4
structure, 0-70% trans-1,4 structure, and 0-90% vinyl structure. This worker
has also described an infrared thin film area method for the analysis of styrene-butadiene copolymers. The integrated absorption area between 6.6 and 7.2μ
has been found to be essentially proportional to total bound butadiene and is
independent of the isomeric type butadiene structure present. This method can
be calibrated for bound styrene contents ranging from 25 to 100 percent.

Sequence distributions of the configurations in high 1,4 polybutadiene have
been determined by infrared spectroscopy. Clark and Chen (1948) determined
the number of dyads in the 1,4 polybutadiene and used the run number concept
of Harwood and Ritchey (1949) to characterize the cis(c) and trans(t) distributions along the chain. They defined a run number R as the average number of
uninterrupted sequences in a polymer chain of 100 units. Thus a sequence of -c
t cc ttc t cc t c ttt ccc tt- would have 60 runs per 100 monomeric units. They
obtained the cis and trans and 1,2 polybutadiene concentrations by the infrared procedure of Silas et al (1950) while the cc and ct contents were calculated from measurements made on bands near 778 cm^{-1} and 1075 cm^{-1} respectively.
All polybutadiene spectra observed have a 778 cm^{-1} band when cis is present
and a 1075 cm^{-1} band when both cis and trans are present. Their analytical
data fit the theoretical observation that %c=%(cc) + ½%(ct) reasonably well.

Some typical spectra are shown in Fig. 145 for a high-cis 1,4 polybutadiene,

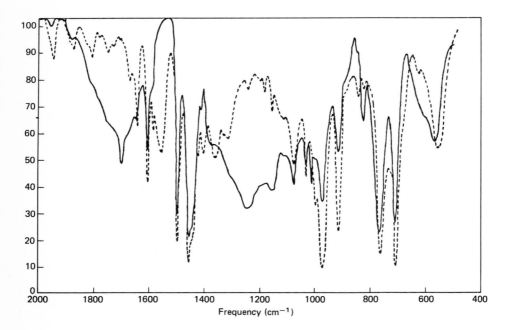

Fig. 144   Spectra of a film of butadiene-styrene copolymer
           and of the film after exposure to bromine vapour.
           ----------- Copolymer of butadiene and styrene
           ——————— Copolymer of butadiene and styrene
           after exposure to bromine vapour.

and a high-trans 1,4 polybutadiene, and a high 1,2 (atactic) polybutadiene,
and other samples of different cis and trans compositions.

Silas et àl (1950) speculated that the 778 cm^{-1} band might be associated with
-CH$_2$CH$_2$- groups in cis-1,4 polybutadiene, while Cornell and Koening (1951) and
Neto and di Lauro (1952) both assign a band near 778 cm^{-1} in the spectrum of
crystalline trans-1,4 polybutadiene to a -CH$_2$- rocking mode absorption of the
-CH$_2$CH$_2$- group. This band disappears in the solution spectrum of trans-1,4
polybutadiene, but it remains in the solution spectrum of cis-1,4 polybutadiene.
Binder also noticed this behaviour (1953). Neto and di Lauro (1952) assigned
a band near 1075 cm^{-1} in the spectrum of trans-1,4 polybutadiene to a mixing
of the -C-C- stretching mode and -CH bending modes of the -C=C-C- structure
but Clark and Chen (1948) did not make any comments on the apparent band shift
that results in the spectra of mixed cis and trans polybutadiene.

Ast et al (1955) and Hummel et al (1954) characterized polybutadienes by metathesis with 4-octene.

Braun and Canji (1956, 1957) carried out microstructure studies on the 1,2 configuration in polybutadiene and Hast and Deur-Siftar (1958) distinguished between cis and trans isomers in polybutadiene.

Canji et al (1959) determined the structures of 1,4 polybutadienes and butadiene-styrene copolymers by metathesis with 3-hexane. Reaction with 1,2 linkages

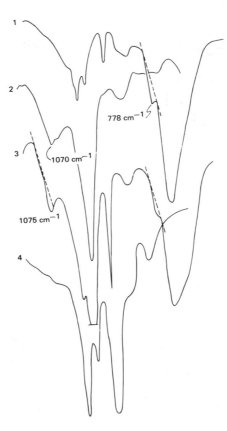

Fig. 145   Spectra of (1) cis-1,4-PBD; (2) trans-1,4-PBD;
            (3) mixed PBD structure; (4) atactic 1,2-PBD.
            Figures displayed vertically for viewing. All
            samples in carbon disulphide solution. 2mm light
            bath cell with NaCl window.

produced 1-butene and with 1,4-1,4 sequences produced 3,7-decadiene. Allara
(1960) used internal reflection infrared to study the oxidation of polybutadi-
ene-1 cast on copper or gold. Only carboxylate ion was formed at the copper
interface whereas bands typical of carbonyl, carboxylic acids, aldehydes, and
ketones appear at the gold interface.

Quack and Fetters (1961) showed the presence of a vinylcyclopentane structure
in polybutadiene apparently resulting from intramolecular cyclization during
propagation at low monomer concentration.

Nuclear Magnetic Resonance Spectroscopy

Infrared (1963, 1965-1968) and pyrolysis-infrared (1964) methods have both been
tried for determination of the composition of vulcanizates but both methods
have serious disadvantages. Carlson and Altenan (1962) investigated nuclear

magnetic resonance spectroscopy as a means of solving this problem. Various
vulcanizates containing two or three of the following, natural rubber, polybutadiene, and butadiene-styrene copolymer, were analyzed in hexachlorobenzene
solution (trimethylsilane internal standard).

Carlson and Altenan (1962) used the equations to calculate polymer compositions
by NMR which have been published by Mochel (1969). The quality of the spectra
was comparable to that of uncompounded polymers. Good agreement was obtained
between the calculated amounts of butadiene, natural rubber, and styrene and
the found values; however, the method will not distinguish between butadiene
from polybutadiene and butadiene-styrene copolymer. The relative amounts of
1,2-butadiene and 1,4-butadiene were found in the cases of polybutadiene and
styrene-butadiene rubber blends. The individual amounts of cis and trans 1,4-
butadiene could not be calculated because of complete overlapping of the two
signals as is also the case in uncompounded polymers.

Determination of the different microstructures in a blend of polybutadiene and/
or styrene butadiene rubber with natural rubber is very difficult because the
signals from the olefinic protons severely overlap one another. Carlson and
Altenan (1962) did not attempt to calculate the percentages of the different
microstructures in these blends of rubber. Natural rubber is essentially 100%
cis 1,4-polyisoprene.

In further work, Carlson et al (1970) discuss a disadvatage of their NMR method connected with its inability to provide only limited microstructural data.
This was due to the lack of resolution of the 60-megacycle NMR. They developed
alternate carbon disulphide extraction techniques to overcome these limitations
and applied infrared methods as described by Binder (1971, 1973).

Table 72 shows infrared analyses of vulcanizates containing either polybutadiene or styrene-butadiene copolymer, or a blend of these as reported by Carlson et al (1970). The relative amounts and types of polybutadiene and styrene-
butadiene rubber varied in these blends. These differences are reflected in
the calculated per cents for the samples.

Table 72   Determination of Polymer Composition.

Calculated compounded formula				Found compounded formula (infrared)			
Styrene	Butadiene			Styrene	Butadiene		
	cis	trans	1,2		cis	trans	1,2
	37	55	8		35	55	10
	14	30	56		16	27	57
15	20	53	12	16	21	51	12
21	13	52	14	23	15	49	13
26	8	53	14	28	12	48	12
17	17	43	23	22	17	39	22

Table 73 shows infrared analyses of vulcanizates containing natural rubber,
polybutadiene and styrene-butadiene rubber. The infrared results indicate that
the concentration of natural rubber in the carbon disulphide solution is considerably higher than that present in the vulcanizate.

Various workers (1974-1977) have discussed use of ^{13}C NMR spectroscopy in establishing sequence distribution in polybutadienes.

Grant and Paul (1978) developed an empirical procedure for calculating ^{13}C NMR

Table 73 Determination of Polymer Composition.

Calculated compounded formula				Found compounded formula (infrared)				
Styrene	Butadiene		Natural rubber	Styrene	Butadiene			Natural rubber
	cis	trans	1,2			cis	trans	1,2	

Styrene	cis	trans	1,2	Natural rubber	Styrene	cis	trans	1,2	Natural rubber
	19	28	4	50		13	18	3	66
	37	55	8	-		37	54	10	-
19	4	22	20	35	14	4	20	12	50
29	6	35	30	-	31	11	39	19	-
14	5	32	9	40	10	5	30	6	49
24	8	54	15	-	30	10	49	11	-
13	4	27	7	50	8	6	22	6	59
24	7	53	14	-	32	9	49	10	-

hydrocarbon chemical shifts. This procedure has been applied to polymers by various workers. Grant and Paul (1978) did not discuss an investigation of the effect of the temperature effect of their parameter values. It was not, in fact, until the work of Randall (1979) that it was shown that the correct- ive terms can be temperature sensitive. He presents results for a hydrogen- ated polybutadiene and various ethylene-1-olefin copolymers and finds that as a result of temperature compensation the standard deviation between calc- ulated and observed chemical shifts has been improved from 1.04 to 0.30 ppm.

Shown in Fig. 146 are ^{13}C NMR spectra at 25.2 MHz of a hydrogenated polybut- adiene that contained 73% 1,2 additions (or 183 branches per 1000 carbon at- oms). The spectra were recorded on a Varian XL-100 NMR spectrometer at temp- eratures of 37 and 125°C in approximately 10 wt% polymer solutions of 1,2,4- trichlorobenzene and perdeuterobenzene. The temperature sensitivity of some of the resonances in Fig. 146 is clearly indicated by the overlap that occurs at one temperature but is resolved at the other. In all, nineteen resonances could be identified even though only eighteen were visible at 37°C and only 16 at 125°C.

Hatada et al (1980) used spin decoupling techniques in NMR spectroscopy to show that the distribution of cis and trans units in polybutadiene was stat- istically random.

The amorphous component of carbon disulphide-wetted poly(4-methyl-1-pentene) and trans-1,4-polybutadiene has been studied by broad-line NMR (1981, 1982).

Chen (1983) compared the ^{13}C NMR spectra of poly(1,4-butadienes) and polypent- enamer with emphasis on methylene peaks.

The ^{13}C NMR spectra of polybutadiene has been interpreted and signals assign- ed to cis-1,4, trans-1,4 units and 1,2 units by Elgert et al (1984, 1985).

Based on ^{19}F NMR and ^{13}C NMR spectra, Toy and Stringham (1986) concluded that the microstructure or polyperfluorobutadiene consisted of mainly 1,4 moities. Based on the infrared spectrum, the infrared inactive trans-1,4 moiety was suggested as the predominant configuration.

The stereosequence distribution in 1,4-poly(2,3-dimethyl-1,3-butadiene) as determined by ^{13}C NMR has been evaluated (1987).

^{13}C NMR studies by Elgert et al (1988) showed that polybutadiene prepared us-

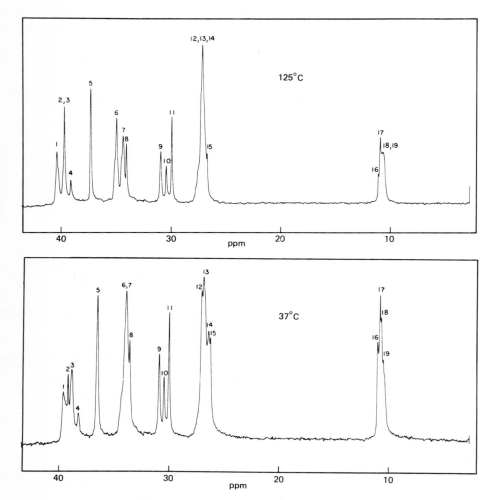

Fig. 146 ^{13}C NMR spectra of a hydrogenated polybutadiene
(73% 1,2-additions) at 37 and 125°C shown with
respect to a trimethylsilane internal standard.

ing butyllithium in tetrahydrofuran at -10°C contained approximately 80% 1,2
units and was predominantly syndiotactic.

Shibata and coworkers (1989) determined the vinyl content in 1,2-polybutadiene
by pulse Fourier transform ^{13}C NMR using acetonitrile as an internal standard.

Shahab and Basheer (1990, 1991) have applied proton NMR spectroscopy to partially hydrogenated and partially deuterated natural rubber, gutta percha and
cis-1,4-polybutadiene.

The ^{13}C NMR spectroscopy of polybutadienes has been studied by Suman and Werstler (1992) and Mockel (1993).

Gemmer and Golub (1994) studied the ^{13}C NMR spectrum of 1,4-polybutadiene.

Tompa et al (1995) studied the wide-line and pulsed NMR spectrum of carboxy terminated polybutadienes.

Brame and Khan (1996) used proton NMR to determine the composition of 2-chloro-butadiene-1,3 and 2,3-dichlorobutadiene-1,3 copolymers and ^{13}C NMR to characterize the sequencing in the polymers and copolymers.

Tanaka et al (1997) determined the cyclicity in cyclized polybutadiene using proton NMR spectroscopy.

Electron Spin Resonance Spectroscopy

Hiraka et al (1998) carried out an electron spin resonance study of the polymerization of butadiene with tris(acetylacetonate)titanium and triethylaluminium.

Pyrolysis-Gas Chromatography

Polybutadiene upon pyrolysis in an open combustion furnace undergoes random chain scission to a mixture of products as shown in the pyrolysis-gas chromatography studies of Cobler and Samsel (1999) shown in Fig. 147. Within a specified pyrolysis temperature range all the various products are found in each chromatogram, even though the amounts of the volatile components formed increase with increasing pyrolysis temperature.

Tamura et al (2000) developed a pyrolysis-molecular weight chromatography-vapour phase infrared spectrophotometric method for the analyses of polymers. When applied to the analysis of 1,4-polybutadiene, the main pyrolysis products were 4-vinyl-1-cyclohexane, 1,3-butadiene, cyclopentane, and 1,3-cyclohexadiene, the relative amounts of which depend upon the cis-trans configurations in the polymer.

The structure of γ irradiated 1,2-polybutadiene has been studied using chromatography-mass spectrometry (2001).

Thin-layer Chromatography

Thin-layer chromatography has been shown to be effective in separations of trans-1,4, cis-1,4 and 1,2 vinyl polybutadienes (2002).

Fractionation and Molecular Weight

Harmon (2003) has compared results obtained in solution and gel permeation methods for the fractionation of cis-1,4-polybutadiene. The gel permeation fractionation method gave an overall wider distribution with lower number- and weight-average molecular weights. The difference may be partially explained by the diffusion of low molecular weight material through the osmotic membranes resulting in high values of $\overline{M}_n$. Isopiestic measurements tended to support lower values of $\overline{M}_n$

Kössler and Vadchnal (2004) have discussed the fractionation of polybutadiene using the Baker and Williams (2005) chromatographic precipitation technique (benzene-methanol fractionation medium). These workers explain the previously observed (2006-2015) anomolous decrease in molecular weights of the last fract-

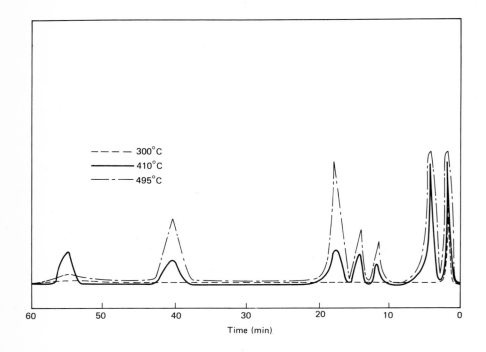

Fig. 147   Gas chromatogram of pyrolyzate of polybutadiene.

ions as a result of oxidation processes leading to the formation of peroxide
groups in the polymer. They found a correlation between the carbonyl group
content of the polymer and the constant k^1 of the Schulz-Blaschke equation.

Functionality distribution of carboxyl- and hydroxy-terminated polybutadienes
has been determined by silica gel elution (2016, 2017).

Polymer mixtures of oligobutadiene have been quantitatively separated by column chromatography on silica gel (2018).

Maley (2019) has applied gel permeation chromatography to the determination
of the molecular weight distribution of polybutadiene.

Additives

Solvent extraction-gas chromatographic methods have been described (2020) for
determining alkylated cresols (2,6 di-t-butyl-p-cresol) and amine antioxidants (N-phenyl-2-naphthylamine, p-phenylene diamine type) and Santoflex (NN'
sec-heptyl phenyl-p-phenylene diamine) in polybutadiene.

Miscellaneous

Beevers (2021) reviewed refractometry for polymer characterization with emphasis on use of molar refractivity as defined by the Lorenz-Lorentz equation.
Critical angle refractometry have been discussed in studies of films of poly-

butadiene. Gel permeation chromatography combined with viscometry has been used
to analyze branching in polybutadiene (2023).

CARBOXY TERMINATED BUTADIENES AND DIHYDROXY BUTADIENES

Chemical Methods

Law (2024) has also described infrared methods for the determination of hydroxy equivalent weight and carboxyl equivalent weight of carboxy and hydroxy
polybutadienes.

A liquid-solid chromatographic technique has been developed, utilizing stepwise elution from activated silica gel (2025, 2026) for effecting polymer separations based on functionality distribution. (Functionality distribution is
the relative proportion of nonfunctional, monofunctional, difunctional, tri-
functional, etc, present in the polymer.) This method was, however, unsatis-
factory for unsaturated polymers due to poor recovery from the column, apparently caused by polymerization on the column. Law (2024, 2029-2031) circum-
vented this problem by using partially deactivated silica gel, and recycling
the unfractionated portion through additional columns of increasing activity
to obtain the desired resolution. Using this technique Law (2024, 2029-2031)
separated carboxy-polybutadienes and hydroxy-polybutadienes according to functionality using stepwise elution from silica gel. Recoveries in the 95-100%
range were achieved. Subjection of the fractions obtained from the silica gel
separation to analysis via gel-permeation chromatography of chloroform solutions and infrared or near-infrared spectroscopy yielded not only functionality
distribution data, but also provided the relationship between molecular weight
distribution and functional type.

Adiscoff and Martin (2027) have discussed a method for the estimation of di-
hydroxy polybutadiene. The method consists of dissolving the polymer in a suitable solvent mixture such as toluene-dimethoxyethane; the addition of a known
amount of acetic anhydride reagent and the analysis of the acetic acid liberated with standard base.

Fractionation

Law (2028) has applied preparative gel permeation chromatography to studies
of low molecular weight carboxy-polybutadienes.

Miscellaneous

Natta et al (2032, 2033, 2034) have applied x-ray fibre optics methods to a
series of crystalline polymers made from various alkyl esters of the trans-
trans isomer of sorbic acid and β- styrylacrylic acid. The monomer units with
1,4 enchainment are of the type -CHR-CH=CH-CHCOOR'-.

BUTADIENE COPOLYMERS

BUTADIENE-OLEFIN COPOLYMERS

Chemical Methods

The 1,2 and 1,4 structures of polybutadiene, polyisoprene and butadiene-propene copolymers have been determined by Hackathorn and Brock (2035) by micro-

ozonolysis followed by gas chromatography. Ozonolysis of polybutadiene containing 1,4-1,2-1,4 sequences gave 3-formyl-1,6-hexanedial in direct proportion to the 1,2 content of the polymers. Ozonolysis of high 1,4-polybutadiene
gave an unusual product, 4-octene-1,8-dial. The ozonolysis of butadiene-propene polymers having an alternating structure yielded 3-methyl-1,6-hexanedial.

Infrared Spectroscopy

Valuev et al (2036) separated oligomeric butadiene-isoprene block copolymers
into fractions containing 2,1 and no terminal hydroxyl groups, respectively,
by thin-layer chromatography. Infrared spectroscopy was used for the determination of hydroxy content.

Shipman and Golub (2037) carried on an infrared study of the reaction of polyisoprene and polybutadiene with sulphur by the use of deuterated polymers.

Nuclear Magnetic Resonance Spectroscopy

Pulsed ^{13}C NMR was used by Carman (2038) to show that butadiene-propene polymers prepared with vanadium or titanium catalysts had perfectly alternating
monomer sequence distribution. ^{13}C NMR data were shown to be more sensitive
to structure modification than PMR.

Pilar et al (2039) carried out an EPR study of the photoinitiated polymerization of butadiene and of isobutylene mixture in the presence of vanadium
tetrachloride.

Hackathorn and Brock (2040) used microozonolysis-gas chromatography to elucidate the structure of butadiene-propylene copolymers. Samples of highly alternating copolymers of butadiene and propylene yielded large amounts of 3-
methyl-1,6-hexanedial when submitted to ozonolysis. The ozonolysis product
from 4-methyl-cyclohexene-1 (i.e. 3-methyl-1,6 hexanedial, O=C-C-C-C-C-C=CO)
                                                                          |
                                                                          C

was used as a model compound for this structure. Ozonolysis of these polymers
occurs as shown in eq. (3)

```
        BD  prop   BD  prop   BD
        |    |     |    |     |
      C-C=C-C-C-C-C-C=C-C-C-C-C-C=C-C   →      triphenylphosphine
        ↑    |     ↑    |     ↑
       O₃    C    O₃    C    O₃

                    O=C-C-C-C-C-C=O              (3)
                            |
                            C
```

The amount of alternation in these polymers can be determined if the amounts
of 1,4- and 1,2-polybutadiene structure and total propylene have been determined by infrared or NMR spectroscopy. Table 74 shows results obtained for
several butadiene-propylene copolymers having more or less alternating structures. Similar polymers have been analyzed by Kawasaki (2041) by use of conventional ozonolysis methods with esters as the final products.

CHLORINATED POLYBUTADIENES

Iida et al (2042) studied the pyrolysis of chlorinated polybutadienes as a

Table 74   Microozonolysis of Butadiene/Propylene Copolymers.

	1,4, %	1,2, %	propylene mole %	Area(from gas chromatography),%				alternating BD/Pr %
				succin- aldehyde	3-methyl -1,6- hexane -dial	3-formyl 1,6- hexane -dial	4-octene -1,8- dial	
A	45	5.7	49.3	5	92	1	2	77
B	47.8	2.2	50	11.5	85	0.5	3	71
C	53.1	3.2	43.7	25	61	6	8	48
D	-	-	30	49	38	1	12	33

model for polyvinyl chloride to investigate the thermal decomposition mechanisms of polyvinyl chloride. Chlorinated polybutadiene in which all double bonds
of polybutadiene are separated with chlorine corresponds to polyvinyl chloride
constructed with head-head and tail-tail linkages for vinyl chloride units;

$$-CH_2-CH-CH-CH_2-CH_2-CH-CH-CH_2-CH_2-CH-CH-CH_2-$$
$$Cl\ Cl \qquad Cl\ Cl \qquad Cl\ Cl$$

These workers used pyrolysis-gas chromatography to investigate the pyrolysis
of chlorinated polybutadienes. Benzene, toluene, ethylbenzene, o-xylene, styrene, vinyltoluene, chlorobenzenes, naphthalene and methyl naphthalenes were
detected in the pyrolysis products above 300°C and no hydrocarbons could be
detected at 200°C. The pyrolysis products from chlorinated polybutadiene were
similar to those from polyvinyl chloride and new products could not be detected. Lower aliphatics, toluene, ethylbenzene, o-xylene, chlorobenzenes and meth-
ylnaphthalenes were released more easily from pyrolysis of chlorinated polybutadiene than from polyvinyl chloride; amounts of benzene, styrene and naphthalene formed were small. These results support the conclusion that recombination of chlorine atoms with the double bonds in the polyene chain takes
place and that scission of the main chain may depend on the location of methylene groups isolated along the polyene chain during the thermal decomposition
of polyvinyl chloride.

POLYISOPRENE

Chemical Methods

Hackathorn and Brock (2644) have used microozonolysis followed by gas chromatography to elucidate the structure of polyisoprene having nearly equal 1,4
and 3,4 structures. The products included large amounts of levulinaldehyde,
succinaldehyde and 2,5-hexanedione, indicating blocks of 1,4 structures in
head-tail, tail- tail, and head-head configurations. Pyrolysis products indicated a structure similar to that of sodium-initiated polyisoprene when compared with earlier data of Hackathorn and Brock (3242).

Infrared Spectroscopy

Various infrared methods (2645-2648) have been reported for the analysis of
polyisoprene with predominantly cis-1,4 and trans-1,4 structural units with
low amounts of 3,4 addition.

Sutherland and Jones (2649) and Saunders and Smith (2650) have applied infrared spectroscopy to the elucidation of the structure of polyisoprenes.

Binder (2651) has carried out a detailed study of the infrared polarization
spectra of various polyisoprenes (Fig. 148).

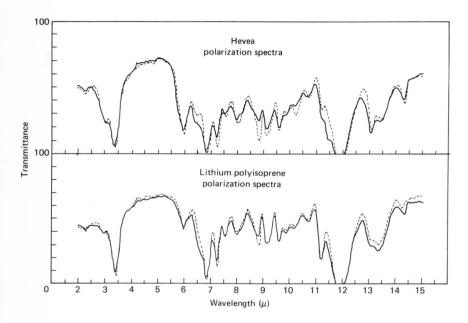

Fig. 148   Polarization spectra of Hevea and lithium poly-
           isoprene at liquid nitrogen temperature. Dotted
           curve parallel to direction of stretch; solid
           curve perpendicular.

In Table 75 are compared the conclusions of Saunders and Smith (2650), Sutherland and Jones (2649) and Binder (2651) regarding the infrared spectra of
Hevea and balata polyisoprenes.

A comparison of the spectra of Hevea and balata shows that, in carbon disulphide solutions, many bands in the spectrum of balata are shifted toward shorter wave-lengths than their counterparts in the spectrum of Hevea. Thus,
for example, the maximum of the 1130 cm^{-1} band in hevea is at 1152 cm^{-1} in
balata. This raised the question of whether these bands are connected with
some vibration of the $-C(CH_3)=CH-$ group. A direct correlation was found between the intensity of the 742 cm^{-1} band and the percent net cis-1, 4 x-total
found per cent, obtained from the microstructure determination, for various
high cis-1,4-polyisoprenes. The polarization spectra of Hevea and lithium
polyisoprene (94% cis-1,4) show that these bands are perpendicularly polarized
with respect to the direction of stretch and hence are due to a group which
is oriented to some extent. The spectrum of hyrdogenated Hevea shows that these
bands have been replaced by one strong band at about 735 cm^{-1} which is, most
likely, due to the $(CH_2)_3$ groups formed in the hydrogenation. The spectrum of
the maleic anhydride adduct of Hevea does not have the same bands either.
Maleic anhydride reacts at the carbon double bond so that the disapp-

Table 75   Bands in Spectra of Polyisoprenes.

Band cm^{-1}	Saunders and Smith (2650)	Sutherland and Jones (2649)	Binder (2651)
3.25 3077			=CH stretch of C(CH$_3$)=CH$_2$
3.33 3000			=CH stretch of C(CH$_3$)=CH-
3.42 2924			CH$_3$ stretch asymmetrical
3.44 2907			CH$_2$ in phase stretch
3.46 2890			CH$_3$ stretch symmetrical
3.53 2833			CH$_2$ out-of-phase stretch
3.68 2717			
6.01 1665	C=C		C=C of C(CH$_3$)=CH-
6.10 1645			C=C of CH$_2$=C(CH$_3$)-
6.90 1450	CH$_3$ antisymmetrical		
7.25 1380	CH$_3$ symmetrical	CH$_3$ symmetrical	
7.35 1361	CH$_2$ wag	CH in-plane def.^{c}	
7.55 1325			=CH of trans-C(CH$_3$)=CH
7.60 1315	CH$_2$ twist		=CH of cis-C(CH$_3$)=CH
7.73 1294	C-H in-plane bend		
8.03 1245	CH$_2$ twist		
8.25 1212			
8.68 1152			C-CH$_3$ of trans-C(CH$_3$)=CH
8.85 1130	CH$_2$ wag	CH$_3$ in-plane def.^{c}	C-CH$_3$ of cis-C(CH$_3$)=CH
9.15 1105	C-CH$_2$ stretch		
9.62 1040	CH$_3$ rock		
9.87 1013	C-CH$_2$ stretch		
10.20 980	C-CH$_3$ stretch		
10.74 931			
11.24 890	CH$_3$ wag		CH of CH$_2$=C(CH$_3$)
11.49 870	CH$_3$ wag out-of-plane		Crystallinity band
11.90 840	CH$_2$. CH$_2$ stretch		CH out-of-plane of cis-C(CH$_3$)=CH.
11.83 845	CH wag out-of-plane		
		CH out-of-plane	CH out-of-plane of trans-C(CH$_3$)=CH
13.12 762			
13.48 742	CH$_2$ rock		C(CH$_3$)=CH

^{c}Def. means deformation.

earance of the bands in the spectrum of the maleic anhydride adduct indicates the removal of the double bonded carbon. The spectrum of an essentially all 3,4-polyisoprene, shown in Fig. 149 does not have these bands either and in this polymer all of the double bonds are in isopropenyl groups CH$_2$=C(CH$_3$)-

While the evidence points to the origins of the 764 cm^{-1} and 742 cm^{-1} bands being in the -C(CH$_3$)=CH- group, Binder (2651) was unable to state what particular kind of vibration is involved. The assignment of Saunders and Smith (2650) in Table 75 to a CH$_2$ rocking seems improbable.

The band at about 833 cm^{-1} is now generally recognized to be due to the CH out-of-plane vibration of the C(CH$_3$)=CH- group in both Hevea and balata spectra.

Saunders and Smith (2650) assigned the other band in this region, at 890 cm^{-1} (11.25μ), to a CH$_3$ wagging vibration. Binder and Ransaw (2652) assigned it to isopropenyl, CH$_2$=C(CH$_3$)-group for the following reasons. If there is an iso-

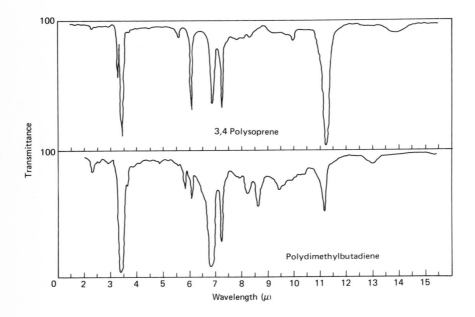

Fig. 149   Infrared spectra of all 3,4-polyisoprene and poly-
           dimethylbutadiene.

propenyl group in Hevea then, besides this 890 cm^{-1} band, there should be one
due to the C=C group near 1645 cm^{-1} and another near 3077 cm^{-1} due to the CH
stretch. Figure 150 shows the infrared spectra of Hevea and two synthetic
polyisoprenes containing different amounts of 3,4 addition (isopropenyl groups)
in the 6μ region.

It is evident that the intensity of the 6.1μ band increases as the amount of
isopropenyl group increases. The magnitude of the change in intensity with
the change in amount of isopropenyl groups is such that if Hevea contains only
about 2% isopropenyl the 6.1μ band would be weak and appear as a shoulder, as
observed.

The absorption at 3077 cm^{-1} due to the =CH stretch vibration of the CH$_2$=C(CH$_3$)
appears in the spectrum of Hevea and also in spectrum of the synthetic polyisoprene. The intensity of this band increases with increasing amount of isopropenyl. The absorption in the Hevea spectrum, although only a slight shoulder on the other strong CH bands, appears at the same place and is real. Therefore, it is concluded that it is also due to the =CH stretch of the CH$_2$=C(CH$_3$)
group. The polarization spectra show some dichroism of the 890 cm^{-1} band and
it disappears in the hydrogenated spectra. These facts reinforce the assignment of the 890 cm^{-1} band to the isopropenyl group and thus it is due to a
hydrogen out-of-plane vibration rather than a CH$_3$ wag.

Golub (2653) has reported the isomerization of Hevea and it might appear from
his spectra that the intensity of the 890 cm^{-1} band decreased as the cis-1,4
decreased. Binder (2651) reported his experiment with purified, deproteinized

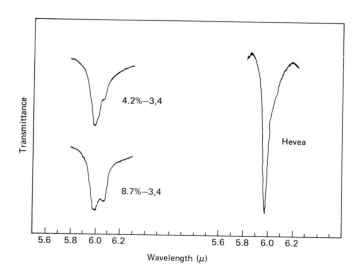

Fig. 150   Infrared spectra of Hevea and synthetic polyisoprenes in the 6μ region.

Hevea. In this experiment the cis-1,4 decreased to about 32% but the 3,4 actually increased, as was plainly evident from the spectrum, to about 6%. It should be pointed out that the 10.4μ band in the Hevea spectrum also increased
somewhat in intensity in the spectrum of the isomerized product. Thus the isomerization experiment of Golub does not invalidate the assignment of the 890
cm^{-1} band to isopropenyl, $CH_2=C(CH_3)-$.

The 1130 cm^{-1} band in the spectrum of Hevea has been assigned by Saunders and
Smith (2650) to a CH_2 wagging vibration but Sutherland and Jones (2649) on
much the same evidence, suggest that it could be due to a CH_3 in-plane deformation. Binder's (2651) results on synthetic polyisoprenes show, however, that
the intensity of this band does change with the amount of cis-1,4 addition,
or cis-$C(CH_3)=CH-$, as does the 1152.5 cm^{-1} band in the spectrum of balata with
the amount of trans-1,4 addition. In synthetic polyisoprenes the number of
CH_2 groups should be constant regardless of the kind of $-C(CH_3)=CH-$ groups
(cis or trans) so that the intensity of this bond should be constant, if due
to CH_2, regardless of the amount of cis- or trans-$C(CH_3)=CH$. Hence, more in
agreement with Sutherland's assignment, it appears that this band is due to
a $C-CH_3$ vibration of the cis-$C(CH_3)=CH-$ group. This conclusion is confirmed by
the hydrogenation spectrum and the maleic anhydride adduct spectrum. In both,
the 1130 cm^{-1} band has disappeared as it should if it is connected with a C=C
group.

A comparison of the polarization spectra of Hevea and lithium polymer, in Fig.
148, confirm some remarks of Sutherland and Jones (2649) about the effect of
crystallinity on the intensity of the 1130 cm^{-1} (8.85μ) band in the polarization spectra of Hevea. The lithium polyisoprene contained 94% cis-1,4 addition and was found, by x-rays, to crystallize less readily than Hevea. It seems
unlikely the small difference in cis-1,4 would produce the difference in the
amount of polarization shown.

As a general rule, infrared vibration bands are symmetrical. From the shapes of the 1130 and 1152 cm^{-1} bands it is evident that there must be another band lying between them. Stronger evidence of this band is found in the spectra of emulsion or sodium or potassium-catalyzed polyisoprenes. In these spectra the band appears to be near 1143 cm^{-1}. It is this band which interferes with determinations of the amounts of cis-1,4 and trans-1,4 additions in these types of polyisoprenes, using intensities measured at 1130 and 1152 cm^{-1} respectively.

The bands at about 1315 cm^{-1} in the spectrum of Hevea and at 1325 cm^{-1} in balata also appear to be connected with the -C(CH$_3$)=CH- group. Golub (2653) has pointed out that the 1325 cm^{-1} band increases in intensity and the 1315 cm^{-1} band decreases during the isomerization of Hevea. His conclusions are supported by the results of Binder (2651) from synthetic polyisoprenes, which also indicate that about 10% trans-C(CH$_3$)=CH is required for the 1325 cm^{-1} band to definitely appear in the spectrum. Again this conclusion is confirmed by the spectra of hydrogenated Hevea (Fig. 151) and of the maleic anhydride adduct (Fig. 152) which show that these bands have disappeared. The polarization spectrum shows that this band is scarcely polarized.

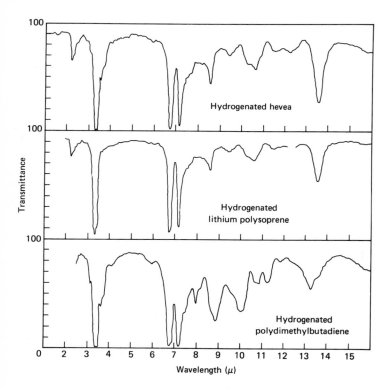

Fig. 151 Infrared spectra of hydrogenated polyisoprenes
and polydimethylbutadiene.

These two bands, 1315 and 1325 cm^{-1} are probably due to a =C-H vibration of the C(CH$_3$)=CH- group in Hevea and balata, respectively, for several reasons. Bands due to the C(CH$_3$) part of the group are assigned. Bands at 1310 and ¹⁻⁻⁻

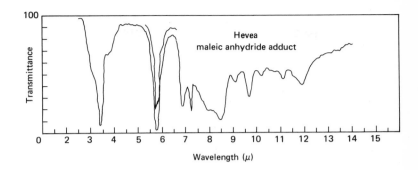

Fig. 152   Infrared spectrum of maleic anhydride adduct of
           Hevea.

1320 cm^{-1} appear in polybutadiene spectra and are connected with the cis- and
trans-CH=CH-,

Binder (2651) concludes that bands at 764, 742, 1130 and 1315 cm^{-1} are connected with vibrations of the cis-C(CH$_3$)=CH- group. Of these the 1130 cm^{-1} band
appears to be due to a vibration of the CH$_3$ or C(CH$_3$) in this group while the
1315 cm^{-1} band is connected with a C-H vibration of this group. Bands at 766,
745, 1152 and 1325 cm^{-1} are due to vibrations of the trans-C(CH$_3$)=CH- group.
Of these the 1152 cm^{-1} is due to a vibration of the CH$_3$ or C(CH$_3$) in this group
and the 1325 cm^{-1} band due to a CH vibration of this group. The 1152 cm^{-1} band
appears in the spectrum of polydimethylbutadiene and is due to a C(CH$_3$) vibration of the trans-C(CH$_3$)=C(CH$_3$) group.

Maynard and Moobel (2654) and Ferguson (2655) have described infrared methods,
respectively, for determining cis-1,4, trans-1,4, 3,4 and 1,2 structures and
1,4 structures in polyisoprene.

Kössler Vodchnal(2656) came to the conclusion that the infrared spectra of
polymers containing cis-1,4, trans-1,4 and 3,4- or cyclic structural units are
not additively composed of the spectra of stereoregular polymers containing
only one of these structures.

It is known that stereoregular cis-1,4-polyisoprene (Hevea) and stereoregular
trans-1,4-polyisoprene (balata) have absorption bands at 1130 and 1150 cm^{-1},
respectively. These workers found that a polymer having a high content of 3,4-
structural units, in addition to the 1,4-structural units, has no absorption
band at either 1130 or 1150 cm^{-1} but does have a band at 1140 cm^{-1}. They att-
ribute this band to the C-CH$_3$ vibration of the -C(CH$_3$)=CH- structural unit
separated by other structural units.

The appearance of the absorption band at 1140 cm^{-1} in some synthetic polyisoprenes has been mentioned several times by Binder (2657-2659) with the comment that the origin of this band is not known (2659).

A similar phenomenon has been discovered by analysis of a polymer having approximately 20% trans-1,4 in addition to about 75% cis-1,4 structural units,
as estimated by an analysis using the absorption bands at 572 and 980 cm^{-1}(2660).

The band at 1130 cm^{-1} was shifted towards higher values. In a mixture of Hevea and balata with the same content of 20% trans-1,4-structural units both the 1130 and 1150 cm^{-1} bands are quite distinct. The behaviour of the 1130 and 1150 cm^{-1} bands is in agreement with the finding of Golub (2653) who has shown that during the cis-trans isomerization of polyisoprene the 8.80μ absorption band appears instead of the 8.88μ band in cis-1,4 isomers or the 8.70μ band in trans-1,4 isomers. The statement of Binder (2654) that the small amount of trans-1,4 structural units may be better detected using the band pair near 7.65μ rather than the bands at 8.84 and 8.68μ is also in good agreement with the findings of Kössler and Vodchnal (2656).

As a consequence of these results, it is possible to conclude that only poly-isoprenes having long sequences of cis-1,4 or trans-1,4 units have the absorp-tion bands at 1130 and 1150 cm^{-1} respectively. It is also evident that the analysis of synthetic polyisoprenes using these absorption bands leads to distorted results. Kössler and Vodchnal (2656) obtained better results using the absorption bands at 572, 980, and 888 cm^{-1} for cis-1,4 and trans-1,4 and 3,4-polyisoprene structural units, respectively. The use of various combin-ations of different absorption bands permits one to conclude whether a poly-mer in question is more of the block copolymer type or a mixture of stereo-regular polymers (2661).

Vodchnal and Kössler (2662, 2663) have reported an infrared method for anal-ysis of polyisoprenes suitable for polymers with a high content of 3,4 add-ition and relatively low amounts of cis-1,4 and trans-1,4 structural units.

Absorptivities of the bands which are commonly used for the determination of 1,4-structural units amount are about 50 times lower than the absorptivity of the band at 888 cm^{-1} which is used for the determination of 3,4-polyisoprene units amount. Therefore, in analyses of samples with high content of 3,4 poly-isoprene units it is necessary to use two concentrations or two cuvettes with different thickness. Application of the 1780 cm^{-1} and 3070 cm^{-1}, bands offers the possibility of using only one cuvette and one concentration. The 1780 cm^{-1} and 3070 cm^{-1} absorption bands do not overlap with absorption bands of other structural forms, the accuracy of analyses thus being increased. Besides ex-act determination of the amount of 3,4 structural units it is possible to estimate an approximate amount of 1,4 addition from the 840 cm^{-1}, 572 cm^{-1} and 600 cm^{-1} absorption bands.

Values of apparent molar absorptivities of 3,4-polyisoprene, Hevea and balata in carbon disulphide solutions for the 572 cm^{-1}, 840 cm^{-1}, 888 cm^{-1}, 1780 cm^{-1}, and 3070 cm^{-1} absorption bands are summarized in Table 76.

Table 76   Apparent Molar Absorptivities k_m for CS_2 Solutions.

k_m, mole^{-1} 1. cm^{-1}

Sample	572 cm^{-1}	840 cm^{-1}	888 cm^{-1}	1780 cm^{-1}	3070 cm^{-1}
Hevea	5.7	16.6[a]	1.72	0	0
Balata	2.7	7.6[a]	0.58	0	0
3,4-polyisoprene	6.5	0	110	3.46	30.6

[a] Measured in absorption maximum.

Absorptivities $k_p(g^{-1}cm^2)$ of the samples in potassium bromide pellet, which were determined similarly as in solution by the base-line method at the 840 cm^{-1}, 888 cm^{-1} and 3070 cm^{-1} bands and by the method of density difference at the 1780 cm^{-1} band, are presented in Table 77. For comparison of their magnitudes with the absorptivities in solutions they are transformed to "molar" according to the relation $k_m = k_p M/1000$ (M denotes molecular weight; concentration is in grams per 1 cm^2 of the pellet surface area). The factor f in Table 77 represents the ratio of both molar absorptivities. In the case of the 888 cm^{-1}, 1780 cm^{-1}, and 3070 cm^{-1} bands its value varies in the range of 0.82 ± 0.05. A large deviation of this factor in the case of the 840 cm^{-1} band (f=0.51) of balata shows that the vibration belonging to this absorption band is to a great extent affected by the state of the sample in the solid. The 840 cm^{-1} absorption band can be used with the potassium bromide method and probably also with films only for a rough estimation of the amount of 1,4 structural units.

Table 77   Absorptivities for KBr Pellet Technique.

k_p (g^{-1} cm^2) and k_m(mole^{-1} 1. cm^{-1})

Sample	840 cm^{-1a}			888 cm^{-1}			1780 cm^{-1}			3070 cm^{-1}		
	k_p	k_m	f	k_p	k_m	f	k_p	k_m	f	k_p	k_m	f
Hevea	230	15.6	0.94	22	1.5	0.87	0	-	-	0	-	-
Balata	57	3.9	0.51	0	-	-	0	-	-	0	-	-
3,4-polyisoprene	0	-	-	1294	88	0.80	40.5	2.8	0.81	354	24.1	0.79

^{a}Measured in absorption maximum.

The results of measurements of the samples in carbon disulphide solutions, obtained using absorptivities from Table 76 are summarized in Table 78. The cuvettes employed were 2 mm thick, concentration of polymer varying from 0.5 to 3 g in 100 g solution. From the value of absorption at 840 cm^{-1} the minimum amount of 1,4 structural units was estimated assuming that all 1,4 units are cis. Analysis using the 572 cm^{-1} and 980 cm^{-1} bands was inapplicable due to the presence of cyclic structures.

Table 78   Results of Analyses Using Various Absorption Bands.

Structure,%

	3,4 units			1,4 units
888 cm^{-1}	3070 cm^{-1}	1780 cm^{-1}	average	840 cm^{-1}
31.6	32.9	-	32	43
37.0	39.0	-	38	42
37.2	40.2	-	39	47
39.0	42.3	-	41	41
-	49.7	51.3	51	-
-	54.6	59.6	58	28

The results of analyses of the samples in potassium bromide pellets are presented in Table 79. In these analyses it was possible to utilize the 572 cm^{-1} and 600 cm^{-1} absorption bands only for an approximate estimation of the relative abundance of 1,4 structural units.

Fraga and Beason (2664) have investigated a thin film infrared method for the analysis of polyisoprene.They emphasize that clear smooth and uniform films are necessary and that these can be cast from a toluene solution of the polymer. Film thickness should be maintained to provide between 0.5 and 0.7 absorbance units at the peak near 7.3μ. When good quality films are used the re-

Table 79   Results of Analyses.

cis-1,4 %	trans-1,4 %	3,4 %		
		888 cm^{-1}	3070 cm^{-1}	1780 cm^{-1}
0	10	-	57	62
60	9	15	11	-
45	17	-	20	-

peatability of the method is excellent. Thinner films will give slightly lower results.

Microstructure studies have been carried out on the cyclic content and cis-trans isomer distribution in polyisoprene (2665).

Nuclear Magnetic Resonance Spectroscopy

Yuki and Okamoto (2667) have conducted nuclear magnetic resonance studies of isoprenyllithium derived from 1,1-diphenyl-n-butyllithium 3,4-d$_5$ and isoprene.

Golub and coworkers (2668) applied proton and ^{13}C NMR and infrared analysis to the study of the thermal oxidation of polyisoprene. The major spectral changes were associated with a loss of the original C(CH$_3$)=CH double bonds and cyclization via the RO$_2$ radicals accompanied by appearance of epoxy, peroxy, and hydroperoxy groups.

Gronski and coworkers (2669) used the chemical shift correction parameters for linear alkanes in the aliphatic region of ^{13}C NMR spectra to determine the relative amounts of 3,4 and cis-1,4 units of polyisoprene.

The diad distribution of cis-1,4 and trans-1,4 units in low molecular weight 1,4-polyisoprene has been determined from the ^{13}C NMR spectra at 350 K by Morese-Seguela et al (2670).

Duch and Grant (2671) assigned the ^{13}C NMR spectra of cis-1,4 and trans-1,4 homopolyisoprenes.

Isoprene polymerizes to give four types of isomeric structures, i.e., cis-1, 4, trans-1,4, 3,4, and 1,2 structures, depending on the catalysts or polymerization conditions. For example, synthetic cis-1,4 polyisoprenes prepared with Zeigler catalysts or lithium catalysts contain a small percentage of the 3,4 structure.

On the other hand, naturally occurring polyisoprenes such as natural rubber (Hevea), gutta-percha, balata, and chicle consist exclusively of the 1,4 structure. The difference between the thermal and mechanical properties of the natural and synthetic polyisoprenes have been attributed to the amount of cis-1,4 units. It is reasonable to expect that the physical properties of polyisoprenes are also affected by the distribution of the isomeric structure units along the polymer chain as well as the composition of the polymers.

Tanaka et al (2672) determined the distribution of cis-1,4 and trans-1,4 units in 1,4-polyisoprenes by using ^{13}C NMR spectroscopy and found that cis-1,4 and trans-1,4 units are distributed almost randomly along the polymer chain in cis-trans isomerized polyisoprenes, and that chicle is a mixture of cis-1,4 and trans-1,4 polyisoprenes. These workers (2673) have also investigated the ^{13}C NMR spectra of hydrogenated polyisoprenes and determined the distribution of 1,4 and 3,4 units along the polymer chain for n-butyllithium catalysed

polymers and have confirmed that these units are randomly distributed along
the polymer chain. The polymers did not contain appreciable amounts of head-
to-head or head-to-tail 1,4 linkages.

Gemmer and Golub (2675) carried out a ^{13}C NMR spectroscopic study of epoxidized
1,4-polyisoprene and 1,4-polybutadiene.

Beebe (2676) and Dolinskaya et al (2677) studied the sequence structure of
polyisoprene with 3,4- and cis, trans-1,4 structural units by ^{13}C NMR.

Katritzky and Smith (2730) also discussed use of ^{13}C NMR in studies of linear
polymers and rubbers.

Campos-Lopez and Palacios (2731) also derived the microstructure of natural
guayule rubber from its 300 MHz nuclear magnetic resonance spectra.

Shahab and Basheer (2732, 2733) investigated the ^{13}C NMR spectra of partially
hydrogenated natural rubber, gutta percha and cis-1,4-polybutadiene.

Kusumoto and Gntowsky (2735) have carried out PMR measurements on natural rubber.

Electron Spin Resonance Spectroscopy

Jamroz et al (2736) used ESR spectrscopy to investigate the formation of free
radicals on compounding rubbers with carbon blacks.

Pyrolysis-Gas Chromatography

Pyrolysis-gas chromatography has been applied to the investigation of the sequence distribution of the 1,4 and 3,4 units in polyisoprenes (2678, 2679).
This method is based on the structural relationiship between the isoprene dimers and the dyad sequences of 1,4 and 3,4 units. It has been pointed out by
Tanaka et al (2680) however, that it is difficult to deduce accurately the
polymer structure by this technique because of two major problems. First, the
dimer fraction is only a minor pyrolytic product (about 30%), the monomer being the major product and other products constituting a small percentage. Second, during the pyrolysis the chain-scission reaction is often complicated
by side reactions which can alter the structure of the products.

A method is given by Feuerberg and Weigel (2737) for the identification and
determination by gas chromatography of the pyrolysis product of elastomers
including natural rubber, butyl rubber, neoprene, Vulkollan, Perbunan, Hyp-
alon, Viton A, silicone rubber, styrene-butadiene and styrene butadiene/natur-
al rubber mixtures.

The examination of synthetic rubbers is described by Kubinova and Mikl (2738).
The pyrolysis gases are passed through absorption columns to remove hydrogen,
air, carbon monoxide and methane and the remaining products then separated on
columns chosen for their suitability for the polymers being examined.

Baba and Tokumaru (2739) examined solutions of raw rubber pyrolyzates to obtain gas characteristic chromatograms for use in the identification of components in blended rubbers.

Krishen (2740) reviewed the application of pyrolysis-gas chromatography to
qualitative and quantitative analysis of polymers. Results for two correlation studies of pyrolysis-gas chromatography of paint vehicles, synthetic rub-

bers, and polymers to standardize the procedures were reported (2741, 2742).

Chih-An (2743) has developed a pyrolysis-gas chromatography method for the analysis of rubbers and other high polymers. A chromatogram which characterizes the volatile additives is followed by a pyrogram which identifies the polymers.

Hydrogenated natural rubber is a completely alternating ethylene/propylene copolymer (i.e. has no ethylene or propylene blocking) and is an interesting substance, therefore, for pyrolysis studies. The surface areas of the main peaks up to C_{13} obtained by Van Schooten and Evenhuis (2744) are given in Fig. 153. The unzipping reaction, which would yield equal amounts of ethylene and propylene in the hydrogenated pyrolyzate, evidently takes place to some extent, but is less important than the hydrogen transfer reactions. The large number of peaks in Fig. 153 reflects the many possible transfer reactions for this polymer. Hydrogen transfer reactions from the fifth carbon atom predominate as indicated by the large butane, 3-methyl hexane, 2-methyl heptane peaks, followed by transfer from the ninth carbon atom, which is shown by the size of the 3-methyl octane, 3,7-methyl decane (C_{10}) and 2,6-dimethyl undecane (C_{11}) peaks. Transfer reactions with the other carbon atoms may also be due to intermolecular hydrogen transfer in the same way as discussed for polyethylene (n-butane) and polypropylene, (2 methyl pentane).

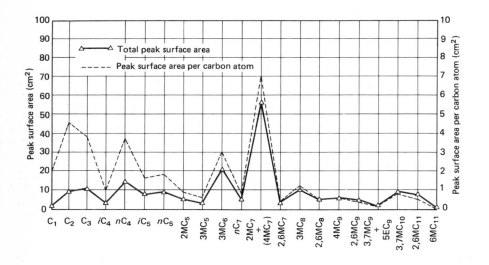

Fig. 153   Surface areas of the main peaks of the pyrogram
           of hydrogenated natural rubber.

Crosslinking

Bristow et al (2745) developed a method for the determination of the degree of crosslinking in natural rubber vulcanizates.

Burfield and Gan (2746) studied the nonoxidative crosslinking reactions in natural rubber.

Fractionation

Kössler and Vadchnal (2681) have discussed the fractionation of polyisoprene using the Baker/Williams (2682) chromatographic precipitation technique (benzene-methanol fractionation medium). These workers explain the previously observed (2683-2692) anomalous decrease in molecular weights of the last fractions as a result of oxidation processes leading to the formation of peroxide groups in the polymer. They found a correlation between the carbonyl group content of the polymer and the constant k' of the Schulz-Blaschke equation.

Molecular Weight

Calderon and Scott (2693) have described a method for the determination of the molecular weight distribution of narrow molecular weight distribution polyisoprenes prepared from the sol dependence on crosslinking. Procedures are described for estimating the breadth of the molecular weight distribution from the sol dependence on crosslinking. These yield the value of 1.05 ± 0.05 for the ratio of the weight-average molecular weight for this polyisoprene. These results demonstrate that it is possible to determine the molecular weight and molecular weight distribution of polyisoprenes, containing cis-1,4, trans-1,4 or moderate amounts of 3,4 addition structures, from the sol dependence on the amount of dicumyl peroxide used in crosslinking.

X-ray Diffraction

Mayer (2694) has investigated the structure of 1,4 polyisoprene using small-angle x-ray diffraction and electron microscopy.

Miscellaneous

Gruber and Elias (2695) have discussed apparatus suitable for the cloud point titration to determine the critical non solvent fraction (φ crit) of polyisoprene. The influence and significance of possible errors are examined. They emphasize that to obtain optimum results one should use good solvents, nonsolvents of moderate precipitating power, and small volumes in the cuvettes. Observing these working conditions, a standard error of ± 0.3% can be obtained for ρ crit. To work within this limit of error, the absolute temperature should not deviate by more than ± 0.15°C. Elias and Gruber (2696) found that a critical volume fraction ρ crit. can be obtained by extrapolation of the volume fractions of non-solvent at the cloud point to 100% polymer concentration. ρ crit depends only slightly on molecular weight. The molecular weight dependence of the slope coefficient K_1 of these curves is stronger. K_1 is approximately proportional to the arithmetic mean of weight and z-average of molecular weight.

The cross-link density of small microtomed samples of rubber elastomers has been determined using a gas chromatographic method for estimating the amount of solvent in the swollen sample (2791).

The identification of a wide range of rubber elastomers has been described by Hulot and Lebel (2792).

Volatiles

Takashima and Okada (2747) have described a method for the determination of benzene and toluene in rubber adhesives. A dioctyl phthalate column is used for the separation and trichloroethylene is the chosen internal standard.

Monomers

Brodsky (2748) has described a method for the determination of monomer in aqueous emulsions and rubber latices in which latex samples are injected into
a gas chromatograph and the monomer estimated from its chromatograph peak size.
For a similar analysis, Nelsen et al (2749) again injects the latex sample
directly into the gas chromatograph but then oxidizes the eluted products,
absorbs the water and finally determines the monomer as carbon dioxide. An example is given of the determination of styrene using a silicone oil column.

Additives

Tables 80 and 81 list various procedures for the determination of antioxidants,
antiozonants, accelerators and plasticizers in polyisoprenes, rubbers and rubber vulcanizates.

Table 80 Antioxidants, Antiozonants, Plasticizers and Accelerators in Rubbers and Polyisoprenes.

Spectroscopic Methods

amine and phenolic antioxidants	ethanolic hydrochloric acid extraction, spectroscopy	2750
phenyl salicylate resor- cinol benzoate	diethylether extraction, spectroscopy	2751
Antioxidants	acetone extraction, spectroscopy	2752, 2753
ketone-amine condensates 2-mercapto-benzinidazole	acetone, extraction, spectrscopy	2754
Antioxidants phenylnapthyl amines	solvent extraction, fluorescence spectro- scopy	2755
rubber accelerators	solvent extraction, polarography	2756-62
Polygard (tris-nonylated phenyl) phosphate	solvent extraction, hydrolysis to nonyl phenol, diazotization and spectrophoto- metric evaluation	2763

Column Chromatography

amine types of antioxidants and antiozonants	solvent extraction, gel permeation chromatography using tetrahydrofuran dilution solvent	2764-2768
antioxidant, plasticizers accelerators	carbon tetrachloride solution of sample put through column chromatography, ultra- violet monitoring of effluent successive elution with various solvents	2769
antioxidants N-N' diethyl aniline N-ethylaniline diphenylamine N-phenyl-2-napthylamine	freeze grinding, diethyl ether extract- ion, column chromatography with refractive index and ultraviolet detectors	2770

Gas Chromatography

antioxidants	solvent extraction, glc	2771-2773

Table 80   (Continued)

amine type	acetone extraction, glc	
1,2 dihydroxy-2,2,4 trimethyl-6-ethoxy quinoline N-isopropyl-N'-phenyl-p-phenylenediamine N,phenyl-2-napthylamine N,N'di-2-octyl-p-phenylene diamine N,N' diphenyl-p-phenylene diamine		
di-2-napthyl-p-phenylene diamine poly(trimethyl-dihydro-quinoline)	solvent extraction, glc	2771

Table 81   Antioxidants, Antiozonants, Accelerators, in Rubber Vulcanizates.

Antioxidants

Gas Chromatography

Amine and phenolic type	solvent extraction, acetylation glc	2774

Column Chromatography

antioxidants and accelerators, guanidine type, thiocarbanilide mercapto-benz-thiazole-2-disulphide 2-benzthiazolyl-N' cyclo-hexyl sulphumamide phenyl napthylamines sym-di-β-napthyl-p-phenylene diamine zinc dialkyldithiocarbamates formaldehyde-aniline condensation products aldol napthylamine condensation products polymerized trimethyl dihydroquinone	solvent extraction, column chromatography Column streaked with detection reagents UV and IR examination of extracts	2775-2777

Thin-Layer Chromatography

amine and phenolic types N-phenylnapthylamine acylated diphenylamines 4,4' dimethyoxydiphenylamine	solvent extraction, TLC for identification	2778
phenolic and amine antioxidants	solvent extraction, TLC solvent extraction, TLC	2778 2779

Paper Chromatography

antioxidants	solvent extraction, paper chromatography	2780

Table 81   (Continued).

		2781
phenolic and amine antioxidants		2782

Spectroscopy

Agerite range of anti-oxidants Stabilite Alba (di-o-tolyl ethylene diamine), Santoflex 66 (N-phenyl-N'-cyclohexyl-p-phenylene diamine), Santoflex DD (6-dodecyl-1,2,-dihydro-2,2,4-trimethyl-quinoline), Santoflex 13, N-phenyl-β-napthylamine, Wingstay S (styrenated phenol), 2,6, di-tertbutyl-p-cresol, N,N' diphenyl-p-phenylene diamine,	acetonitrile extraction - then mass spectrometry, infrared spectroscopy and NMR	2783

Antiozonants
Thin-Layer Chromatography

antiozonants	solvent extraction, TLC	2778

Paper Chromatography

antiozonants	solvent extraction, paper chromatography for identification	2782, 2784- 2786

Accelerators
Thin-Layer Chromatography

accelerators	solvent extraction, TLC	2787 2788 2789
guanidines, thiuram type carbamate type e.g. tetramethylthiuram monosulphide, dipentamethylene thiuram-tetrasulphide, cyclic thiuram piperidinium pentamethylene-dithiocarbamate, zinc (bismuth and cadmium)-dimethylthiocarbamate, 2-benzothiazyl N,N' diethyl-thiocarbonyl sulphide, 2-mercaptobenzthiazole, benzothiazyl disulphide,	solvent extraction, TLC for identi-fication	2778

Paper Chromatography

<u>Table 81</u>   (Continued).

accelerators   solvent extraction, paper chromatography 2784,2786
                                                                          2782

<h2>Mass Spectrmetry</h2>

2-mercapto-benzothiazole accelerator   volatile preconcentration/ 2790
sulphunamide types,   mass spectrometry down to
e.g. N-tertbutyl-2-   0.2%
benzothiazole sulphenamide,
2-(4 morpholinothio)benzothiazole,
2-(2,6-dimethyl morpholinothio)benz-
othiazole,
N,N-diisopropyl-2-sulphenamide,
N,N-dicyclohexyl-2-benzothiazole-
sulphenamide

MISCELLANEOUS RUBBERS

POLYCHLOROPRENE

<h2>Infrared Spectroscopy</h2>

Ferguson (2697) used infrared and NMR in his study of the microstructure of polychloroprenes.

Coleman and Painter (2698) separated the infrared bands for the preferred conformation of semi-crystalline trans-1,4-polychloroprene from the total infrared spectrum.

<h2>Raman Spectroscopy</h2>

Painter and Koenig (2699) have reported the applications of computerized Raman spectroscopy to the analysis of polymer molecular structure using trans-1, 4-polychloroprene, neoprene rubber, and butadiene rubber as models. The longitudinal acoustical mode of Raman spectroscopy was found to have a frequency sensitive to chain length and therefore related to lamellar thickness of crystalline polymers (2700).

Coleman et al (2701) used Raman to identify the crystalline and amorphous regions of trans-1,4-polychloroprene.

<h2>Nuclear Magnetic Resonance Spectroscopy</h2>

The microstructure of polychloroprene has been characterized by ^{13}C NMR (2702) and by high resolution NMR. Lines associated with diad and triad sequences involving trans-1,4; cis-1,4; 1,2-; isomerized 1,2-; and 3,4-structural irregularities were identified and assigned.

Okada and Ikushige (2674) have reported that in the proton NMR spectrum of polychloroprene dissolved in deutrobenzene, the =CH proton signal was separated into two triplet peaks. These triplet signals were assigned to the =CH proton in the trans-1,4 and cis-1,4 isomers by measurement of proton NMR spectra of 3-chloro-1-butene and a mixture of trans- and cis-2-chloro-2-butene as model compounds for the 1,2, trans-1,4 and cis-1,4 isomers. In proton NMR sp-

ectra (220 Mcps) of chloroprene dissolved in deuterobenzene, two triplet signals were separated completely from which the relative concentrations of trans-
1,4 and cis-1,4 isomers could be obtained quantitatively.

Fractionation

Kössler and Vodchnal (2703) have discussed the fractionation of polychloroprene using the Baker/Williams (2704) chromatographic precipitation technique
(benzene-methanol fractionating medium). These workers explain the previously
observed (2705-2714) anomalous decrease in molecular weights of the last fractions as a result of oxidation processes leading to the formation of peroxide
groups in the polymer. They found a correlation between the carbonyl group
content of the polymer and the constant K_1 of the Schulz-Blaschke equation.

Miscellaneous

Coleman et al (2715) discussed Fourier transform methods for determining crystalline vibrational bands, illustrating the approach by study of trans-1,4-
polychloroprene.

Kawasaki and Hashimoto (2716) have used swelling data to obtain the quantity
of crosslinking units in polychloroprene. Large has studied the effects of
solvent and wavelength of light in light scattering measurements on polychloroprene (2717).

BUTYL AND CHLOR-BUTYL RUBBERS

BUTYL RUBBER

Gallo et al (2718) and Lee et al (2719) have measured terminal unsaturation
in butyl rubber, but their hydrogenation was limited to butyl rubbers having
relatively high unsaturation in the 0.5 to 5.0 mole per cent range.

A radiochemical technique has been described by McNeill (2720) in which butyl
rubbers with unsaturations from 0.5 to 2.0 mole-% were reacted with radiochlorine (^{36}Cl). The unsaturation values obtained were in agreement with those
found by other procedures.

CHLOROBUTYL RUBBER

Falcon et al (2721) have described a method based on oxygen flask combustion
followed by a turbidimetric finish for determining chlorine at low levels in
chlorobutyl and other chlorine-containing polymers.

Chen and Field (2726) have used time averaged nuclear magnetic resonance spectroscopy for the detection of olefinic resonance at about 5 r in butyl rubber.

NITROSO RUBBERS

Clark et al (2727) have investigated the electron microscopy for chemical analysis (ESCA) of nitroso rubbers.

PVC-RUBBER BLENDS

Zitck and Zelingert (2728) have investigated the structure of PVC-rubber blends.

McNeill and Straiton (2729) have studied the degradation of blends of PVC with
chlorinated rubber.

POLY 1,3 PENTADIENE

NMR Spectroscopy

Beebe et al (882) have carried out a micro-structure determination on this
polymer by a combination of infrared, 60-MHz NMR, 300 MHz NMR, and x-ray diffraction spectrscopy.

The micro-structure of poly-trans-1,3-pentadiene prepared in heptane at 80°C
with butyllithium catalyst has been found by NMR spectroscopy at 90.51 MHz to
contain 15% 1,2 and 85% 1,4 units; 30% of the 1,4 units were cis and 70% were
of trans configuration (883).

Hirai et al (884) have carried out an electron spin resonance study on polymerization of conjugated dienes produced using an homogeneous catalyst derived
from n-butyl titanate and triethylaluminium.

Takeda et al (885) carried on electron spin resonance study of radiation induced solid-state polymerization on conjugated dienes.

Dannals (886) measured the number of crosslinked monomer units per weight-
average primary chain at the gel-point in the emulsion polymerization 1,3-
dienes.

Malone (887) has discussed a new approach for estimating the limiting refractive index (n) and tacticity of polymers based on structural knowledge and
derived an equation relating solution end-to-end distance with polymer solution refractive index. Correlations have been reported (888) between n and
structure of homo and copolymers of dienes and vinyl monomers.

CHAPTER 8

POLYESTERS AND POLYETHERS

POLYETHYLENE TEREPHTHALATE

Chemical Methods

Low temperature ashing has been used by Narasaki and Umezawa (2043) for controlled decomposition of polyethylene terephthalate prior to the determination of trace metals.

Carbon black in pigmented polyethylene terephthalate sheets has been determined by turbidimetry with a reproducibility of ± 0.01% and a variation coefficient of ± 3% (2044).

Pohorelsky and Heran (2045) determined 6-caprolactam in poly(ethylene terephthalate), after hydrolysis in 1 N hydrochloric acid colorimetrically with
ninhydrin.

Oprea and Pogorevici (2046) have reported on the gas chromatography determination of ethylene glycol, dimethyl terephthalate, dimethyl isophthalate, and
dimethyl adipate in ethylene glycol-isophthalic acid-terephthalic acid copolymers and in adipic acid-ethylene glycol-terephthalic copolymers.

Hydroxyl groups and ethylene glycol contents during poly(ethylene terephthalate) synthesis have been measured by acetylation and periodate methods respectively (2047).

Nissen and coworkers have described a method for carboxyl and groups in poly
(ethylene terephthalate). Hydrazinolysis led to formation of terephthalomono-
hydrazide from carboxylated terephthalyl residues to provide a selective analysis for carboxyl groups via ultraviolet absorbance at 240 nm. (2048)

Infrared Spectroscopy

Alter and Bonart (2050) showed that poly(1,4-butylene terephthalate) has a
kinked chain conformation with a tri-clinic unit cell.

348

The infrared absorbance substraction technique has been used to isolate bands, in composite samples of annealed melt-quenched and solution case samples of polyethylene terephthalate according to their crystalline or amorphous character (2051). The amorphous trans bands were found to shift approximately 1 to 3 cm^{-1} from their positions in the crystalline trans spectrum.

Stas'kov (2052, 2053) used polarization infrared to determine the character of molecular orientation in subsurface layers of rigid polymers.

Application of alternated total reflectance to infrared analysis of blocks, filaments, and films of polyethylene terephthalate has been discussed by Gusev and Glovachev (2054). Formation of a crystalline phase in polyethylene terephthalate led to absorption bands at 235, 138, and 85 cm^{-1} (2055).

Bell and coworkers used infrared dynamic mechanical and molecular weight measurements in studies of morphology of uni-axially oriented polyethylene terephthalate (2056) and of chain folding in annealed polymer (2057).

Raman Spectroscopy

Purvis and Bower (2058) studied molecular orientation in poly(ethylene terephthalate) by means of laser Raman spectroscopy.

Boerio et al (2059) discussed the band assignments in the Raman spectra of polyethylene terephthalate.

Bahl and coworkers (2060) and Boerio and Bailey (2061) discussed Raman spectra of polyethylene terephthalate and interpreted the nature of the bands at 1096,1000, 857 and 278 cm^{-1} and compared them with infrared absorption.

Raman spectrum of polyethylene terephthalate and its relation to molecular order has been reported by Derouault et al (2062, 2063).

Molecular orientation in polyethylene terephthalate was studied by polarized Raman scattering (2064, 2065).

Tirpak and Sibilia (2066) have described a new technique for obtaining infrared spectra of fine fibres of poly(ethylene terephthalate).

Nuclear Magnetic Resonance Spectroscopy

Several methods exist for the determination of the copolymerization ratio in polyethylene terephthalate-isophthalate using paper chromatography (2068, 2069), ascending chromatography (2070) gas chromatography (2071), infrared spectroscopy (2072) and the gravimetric method (2067). Although these methods can be used for the determination of the copolymerization ratio of polyethylene terephthalate-isophthalate after its hydrolysis, they are troublesome and not very accurate. Recently, the polarographic determination of isophthalate content in nitrated polyethylene terephthalate-1 polymers containing less than 30% isophthalate has been reported by Benisek (2073). Nuclear magnetic resonance spectroscopy has become useful for the determination of the copolymerization ratio of high polymers (2074).

Murano et al (2075) have described an accurate NMR method for the determination of isophthalate in polyethylene terephthalate isophthalate dissolved in 5% trichloroacetic acid. The NMR spectra of these polymers were measured on a High Resolution NMR spectrometer at 80°C. A singlet at 7.74 ppm is due to the four equivalent protons attached to the nucleus of the terephthalate unit. The complicated signals which appear at 8.21, 7.90, 7.80, 7.35, 7.22 and 7.10 ppm are due to the four protons attached to the nucleus of the isophthalate unit. The content of the isophthalate unit can be calculated from the integrated intensities of these peaks.

Structural information from NMR data on strongly extended noncrystalline chains in polyethylene terephthalate and in nylon 66 have been reported (2076).

High resolution NMR (2077) has been used for the qualitative and quantitative analysis of binary polyesters from phthalic and maleic anhydride and ethylene glycol, diethylene glycol, propylene glycol, and dipropylene glycol.

Electron Spin Resonance Spectroscopy

Cambell et al (2078) and Cambell and Turner (2079) have carried out ESR studies of free radicals formed by gamma irradiation of polyethylene terephthalate.

Pyrolysis-Gas Chromatography

Janssen et al (2080) have described a reaction-gas chromatographic method for the determination of 0-5% diethylene glycol incorporated in the polyethylene terephthalate chain. The method consists in a trans-esterification at high temperature (250°C) under pressure with an excess of ethanol. The diethylene glycol in the reaction mixture was determined by gas chromatographic analysis using a flame ionization detector. (Fig. 154).

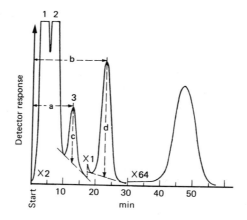

Fig. 154   Gas chromatogram of the trans-esterification mixture 1: ethanol; 2: glycol; 3: diethylene glycol (D.E.G.); 4: internat. standard (T.E.G.D.M.E.); 5: diethyl terephthalate; X2, etc; sensitivity factor F.I.D.

Thermal Methods

Schulkin et al (2081) have used differential thermal analysis to study crystallization, melting and glass transition temperatures in polyethylene terephthalate and poly 1,4-cyclohexylene dimethyl terephthalate.

Millen (2083) discussed multiple melting peaks in polyethylene terephthalate obtained by differential thermal analysis-differential scanning calorimetry.

The thermal degradation of poly(ethylene terephthalate) has been investigates by Troitskii and Varyukhin (2084). The kinetics of the polyethylene terephthalate degradation were affected by diethylene glycol content.

Gilland and Lewis (2085) found that cyclic trimer. tetramer, pentamer and hexamer were produced when terephthalate were heated at 240, 280, 330, and 360°C respectively.

Thin-layer and Column Chromatographic Techiques

Peebles et al (2086) have described thin-layer and column chromatographic techniques which can be used to isolate and identify the linear and the cyclic oligomers of poly(ethylene terephthalate). Extraction of the oligomers from high molecular weight polymer produces at least eight different cyclic species, some of which, they isolated and identified.

Cyclic oligimers in poly(ethylene terephthalate) have been determined by thin-layer chromatography using chloroform-ether (9:1) as a solvent (2087).

Zaborsky (2088) has described a high performance liquid chromatographic procedure for the determination of poly(ethylene terephthalate) prepolymer oligomers containing from 1-7 terephthaloyl repeat units.

Shiono (2089) has described a procedure for the determination of poly(ethylene terephthalate) oligomers in refrigeration oils by absorption column chromatography-gel permeation chromatography.

X-ray Diffraction

Lindner (2090) used x-ray diffraction to characterize crystalline, intermediate, and amorphous phases in poly(ethylene terephthalate). Wide and small-angle x-ray have been used in morphological studies of solvent induced crystallization of polyethylene terephthalate (2091). Small-angle measurements aided in establishing a relationship between chain fold period and crystallization temperature in polyethylene terephthalate (2092) and on its quantitative morphological characterization (2093),

Kashimaga et al (2094) gave a description of molecular orientation in polyethylene terephthalate films based on NMR, optical, and x-ray diffraction measurements.

The crystal density of semi-crystalline polymers such as poly(ethylene terephthalate) has been calculated from unit cell dimensions obtained by x-ray diffraction (2095).

Fractionation

Gel permeation chromatography using a nitrobenzene-tetrachloroethane solvent mixture has been applied to the fractionation of polyethylene terephthalate. (2096)

Gel permeation chromatography has been used to determine the oligomer composition in polymerization mixture of terephthalic acid and ethylene glycol (2097).

Van der-Maeden and coworkers (2098) used gradient elution high performance

liquid chromatography to analyze oligomeric mixtures of low molecular weight
resins, prepolymers and polymer extracts. The separation of epoxy resins, novo-
laks, poly(ethylene terephthalate) and nonionic surfactants are described.

Fractionation of poly(ethylene terephthalate) has been carried out from
phenol or phenol-tetrachloroethane solution (2099, 2100). Vasile et al proposed a turbidimetric procedure for determining the molecular weight distribution and the average molecular weight of polyethylene terephthalate which included viscosity measurements. Cooper and Samlyan (2101) described methods
for extracting cyclic oligomers from polyethylene terephthalate and analyzing
the extracts by gel permeation chromatography.

Shiono (2102) has used gel permeation chromatography to separate and identify
the oligomers of this polymer.

Molecular Weight

A nitrobenzene-tetrachloroethane solvent system has been developed for the molecular weight distribution analysis of poly(ethylene terephthalate) at room
temperature. Advantages of this solvent over the common solvent m-cresol at
110-135°C are discussed (2103).

The molecular weight of poly(ethylene terephthalate) has been determined from
intrinsic viscosity and from end-group analysis (2104).

Steinke and Vogel (2105) described how the solution viscosity of small poly
(ethylene terephthalate) samples could be derived from viscosity of the mixture
of the sample and a standard polyethylene terephthalate of known viscosity.

Walent et al (2106) reported a continuous fractionation procedure suitable for
molecular weight distribution studies on polyethylene terephthalate.

The molecular weight of polyethylene terephthalate has been determined by gel
permeation chromatography (2101).

Monomers and Water

A method involving solution in a solvent followed by gas chromatography has
been described for the determination of water (2108, 2109) and diethylene glycol (2107) in polyethylene terephthalate.

Miscellaneous

The application of fluorescence polarization to the measurement of orientation
in amorphous regions in polyethylene terephthalate has been discussed by McGraw
(2110). Burow et al (2111) discussed the yields of gases and carboxyl groups
produced upon gamma irradiation of polyethylene terephthalate.

Based on small-angle measurements, Mueller (2112) observed that the number of
crystallites and the maximum crystalline volume of amorphous poly(ethylene
terephthalate) annealed at 100-80°C was a function only of temperature. The
kinetics of reodering of poly(ethylene terephthalate) from the glassy amorphous state and from the melt were recorded as a function of temperature (2113).
Two crystalline modifications in poly(ethylene terephthalate) were indicated
by x-ray diffractograms prepared by Razumova et al (2114).

Berghmans and coworkers (2115) studied low frequency motions in poly(ethylene
terephthalate) using neutron inelastic scattering.

POLYCARBONATES

Chemical Methods

Terminal hydroxy groups were measured in polycarbonate following complexation with ceric ammonium nitrate; absorbance has been measured at 500 nm and at 530-540 (2248).

The reactivity of initial phenols and intermediate bisphenols in polycarbonate synthesis has been followed by photometric titration with sodium isopropoxide observing the equivalence point at 290-310 nm (2249).

Senetskaya et al (24) have described a spectrophotometric method for the determination of bisphenols of the triphenylmethane series in polycarbonate. The method is based on direct acid-base titration by a sodium isopropylate with the use of spectrophotometric titration. The method is also suitable for the separate determination of components in mixtures of phenol and bisphenols of the triphenylmethane series, used as initial and intermediate products in the synthesis of polycarbonates.

Infrared Spectroscopy

Gedemer (2250) has used attenuated total reflectance spectroscopy in conjunction with infrared spectroscopy to study the rate of degradation by ultraviolet light of polycarbonates. Since infrared spectral bands can, in many cases, be directly correlated with molecular structure, changes in sample composition due to ultraviolet attack become observable. Gedemer (2250) concludes that degradation of polycarbonates almost certainly includes scission of the ester linkages and transformation of the gem-dimethyl groups to terminal olefins in conjunction with the aromatic rings.

Infrared spectrscopy has been shown to distinguish between polycarbonate prepared from bisphenol A and 2,2-bis(4-hydroxyphenyl)-1,1-dichloroethylene or 2,2-bis(4-hydroxy-3,5-dichlorophenyl)-propane (2251).

Jansson and Yannas (2252) measured the infrared dichroism of polycarbonates at different strain levels. Below 0.6% strain, the dichroism is negligible while above this level the dichroism increases linearly.

Structural transitions and relaxation phenomena of polycarbonates have been followed by plotting the absorbances at 1230 cm^{-1} (stretching vibration of C-O-C groups) and 940 cm^{-1} (P-O-C group) against temperature (2253).

Vorob'yev and Vettegren (63) used infrared spectroscopy to determine temperature transitions in polycarbonate. Thermal transitions in polycarbonates were determined from the concentration variation of the residual solvent or plasticizer.

Difference infrared spectroscopy (100) has been used for the accurate measurement of small-induced frequency shifts in polycarbonate.

Nuclear Magnetic Resonance Spectroscopy

Stefan and Williams used pulsed NMR techniques to study molecular motions in bisphenol-A polycarbonate in homopolymers and copolymers (2258), and in blends, block copolymers, and composites (2259).

Phase separation in binary polymer systems have been studied by Nagumanova et

al (2260) using broad-line NMR. A non-additive increase in the second moment
(M_2) of the NMR line was observed in the region of critical concentration of
the second polymer. For polycarbonates, the non-additive changes in M_2 were
explained by an increase in the degree of ordering of the polycarbonate upon
transition from a single phase to a two-phase system. The composition of mixed polycarbonates formed by the polycondensation of bisphenol with phosgene
and 5-azidoisophthaloyl chloride has been determined by Smirnova et al (2261).

Jones and Bisceglia (164) have interpreted the proton and C^{13} spin-lattice
relaxation in methylene chloride solutions of polycarbonate. Carbon-13 and
proton spin-lattice relaxation times, the latter at two static field strengths,
were determined as a function of temperature for a bisphenol A polycarbonate
in methylene chloride solution: The results correlated well with those obtained from dielectric, wide-line and high resolution solid-state carbon-13 NMR
and dynamic mechanical analysis on solid polycarbonates

Electron Spin Resonance Spectroscopy

Placek et al (2262) conducted an electron spin resonance study of the macroradicals produced in polycarbonate by thermal decomposition of benzoyl peroxide
at high pressure.

Hama and Shinohara (2263) carried out electron spin resonance studies of polycarbonate irradiated by γ-rays and ultraviolet light.

Thermal Methods

Differential thermal analysis has been used (2264) to identify polycarbonates.

Differential scanning calorimetry has been used to examine polycarbonate (2265,
165).

Gel-Permeation Chromatography

Robertson et al (166) used gel-permeation chromatography to examine polycarbonates produced by different synthetic routes. They concluded that this method
is superior to end-group analysis. Hoore and Hillman (193) carried out fractionations of polycarbonate by gel-permeation chromatography. The polycarbonate
was fractionated from methylene chloride solution by a progressive precipitation technique. The intrinsic viscosities of these fractions were obtained
in methylene chloride solution and corresponded to a viscosity average molec.
wt. range of 2,000 to 80,000. The fractions were used to calibrate a gel-permeation chromatograph and an experimental Q-factor of 23.8 was found. The calibration was further confirmed by membrane osmometry and light scattering experiments. The experimental viscosity data demonstrated that the Kurata-Stock-
mayer-Roig relationship is suitable for interpreting the molecular expansion
of polycarbonate in methylene chloride.

X-ray Diffraction

X-ray diffraction and electron microscopy results have indicated that a molyb-
denum sulphide filler did not interfere with the crystallization of bisphenol
A polycarbonate (2266).

Volatiles

Di Pasquals et al (2267) have developed a head space technique for the determination of trace amounts of dichloromethane in polycarbonates.

Additives

A colorimetric method has been described for the determination of p-tert-butyl phenol in bisphenol A polycarbonates employing diazotized p-nitroanaline (2268).

POLYESTERS, (MISCELLANEOUS)

Chemical Methods

Urbanski (2116) measured small amounts of propylene oxide and epichlorohydrin in polyesters colorimetrically after reaction with a chloroform solution of 2,4 dinitrosulphonic acid.

Radovici et al (2117) reported on a polarographic method for the microdeter-mination of zinc in polyesters. Phthalic acid and chloride, bromide, iodide and nitro phthalic acids in polyesters were determined by conductometric tit-ration with triethanolamine in isopropanol (2118).

Kalinina and Doroshina (2119) reviewed chemical methods for qualitative and quantitative analyses of polyesters.

Moiseeva et al (2120) noted two inflections in the potentiometric titration of 1,4-diazobicycle (2.2.2) octane, a catalyst used in foaming of polyesters and polyethers. The titrant was 0.1 N hydrochloric acid in methyl ethyl ketone. The method permitted determination of the above catalyst in the presence of other amines, sodium acetate, or potassium hydroxide.

King and coworkers (2121) determined polyester in cotton-polyester blend fab-rics by dissolving the polyester in boiling monoethanolamine and weighing the cotton residue.

Infrared Spectroscopy

Near infrared absorption of irradiated polymers has been discussed by Vigneron and Deschreider (2122); included were polyesters, polystyrene and cellophane.

The infrared spectra of polyesters in the C-O-R stretching region has been discussed (2123). Infrared and Raman band assignments were made by Holland Moritz and coworkers (2124-2126) on linear aliphatic polyesters including in-fluence of the ester and CH_2 groups. The α and β forms of crystalline poly (ethylene glycol adipate) has been explained by infrared spectroscopy (2127). Funke and Schuh (2128) used infrared in studies of the chemical structure of radiation cured polyester resins.

The composition of cotton-polyester blends has been determined from measure-ments of the polyester C=O stretching at 1725 cm^{-1} (2129).

Vance et al (2130) have determined the oil content of oil modified o-phthalic polyester resins by infrared spectroscopy.

Nuclear Magnetic Resonance Spectroscopy

Yamadira and Murano (2131) have determined the randomness of copolyesters by high resolution NMR.

Murano (2132) has carried out NMR studies on terpolyesters. Olefinic proton resonance data have been reported on unsaturated polyesters of fumaric acid with ethylene glycol and sebacic or adipic acid (2133).

Slonim and coworkers (2134) described NMR methods to determine chain structure, composition and molecular weight of unsaturated polyesters.

Computer programmes have been reported for automated analyses of polyester NMR spectra, and cis-trans isomerization kinetics; results from analysis of polyethylene glycol adipate-sebacate having butoxy end-groups were given (2134).

Hydrolytic degradation of polyester plasticizers has been studied by NMR spectroscopy; observations were reported for plasticizers in neutral media and in carboxy-terminated adipic acid-ethylene glycol and adipic acid-neopentyl glycol copolymers (2135).

Cumene hydroperoxide in hardened polyester resins has been determined by decomposition with cobalt naphthalate and reaction of the free radical with diphenylamine to form a stable radical measurable ESR spectrum (2136).

A ^{13}C NMR and proton NMR study of the distribution of monomer units in copolyesters such as ethylene glycol sebacic acid-terephthalic copolymers has been reported by Urman et al (2137).

High resolution magnetic resonance spectroscopy has been used for the determination of composition and molecular weight of polyester urethanes(2138).

Cellarelli et al (197) prepared NMR spectra of polyesters derived from bridged bicyclic lactones. From a comparison with the spectrum of the monomer and by hydrolysis of the polymer, it is possible to assign to the polymer, obtained using n-butyllithium polymerization catalyst, a microstructure in which the ester groups on the 1 and 4 positions of the cyclohexane unit are cis to each other. Polymers obtained using sodium tert-butoxide and sodium/potassium alloy as polymerization initiators show spectral differences associated with the presence of another isomeric structural unit. Decreasing relaxation times of the cis unit by increasing the concentration of this new unit is reasonably explained by assuming for the latter a trans structure linked with the cis unit to form a stereocopolymer.

Pyrolysis-Gas Chromatography

de Anglis et al (2139) have investigated the pyrolysis in an oxygen-free atmosphere of polyester resin at temperatures not above 550°C. Pyrolysis gases were identified by gas chromatography. These workers quote reproducibility data and describe their pyrolysis apparatus.

Nelson et al (2140) identified polyester resins by pyrolysis-gas chromatography.

Esposito (2141) identified polyhydric alcohols in resins by programmed temperature gas chromatography.

Gas Chromatography

Esposito and Swann (2142) published a technique involving methanolysis of a polyester resin with lithium methoxide as a catalyst: the methyl esters formed were separated from the polyols and identified by gas chromatography. This method was recently improved by Percival (2143) by using sodium methoxide as a catalyst and injecting the reaction mixture of the transesterification directly into the gas chromatograph (without any preliminary separation).

Allen and coworkers (2144) hydrolyzed polyesters and converted the product acids and glycols including 1,4 cyclohexane-di-methanol and isophthalic acid to the corresponding trimethylsilyl esters and ethers which were then analyzed by gas chromatography.

Miejnek (2145) has described a method for identification of the basic components of polyesters consisting of splitting the polymers into the parent monomers by either hydrazinolysis or aminolysis followed by gas chromatography.

Aliphatic dicarboxylic acid esters such as adipates, sebacates, and azelates have been analyzed by coupling thin-layer chromatography with programmed gas chromatography (2146).

Tsarfin and Kharchenkova (2147) reported on a gas chromatographic method for the determination of diethylene glycol, glycerol, and trimethylolpropane in complex polyesters.

Resole polyesters have been analyzed after treatment with excess sodium sulphite at pH 9.9 to prevent hemiformal formation, followed by drying and acetylation with acetic anhydride-pyridine. The acetylated resole components were analyzed by gas chromatography (2148).

Thermal Methods

Differential thermal analysis has been used to identify polyesters (2149).

A directly coupled thermal balance and quadropole mass spectrometer system has been applied to polyesters (2150).

Direct pyrolysis of polyesters and copolyesters of lactic acid and glycolic acid yielded cyclic oligomers which were further degraded by an electron impact induced mechanism with the elimination of formaldehyde, acetaldehyde and carbon dioxide (2151).

Ng and Williams (210) carried out a thermochemical analysis of linear aromatic polyesters.

Fractionation

Both high performance liquid chromatography (2152) and gel permeation chromatography (2153) have been applied to the analysis of polyester prepolymer oligomers and resoles, respectively.

X-ray Diffraction

Vancso-Szmersanyi and Bodor (231) used x-ray diffraction to investigate the structure of polyesters with a molecular weight of 700 to 6000 and of various chemical compositions. The polyesters were investigated in solutions based on a number of solvents.

Molecular Weight

Gilbert and Hybart (2154) have discussed the osmometric determination of the molecular weight of linear polyesters with molecular weights between 900 and 20,000. O-chlorophenol tetrachloroethylene (60:40) was a suitable solvent. Using this method, number average molecular weights were measured for the aromatic polyesters; polyethylene terephthalate, polytetramethylene terephthalate, polypentamethylene terephthalate, polyhexamethylene terephthalate and polytetra-

358   Analysis of Plastics

methylene isophthalate.

Aliphatic polyesters such as polytetramethylene adipate and polyhexamethylene adipate with similar molecular weights were measured with the same membranes but with chloroform as the solvent at 25°C.

Eremeeva et al (2155) used absorption chromatography to determine the molecular weight distribution and the functionality of oligomeric adipic acid-trimethylol propane copolymer and adipic acid-diethylene glycol-trimethylol propane copolymer.

Additives

Iodometric methods have been described (2156) for the determination of organic peroxides in polyesters.

Gas chromatographic methods have been described for the determination of volatile constituents in polyester coated film (232) and in polyesters (248). Welsch et al (232) used gas chromatography combined with accumulated dosage. Volatiles from a heated sample are transferred in a carrier gas stream to a cooled initial part of the GC column, and are subsequently discharged onto the column by heating. The products were identified by mass spectrometry.

Tengler and Von Falkai (248) determined diols, adipic acid and volatile cyclo-mono-diol adipates qualitatively and quantitatively in aliphatic polyesters based on adipic acid. The identification of the cyclic mono-esters is accomplished with the help of coupling a gas chromatograph with a mass spectrometer, the diols are classified by gas chromatographic retention volume.

Surface tris(2,3-dibromopropyl)phosphate has been determined on the surface of retardent polyester fabrics (252). The technique used to determine the surface TDP levels involved extraction of the fabric with an organic solvent followed by analysis of the solvent by x-ray fluorescence for surface bromine and by high pressure liquid chromatography for molecular tris(2,3-dibromopropyl) phosphate.

Water

Moisture in polyesters has been reported to enhance the x-ray absorption coefficient but appears to have no influence on the crystal lattice (2157).

Miscellaneous

The fluorescence spectra of polyesters with a terephthalate main chain and a pendent ω-carbazylbutyl group has been explained on the basis of inter- and intramolecular interaction (2158).

Skorokhadov has reported structure studies on copolymers of unsaturated benzyl esters and dimethoxyethylene (2159).

The Bushuk and Benoit theory of light scattering in copolymer systems has been evaluated on polyesters relative to compositional heterogeneity (2160).

POLYGLYCOLS

POLY(ETHYLENE GLYCOL)

Chemical Methods

Spirin and Yatsimirskaya (2560) developed a sensitive dilatometric method for determining primary and secondary hydroxyl groups in oligomers based on reaction rate differences with phenyl isocyanate reagent.

Groom and coworkers (2561) employed trichloroacetyl isocyanate and trifluoro acetic anhydride acetylations to determine hydroxy end-groups in polyester polyols. The isocyanate reagent measured proton resonance and was best suited for samples having molecular weights of less than 4500; the anhydride method was more sensitive and applicable to higher molecular weights than is a ^{19}F NMR method.

Fritz et al (2562) have described a method for the determination of hydroxy groups in poly(ethylene glycols). The method has a sensitivity of 10^{-4} mol hydroxyl per kg polymer which is achieved by using silylation with arylsilyl-amines and subsequent photometric measurement of the silylated polymer.

Kaduji and Rees (432) have described a direct injection enthalpimetric method to measure the hydroxy valve of glycerol alkylene oxide polyethenes and butane-1,4-diol-adipic acid polyesters.

Infrared Spectroscopy

Infrared spectroscopy has given evidence for inversion ring opening in ethylene oxide polymerization based on comparisons of non-deuterated and deuterated species (2563).

The nature of bonding of adsorbed poly(ethylene glycol) to silica has been established from extinction of the 3300 cm^{-1} band (2564).

Infrared band assignments in the 4000-400 cm^{-1} region have been made for crystalline poly(ethylene glycol dimethyl ether) (2566).

Raman Spectroscopy

The longitudinal acoustic mode fundamental and the 3rd harmonic in the Raman spectrum of poly(ethylene oxide) have been observed to be a function of the oligomer (molecular weight about 200) content (2567).

Hartley et al (2565) used Raman scattering from the longitudinal acoustic mode in crystalline poly(ethylene oxide) to study crystal structure. Results from α, ω-hydroxy poly(ethylene oxides) with C_1-C_{18} alkoxy groups indicated that samples with lower and higher alkoxyl groups crystallized in chain-extended and folded forms respectively. The sample with C_7 alkoxyl groups formed both types of crystals.

Nuclear Magnetic Resonance Spectroscopy

Cross and Mackay (2568) analyzed alkyl ethoxylates for average alkyl chain length and average number of ethylene oxide units by examination of the NMR spectra of their trimethylsilylated derivatives. Hydrate formation in poly (ethylene oxides) has been followed by NMR (2569).

Slonim et al (2570) used Carbon-13 NMR to determine the molecular weight distribution of poly(methyl glycol) in aqueous formaldehyde solutions from the equilibrium constants of formaldehyde hydration.

Matsuzaki and Ito (2571) have carried out nuclear magnetic resonance studies

on the mechanism of ring opening in ethylene oxide polymerization, and Okada
(2572) has carried out proton NMR spectra measurements of poly(ethylene glycols) in the presence of shift reagent.

Holmes and Moniz (466) carried out a C^{13} NMR study of polyethylene glycol in
solution. They investigated polymer-solvent interactions in polyethylene glycol solutions in water, methanol, benzene, carbon tetrachloride and potassium
sulphate. A temperature study of the carbon-13 chemical shifts and spin-lattice relaxation times in polyethylene glycol of molecular weight 1000 in carbon tetrachloride and potassium sulphate was undertaken in an attempt to clar-
ify these interactions.

Electron Spin Resonance Spectroscopy

Thyron and Baijal (2573) have carried out an electron spin resonance study of
polyglycols.

Concentrations of free radicals produced by γ-irradiation of poly(ethylene
glycol) have been determined by an electron spin echo method at 77 and 4.20K,
(2574).

Gas Chromatography

Using programmed temperature gas chromatography, esters prepared from polyols
present in synthetic resins have been identified by Esposito and Swann (2575).
The polyols are first treated with butylamine and then esterified with acetic
anhydride. The esters, as a chloroform extract, are resolved on a Carbowax
20-M column over a temperature range of 50°-$225^{o}C$.

Oxidative Degradation-Gas Chromatography

Schole et al (2576) have reported on the characterization of polymers using
an oxidative degradation technique. In this method the oxidation products of
the polymers are produced in a short pre-column maintained at 100 to $600^{o}C$
just ahead of the separation column in a gas chromatograph. The oxidation
products are swept on to the separation column and detected in the normal manner.

Figure 155 shows the chromatograms of Carbowax 20M and UCON Polar polyglycols.
The compnonents are the oxidation products produced in the pre-column by the
oxygen. The pre-columns were thermostated at 203° and $202.5^{o}C$ respectively.

Fractionation

Polymer mixtures of polyethylenes, polyoxypropylene, polytetrahydrofuran, polyepichlorohydrin, or oligobutadiene have been quantitatively separated by column chromatography on silica gel (2577).

Kato and coworkers (2578) have used a high speed aqueous gel permeation chromatography to demonstrate good resolution of polyethylene glycol oligomers and
small molecules on TSK-Gel type PW packing.

Molecular Weight

Molecular weight distributions of poly(ethylene glycols) have been determined
by liquid-liquid extraction (2579), paper chromatography (2580), thin-layer
chromatography (2581), gas chromatography (2582-2583), fractional precipitation (2584) and gel permeation chromatography (2585). Gas chromatography,

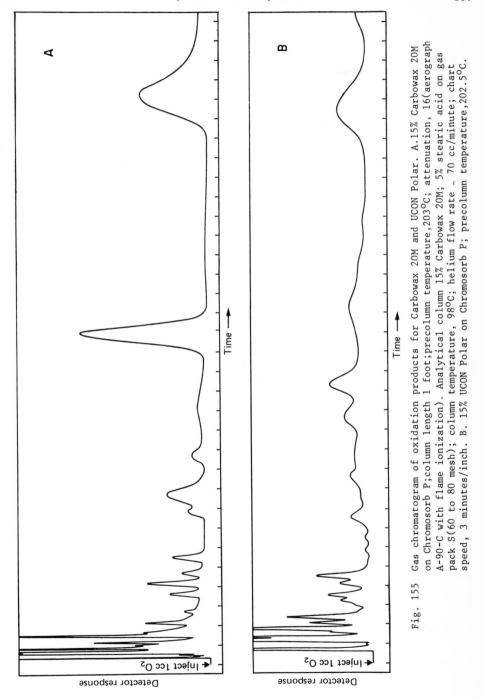

Fig. 155 Gas chromatogram of oxidation products for Carbowax 20M and UCON Polar. A.15% Carbowax 20M on Chromosorb P;column length 1 foot;precolumn temperature,203°C;precolumn temperature,203°C; attenuation, 16(aerograph A-90-C with flame ionization). Analytical column 15% Carbowax 20M; 5% stearic acid on gas pack S(60 to 80 mesh); column temperature, 98°C; helium flow rate ~ 70 cc/minute; chart speed, 3 minutes/inch. B. 15% UCON Polar on Chromosorb P; precolumn temperature,202.5°C.

when applicable, is much simpler and considerably faster than other methods
for polyethylene glycols.

Mikkelsen (2582) reported a gas chromatographic analysis of polyethylene glycol 400; the original sample is injected directly into the gas chromatograph,a
technique also employed by Puschmann (2583). Celades and Pacquot (2584) converted the polyethylene glycols into methyl ethers in a reaction with dimethyl
sulphate before injection into the gas chromatograph.

The rapid formation of trimethylsilyl ether derivatives of polyhydroxy compounds, followed by their separation and estimation by gas chromatography, has
been described by Sweeley et al (2586). Crabb (2587) applied this most useful
technique to the liquid poly(ethylene glycols) 200, 300 and 400. Fletcher and
Persinger (2588) studied this application using poly(ethylene glycols) 200,
3oo and 400 and reported the determination of response factors relative to
ethylene glycol for conversion of peak areas to weight per cent when a thermal
conductivity detector is used. A typical chromatogram of PEG 400 is shown in
Fig. 156.

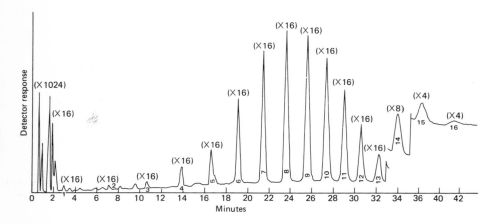

Fig. 156 Gas chromatogram of trimethylsilyl derivative of
          PEG 400. Dual columns, 3ft. by 0.25 in o.d. 5%
          SE-30 on Chromosorb G. Temperature programmed
          from 100 to 370°C at 10°C/min. and then isother-
          mal at 370°C.

Gruber and Elias (2589) have discussed apparatus suitable for the cloud point
titration to determine the critical non-solvent fraction (φ crit) of polyethylene glycol. The influence and significance of possible errors are examined.
They emphasize that to obtain optimum results one should use good solvents,
non-solvents of moderate precipitating power, and small volumes in the cell
Observing these working conditions, a standard error of ± 0.3% can be obtained
for φ crit. To work within this limit of error, the absolute temperature should
not deviate by more than ± 0.15°C.

Turbidimetric titration to determine the molecular weight has been demonstrated with poly(ethylene oxide) (2590).

Moore and Hendrickson (2591) applied gel permeation chromatography to the determination of the molecular weight distribution of polyglycols.

Kalinina and Motorina (2592) developed a method for the determination of the
molecular weight of compounds having terminal hydroxyl groups, e.g. polyethylene glycols. This was estimated from treatment with ceric ammonium nitrate
reagent in nitric acid followed by spectrophotometric measurement of the coloured complex.

Ingham and Lawson (2593) have established refractive index-molecular weight
relationships for polyethylene oxide.

Kuresavic and Samsa (2594) have determined the molecular weight of polyethylene glycols in the range 400-6000 by thermoelectric vapour pressure osmometry
in methyl alcohol or chloroform solution.

Suzuki and Leonis (2595) have used sedimentation data to estimate the molecular weight of polyglycols and polyesters.

X-ray Diffraction

Poly(ethylene oxide) swoolen with diethyl phthalate has been examined by small-
angle x-ray and differential scanning calorimetry (2596).

Miscellaneous

Studies of molecular motions in partially crystalline polymers have been reported by Bergmann (2597) on poly(ethylene oxide).

Steigen has described instrumentation for determining cloud points of polymer
systems, such as polyoxyethylenated non-ionic surfactants (2598).

Takahashi and coworkers (2599) reported on a new crystal modification in poly
(ethylene oxide) which has been stretched at room temperature.

Klein and Widdecke (2600) discussed the use of gas chromatography for characterizing poly(ethylene oxide) and polystyrene. Retention indexes were determined
for benzene, ethanol, ethyl acetate, methyl nitrate and pyridine as polar test
molecules.

Lee and Segwick (2601) found that the mass spectra of linear copolymers of
ethylene and propylene oxides showed features characteristic of the two monomers permitting identification of random and block structures.

POLYPROPYLENE GLYCOL

Chemical Methods

Stetzler and Smullin (2603) found that for polypropylene glycols the classical phthalation procedure gave consistently low results as did the perchloric
acid catalyzed acetylation procedure, developed by Fritz and Schenk (2604).
In addition to its greater intrinsic accuracy, the advantages claimed for the
Stetzler and Smullin (2603) toluene p-sulphonic acid catalyzed procedure over
phthalation include shorter reaction time and lower reaction temperature (en-
abling reaction to be carried out in a closed vessel), also good reproducibility, indicating no reaction with the ether groups.

In Table 82 are compared hydroxy values obtained by three methods. It is seen

Table 82 Comparison of Results Obtained on Polypropylene-
glycol by Catalyzed Acetylation,Phthalation and
Phenyl Isocyanate Reaction.

Mol. wt.	Stetzler and Smullin (2603) PTSA Method Average OH No. mg/KOH/g	Phthalation Method Average OH No. mg/KOH/g	Reaction with phenyl isocyanate Average OH No. mg/KOH/g
700	353	346	
1260	276	265	
5000	34.9	34.0	34.9
5000	34.3	33.8	35.1
2890	58.3	57.0	

that in every case the acid catalyzed acetylation method gives a higher result
than that obtained by phthalation, the average difference between the two
methods being 2.5% of the determined value. It will be seen from the Table
that the agreement between the p-toluene sulphonic acid method and the phenyl
isocyanate method is good.

The Stetzler and Smullin (2603) method is applicable to polyoxyethylene, polyoxypropylene, ethylene oxide tipped glycerol/propylene oxide condensates. Compounds of this type in the molecular weight range 500 to 5000 can be analyzed
by this procedure.

Earlier work by Price and Osgan (2605) suggested that the origin of non-cry-
stallizatine fractions of high molecular weight poly(propylene oxide) from
optically active monomer with stereoselective coordination catalysts was partial racemization of the asymmetric centre through carbonium intermediates
producing partially atactic polymer. More recent work (Price et al 2632) has
shown that this interpretation is generally incorrect and that the non-crystalline fractions owe their origin to positional isomerism. This work showed
that by extensive ozonation and lithium aluminium hydride reduction, poly(propylene oxide) can be converted to material containing a dipropylene glycol.
Since the diprimary, disecondary, and primary-secondary isomers are readily
separable by gas chromatography, it was possible to show that some (but not
all) non-crystalline polypropylene oxide samples contain many head-to-head,
tail-to-tail monomer units. Correlation of the fraction of head-to-head units
with the optical rotation of non-crystalline fractions from optically active
monomer, indicates that there is one asymmetric centre inverted for every unit
inserted head-to-head.

The ozone degradation of polyethers (2633) has been extended so as to produce
three structurally isomeric dipropylene glycols, diprimary (I), disecondary
(II), and primary-secondary (III), and for isomer I, the two stereoisomeric
forms were readily separable.

$$(HOCH_2CH)_2O \qquad\qquad (HOCHCH_2)_2O \qquad\qquad HOCH_2CHOCH_2CHOH$$
$$\,\,\,\,\,\,\,| \qquad\qquad\qquad\qquad\quad | \qquad\qquad\qquad\qquad\quad | \qquad\quad\, |$$
$$CH_3 \qquad\qquad\qquad\qquad CH_3 \qquad\qquad\qquad\quad CH_3 \quad CH_3$$

I   II   III

(two isomers)

For the non-crystalline fractions of polypropylene oxide substantial amounts of dimer were isomers I and II. The presence of these two isomers must have arisen from head-to-head and tail-to-tail units in the polymer chain.

$$
\underset{\text{(1)}}{-\overset{\displaystyle a}{|}\!\!\!\overset{|}{O}\text{-}CH_2CH\text{-}\overset{\displaystyle b}{|}\!\!\!\overset{|}{O}\text{-}CH_2CH\text{-}} \underset{\text{(2)}}{\overset{\displaystyle c}{|}\!\!\!\overset{|}{O}\text{-}} \underset{\text{(3)}}{\overset{\displaystyle a}{|}\!\!\!\overset{|}{}CHCH_2\text{-}O\text{-}} \underset{\text{(4)}}{\overset{\displaystyle b}{|}CH_2CH\text{-}O\overset{\displaystyle c}{|}\!\!\!\overset{|}{}\text{-}}
$$

```
      a |         b |        c | a |       b           c
  - -|O-CH2CH-|-O-CH2CH-|-O--|-CHCH2-O-|-CH2CH-O-|-
      |  |        |  |     | |         |        |
      CH3          CH3      CH3         CH3

     (1)       (2)          (3)         (4)
```

The insertion of a single unit, (3) in the chain, in the inverse structural sense would provide the opportunity for producing the various diglycol isomers observed. Ozone cleavage at aa would produce III; cleavage at bb would produce I, at cc, II. The oxone cleavage procedure employed by Price et al (2605) cannot necessarily give information on the stereochemistry of adjacent units in the chain, as can the butyllithium procedure used by Vandenberg (2606). The ozonation produces ester, ketone, aldehyde, and carbonyl groups, all reduced back to alcohol by lithium aluminium hydride. It is highly likely that many secondary alcohols were ketones before reduction, and thus would have lost their original configuration.

In Table 83 are correlated the results of the structural isomerism in various samples of polypropylene oxide with optical data obtained for similar polymer samples made from optically pure monomer. The data are very well interrelated if one assumes that whenever a monomer unit was inserted inverse to the normal opening at the primary carbon, it not only produced head-to-head and tail-to-tail sequences, but must have opened the epoxide ring at the asymmetric centre with inversion of configuration.

These data are thus in full accord with those of Vandenberg on cis- and trans-2-butene oxides (2606).

Table 83   Diglycols from Ozonation of Poly(propylene Oxide).

PPO	III,%[a]	I and II,%		I and II (calcd.),[e] %
Al(i-PrO)$_3$ZnCl$_2$ (isotactic)[c]	>95	<5	-25°	-
Commercial (General Tire)	>95	<5	-	-
KOH	~90	~10	~-20°	10
AlEt$_3$ H$_2$O[d]	76	24	-18°	14
FeCl$_3$(PO)$_4$[d]	71	29	-10°	30
Et$_2$Zn H$_2$O[d]	67	33	-5°	40

a By vapour-phase chromatographic analysis
d Amorphous fractions
e Calculated from $(\alpha)_D^{20}$ assuming an inversion of configuration for every head-to-head unit.

Figure 157 is an example of a gas chromatogram obtained for the ozonolysis product of an amorphous polypropyleneoxide showing substantial amounts of I, II and III. Identifications of ozonolysis products of polypropylene oxide were compared by comparing their infrared spectra with those of authentic compounds.

Belen'kii et al (2607) determined the number of hydroxy groups in poly(prop-

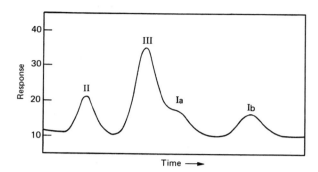

Fig. 157   Gas liquid chromatogram of dipropylene glycol from
amorphous poly(propylene glycol), $ZnEt_2$. H_2O catalyst.

ylene oxide) and hydroxy containing impurities such as alcohol by thin-layer
chromatography on silica gel. Water-saturated ethyl acetate with 2% methyl
ethyl ketone was used as the eluent.

Infrared Spectroscopy

The monoconfigurational sequence distribution for poly(propylene oxide) has
been determined from the band shape analysis of the infrared spectra within
the range of 1240 to 1300 cm^{-1} (2608).

Nuclear Magnetic Resonance Spectroscopy

Stereoregularity in polypropylene oxide has been measured by ^{13}C NMR, together with mechanism of polymerization (2609, 2610). Schaefer (2611) used
noise-modulated partial decoupling to aid in interpreting ^{13}C NMR spectra of
polypropylene oxide.

Pyrolysis-Gas Chromatography

The ethylene oxide and propylene oxide content of ethylene oxide-propylene
oxide copolymers has been calculated from the ethene and propene formed on
pyrolysis of the copolymer at 700-800°C (2612).

Fractionation

Vakhtina et al (2613) have reported on the thin-layer chromatographic fractionation of poly(oxypropylene) glycols on the basis of molecular weight and
composition.

Bode et al (2614) have used a dielectric constant detector for ion-exclusion
chromatography (SEC) of polypropylene oxide-polypropylene oxide oligomers.

Molecular Weight

Vakhtina (2615) has reported on the determination of functionality, molecular
weight and polydispersity of poly(propylene glycol) by gel permeation chromatography.

Oligomers of poly(ethylene glycol) and poly(propylene glycol) have been characterized by gel permeation chromatography (2616).

Miscellaneous

The linear relation between the density difference of polymer and solvent and
polymer concentration has been demonstrated for solutions of poly(propylene
oxide) in dimethyl formamide and chloroform (2617),

Velocity and attenuation coefficient of longitudinal hypersound measurements
of poly(propylene glycol) by Huang and Wang (2618) showed the presence of molecular relaxations between 250 and 390 K. The observation of entire relaxation
in the sound attenuation vs. temperature curve appears to be the first of its
kind ever to have been made by Brillouin scattering measurements.

ETHYLENE OXIDE-PROPYLENE OXIDE CONDENSATES

Chemical Methods

Cervenka and Merrall (2619) have investigated the application of acidic dehydration of ethylene oxide-propylene oxide condensates in α-bromonapthalene in
the presence of p-toluene sulphonic acid to methyldioxane-1,4 to the elucidation of the molecular structure and monomer sequence of these polymers. Gas
chromatography was used to determine dehydration products.

Studies on poly(ethylene glycol) and poly(propylene glycol) homopolymers showed
that dioxane and its derivatives are not the only reaction products. Dehydration of poly(ethylene glycol) gave three products (E_1-E_3); that of poly(propylene glycol) six products (P_1-P_6). The majority of them were identified (Table
84).

Data obtained on dimers, trimers, and poly(alkylene oxides) of molecular weight
400 to 1500 show that in the case of ethylene oxide chains the amount of dioxane decreases very slightly with decreasing molecular weight, and this is
compensated by an increase in the proportion of dioxolane. The acetaldehyde
concentration is independent of the molecular weight. The reverse and more
pronounced trend has been obtained for dimethyldioxanes in the case of propylene oxide based materials. The concentration of acetone and mesityl oxide
formed by the aldolization and dehydration thereof decreases with decreasing
molecular weight.

Cervenka and Merrall (2619) conclude that results on homopolymers, their blends,
and model copolymers of different chain architectures demonstrate that acidic
dehydration is capable of distinguishing ethylene oxide-propylene oxide copolymers of different structures, giving correct absolute values of overall
monomer contents and also ranking polyols according to their degrees of randomness. In comparison with other techniques suitable for characterization of
ethylene oxide-propylene oxide copolymers, e.g. infrared spectroscopy (2620,
2621) chemical fission (2622, 2623) and pyrolysis (2624, 2625) (all methods
giving overall monomer composition only) they believe that their method yields
a deeper insight into the structure.

Nuclear Magnetic Resonance Spectroscopy

Diad and triad sequences in ethylene oxide-propylene oxide copolymers were
determined from ^{13}C NMR intensities (2626).

Pyrolysis-Gas Chromatography

Neumann and Nadeau (2627) used pyrolysis in an evacuated vial at 360-410°C
(2628) and gas chromatography to determine the quantitative composition of various ethylene oxide-propylene oxide copolymers.

Table 84   Products from the Acidic Dehydration of Homopoly-
           mers and Their Blends.

Code	Product
E_1	Acetaldehyde
E_2	2-Methyldioxolane
E_3	Dioxane-1,4
P_1	Acetone
P_2	-
P_3 }	Isomers of dimethyl-dioxane-
P_4 }	1,4; substitution in the
P_5 }	positions, 2,5 and 2,6
P_6	Mesityloxide
X_1	dimethyldioxolane
X_2	Methyldioxane-1,4

In all of the polymers studied by Neumann and Nadeau (2627) there was no significant differences in the pyrolysis chromatograms of samples heated for ½
to 2 hours (Table 85 shows the differences between a ½ and a 2 hour pyrolysis).
It appears that the only effect of time of pyrolysis is the total amount of gas
produced. The relative concentrations of components is not significantly chan-
ged. The temperature of pyrolysis, however, does play a large part in both the
amount and type of components in the volatile gases. Between 390° to 410°C,
there was no noticeable change in products.

Table 85   Effect of Pyrolysis Time on the Pyrolysis Pattern.

	CH_4,%	C_2H_4, %	C_2H_6, %	C_3's, %
Polypropylene glycol (½ hr. at 360°C)	15	0.5	8	77
Polypropylene glycol (2 hr. at 360°C)	15	0.5	10	74

Neumann and Nadeau (2627) pyrolyzed several simple glycols and their derivatives in order to obtain information on the degradation products. The pyrolysis
products are listed in Table 85.

Figure 158 is a plot of the per cent ethylene as a function of the ethylene
oxide content of the condensate. As the ethoxy character of the pluronic increases, so does the amount of ethylene produced. The curve is linear up to
about 50% ethylene oxide, then turns sharply upward to an ethylene content of
38.6% for pure polyethylene glycol.

The relative contents of ethylene oxide and propylene oxide in polyethylene-
polypropylene glycols has been determined using combined pyrolysis-gas chromatography calibrated with polyethylene glycol and polypropylene glycol standards (2629).

GLYCEROL-PROPYLENE OXIDE CONDENSATES

Chemical Methods

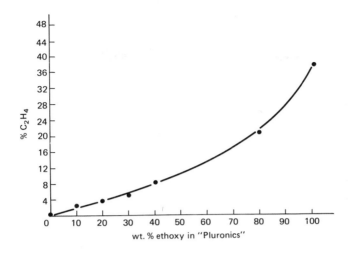

Fig. 158   Per cent C_2H_4 as function of ethoxy content of
Pluronics, listed in Table 85.

Crompton (2630) has developed a kinetic method for the evaluation of the reactivity of polyols in the range 3000-5000. This procedure may be used to determine the degree of 'tipping' produced by the addition of ethylene oxide to
glycerol/propylene oxide condensates. The method was developed to measure the
relative reactivity with isocyanates of such condensates as a function of their
ethylene oxide tipping content. The method is based on the observation that
primary hydroxyl groups react with phenyl isocyanate to form a urethane faster
than do secondary hydroxyl groups. Hence a 'tipped' polyol will react more
completely in a given time with an equivalent amount of phenyl isocyanate than
an 'untipped' polyol reacting under the same conditions. The greater the amount
of ethylene oxide tipping the greater its rate of reaction with phenyl isocyanate.

The reaction is carried out under standard conditions in which a calculated
weight of the polyol (depending on its hydroxyl number) is reacted under standard conditions with a standard toluene solution of phenyl isocyanate in the
presence of a basic catalyst. Reactivity is calculated by a kinetic procedure
in which the percentage of the original phenyl isocyanate addition, which reacts with the sample in a given time, is taken as an index of its reactivity.
A calibration graph can be prepared in which the determined phenyl isocyanate
reactivity is plotted against the ethylene oxide content and this enables a
determination to be made of the ethylene oxide content of unknown samples from
the calibration graph by interpolation.

Crompton applied this method to a range of glycerol/propylene oxide adducts
containing various accurately known amounts of ethylene oxide tipping, up to
5.3 moles, (Table 86). Figure 159 shows the reaction curves for each of the
standard samples analyzed. It is seen that increasing the ethylene oxide tip
content of a polyol leads to a distinct increase in the reactivity of the polyol with phenyl isocyanate. As expected, "untipped" polyols which are relatively free from primary hydroxyl groups, i.e. curves A and B, react comparatively
slowly with phenyl isocyanate. Decinormal solutions of 'untipped' glycerol/

Table 86   Application of Reactivity Method to Standard Ethylene Oxide Tipped Polyols.

Sample identification	Approximate molecular weight	Hydroxyl number mg KOH/g polyol	Ethylene oxide tipping moles ethylene oxide/ mole glycerol (by weight addition)	Reactivity, i.e. % of original phenyl isocyanate addition consumed in:-	
				60 minutes reaction	100 minutes reaction period
A	3000	59.0	0.0	25.2	35.2
B	5000	34.9	0.0	27.1	36.8
C	5000	34.3	3.0	41.5	46.6
D	5000	35.9	3.5	44.4	51.0
E	5000	36.0	4.3	47.5	54.1
F	5000	33.3	5.3	51.9	57.4

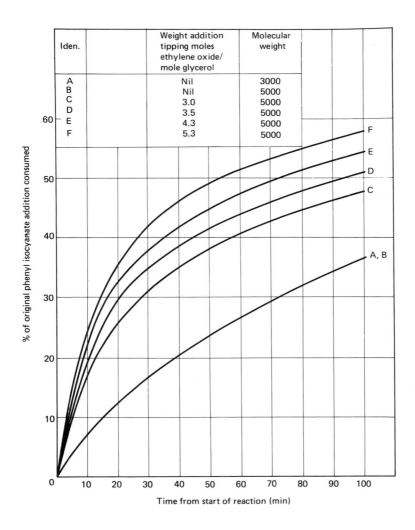

Fig. 159 Standard polyol samples-reaction curves.

propylene oxide condensates of molecular weight 3000 and 5000 had an identical rate of reaction with phenyl isocyanate (Fig. 159 curves A and B). Thus the rate of reaction with phenyl isocyanate of the terminal isopropanol end-groups in polyols is independent of molecular weight in the molecular weight range 3000 to 5000 and depends only on proportions of primary and secondary end-groups present.

A calibration curve is prepared (Fig. 160) by plotting moles ethylene oxide per mole of glycerol for the range of standard tipped polyols of known ethylene oxide content against % of original phenyl isocyanate addition consumed after 60 and 100 minutes, i.e. P60% and P100%. This curve can be used to obtain from

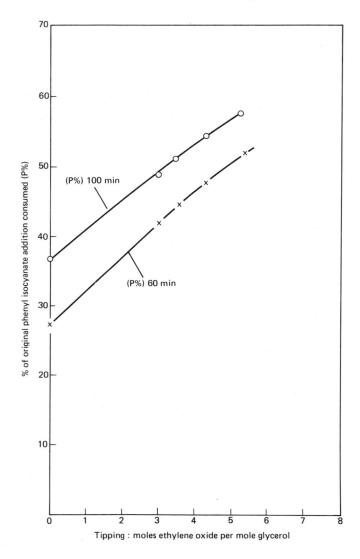

Fig. 160   Calibration plot. Ethylene oxide glycerol-propy-
            lene oxide polyols.

P60% and P100% data obtained for tipped glycerol for propylene oxide polyols
of unknown composition their tipped ethylene oxide contents (in moles ethylene oxide per mole glycerol).

Gas Chromatography

Gas chromatography (2631), has been applied to the analysis of propylene oxide glycerol condensates of molecular weight up to 300. The condensates are first acetylated by a rapid procedure using p-toluene sulphonic acid as catalyst. Up to seven peaks were observed with initial retention distances (DRI) and identities as given in Table 87. Graphs relating log DRI and molecular weights gave straight line relationships.

Table 87   Initial Retention Distances

Peak	Identity	on "Reoplex' 400 column Column temp. 240°C Column inlet pressure (psig			on Silicone E301 column Column temp. 200°C
		5	10	15	
		D_{RI}(mm.)			
A	ethyl acetate	0	0	0	0
B	glycerol triacetate	5	2½	1½	1¼
C	triacetate 4 1:1 glycerol/propylene oxide adduct	10	5	3	3½
D	triacetate of 1:2 glycerol/propylene oxide adducts	21½	10½	6½	10
E	triacetate of 1:3 glycerol/propylene oxide adducts	44½	22	14	26
F	triacetate of 1:4 glycerol/propylene oxide adducts	93½	45½	29	72
G	triacetate of 1:5 glycerol/propylene oxide adducts	195	95½	60½	-

POLYETHERS, MISCELLANEOUS

Ultraviolet Spectroscopy

Barzykina et al (2161) have determined the concentration of active centres in the cationic polymerization of cyclic ethers by ultraviolet spectroscopy.

Raman Spectroscopy

The rate constant activation energies and entropies of activation for reactions of diglycidyl ethers with diols have been determined by following changes in the intensity of the band at 917 cm^{-1} (stretching vibration of the epoxy ring) (2162). Matsui et al (2163) have prepared Raman spectra of polyethers.

Nuclear Magnetic Resonance Spectroscopy

Yuki et al (2164) have discussed the NMR spectra of vinyl ethers.

High resolution NMR has been used in studies of the solution polymerization of cyclic ethers (2165).

Kawamura and Uryu et al (2166) have carried out ^{13}C nuclear magnetic resonance studies on the stereochemical configuration of 2,4-dimethoxypentane and poly (alkyl vinyl ethers).

Diad, triad and pentad sequences in poly(methyl vinyl ether) have been assigned from ^{13}C NMR data; 2,4-dimethoxypentane served as model compound in these studies which included effects of solvent and temperature on chemical shifts (2167).

The mechanism of polymerization and chemical structure of poly(vinyl phenyl ether) have been established from NMR and IR data (2168).

Hiyashimura et al (2169) have prepared the proton and ^{13}C NMR spectra of poly (propenyl ethers).

Matsuzaki et al (2170) have investigated the stereoregularity of poly(α-methyl-vinyl alkyl ethers) determined by proton and ^{13}C NMR.spectroscopy.

Miscellaneous

Wiley (2171) has investigated the mass spectral characteristics of poly(2,6-dimethyl-1,4-phenylene ether).

High pressure viscosity measurements on a polyphenyl ether have been made using a concentric-cylinder or Couette viscometer (2172).

POLYPHENYLENES AND POLYPHENYLENE OXIDES

Electron Spin Resonance Spectroscopy

Lerner (2173) applied ESR and chemical analysis in his study of p-phenylene formed using aluminium trichloride and cuprous chloride catalysts.

Montaudo and coworkers (2174) established the thermo degradation mechanisms for poly(oxy-1,4 phenylene), poly(thio-1,4-phenylene), and poly(dithio-1,4-phenylene) by direct pyrolysis of the polymers in the ion source of a mass spectrometer. The electron impact induced fragmentation of the pyrolytic fragments was suppressed by using low ionization energy.

Thermal Degradation

Ehlers et al (2175-2177) have conducted studies of the thermal degradation of various polymers with phenylene units in the chain including polyphenylenes, polyphenylene oxides, sulphur containing polyarylenes and polyarylates.

Molecular Weight

Lanikova (2178) has determined the molecular weight distribution of polyphenyl-oxide using turbidimetric titration.

The molecular weight of poly(phenylene oxide) has been calculated from the absorbance of the terminal hydroxy groups measured at 3615 cm^{-1} in carbon tetrachloride solution (2179).

Structural modifications of polydioxolane and polyoxymethylene were determined by infrared and x-ray analysis; a spiral form of the latter was found characterized by absorption bands at 2900, 1125, 1028, 980, 970 and 938 cm^{-1} (2180).

REFERENCES

1.   Gouveneur, P., Kuijper, E. and Schonebaum, R.C., Shell Chemical Co.
        Ltd., Amsterdam, private communication.
2.   Crompton, T.R., private communication.
3.   Colman, D.M., Determination of low chlorine concentration in
        plastics. Radiation Laboratory, University of California,
        Livermore, California, USA (1958).
4.   Ubaldini, I., Capizzi Maitan, Chim. e Ind., 37, 799 (1955).
5.   Seefield, E.W. and Robinson, J.W., Anal. Chem. Acta, 23, 301 (1960).
6.   Zall, D.M., Fischer, D. and Gardner, M.A., Anal. Chem., 38, 1665
        (1956).
7.   Vecera, M. and Spevak, A., Chem. Listy, 51, 2037 (1957).
8.   Lefferts, D.T., Microchem. J., 2, 257 (1958).
9.   Urbanski, J. and Iwanska, S., Chem. Anal. (Warsaw), 7, 1129 (1962).
10.   Hernandez, H.A., International Laboratory, 84, (Sept. 1981).
11.   Sokolov, D.N., Nesterenko, G.N. and Golubeva, L.K., Zavod. Lab.,
        39, (8), 939 (1973).
12.   Gorsuch, T., Analyst, 87, 112 (1962).
13.   Gorsuch, T., Analyst, 84, 135 (159).
14.   Kalinina, L.S. and Doroshina, L.I., Metody Ispyt. Kontr. Issled.
        Mashinostrolt, Mater., 3, 5 (1973); Chem. Abstr., 80, 100 (1974).
15.   Hausler, K.G., Klemm, E., Fricke, E. and Benn, I., Plast. u. Kaut.,
        27, 678 (1980).
16.   D-ASTM-2283-64T, Tentative Methods of Test for the Absorbance of
        Polyethylene Due to Methyl Groups at 1378 cm^{-1}, Method A.
17.   Neilson, O., and Holland, O., J. Mol. Spect., 4, 448 (1960).
18.   Stanescu, G., Revista de Chemie Bucharest, 64, 42 (1963).
19.   Boyle, D.A., Simpson, W. and Waldron, J.D., Polymer, 2, 323 (1961).
20.   Willbourn, A.H., J. Polymer Science, 34, 509 (1959).
21.   Salovey, R. and Pascale, J.V., J. Polymer Science, 2, 2041 (1964).
22.   Luongo, J.P. and Salovey, R., J..Appl. Polymer Science, 7, 2307 (1963).
23.   Lindberg, J., Stenman F. and Laipio, I., J. Polym. Sci. Symp. No. 42,
        925 (1973).
24.   Senetskaya, L.P., Smirnova, O.V. and Domozhakova, L.M., Plast.
        Massy No. 1, 68 (1974).
25.   Tasumi, M., Shimanouchi, T., Kenjo H. and Ikeda, S., J. Polymer
        Science A 4, 1011 (1966).
26.   Nerheim, A.G., Anal. Chem., 47   1128   (1975).
27.   Laiber, M., Opera, H., Toader, M. and Laiber, V., Rev. Chim.
        (Bucharest), 28(9), 881-5 (1977); Chem. Abstr. 88, 23539 p (1978).
28.   Barral, M., J. Polymer Science, 13, 1515 (1975).
29.   Luongo, J.P., J. Polym. Sci., Polym. Chem. Ed. 12, 1203 (1974).
30.   Zerbi, G., Phonons, Proc. Int. Conf., 248 (1971).
31.   Kveder, H. and Ungar, G., Nafta (Zagreb), 24(2), 85 (1973).
32.   Pentin, Yu. A., Tarasevich, B.N. and El'tsefon, B.S., Vestn. Mosk.

Univ., Khim., 14(1), 13 (1973).

33. Miller, P.J., Jackson, J.F. and Porter, R.S., J. Polym. Sci., Polym. Phys. Ed., 11, 2001 (1973).

34. Kobayashi, M., Kagaku (Kyoto), 27(3), 247 (1972).

35. Chan, M.G. and Allara, D.I., Polym. Eng. Sci., 14(1), 12 (1974).

36. Bezdadea, E., Braun, D., Buruiana, E., Caraculacu, A. and Istrate-Robila, G., Angew Makromol. Chem. 37, 35 (1974).

37. Dankovics, A., Muanyag Gumi, 11(12), 380 (1974); Chem. Abstr. 82, 86773g.(1975).

38. Rueda, D.R., Balta-Calleja, F.J. and Hidalgo, A., Spectrochim. Acta, Part A, 30(8), 1545 (1974).

39. Seegar, M. and Barrall, E.M., J. Polym. Sci., Polym. Chem. Ed 13(7), 1515 (1975).

40. Tabb, D.L., Sevcik, J.J. and Koenig, J.L., J. Polym. Sci., Polym. Phys. Ed. 13(4), 815 (1975).

41. McRae, M.A., Maddams, W.F..and Preedy, J.E., J..Mater. Sci. 11(11), 2036 (1976).

42. Painter, P.C., Havens, J., Hart, W.W. and Koenig, J.L., J. Polym. Sci., Polym. Phys. Ed. 15(7), 1237 (1977).

43. Painter, P.C., Havens, J., Hart, W.W. and Koenig, J.L., J. Polym. Sci., Polym. Phys. Ed. 15(7), 1223 (1977).

44. Rueda, D.R., Balta-Calleja, F.J. and Hidalgo, A., Spectrochimica Acta, 30A, 1545 (1974).

45. Stas'kov, N.I. and Gusev, S.S., Vestsi Akad. Navuk BSSR, Ser, Fiz.-Mat. Nuvuk (1), 124 (1978); Chem. Abstr. 88, 170877a (1978).

46. Krimm, S., Proc. Int. Symp. Macromol, 107 (1974), (pub. 1975).

47. Frank, W. and Wulff, W., Colloid Polym. Sci., 254(5), 534 (1976).

48. Fraser, G.V., Hendra, P.J., Cudby, M.E.A. and Willis, H.A., J. Mater. Sci. 9(8), 1270 (1974).

49. Hendra, P.J., Jobic, H.P., Marsden, E.P. and Bloor, D., Spectrochim. Acta, Part A, 33(3-4), 445 (1977).

50. Willis, H.A. and Cudby, M.E.A., Struct. Stud. Micromol. Spectrosc. Methods, (Meet.) 81-9 (1974); (Pub. 1976), Ed. K.J. Ivin, Wiley, Chichester, Enal.

51. Brandmueller, J. and Schroetter, H.W., Fortschr. Chem. Forsch. 36, 85 (1972).

52. Strobl, G.R. and Hagedorn, W., J. Polym. Sci., Polym. Phys. Ed., 16(7), 1181 (1978).

53. Maxfield, J., Stein, R.S. and Chen, M.C., J. Polym. Sci., Polym. Phys. Ed., 16(1), 37 (1978).

54. Woodward, A.E., J. Polymer Science, C8, 137 (1965).

55. Axelson, D.E. and Mandelkern, L., J. Polym. Sci., Polym. Phys. Ed., 16(6), 1135 (1978).

56. Beatty, C.L. and Froix, M.F., Polym. Prepr., Am. Chem. Soc., Div. Polym. Chem. 16(2), 628 (1975).

57. Smith, J.B., Manuel, A.J. and Ward, I.M., Polymer, 16(1), 57 (1975).

58. Sobottka, J., Wunderlich, K. and Priedel, B., Faserforsch. Textiltech., 27(7), 343 (1976).

59. Horii, F. and Kitamaru, R., J. Polym. Sci., Polym. Phys. Ed., 16(2), 265 (1978).

60. Pembleton, R.G., Wilson, R.C. and Gerstein, B.C., J. Chem. Phys. 66(11), 5133 (1977).

61. Hayakawa, N. and Kurijama, I., J. Polymer Science A-1 14, 1513 (1976).

62. Kitamura, R., Horri, F. and Hyon, S.H., J. Polym. Sci., Polym. Phys. Ed. 15(5), 821 (1977).

63. Vorob'yev, V.M. and Vettegren, V.T., Polym. Sci., USSR 17, 520 (1975).

64. Zetta, L. and Gatti, G., Macromoleculaes, 5(4), 535 (1972).

65. Nishioka, A., Goto, K., Kambe, M., Udagawa, K., Kambe, S., Nishizawa, K. and Watanable, S., Kobunski Kagaku, 29(10), 673 (1972).

66. Dorman, D.E., Otocka, E.P. and Bovey, F.A., Macromolecules, 5(5), 574 (1972).
67. Randall, J.C., J. Polym. Sci., Polym. Phys. 11, 275, Ed. (1973).
68. Zachmann, H.G., J. Polym. Sci., Symp. No. 42, 693 (1973).
69. Tsuji, K., J. Polym. Sci., Polym. Chem. Ed., 11(2), 467 (1973).
70. Tsuji, K., J. Polym. Sci., Polym. Chem. Ed., 11, 1407 (1973).
71. Nagamura, T., Kusumoto, N. and Takayanagi, M., J. Polym. Sci., Polym. Phys. Ed., 11, 2357 (1973).
72. Kuznetsov, A.N. and Radtsig, V.A., Chem. Phys. Lett., 17, 377 (1972).
73. Kawashima, T., Shimada, S., Kashiwabara, H. and Sohma, J., Polym. J. 5(2), 135 (1973).
74. Ahmad, S.R. and Charlesby, A., Eur. Polym. J. 11(1), 91 (1975).
75. Cudby, M.E.A. and Bunn, A., Polymer, 17(4), 345 (1976).
76. Hama, T., Suzuki, T. and Kosaka, K., Kobunshi Ronbunshu 32(2), 91 (1975); Chem. Abstr. 82, 156917h. (1975).
77. Nishioka, A., Ando, I. and Matsumoto, J., Bunseki Kagaku, 26(5), 308 (1977); Chem. Abstr. 87, 102729h (1977).
78. Cutler, D.J., Hendra, P.J., Cudby, M.E.A. and Willis, H.A., Polymer, 18(10), 1005 (1977).
79. Randall, J.C., J. Appl. Polym. Sci., 22(2), 585 (1978).
80. Bennett, R.L., Keller, A. and Stejny, J., J. Polymer Science A-1 14, 3021 (1976).
81. Cudby, M.E.A. and Bunn, A., Polymer, 17, 345 (1976).
82. Randall, J.C., J. Polym. Sci., Polym. Phys. Ed., 11, 275 (1973).
83. Dorman, D.E., Otocka, E.P. and Bovey, F.A., Macromolecules, 5, 574 (1972).
84. Bovey, F.A., Schilling, F.C., McCrackin, F.L. and Wagner, H.L., Macromolecules, 9, 76 (1976).
85. Randall, J.C., J. Appl. Polymer. Sci. 22, 585 (1978).
86. Miller, M.L., The Structure of Polymers, Reinhold, New York, p 118, (1966).
87. Willbourn, A.H., J. Polym. Sci., 34, 569 (1959) Nottingham Symposium.
88. Otocka, E.P., Roe, R.J., Hellman, M.Y. and Muglia, P.M., Macromolecules, 4, 507 (1971).
89. Dorman, D.E., Otocka, E.P. and Bovey, F.A., Macromolecules, 5, 574 (1972).
90. Randall, J.C., J. Polym. Sci., Polym. Phys. Ed., 11, 275 (1973).
91. Axelson, D.E., Levy, G.C. and Mandelkern, L., Macromolecules, 12, 41 (1979).
92. Cudby, M.E.A. and Bunn, A., Polymer, 17, 345 (1976).
93. Randall, J.C., J. Appl. Polym. Sci., 22, 585 (1978).
94. Bovey, F.A., Schilling, F.C., McCrackin, F.L. and Wagner, H.L., Macromolecules, 9, 76 (1976).
95. Foster, G.N., Polymer Preprints, 20, 463 (1979).
96. Axelson, D.E. and Knapp, W.C., (in press).
97. Hogan, J.P., Levett, C.T. and Werkman, R.T., S.P.E. Journal 23(11), 87 (1967).
98. Kraus, G. and Stacy, C.J., J. Polym. Sci., Symposium No. 43, 329 (1973).
99. Randall, J.C., Reprint ACS Symposium series No. 142 Polymer Characterization by ESR and NMR, American Chemical Society (1980).
100. Rueda, D.R., Balta Calleja, F.J. and Hidalgo, A., Spectrochimica Acta, 35A 847 (1979).
101. Barlow, A., Wild, L. and Ranganath, R., J. Appl. Polym. Sci., 21, 3319 (1977).
102. Wild, L., Ranganath, R. and Barlow, A., J. Appl. Polym. Sci., 21, 3331 (1977).
103. Grant, D.M. and Paul, E.G., J. Am. Chem. Soc., 86, 2984 (1964).
104. Olnishi, S.I., Sugimoto, S.I. and Nitta, I., J. Polymer Science, Part A, - 1, 605 (1963).

105.   Ono, Y. and Keii, T., J. Polymer Science, A-I 4, 2429 (1966).
106.   Hama, Y., Hosano, Y. and Shinohara, K., J. Polymer Science A9,
       1411 (1971).
107.   Seiki, T. and Takeshita, T., J. Polymer Science A-I 10, 3119 (1972).
108.   Tsuji, K., J. Polymer Science A-I, 11, 467 (1973).
109.   Tsuji, K., J. Polymer Science, A-I, 11, 1407 (1973).
110.   Hori, Y., Shimada, S. and Kashiwabara, H., Polymer, 18(6), 567 (1977).
111.   Hori, Y., Shimada, S. and Kashiwabara, H., Polymer, 18(11), 1143 (1977).
112.   Seguchi, T. and Tamura, N., J. Polymer Science A-I 12, 1671 (1974).
113.   Seguchi, T. and Tamura, N., J. Polymer Science A-I 12, 1953 (1974).
114.   Browning, H.L., Ackermann, H.D. and Patton, H.W., J. Polymer Science
       A 4, 1433 (1966).
115.   Cross, L.H., Richards, R.B. and Willis, H.A., Discussions Faraday
       Soc., 9, 235 (1950).
116.   Rugg, R.M., Smith, J.J. and Bacon, R.C., J. Polymer Sci., 13, 535
       (1954).
117.   Pross, A.W. and Black, R.M., J. Soc. Chem. Ind. (London), 69, 115
       (1950).
118.   Beachell, A.C. and Nemphos, S.P., J. Polymer Sci., 21, 113 (1956).
119.   Beachell, H.C. and Tarbet, G.W., J. Polymer Sci., 45, 451 (1960).
120.   Luongo, J.P., J. Polymer Sci., 42, 139 (1960).
121.   Cooper, G.D. and Prober, X., J. Polymer Sci., 44, 397 (1960).
122.   Thompson, H.W. and Tarkington, P., Proc. Roy. Soc. (London), 184A,
       3 (1945).
123.   Pross, A.W. and Black, R.M., J. Soc. Chem. Ind., 69, 113 (1950).
124.   Cross, L.H., Richards, R.B. and Willis, H.A., Discussions Faraday
       Soc., 9, 235 (1950).
125.   Rugg, F.M., Smith, J.J. and Bacon, R.C., J. Polymer Sci., 13, 535
       (1954).
126.   Cooper, G.D. and Prober, M., J. Polymer Sci., 44, 397 (1960).
127.   Heacock, J.F., J. Appl. Polymer Science, 7, 2319 (1963).
128.   Das, N. and Palit, S.R., J. Polym. Sci., Polym. Chem. Ed. 11, 1025
       (1973).
129.   Margolin, A.L., Postnikov, L.M., Bordakov, V.S. and Vichutinskaya,
       E.V., Vysokomol. Soedin., Ser. A, 14(7), 1586 (1972).
130.   Kruger, R.A., J. Polym. Sci., 17, 2305 (1973).
131.   Hedvig, P. and Somogyi, A., Proc. Tihany Symp. Radiat. Chem., 3rd
       1971, 1, 907 (Pub. 1972).
132.   Bair, H.E., Polym. Eng. Sci., 13, 435 (1973).
133.   Roe, R.J., Bair, H.E. and Gleniewski, C., J. Appl. Polym. Sci., 18,
       843 (1974).
134.   Kato, K., J. Appl. Polym. Sci., 18, 3087 (1974).
135.   Guseinov, T.I., Bagirov, M.A., Bolchenkov, E.Ya. and Abasov, S.A.,
       Elektron. Obrab. Mater., 6, 42 (1974); Chem. Abstr. 83, 89403t. (1974).
136.   Le Poidevin, G.J., Electr. Council Res. Cent., (Note) ECRC/N1062
       (1977); Chem. Abstr. 87, 136690d (1977).
137.   Brauer, G.M., J. Polymer Science, C8, 3 (1965).
138.   Cieplinski, E.W., Ettre, L.S., Kolb, B. and Kemner, G., Z. Anal. Chem.,
       205, 357 (1964).
139.   Wall, L.A., J. Elastoplast., 5, 38 (1973).
140.   Voigt, J., Kunststoffe, 54, 2 (1964).
141.   Kiran, E. and Gilham, J.K., J. Macromol. Sci., Chem., 8(1), 211 (1974).
142.   Van Schooten, J. and Evenhuis, J.K., Polymer (London) 561, Nov. (1965).
143.   Van Schooten, J. and Evenhuis, J.K., Polymer (London) 6, 353 (1965).
144.   Harlen, F., et al, J. Polymer Sci., 18, 589 (1955).
145.   Rodel, M.J., J. Am. Chem. Soc., 75, 6110 (1953).
146.   Wright, B., private communication.
147.   Shell Chemical Co., Emeryville, California, private communication,
       (1964).

148. Willbourn, A.H., J. Polym. Sci., 34, 569 (1959).
149. Nakajima, A., Hamada, H. and Hayaski, S., Makromol. Chem., 95, 40
      (1966).
150. Boccato, G., Rigo, A., Talamini, G. and Zilo-Grandi, F., Makromol.
      Chem., 108, 218 (1967).
151. Burnett, G.M., Ross, F.L. and Hay, J.N., J. Polym. Sci., A-1, 5,
      1467 (1967).
152. Rigo, A., Palma, G. and Talamini, G., Makromol. Chem., 153, 219 (1972).
153. Baker, C., Maddams, W.F., Park, G.S. and Robertson, B., Makromol.
      Chem., 165, 321 (1973).
154. Salovey, R., J. Polym. Sci., A, 2, 2041 (1964).
155. Harlan, F., Simpson, W., Waddington, F.B., Waldron, J.D. and Baskett,
      A.C., J. Polym. Sci., 18, 489 (1955).
156. Kamath, P. and Barlow, A., J..Polymer Sci., A-1, 5, 2023 (1967).
157. Kolb, B. and Kaiser, H., J. Gas Chrom., 2, 233 (1964).
158. Van Schooten, J. and Evenhuis, J.K., Polymer, 6, 343 (1965).
159. Van Schooten, J. and Evenhuis, J.K., Polymer, 6, 561 (1965).
160. Michajlov, L., Zugenmaier, P. and Cantow, H.J., Polymer, 9, 325 (1968).
161. Seeger, M. and Barrall, E., II, J. Polym. Sci., A-1, 13, 1515 (1975).
162. Seeger, M., Exner, J. and Cantow, H.J., paper presented at 23rd
      Internat. Congress of Pure and Appl. Chem., Boston, 1971, Macro-
      molecular Preprints, II, 739.
163. Exner, J., Seeger, M. and Cantow, J.J., Angew Chem. (Internat. Ed.),
      10, 346 (1971).
164. Jones, A.A. and Bisceglia, M., Polym. Preprints, 20, 227 (1979).
165. Gedemer, I.J., Plast. Engng., 31, 28 (1975).
166. Robertson, A.B., Cook, J.A. and Gregory, J.T., Kinetics Symposium,
      Boston, Mass. P258-73 (1972).
167. Wall, L.A., Madorsky, S.L., Brown, D.W., Straus, S. and Simka, R.,
      J. Amer. Chem. Soc., 76, 3430 (1954).
168. Kolb, B. and Kaiser, H., J. Gas Chromatography, 2, 233 (1964).
169. Seeger, M., Exner, J. and Cantow, H.J., paper presented at the 23rd
      Internat. Congress of Pure and Applied Chemistry, Boston, 1971,
      Macromolecular Preprints, II, 739.
170. Exner, J., Seeger, M. and Cantow, J.J., Angew Chem. (Internat. Ed.)
      10, 346 (1971).
171. Seeger, M. and Barrall, E., J. Polymer Science, A-1 13, 1515 (1975).
172. Van Schooten, J. and Evenhuis, J.K., Polymer 6, 561 (1965).
173. Seeger, M. and Barrall, E.M., J. Polymer Science, Polymer Chemistry
      Ed., 13, 1515 (1975).
174. Michajlov, L., Zugenmaier, P. and Cantow, H.J., Polymer 9, 325 (1968).
175. Michajlov, L., Cantow, H.J. and Zugenmaier, P., Polymer 12, 70 (1971).
176. Seeger, M., Exner, J. and Cantow, H.J., paper presented at 23rd
      Internat. Congress of Pure and Appl. Chem. Boston, 1971;
      Macromolecular Preprints, III, 739.
177. Seeger, M., Exner, J. and Cantow, H.J., paper presented at Internat.
      Symposium on Macromolecules (IUPAC), Helsinki, 1972; Preprints, 5,
      177.
178. Seeger, M., Barrall, E.M. and Shen, M., J. Polymer Science, Polymer
      Chemistry Edition, 13, 1541 (1975).
179. Michajlov, L., Zugenmaier, P. and Cantow, H.J., Polymer, 9, 325 (1968).
180. Shen, M., Kobayashi, H. and Bell, A.T., J. Appl. Polym. Sci., 17,
      885 (1973).
181. Ahlstrom, D.H., Leibman, S.A. and Abbas, K.P., J. Polymer Science,
      Polymer Chemistry Edition, 14, 2479 (1976).
182. Tsuchiya, Y. and Sumi, K., J. Polymer Sci., B, 6, 356 (1968).
183. Tsuchiya, Y. and Sumi, K., J. Polym. Sci. A-1, 6, 415 (1968).
184. Rugg, F., Smith, J. and Wartman, L., J. Polymer Sci., 11, 1 (1954).

185. Rugg, F., Smith, J. and Atkinson, J., J. Polymer Sci. 9, 579 (1952).
186. Holmes, D., Miller, R., Palmer, R. and Bunn, C., Nature, 171, 1104
      (1953).
187. Aggarwal, S.L., unpublished research.
188. Bryant, W., J. Polymer Sci., 2, 547 (1947).
189. Wiener, O., Abhandl. sächs. Ges. Wiss. (Math. Phys. Kl.), 32, 507
      (1912).
190. Tobin, M.C. and Carrano, M.J., J. Polymer Science, 24, 93 (1957).
191. Aggarwal, S.L. and Tilley, G.P., J. Polymer Sci., 18, 17 (1955).
192. Nichols, J.B., J. Appl. Phys. 25, 840 (1954)
193. Hoore, J. and Hillman, D.E., Brit. Polym. J., 3, 259 (1971).
194. Stein, R.S., Tanaka, A. and Finklestein, R.S., Appl. Polym. Symp. No.
      20, 235 (1973).
195. Tanaka, A., Chang, E.P., Delf, B., Kimura, I. and Stein, R.S., J.
      Polym. Sci., Polym. Phys. Ed., 11, 1891 (1973).
196. Ginzburg, B.M., Kurbanov, K.B. and Rashidov, D., Mekh. Polim., 2,
      375 (1974).
197. Cellarelli, G., Andruzzi, F. and Paci, M., Polymer, 20, 605 (1979).
198. Gerasimov, V.I., Genin, Ya. V. and Tsvankin, D. Ya., J. Polym. Sci.,
      Polym. Phys. Ed. 12, 2035 (1974).
199. Roe, R.J. and Gieniewski, C., Macromolecules, 6, 212 (1973).
200. Marx, C.L. and Cooper, S.L., Makromol. Chem., 168, 339 (1973).
201. Markova, G.S., Ovchinnikov, Yu. K. and Bokhyan, J. Polym. Sci., Polym.
      Symp. 42 pt. 2, 671 (1973).
202. Perepechko, I.I., Startsev, O.V. and Mirzakarimov, A., Plast. Massy.
      4, 71 (1973).
203. Brown, D.S., Fulcher, K.U. and Wetton, R.E., Polymer, 14, 379 (1973).
204. Takaynagi, M., Nippon Butsuri Gakkaishi, 28(8), 619 (1973).
205. Marx, C.L., Caulfield, D.F. and Cooper, S.L., Macromolecules, 6(3),
      344 (1973).
206. Scultz, J.M. and Long, T.C., J. Mater. Sci., 10(4), 567 (1975).
207. Das Gupta, D.K. and Noon, T., Inst. Phys. Conf. Ser. 27(Static.
      Electrif.) 122 (1975).
208. Rybnikar, F., Plasty Kauc. 12(11), 326 (1975); Chem. Abstr. 84:
      90668k.(1975).
209. Ng, T.H., and Williams, H.L., Makromol. Chem., 182, 3323 (1981).
      Res., Kyoto Univ. 52(2), 359 (1974).
210. Ng, T.H., and Williams, H.L., Makromol. Chem., 182, 3323 (1981).
211. Pae, K.D., Newman, B.A. and Sham, T.P., J. Mater. Sci., 12(9), 1793
      (1977).
212. Strobl, G.R. and Eckel, R., Prog. Colloid Polym. Sci., 62, 9 (1977).
213. McRae, M.A. and Maddams, W.F., Polymer 18(5), 524 (1977).
214. Baczek, S.K., Diss. Abstr. Int. B. (Univ. Michigan:Ann Arbor), 38(6),
      2697 (1977); Chem. Abstr., 88, 51300x (1978).
215. Rozkuszka, K.P., Diss. Abstr. Int. B. (Univ. Michigan:Ann Arbor),
      39(1), 258 (1978).
216. Clampett, B.H., Anal. Chem., 35, 1834 (1963).
217. Clampett, B.H., J. Polymer Science A3, 671 (1965).
218. Holden, H.W., J. Polymer Science C 6, 53 (1964).
219. Paulik, J. and Paulik, F., GIT(Glas-Instrum.-Tech.) Fachz. Lab.
      16(9), 1043 (1972).
220. Bosch, K., Beitr. Gerichtl. Med., 33, 280 (1975).
221. Harland, W.G., Khadr, M.M. and Peters, R.H., Polymer, 15, 81 (1974).
222. Mehta, A. and Wunderlich, B., Makromol. Chem. 175, 977 (1974).
223. Southern, J.H. and Wilkes, G.L., J. Polym. Sci., Polym. Lett. Ed.
      11, 555 (1973).
224. Caillot, C., Fournie, R., Audoux, C. and Brive, R., Rev. Gen. Therm.,
      11, (125), 461 (1972).

225. Spencer, L.R., Food Prod. Develop., 7(1), 46 (1973).
226. Marx, C.L. and Cooper, S.L., Makromol. Chem., 168, 339 (1973).
227. Schmitt, C.R., J. Fire Flammability, 3, 303 (1972).
228. Nakajima, A., Chem. High Polymers, Japan, (Kobunski Kagaku), 7,
      64 (1950).
229. Nakajima, A. and Fujiwara, H., High Polymers Japan (Kobunski Kagaku),
      37, 909 (1964).
230. Wijga, P.W.O., Van Schooten, J. and Boerma, J., Makromol. Chemie,
      36, 115 (1960).
231. Vansco-Szmersanyi, I., Bodorg. Arbeits. Verst. Kunst. Internationale
      Tagung 14 O Heutliche Freudenstadt, P42/1-42/4 (1977).
232. Welsch, T., Engewald, W. and Kowash, E., Plaste. u Kaut. 23, 584
      (1976).
233. Natta, G., Corradini, P. and Cesara, M., Lincei-Rend. Sci. fis. mat.
      a Mat., 22, 11 (1957).
234. Natta, G., Danusso, F. and Moraglio, G., Angew. Chem., 69, 686 (1957).
235. Natta, G., J. Polymer. Sci., 34, 531 (1959).
236. Flory, P.J., J. Chem. Phys., 17, 223 (1949).
237. Danusso, F., Moraglio, G. and Flores, E., Lincei-Rend. Sci. fis. mat.
      e nat., 25, 520 (1958).
238. Kenyon, A.S., Salyer, I.O., Kurz, J.E. and Brown, D.R., J. Polymer
      Sci., C8, 205 (1965).
239. Kolke, V. and Billmeyer, F.W., J. Polymer Science, C8, 217 (1965).
240. Williamson, G.R. and Cervenka, A., Eur. Polym. J., 8(8), 1009 (1972).
241. Nakajima, N., Advan. Chem. Ser., 125, 98 (1971) (Pub. 1973).
242. Eldarov, E.G., Goldberg, V.M., Pankratova, G.V., Akutin, M.S. and
      Toptygin, D. Ya., Zavod. Lab., 40(3), 269 (1974).
243. Lovric, L., Nafta (Zagreb), 23(12), 606 (1972).
244. Lovric, L., Nafta (Zagreb), 24(1), 47 (1973).
245. Peyrouset, A., Prechner, R., Panaris, R. and Benoit, H., J. Appl.
      Polym. Sci., 19(5), 1363 (1975).
246. Akutin, M.S., Goldberg, V.M. and Lavrushin, F.G., Vysokomol. Soedin.,
      Ser. A 19(5), 1113 (1977).
247. Gianotti, G., Gaita, A. and Romanini, D., Polymer 21, 1087 (1980).
248. Tengler, H. and Von Falkai, B., Kunststoffe, 62, 759 (1972).
249. Taylor, W.C. and Tung, L.H., paper presented at 140th Meeting American
      Chemical Society, Chicago, Illinois, Sept. 1961. Also SPE Trans-
      actions page 119 April 1962.
250. Moray, D.R. and Tamblyn, J.W., J. Appl. Phys, 16, 419 (1945).
251. Clacsson, S., J. Polymer Sci., 16, 193 (1955).
252. Smith, T.L. and Whelihan, B.N., Text. Chem. Color. 10, 35 (1978).
253. Gamble, L.W., Nipke, W.T. and Lane, T.L., J. Appl. Polymer Science,
      9, 1503 (1965).
254. Wesslau, H., Makromol. Chem., 20, 111 (1956).
255. Chiang, R., J. Polymer Science A 3, 3679 (1965).
256. Schreiber, H.D. and Waldman, M.H., J. Polymer Science A 2, 1655 (1964).
257. Cuesta De La and Billmeyer, F.W., J. Polymer Science A 1, 1721 (1963).
258. Das, N. and Palit, S.R., J. Polymer Science A-1, 11, 1025 (1973).
259. Chien, J.C.W., J. Polymer Science, A 1, 1839 (1963).
260. Wagner, H.L., Advan. Chem. Ser., 17, No. 125 (1971) (Pub. 1973).
261. Maley, J. Polymer Science C8, 253 (1965).
262. Platonov, J.P., Belyaev, V.M. and Grigor'eva, F.P., Plast. Massy.,
      4, 77 (1975).
263. Starck Pal and Kantola, P., Kem-Kemi 3(2), 100 (1976).
264. Lanikova, J. and Hlousek, M., Chem. Prum. 27(12), 628 (1977).
265. Ross Jr. J.H. and Shank, R.L., Advan. Chem. Ser., 125, 108 (1971).
      (Pub. 1973).
266. Nikolskii, V.G., Zlatkevich, L. Yu., Konstantinopolskaya, M.B.,

Osintseva, L.A. and Sokolskii, V.A., J. Polym. Sci., Polym. Phys.
Ed., 12, 1267 (1974).

267.   Looyenga, H., J. Polym. Sci., Polym. Phys. Ed., 11, 1331 (1973).
268.   Stein, R.S., J. Makromol. Sci., Phys., 8(1), 29 (1973).
269.   Kovak'chuk, V.M., Ivanov, A. and Samenenko, E.I., Khira Farm Zn 10,
        114 (1976).
270.   Crewther, W.G., Stacy, C.J. and Arnett, R.C., J. Polymer Science,
        A 2, 167 (1964).
271.   Olsen, D.A. and Osteraas, A.J., J. Polymer Science A-1, 7, 1927 (1969).
272.   Ranicar, J.H. and Fleming, R.J., J. Polym. Sci., Part A-2, 10 1979
        (1972).
273.   Onogi, S. and Asada, T., Progr. Polym. Sci., 2, 261 (1971).
274.   Schelten, J., Wignall, G.D., Ballard, D.G. and Schmatz, W., Colloid
        Polym. Sci., 252(9), 749 (1974).
275.   Bergmann, K., Kolloid-z. Z. Polym., 251(11), 962 (1973).
276.   Ballard, D.G.H., Cunningham, A. and Schelten, J., Polymer, 18(3),
        259 (1977).
277.   Casper, R., Biskup, U., Lange, H., and Pohl, U., Makromol. Chem. 177(4),
        1111 (1976).
278.   Clark, D.T., Feast, W.F., Musgrave, W.K.R. and Ritchie, I., J. Polym.
        Sci., Polym. Chem. Ed., 13(4), 857 (1975).
279.   Clark, D.T., Feast, W.F., Musgrave, W.K.R. and Ritchie, I., J. Polym.
        Science A-1, 13 837 (1975).
280.   King, J.S., Summerfield, G.C. and Ullman, R., Polym. Prepr. Am. Chem.
        Soc., Div. Polym. Chem. 16(2), 410 (1975).
281.   Schelten, J., Wignall, G.D., Ballard, D.G.H. and Longman, G.W.,
        Polymer, 18(11), 1111 (1977).
282.   Picot, C., Duplessix, R., Decker, D., Benoit, H., Boue, F., Cotton,
        J.P., Daoud, M., Farnoux, B. and Jannink, G., Macromolecules, 10(2),
        436 (1977).
283.   McCubbin, W.C. and Weeks, J.C., J. Applied Physics, 3644 August (1966).
284.   Anderson, F.R., J. Polymer Science, C 8, 275 (1965).
285.   Schreiber, H.P. and Bagley, E.B., J. Polymer Science 58, 29 (1962).
286.   Slovokhotova, N.A., Koritskii, A.T., Kargin, V.A., Buben, N. Ya.,
        Bibikov, V.V., Il'icheva, Z.F. and Rudnaya, G.V., Vysokomol. soyed.
        5, 568 (1963).
287.   Bursfield, D.R., J. Polymer Science A-1, 16, 3301 (1978).
288.   Braun, J.M. and Guillet, J.E., J. Polymer Science, Polymer Chem.,
        Ed. 13, 1119 (1975).
289.   Jeffs, A.R., Analyst 94, 249 (1969)..
290.   Crompton, T.R. and Myers, L.W., Plastics and Polymers, 205, June (1968).
291.   Crompton, T.R., Myers, L.W. and Blair, D., British Plastics,
        December (1965).
292.   Schroder, E. and Hagen, E., Plaste Kautsch, 15, 625 (1968).
293.   Wandel, M., Tengler, H. and Ostromov, H., Die Analyse von Weich-
        machern, Springer Verlag, Berlin, Heidelberg, New York (1967).
294.   Kardos, E., Kosljar, V., Hudec, J. and Ilka, P., Chemicke, Zvesti,
        22, 768 (1968).
295.   Buttery, R.G. and Stuckey, B.N., J. Agr. Food Chem. 9, 283 (1961).
296.   Schroeder, E. and Randolph, G., Plast. Kautschuk 1, 22 (1963).
297.   Denning, J.A. and Marshall, J.A., Analyst 97, 710 (1972).
298.   Lappin, G.R. and Zannucci, J., Anal. Chem. 41, 2076 (1969).
299.   Wheeler, D.A., Talenta 15, 1315 (1968).
300.   Roberts, C.B. and Swank, J.D., Anal. Chem., 36, 271 (1964).
301.   Nosikov, Yu. D. and Vetchinkina, V.N., Neftckhimiya, 5, 284 (1965).
302.   Helf, C. and Bockwan, D., Plaste Kauteshuk, 11, 624 (1964).
303.   Pocaro, P.J., Anal. Chem. 36, 1664 (1964).
304.   Duvall, A.H. and Tully, W.F., J. Chromatograph 11, 38 (1963).

305. Kbight, H.S. and Siegel, H., Anal. Chem. 38, 1221 (1966).
306. Long, R.E. and Guvernator, G.C., Anal. Chem. 39, 1493 (1967).
307. Seldar, J., Feniokova, E. and Pac, C., Analyst, 99, 50 (1974).
308. Denning, J.A. and Marshall, J.A., Analyst. 97. 710 (1972).
309. Mayer, H., Deut, Lebensun Rundshau, 57, 170 (1961).
310. Glavind, J., Acta. Chem. Scand., 17, 1635 (1963).
311. Blois, M.S., Nature, 181, 1199 (1958).
312. Kharasch, M.S. and Joshi, B.S., J. Org. Chem., 22, 1439 (1957).
313. Stafford, C., Anal. Chem., 34, 794 (1962).
314. British Standard 2782, Part 4 Method 405 D (1965).
315. Metcalf, K. and Tomlinson, R.F., Plastics, 25, 319 (1960).
316. Hilton, C.L., Anal. Chem., 32, 383 (1960).
317. British Standard 2782, Part 4 Method 405 B (1965).
318. Crompton, T.R., J. Appl. Polymer Science, 6, 538 (1962).
319. Ruddle, L.H. and Wilson, J.R., Analyst, 94, 105 (1969).
320. Haslam, J. and Willis, H.A., Identification and Analysis of Plastics, The Iliffe Group, London, p.307 (1965).
321. Wexler, A.S., Anal. Chem., 35, 1926 (1963).
322. Ruddle, L.H. and Wilson, J.R., Analyst, 94, 105 (1969).
323. Hilton, C.L., Rubber Age, 84, 263 (1958).
324. Burchfield, H.P. and July, J.N., Anal. Chem., 19, 383 (1960).
325. Cieleszky, C. and Nagy, F., Lebensm Unters-Forsch, 114, 13 (1961).
326. Spell, H.L. and Eddy, R.D., Anal. Chem., 32, 1811 (1960).
327. Parks, C.A., Anal. Chem., 37, 140 (1961).
328. Drushel, H.V. and Sommers, A.L., Anal. Chem., 36, 836 (1964).
329. Major, J. and Kocmanova, V., Chem. Prum., 17, 372 (1967).
330. Spell, H.L. and Eddy, R.D., Anal. Chem., 32, 1811 (1960).
331. Szalkowski, C. and Garber, J., J. Agr. Food Chem., 10, 110 (1962).
332. Hilton, C.L., Rubber Age, 84, 263 (1958).
333. Burchfield, H.P. and July, J.N., Anal. Chem., 19, 383 (1960).
334. Drushel, H.V. and Sommers, A.L., Anal. Chem., 36, 836 (1964).
335. Webster, P.V. and Franks, M.C., J. Inst. Petrol, London, 56, 50 (1970).
336. Wake, W.C., The Analysis of Rubber and Rubber Like Polymers, McLaren, London (1969).
337. Carlson, D.W., Hayes, M.W., Bansaw, H.G., McFadden, A.S. and Altenan, A.G., Anal. Chem., 43, 1874 (1971).
338. Campbell, R.H. and Wise, R.W., J. Chromat., 12, 178 (1963).
339. Slonaker, D.F. and Sievers, D.C., Anal. Chem., 36, 1130 (1964).
340. Van der Heide, R.F. and Wouters, O., Lebensm. Untersuch, Forsch, 117, 129 (1962).
341. Schroder, E. and Rudolph, G., Plaste Kautschuk, 10, 22 (1963).
342. Metcalf, K. and Tomlinson, R., Plastics (London) 25, 319 (1960).
343. Waggon, H., Korn, O. and Jehle, D., Nahrung, 9, 495 (1965).
344. British Standard 2782, Part 4, Method 405 D (1965).
345. British Standard 2782, Part 4, Method 405 B (1965).
346. Yusherichyute, S.S. and Shlyapnikov, Yu. A., Plasticheskie Massy, No. 1, 54 (1967).
347. Stafford, C., Anal. Chem., 34, 794 (1962).
348. Williamson, F.B. and Rubber, S., Intern. Plastics, 148 (No. 2), 24 (1958).
349. Zijp, J.W.H., Rec. Trav. Chim., 75, 1155 (1956).
350. Zijp, J.W.H., Dissertation, Technical University, Delft (1955).
351. Zijp, J.W.H., Rec. Trav. Chim., 75, 1129 (1956).
352. Auler, H., Rubber Chem. Technol., 37, 950 (1964).
353. Zijp, J.W.H. and Kautschuk, U., Gummi, 10, 14 (1957).
354. Kreiner, J.G., Rubber Chem. Technol., 44, 381 (1971).
355. Kreiner, J.G. and Warner, W.C., J. Chromatography, 44, 315 (1969).
356. Auler, J., Identification of Antioxidants, Antiozonants and

Accelerators by means of Thin-Layer Chromatography ETDC Aachen,
Materials and Research Dept. (1967).

357. Kreiner, G. and Warner, W.C., J. Chromatography, 44, 315 (1969).

358. Van der Heide, R.F. and Wouters, O., Lebensm. Untersuch. Forsch,
117, 129 (1962).

359. Waggon, H., Korn, O. and Jehle, D., Nahrung, 9, 495 (1965).

360. Simpson, D. and Curnell, B.R., Analyst, 96, 515 (1971).

361. Kirchner, J.G., Thin-layer Chromatography, Interscience, New York,
pp 168-169, 679-684 (1966).

362. Stahl, E., Thin-layer Chromatography, Academic Press, New York,
pp 349-352, 498 (1965).

363. Randerath, K., Thin-layer Chromatography, Academic Press, New York,
pp 179-181 (1966).

364. Robbitt, J.M., Thin-layer Chromatography, Reinhold, New York, p180
(1963).

365. Zijp, J.W.H., Rec. Trav. Chim., 76, 313, 317 (1957).

366. Zijp, J.W.H., Rec. Trav. Chim., 77, 129 (1958).

367. Schroder, E. and Hagen, E., Plaste Kautsch, 15, 625 (1968).

368. Van der Heide, R.F., Ernahrungforschung, 24, 239 (1966).

369. Cumpelik, B.M., Drug Cosmet. Ind., 113, 44 (1973).

370. Waggon, H., Korn, O. and Jehle, D., Die Nahrung, 4, 495 (1965).

371. Sokolowska, R., Rocza panst. Zakl. Hig., 20, 661 (1969).

372. Novitskaya, L. and Kararinova, N., Zhur. analit. Khim., 28, 1233
(1973).

373. Simpson, D. and Curnell, B.R., Analyst, 96, 515 (1971).

374. Dobies, R.S., J. Chromatography, 40, 110 (1969).

375. Berger, K.G., Sylvester, N.D. and Haines, D.M., Analyst, 85, 341
(1960).

376. British Standard 2782 Part 4 Method 405 D (1965).

377. Van der Neut, J.H. and Maagdenberg, A.C., Plastics, 31, 66 (1966).

378. Majors, R.E., J. Chromatographic Science, 8, 338 (1970).

379. Coupek, J., Pokorny, S., Protivova, J., Holak, J., Korvas, H. and
Posposil, J., J. Chromatography, 65, 279 (1972).

380. Campbell, R.H. and Wize, R.W., J. Chromatography, 12, 178 (1963).

381. Majors, R.E., J. Chromata. Sci., 8, 338 (1970).

382. Schroder, E. and Rudolph, G., Plaste Kautschuk, 10, 22 (1963).

383. Morgenthalen, L.P., unpublished work.

384. Neureitter, N.P. and Bown, D.E., Ind. Eng. Chem. Prod. Res. Dev.,
1, 236 (1962).

385. Kellum, G.E., Anal. Chem., 43, 1843 (1971).

386. Kirchner, J.G., Thin-layer Chromatography, Interscience, New York,
pp 168-169, 679-684 (1966).

387. Simpson, D. and Currell, B.R., Analyst, 96, 515 (1971).

388. Stahl, E., Thin-layer Chromatography, Academic Press, New York,
pp 349-352, 498 (1965).

389. Randerath, K., Thin-layer Chromatography, Academic Press, New York,
pp 179-181 (1966).

390. Robbitt, J.M., Thin-layer Chromatography, Reinhold, New York, p 180
(1963).

391. Zijp, J.W.H., Rec. Trav. Chim., 76, 313 (1957).

392. Zijp, J.W.H., Rec. Trav. Chim., 77, 129 (1958).

393. Wandel, M. and Tengler, H., Fette. Seifen Anstrichmittell, 66, 815
(1964).

394. Dobies, R.S., J. Chromatography, 35, 370 (1968).

395. Cumpelik, B.A., Drug Cosmet. Ind., 113, 44 (1973).

396. Schroder, E..and Hagen, E., Plaste Kautsch, 15, 625 (1968).

397. Van der Heide, R.F., Ernahrungforschung, 24, 239 (1966).

398. Jentzsch, D., Kruger, H., Lebricht, G. and Deucks, G., Gut. I. Z.

Anal. Chem. 236, 96 (1968).

399.   Denning, J.A. and Marshall, J.A., Analyst, 97, 710 (1972).
400.   Romano, S.J., Renner, J.A. and Leitner, P.H., Anal. Chem., 45, 2327
       (1973).
401.   Davis, J.T. and Denham, B.H., Analyst, 93, 336 (1968).
402.   Schroder, E. and Hagen, E., Plaste Kautsch., 15, 625 (1968).
403.   Kycera, J., Collect. Czeck. Chem. Commun., 28, 1344 (1963).
404.   Wandel, M., Tengler, H., Ostromov, H., Die Analyse von Weichmachern,
       Springer Venlag, Berlin, Heidelberg, New York, (1967).
405.   Kréiner, J.G., Rubber Chem. Technol., 44, 381 (1971).
406.   Kreiner, J.G. and Warner, W.C., J. Chromatography, 44, 315 (1969).
407.   Auler, J., Identification of Antioxidants, Antiozonants and Accel-
       erators by means of Thin-layer Chromatography. ETDC - Aachen
       Materials and Research Dept. (1967).
408.   Delves, R.B., J. Chromatography, 26, 296 (1967).
409.   Gororyova, A., Elekt. Kabl. Tech., 23, 207 (1970).
410.   Fiorenza, M., Bonomi, G. and Saradi, A., Rubber Chem. Technol., 41,
       630 (1968).
411.   Kreiner, G. and Warner, W.C., J. Chromatography, 44, 315 (1969).
412.   Denning, J.A. and Marshall, J.A., Analyst (London), 97 (1158), 710
       (1972).
413.   Reed Jr., P.R. and Warren, P.L., Tech. Pap. Reg. Tech. Conf. - Soc.
       Plast. Eng., 11(12), 139 (1975).
414.   Dengreville, M., Analusis 5(4), 195 (1977); Chem. Abstr. 87,
       24, 107n (1977).
415.   Dengreville, M., Analusis, 5, 195 (1977).
416.   Narasaki, H., Miyaji and Unno, A., Bunseki Kagaku, 22 (5), 541 (1973).
417.   Bua, E. and Manaresi, P., Anal. Chem., 31, 2022 (1959).
418.   Van Schooten, G. and Duck, E.W., Berkenbosch R. Polymer, 2, 357 (1961).
419.   Voigt, J., Kunststoffe, 54, 2 (1964).
420.   Voigt, J., Kunststoffe, 51, 18 (1961).
421.   Voigt, J., Kunststoffe, 51, 314 (1961).
422.   Brauer, G.M., J. Polymer Science, 8, 3 (1965).
423.   Strassburger, J., Brauer, G.M., Tryon, M. and Forziati, A.F., Analyt.
       Chem., 32, 454 (1960).
424.   Cobler, J.G..and Samsel, E.P., S.P.E. Trans., 2, 145 (1962).
425.   Stanley, C.W. and Peterson, W.R., S.P.E. Trans., 2, 298 (1962).
426.   Groten, B., Analyt. Chem., 36, 1206 (1964).
427.   Van Schooten, J., Duck, E.W. and Berkenbosch, R., Polymer (London),
       2, 357 (1961).
428.   Van Schooten, J. and Mostert, S., Polymer (London), 4, 135 (1963).
429.   Van Schooten, J. and Evenhuis, J.K., Polymer (London), 561, Nov.
       (1965).
430.   Van Schooten, J. and Evenhuis, J.K., Polymer (London), 6, 343 (1965).
431.   Barlow, A., Lehrle, R.S. and Robb, J.C., Polymer, (London), 2, 27
       (1961); S.C.I. Monogr. No. 17 'Techniques of Polymer Science',
       p 267 London, (1963).
432.   Kaduji, I.I. and Rees, J.H., Analyst, 99, 435 (1974).
433.   Van Schooten, J., Duck, E.W. and Berkenbosch, R., Polymer (London),
       2, 357 (1961).
434.   Van Schooten, J. and Mostert, S., Polymer (London), 4, 135 (1963).
435.   Private Communication
436.   Eggertson, F.T., Dimbat, M. and Stross, F.H., Shell Chemical Co.,
       Emeryville, California, Private Communication.
437.   Maddox, L., Dimbat, L. and Anderson, E.H., Shell Chemical Co.,
       Emeryville, California, Private Communication.
438.   McMurray, H.L. and Thornton, V., Analytic. Chem., 24, 318 (1952).
439.   Van Schooten, J., Duck, E.W. and Berkenbosch, R., Polymer, 2, 357
       (1961).

440. Bucci, G. and Simonazzi, T., J. Polymer Sci., C7, 203 (1964).
441. Bucci, G. and Simonazzi, T., Chimica & Industria, 44, 262 (1962).
442. Natta, G., Mazzanti, G., Valvassori, A., Sartori, G. and Morero, D.,
      Chim. e Ind. (Milano) 42, 125, 132 (1960).
443. Veerkamp, Th. A. and Veermans, A., Makromolekulare Chem., 50, 147
      (1961).
444. Van Schooten, J. and Mostert, S., Polymer, 4, 135 (1963).
445. Natta, G., Mazzanti, G., Valvassori, A. and Pajaro, A., Chim. Ind.
      (Milan), 39, 733 (1957).
446. Drushel, H.V. and Iddings, F.A., Anal. Chem., 35, 28 (1963).
447. Drushel, H.V. and Iddings, F.A., Division of Polymer Chemistry,
      142nd Meeting, ACS, Atlantic City, September 1962.
448. Wei, P.E., Anal. Chem., 33, 215 (1961).
449. Gössl, T., Makromol. Chem., 42, 1 (1961).
450. Corish, P.J. Anal..Chem., 33, 1798 (1961).
451. Small, R.B.M., Anal. Chem., 33, 1798 (1961).
452. Bau, E. and Manaresi, P., Anal. Chem., 31, 2022 (1959).
453. Natta, G. et al., Chim. e Industr., 42, 125 (1960).
454. Sheppard, N. and Sutherland, G.B.B.M., Nature, Lond., 159, 739 (1947).
455. Rugg, F.M., Smith, J.J., and Wartman, L.H., Ann. N.Y. Acad. Sci.,
      57, 398 (1953); J. Polym. Sci., 11, 1 (1953).
456. Sutherland, G.B.B.M., Disc. Faraday Soc., 9, 279 (1950).
457. Ciampelli, F., Bucci, G., Simonazzi, A. and Santambrogio, A., La
      Chimicae l'Industria, 44, 489 (1962).
458. Bucci, G. and Simonazzi, T., La Chimica et l'Industria, 44, 262 (1962).
459. Van Schooten, J., Duck, E.W., and Berkenbosch, R., Polymer, Lond.,
      2, 357 (1961).
460. Van Schooten, J. and Mostert, S., Polymer, 4, 135 (1963).
461. Natta, G. et al, Kolloidzschr, 182, No. 1-2, 50 (1962).
462. Veerkamp, Th. A. and Veermans, A., Makromol. Chem., 50, 147 (1961).
463. Drushel, H.V. and Iddings, F.A., Anal. Chem., 35, 28 (1963).
464. Natta, G., Mazzanti, G., Valvassori, A. and Pajoro, G., Chim. Ind.,
      (Milan), 39, 733 (1957).
465. Stoffer, R.L. and Smith, W.E., Anal. Chem., 33, 1112 (1961).
466. Holmes, B.S. and Moniz, W.B., Polym. Preprints, 20, 389 (1979).
467. Staszewska, D., Polim. Tworz. Wielk. 16, 506 (1971).
468. Yamaura, K., Yanagisawa, H. and Matsuzawa, S., Koll. - Z.Z. Polym.,
      248, 883 (1971).
469. Yamaura, K. Yanagisawa, H. and Matsuzawa, Koll.-Z.Z. Polym., 248,
      883 (1971).
470. Pappas, N.A. and Merrill, E.W., J. Appl. Polym. Sci., 20, 1457 (1976).
471. Radhakrishvan, N.G. and Padhye, M.R., Angew. Makromol. Chem., 43,
      177 (1975).
472. Urbanski, J., "Handbook of Analysis of Synthetic Polymers and Plastics"
      Chichester, E., Horwood p388-402 (1977).
473. Liang, C.Y., Lytton, M.R. and Boone, C.J., J. Polymer Sci., 54, 523
      (1961).
474. Liang, C.Y. and Watt, W.R., J. Polymer Sci., 51, S14, (1961).
475. Baldwin, F.P., Ivory, J.E. and Anthony, R.L., J. Appl. Phys., 26,
      750 (1955).
476. Boonstra, B.B.S.T., Ind. Eng. Chem., 43, 362 (1951).
477. Coleman, B.D., J. Polymer Sci., 31, 155 (1958).
478. Flory, P.J., Trans. Faraday Soc., 51, 848 (1955).
479. Veerkamp, T.A. and Veermans, A., Makromol. Chem., 50, 147 (1961).
480. Bucci, G. and Simonazzi, T., Chim. Ind. (Milan), 44, 262 (1962).
481. McMurry, H.L. and Thornton, V., Anal. Chem., 24, 318 (1952).
482. Van Schooten, J., Duck, E.W. and Berkenbosch, R., Polymer, 2, 357
      (1961).

483. Van Schooten, J. and Mostert, S., Polymer, 4, 135 (1963).
484. Natta, G., Dall'Asta, G., Mazzanti, G. and Ciampelli, F., Kolloid Z.,
      182, 50 (1962).
485. Bucci, G. and Simonazzi, T., J. Polymer Sci., C7, 203 (1964).
486. Lomonte, J.N. and Tirpak, G.A., J. Polymer Sci., A2, 705 (1964).
487. Tosi, G. and Simonazzi, T., Die Angemandte Makromoleculare Chemie,
      32 153 (1973).
488. Tosi, G. and Ciampelli, F., Advances in Polymer Science, 12, 21 (1973).
489. Corish, P.J. and Tunnicliffe, M.E., J. Polymer Science, C7, 187 (1964).
490. Corish, P.J. and Tunnicliffe, M.E., J. Polym. Sci., C7, 187 (1964).
491. Davison, S. and Taylor, G.L., Brit. Polym. J4, 65 (1972).
492. Drushel, H.V. and Iddings, F.A., Anal. Chem., 35, 28 (1963).
493. Chapman, N.B. and Stubbs, P.H., Hosiery and Allied Trades Research
      Assoc., Research Report, 40 pp6 (1976).
494. Kissin, Yu. V. and Chirkov, N.M., Private communication (1970).
495. Bucci, G. and Simonazzi, T., Chim. Ind. (Milano), 44, 262 (1962).
496. Toni, C. and Simonazzi, T., Die Angemandte Makromolecular Chemie,
      32, 153 (1973).
497. Brown, J.E., Tyron, M. and Mandel, J., Anal. Chem., 35, 2173 (1963).
498. Smith, D.C., Ind. Eng. Chem., 48, 1161 (1956).
499. Cross, L.H., Richards, R.B. and Willis, H.A., Discussions Faraday Soc.,
      9, 235 (1950).
500. Tyron, N., Horowitz, E. and Mandel, J., J. Res. Natl. Bur. Std., 55,
      219 (1955).
501. Fraser, G.V., Hendra, P.J., Walker, J.H., Cudby, M.E.A. and Willis,
      H.A., Makromol. Chem., 173, 205 (1973).
502. Tosi, C., Makromol. Chem., 170, 231 (1973).
503. Knott, J. and Rossbach, V., Angew. Makromol. Chem., 86, 203 (1980).
504. Varnell, D.F., Runt, J.F. and Coleman, M.M., Org. Coatings Plast.
      Chem., 45, 170 (1981).
505. Popov, V.P. and Duvanova, A.P., Zh. Prikl. Spektrosk., 18 (6), 1077
      (1973).
506. Morimoto, M. and Okamoto, Y., J. Appl. Polym. Sci., 17, 2801 (1973).
507. Seeger, M., Exner, J. and Cantow, H.J., Quad. Ric. Sci., 84, 102 (1973);
      Chem. Abstr., 81, 50225 (1973).
508. Seno, H., Tsuge, S. and Takeuchi, T., Makromol. Chem., 161, 195 (1972).
509. Porter, N., Nicksie, O. and Johnson, P., Anal. Chem., 35, 1948 (1963).
510. Wilkes, C.E., Carmen, C.J. and Harrington, R.A., J. Polymer Sci., 43,
      237 (1973).
511. Carmen, C.J., Harrington, R.A. and Wilkes, C.E., Macromolecules, 10,
      536 (1977).
512. Ray, G.J., Johnson, P.E. and Knox, J.R., Macromolecules, 10, 773 (1977).
513. Randall, J.C., Macromolecules, 11, 33 (1978).
514. Crain Jnr., W.O., Zambelli, A. and Roberts, J.D., Macromolecules, 4,
      330 (1971).
515. Carman, C.J. and Wilkes, C.E., Rubber Chem. Technol., 44, 781 (1971).
516. Schaefer, J., Macromolecules, 4, 107 (1971).
517. Grant, D.M. and Paul, E.G., J. Amer. Chem. Soc., 86, 2984 (1964).
518. Tanaka, T. and Hatada, K., J. Polymer Science, 11, 2057 (1973).
519. Randall, J.C., J. Polymer Sci. Polymer Phys. Ed., 11, 275 (1973).
520. Paxton, J.R. and Randall, J.C., Anal. Chem., 50, 1777 (1978).
521. Goedhart, D.J., Hussem, J.B. and Sweets, B.P.M., "Liquid Chromato-
      graphy of Polymers and Related Materials" Ed. by Copez, Z., X
      Delamare 4th International Liquid Chromatography Symposium,
      Strazbourg, p203-213 Oct. 24th   (1979).
522. Majewska, F., "Handbook of Analysis of Synthetic Polymers and Plastics"
      Urbanski, J., Chichester, E., Horwood, p273-294 (1977).
523. Schaefer, J. and Natusch, D.F.S., Macromolecules, 5, 416 (1972).

References

524. Inoue,Y., Nishioka, A. and Chujo, R., J. Polymer Sci. Polymer Phys. Ed., 11, 2234 (1973).
525. Sanders, J.M. and Komoroski, R.A., Makromolecules, 10, 1214 (1977).
526. Stehling, F.C., J. Polymer Science, Part A-1, 4, 189 (1966).
527. Randall, J.C., Macromolecules, 11, 33 (1978).
528. Sanderson, J.M. and Komoroski, R.A., Macromolecules, 10(6), 1214 (1977).
529. Ray, G.J., Johnson, P.E. and Knox, J.R., Macromolecules, 10(4), 773 (1977).
530. Dudek, J.J. and Buesche, F.J., Polymer Science A-2, 811 (1964).
531. Porter, R.S., J. Polymer Science, Part A-1, 4, 189 (1966).
532. Porter, R.S., Nicksie, J.W. and Johnson, J.F., Anal. Chem., 35, 1948 (1963).
533. Barrall, II, E.M., Porter, R.S. and Johnson, J.F., paper presented to Division of Polymer Chemistry, 148 Natl. Meeting American Chemical Society, Chicago, September 1964; Preprints, 5, No. 2, 816 (1964); J. Appl. Polymer Sci., 9, 3061 (1965).
534. Satah, S.R., Chujo, T., Ozchi, T. and Nagai, E., J. Polymer Sci., 62, 510 (1962).
535. Kamath, P.M. and Barlow, A., J. Polymer Science, A-1, 5, 2023 (1967).
536. Bellamy, L.J., Infra-Red Spectra of Complex Molecules, Wiley, New York, 27, 13 (1958).
537. Willbourn, A.J., J. Polymer Sci., 34, 569 (1959).
538. Bassi, I.W. and Scordamaglia, R., Makromol. Chem., 176(5), 1503 (1975).
539. Kenyon, A.C., Salyer, I.O., Kurz, J.E. and Brown, D.R., J. Polymer Science, C8, 205 (1965).
540. Ogawa, T. and Inaba, T., J. Polymer Sci., 21(11), 2979 (1977).
541. Gol'denberg, A.L., Severova, N.N., Kosmatykh, K.I., Andreeva, I.N., Lobanov, A.M. and Erofeev, B.V., Vesti Akad. Navuk Belarus, SSR, Ser. Khim. Navuk, 1, 29 (1973).
542. Ogawa, T. and Inaba, T., J. Polym. Sci. Polym. Phys. Ed., 12, 785 (1974).
543. Ogawa, T., Tanaka, S. and Inaba, T., J. Appl. Polym. Sci., 17, 319 (1973).
544. Ogawa, T. and Inaba, T., J. Appl. Polym. Sci., 18(11), 3345 (1974).
545. Sitnikova, T.A., Strakhov, V.V., Golubev, V.M. and Pestova, M.B., Vysokomol. Soedin., Ser. A, 17(1), 192 (1975).
546. Taylor, W.C. and Graham, J.P., Polymer Letters, 2, 169 (1964).
547. Taylor, W.C. and Tung, L.H., SPE (Soc. Plastic Engineers) Trans 2, 119 (1963).
548. Gamble, L.W., Wipke, W.T. and Lane, T., Am. Chem. Soc. Polymer Chem. Preprints, 4, No. 2, 162 (1963).
549. Wesslau, H., Makromolekulare Chemie, 20, 111 (1956).
550. Bushick, R.D., J. Polymer Science, A3, 2047 (1965).
551. Semura, O., J. Polymer Science, A1, 12, 2631 (1974).
552. Hank, R., Rubber Chem. Technol., 40, 936 (1967).
553. Kemp, A.R. and Miller, G.S., Ind. Eng. Chem., Anal. Ed. 6, 52 (1943).
554. Kemp, A.R. and Peters, H., Ind. Eng. Chem., Anal. Ed., 15, 52 (1943).
555. Rehner, J., Ind. Eng. Chem., 36, 118 (1944).
556. Lee, T.S., Kolthoff, I.M. and Johnson, E., Anal. Chem., 22, 995 (1950).
557. Hank, R., Rubber Chem. Technol., 40, 936 (1967).
558. Sewell, P.R. and Skidmore, D.W., J. Polymer Science A-1, 6, 2425 (1968).
559. Cooper, W., Eaves, D.E., Tunnicliffe, M.E. and Vaughan, G., European Polymer J., 1, 121 (1965).
560. Tunnicliffe, M.E., MacKillop, D.A. and Hank, R., European Polymer J., 1, 259 (1965).
561. Altenau, A.G., Headley, L.M., Jones, S.O. and Ronsaw, H.C., Anal. Chem., 42, 1280 (1970).
562. Lee, T.S., Kolthoff, I.M. and Johnson, E., Anal. Chem., 22, 995 (1950).

563. Sewell, P.R. and Skidmore, D.W., J. Polymer Sci., A-1, 6, 2425 (1968).
564. Hank, R., Rubber Chem. Technol., 40, 936 (1967).
565. Van Schooten, J. and Evenhuis, J.K., Polymer (London), 6, 561 (Nov.
      1965).
566. Van Schooten, J. and Evenhuis, J.K., Polymer (London), 6, 343 (1965).
567. Boer, H. and Kooyman, E.C., Analyt. Chim. Acta, 5, 550 (1951).
568. McMurry, H.L. and Thornton, V., Anal. Chem., 24, 318 (1952).
569. Van Schooten, J., Duck, E.Q. and Berkenbosch, R., Polymer, 2, 357
      (1961).
570. Bucci, G., Simonazzi, T., J. Polymer Science C7, 203 (1964).
571. Natta, G., Dall'Asta, G., Mazzanti, G. and Fiampelli, F., Kolloid Z.,
      182, 50 (1962).
572. Gol'denberg, A.L., Zh. Prikl. Spektrosk, 19,(3), 510 (1973).
573. Yur'eva, F.A., Guseinova, F.O., Portyanskii, A.E., Seidov, N.M.,
      Mamedova, V.M., Abasov, A.I. and Malova, M.G., Vysokomol. Soedin.,
      Ser. A 19(10), 2401 (1977); Chem. Abstr., 87, 2023 15p (1977).
574. Popov, V.P. and Duvanova, A.P., Zh. Prikl. Spectrosk, 22(6), 1115
      (1975); Chem. Abstr., 83, 115221d (1976).
575. Brown, A.L., Shell Chemical Co., Carrington, Cheshire, U.K. private
      communication.
576. Willbourne, J., Polymer Science, 34, 569 (1959).
577. Witek, E and Wisdarczyk, M., Polim. Tworz. Wielk, 19, 81 (1974).
578. Meumann, E.W. and Nadeau, H.G., Ananl. Chem., 10, 1454 (1963).
579. Gray, A.P., Olin Mathieson Chemical Corp., New Haven, Conn.,
      unpublished Du Pont of Canada Method, (1957).
580. Boyle, D.A., Simpson, W. and Waldron, J.D., Polymer, 2, 323 (1961).
581. Madorsky, S.L. and Straus, S., J. Res. Natl. Bur. Std., 53, 361 (1954).
582. Van Schooten, J. and Evenhuis, J.K., Shell Chemical Company,Amsterdam,
      private communication.
583. Seeger, M. and Barrall, E.M., J. Polymer Science, Polymer Chemistry,
      Ed., 13, 1515 (1975).
584. Willbourn, A.H., J. Polym. Sci., 34, 569 (1959).
585. Willbourn, A.H., J. Polym. Sci., 34, 592 (1959).
586. Cross, L.H., Richards, R.B. and Willis, H.A., Disc. Faraday Soc., 9,
      235 (1950).
587. Harlen, F., Simpson, W., Waddington, F.B., Waldron, J.D. and Baskett,
      A.C., J. Polymer Sci., 18, 589 (1955).
588. Boyle, D.A., Simpson, W. and Waldron, J.D., Polymer, Lond., 2, 323,
      335 (1961).
589. Roedel, M.J., J. Amer. Chem. Soc., 75, 6110 (1953).
590. Bryant, W.M.D. and Voter, R.C.J., J. Amer. Chem. Soc., 75, 6113 (1953).
591. Schnell, G., Struktur und Physikalisches Verhalten der Kunststoffe,
      p 609, Edited by Wolf, K.A., Springer: Berlin/Göttingen/Heidelberg,
      (1962).
592. Van Schooten, J. and Evenhuis, J.K., private communication.
593. Porter, R.S., Nicksic, S.W. and Johnson, J.F., Anal..Chem., 35, 1948
      (1963).
594. De Abajo, J., Dela Campa, J.G. and Nieto, J.L., Makromol. Chem.
      Rapid Commun., 3, 505 (1982).
595. Barrall, E.M., Porter, R.S. and Johnson, J.F., Anal   Chem., 35,
      73 (1963).
596. Porter, R.S., Hoffman, A.S. and Johnson, J.F., Anal. Chem., 34,
      1179 (1962).
597. Bambaugh, K.J. and Clampitt, B.H., J. Polymer Science, A 3, 805
      (1965).
598. Bambaugh, K.J., Cook, C.E. and Clampitt, B.H., Anal. Chem., 35, 1834
      (1963).
599. Strassburger, J., Brauer, M.T. and Froziate, A.F., Anal. Chem., 32,454
      (1960).

600.   Barrall, E.M. II, Porter, R.S. and Johnson, J.F., Anal.   Chem., 35,
       73 (1963).
601.   Ettre, K. and Varadi, F., Anal. Chem., 34, 753 (1962).
602.   Ettre, K. and Varadi, P.F., Anal. Chem., 35, 69 (1963).
603.   Lehmann, F.A. and Brauer, G.M., Anal. Chem., 33, 673 (1961).
604.   Denisov, V.M., Svetlichnyi, V.M., Gindin, V.A., Zubkov, V.A.,
       Kol'tsov, A.I., Koton, M.M. and Kudryavtsev, V.V., Polym. Sci.,
       USSR, 21, 1644 (1979).
605.   Wunderlich, B. and Poland, D., J. Polym. Sci., A1, 357 (1963).
606.   Kretzschmar, H.J. and Gross, D., Kunststoffe, 65, 92 (1975).
607.   Haken, J.K. and Obita, J.A., J. Oil and Colour Chemist. Assoc., 63,
       194 (1980).
608.   Aydin, O., Kaczmar, B.U. and Schulz, R.C., Angew, Makromol. Chem.,
       24, 171 (1972).
609.   Wojcik, Z. and Sporysz, W., Polimery, 18(4), 203 (1973).
610.   Matsumoto, T., Nakamae, K. and Chosokabe, J., Nippon Setchaku Koyaki
       Shi 11(5), 249 (1975); Chem. Abstr. 84:90856v.(1976).
611.   Majer, J. and Sodomka, J., Chem. Prum. 25(11), 601 (1975); Chem.
       Abstr. 84:45101.(1976).
612.   Munteanu, D. and Savu, N., Rev. Chim. (Bucharest), 27(10), 902 (1976);
       Chem. Abstr., 86, 121973d (1977).
613.   Leukroth, G., Gummi. Asbest..Kunstst. 29(9), 585 (1976); Chem. Abstr.
       86, 44183s (1977).
614.   Munteanu, D. and Toader, M., Mater. Plast. (Bucharest), 13(2), 97
       (1976); Chem. Abstr. 85, 193245v (1976).
615.   Siryuk, A.G. and Bulgakova, R.A., Vysokomol. Soedin., Ser. B, 19(2),
       152 (1977); Chem. Abstr. 86, 140597a (1977).
616.   Munteanu, D., Toader, M. and Laiber, M., Rev. Chim. (Bucharest) 29(1),
       67 (1978); Chem. Abstr., 89, 6701p (1978).
617.   Ibrahim, B., Katritzky, A.R., Smith, A. and Weiss, D.E., J. Chem.
       Soc., Perkin Trans. 2 (13), 1537 (1974).
618.   Sobottka, J., Keller, F. and Wunderlich, K., Faserforsch. Textiltech.,
       25(8), 352 (1974).
619.   Keller, F., Plaste, Kautsch., 22(1), 8 (1975).
620.   Okada, T. and Ikushige, T., Polym. J., 9(2), 121 (1977).
621.   Foerster, G. and Brand, P., Plaste. Kautsch., 21(2), 99 (1974).
622.   Mitra, B.C. and Katti, M.R., Pop. Plast., 18(10), 15 (1973).
623.   Munteanu, D., Stud. Cercet. Chim., 21(8), 911 (1973).
624.   Mlejnek, O., Cveckova, L., Cech, K. and Gomory, I., Elektrolzolacna
       Kablova Tech. 30(1), 15 (1977); Chem. Abstr. 87, 185332x (1977).
625.   Mlejnek, O., Cveckova, L., Cech, K. and Gomory, I., Sb. Prednasek,
       Makrotest. Celoststni Konf. 4th, 1, 95 (1976); Chem. Abstr. 86,
       56018e (1977).
626.   Ryasnyanskaya, A. Ya., Lyubimova, S.L., Kalashnikov, V.V., Terteryan,
       R.A. Monastyrskii, V.N. and Gryaznov, B.V., Neftepererab. Neftekhim
       (Moscow), 10, 69 (1973).
627.   Belyaev, V.M., Budtov, V.P., Frenkel, S. Ya. and Daniel, N.V.,
       Vysokomol, Soedin., Ser. A 14(11), 2335 (1972).
628.   German, A.L. and Heikens, D., J. Polymer Science, A-1, 9, 2225 (1971).
629.   Braun, J.M. and Guillet, J.E., J. Polymer Science, Polymer Chemistry
       Ed., 13, 1119 (1975).
630.   Sedlar, J., Feniokova, E. and Pac, J., Analyst, 50, 99 (1974).
631.   Grasley, M.H. and Barnum, E.R., Martinez Research Laboratory, Shell
       Chemical Co. Ltd., private communication.
632.   Bombaugh, K.J., Cook, C.E. and Clampitt, B.H., Anal. Chem., 35,
       1834 (1963).
633.   Smith, O.F., Anal. Chem., 35, 1835 (1963).
634.   Schulz, R.C., Kaiser, E. and Kern, W., Makromolekulare Chem., 76, 99
       (1964).

635. Tanaka, O., J. Polymer Science A-1, 11, 2069 (1973).
636. Wu, T.K., Macromolecules, 6,(5), 737 (1973).
637. Keller, F. and Muegge, C., Faserforsch. Textiltech. 27(7), 347 (1976).
638. Radowanow, L. and Sobottka, J., Plaste Kaut. 21(4), 266 (1974).
639. Delfini, M., Segre, A.L. and Conti, F., Macromolecules, 6, 456 (1973).
640. Wu, T.K., Ovenall, D.W. and Reddy, G.S., J. Polym. Sci., Polym. Phys.
      Ed., 12, 901 (1974).
641. Wu, T.K. and Ovenall, D.W., Polym. Prepr., Am. Chem. Soc., Div. Polym.
      Chem. 17(2), 693 (1976).
642. Morishima, Y., Takizawa, T. and Murahashi, S., Eur. Polym. U., 9,
      669 (1973).
643. Braun, J.M. and Guillet, J.E., J. Polymer Science, Polymer Chemistry
      Ed., 13, 1119 (1975).
644. Ishibashi, M., J. Polym. Sci., A2, 3657 (1964).
645. Bahrani, M.L., Chakravarty, N.K. and Chopra, S.C., Indian J. Technol.
      13(12), 576 (1975).
646. Heunische, G.W., Anal. Chim. Acta, 101, 221 (1978).
647. Quenum, B.M., Berticat, P. and Vallet, G., Polym. J., 7(3), 277 (1975).
648. Lindberg, J.J., Stenman, F. and Laipio, I., J. Polym. Sci., Polym.
      Symp., No. 42, pt. 2, 925 (1973).
649. Oswald, H.J. and Kubu, E.T., Soc. Plastics Eng. Trans. 3, 168 (1963).
650. Nambu, K., J. Appl. Polym. Sci., 4, 69 (1960).
651. Fuchs, W. and Louis, D., Makromolekulare Chem., 22, 1 (1957).
652. Keller, F. and Muegge, C., Plaste. Kautsch., 24(2), 88 (1977); Chem.
      Abstr. 86, 121915m (1977).
653. Keller, F. and Muegge, C., Plaste. Kautsch., 24(4), 239 (1977); Chem.
      Abstr. 86, 190575n (1977).
654. Keller, F., Faserforsch. Textiltech. 28(10), 515 (1977); Chem..Abstr.
      88, 105888r (1978).
655. Kleinpeter, E. and Keller, F., Z. Chem. 18(6), 222 (1978); Chem.
      Abstr. 89, 130030h (1978).
656. Isa, I. Abu., J. Polym. Sci., A-1 10, 881 (1972).
657. Isa, I. Abu. and Meyers, M.E., J. Polym. Sci., A-1 11, 2125 (1973).
658. Brame, E.G., J. Polym. Sci., A-1 9, 2051(1971).
659. Frensdorff, H.K. and Ekiner, O., J. Polym. Sci., A-1, 5, 1157 (1967).
660. Ogure, H., Bunseki Kagaku, 24 (12), 197 (1975); Chem. Abstr., 85,
      33784s (1975).
661. Narasaki, H. and Umezawa, K., Kobunshi Kagaku, 29(6), 438 (1972).
662. Citovicky, P., Simek, I., Mikulasova, D. and Chrastova, V., Chem.
      Zvesti, 30(3), 342 (1976); Chem. Abstr., 88, 105960h (1978).
663. Slovokhotova, N.A., Il'iccheva, Z.F., Vasiliev, L.A. and Kangin, V.A.,
      Karpov Physicochemical Scientific Research Institute, unidentified
      publication (1963).
664. McMurry, H.L. and Thornton, V., Anal. Chem., 24, 318 (1952).
665. Slovokhoyova, N.A., Marupov, M.A. and Kargin, V.A., Sb. Khimicheskiye
      svoistva i modifikatsiya polimerov. (Collected Papers, The Chemical
      Properties and Modification of Polymers), Izd. Nauka, (1964).
666. Binder, Z.Z. and Ransow, H.C., Anal. Chem., 29, 503 (1957).
667. Fisher, H. and Hellwege, R.H., J. Polymer Sci., 56, 33 (1962).
668. Adams, J.H., J. Polymer Science A-1 8, 1279 (1970).
669. Wool, R.P. and Statton, W.O., J. Polym. Sci., Polym. Phys. Ed., 12,
      1575 (1974).
670. Goldstein, M., Seeley, M.E., Willis, H.A. and Zichy, V.J.I., Polymer,
      14(11), 530 (1973).
671. Schleueter, D.D. and Siggia, S., Anal. Chem., 49, 2349 (1977).
672. Lanska, B. and Subendo, J., Chem. Prum., 30, 421 (1980).
673. Fraser, G.V., Hendra, P.J., Watson, D.S., Gall, M.J., Willis, H.A.
      and Cudby, M.E.A., Spectrochim. Acta. Part A, 29(7), 1525 (1973).

674.   Bailey, R.T., Hyde, A.J., and Kim, J.J., Advan. Raman Spectrosc., 1,
       296 (1972).
675.   Bailey, R.T., Hyde, A.J. and Kim, J.J., Spectrochim. Acta, Part A,
       30(1), 91 (1974).
676.   Chalmers, J.M., Polymer, 18(7), 681 (1977).
677.   Willis, H.A. and Cudby, M.E.A., Struct. Stud. Micromol. Spectrosc.,
       Methods (Meet) 81-9 (1974),(Pub. 1976); Ed. Ivin, K.J., Wiley,
       Chichester, Eng.
678.   Chen, T.L., Hua Hsueh Tung Pao, 1, 47 (1977); Chem. Abstr. 87, 6569k
       (1977).
679.   Majer, J. and Brettschneiderova, J., Chem. Prum., 25(6), 311 (1975);
       Chem. Abstr., 83, 132 150x. (1976).
680.   Painter, P.C., Watzek, M. and Koenig, J.L., Polymer, 18(11), 1169
       (1977).
681.   Holland-Moritz, K., Colloid Polym. Sci., 253(11), 922 (1975).
682.   Blais, P., Carlsson, D.J. and Wiles, D.M., J. Polym. Sci., A7 10,
       1077 (1972).
683.   Natta, G. and Danusso, F., J. Polym. Sci., 34, 3 (1959).
684.   Bovey, F.A., Polymer Conformation and Configuration, Academic Press,
       New York, N.Y., p.8 (1969).
685.   Bovey, F.A., Polymer Conformation and Configuration, Academic Press,
       New York, P32 (1969).
686.   Stehling, F.C. and Knox, J.R., Macromolecules, 8, 595 (1975).
687.   Randall, J.C., ACS Symposium Series No. 103, Carbon 13NMR Polymer
       Science, Wallace M. Pasika, Editor, American Chemical Society (1979).
688.   Grant, D.M. and Paul, E.G., J. Amer. Chem. Soc., 86, 2984 (1964).
689.   Zambelli, A., Locatelli, P., Bajo, G. and Bovey, F.A., Macromolecules,
       8, 687 (1975).
690.   Schaefer, J. and Natusch, D.F.S., Macromolecules, 5, 416 (1972).
691.   Axelson, D.E., Mandelkern, L. and Levy, G.C., Macromolecules, 10,
       557 (1977).
692.   Zambelli, A., Locatelli, P., Bajo, J. and Bovey, F.A., Macromolecules,
       8, 687 (1975).
693.   Randall, J.C., J. Polymer Sci., Polym. Phys. Ed., 12, 703 (1974).
694.   Provasoli, A. and Ferre, D.R., Macromolecules, 10, 874 (1977).
695.   Stehling, F.C. and Knox, J.R., Macromolecules, 8, 595 (1975).
696.   Zambelli, A., Locatelli, P., Bajo, G. and Bovey, F.A., Macromolecules,
       8, 687 (1975).
697.   Bovey, F.A., Macromolecules, 8, 687 (1975).
698.   Randall, J.C., Polym. Sci., Polym. Phys. Ed., 12, 703 (1974).
699.   Provasoli, A. and Ferro, D.R., Macromolecules, 10, 874 (1977).
700.   Crain, Jr., W.D., Zambelli, A. and Roberts, J.D., Macromolecules, 3,
       330 (1970).
701.   Inoue, Y., Nishioka, A. and Chûjô, R., Makromol. Chem., 152, 15 (1972).
702.   Zambelli, A., Dormann, D.E., Richard Brewster, A.I. and Bovey, F.A.,
       Macromolecules, 6, 925 (1973).
703.   Bovey, F.A., Polymer Conformation and Configuration, Academic Press,
       New York, N.Y., p16 (1969).
704.   Randall, J.C., J. Polym. Sci., Polym. Phys. Ed., 14, 2083 (1976).
705.   Illers, K.H., Eur. Polym. J., 10, 911 (1974).
706.   Foli, S., Luederwald, I., Montaudo, G. and Przybylski, M., Angew.
       Makromol. Chem., 62, 215 (1977).
707.   Caccawese, S., Przybylski, M., Maravigna, P., Montaudo, G., Recca, A.,
       Luederwald, I. and Przybylski, M, J. Polym. Sci. Polym. Chem., 15,
       5 (1977).
708.   Varma, I.K., Sundari, V.S. and Varma, D.S., J. Appl. Polym. Sci., 22,
       2857 (1978).
709.   Woodward, A.E., J. Polym. Sci., C 8, 137 (1965).

710. Stehling, F.C., J. Polym. Sci., A 2, 1815 (1964).
711. Brosio, E., Delfini, M. and Conti, F., Nuova Chim., 48(11), 35 (1972).
712. Cavalli, L., Relaz. Corso Teor.-Prat. Risonanza Magn. Nuci, 351 (1973).
713. Mitani, K., J. Macromol. Sci., Chem. A 8(6), 1033 (1974).
714. Inoue, Y., Nishioka and Chujo, R., Makromol. Chem., 152, 15 (1972).
715. Zambelli, A., Dorman, D.E., Brewster, A.I.R. and Bovey, F.A., Macromolecules, 6, 925 (1973).
716. Randall, J.C., J. Polym. Sci., Polym. Phys. Ed., 12, 703 (1974).
717. Randall, J.C., J. Polym. Sci., Polym. Phys. Ed., 14, 2083 (1976).
718. Randall, J.C., J. Polym. Sci., 14, 1693 (1976).
719. Randall, J.C., J. Polym. Sci., Polym. Phys. Ed., 14(11), 283 (1976).
720. Asakura, T., Ando, I., Nishioka, A. Doi, Y. and Keii, T., Makromol. Chem., 178(3),791 (1977).
721. Mitani, K., Ogata, T. and Iwasaki, M., J. Polym. Sci., Polym. Chem. Ed., 12, 1653 (1974).
722. Reilly, G.A., Shell Chemical Company, Emeryville, California, private communication (1964).
723. Loy, B.R., J. Polym. Sci., Part A, 1, 2251 (1963).
724. Wall, L.A., J. Polym. Sci., 17, 141 (1955).
725. Buch, T., PHD Thesis Norwestern University 1960, Microfilms MIC 60-4761 Ann Arbor, Michigan.
726. Kusumoto, N., Matsumoto, K. and Tabayagni, M., J. Polym. Sci., A-1, 7, 1773 (1969).
727. Forrestal, L.J. and Hodgson, W.G., J. Polym. Sci., A2, 1275 (1964).
728. Ohnishi, S., Sugimoto, S. and Nitta, I., J. Poly. Sci., A1, 625 (1963).
729. Stehling, F.C., Knox, J.R., Makromolecules, 8(5), 595 (1975).
730. Ooi, T., Shiotsubo, M., Hama, Y. and Shinohara, K., Polymer, 16(7), 510 (1975).
732. Wall, L.A. and Straus, S., J. Polym. Sci., 44, 313 (1960).
733. Grassie, N., Chemistry of Vinyl Polymer Degradation Processes, Butterworth, London, (1956).
734. Jelliner, H.H.G., The Degradation of Vinyl Polymers, Academic Press, New York, (1955).
735. Simha, R. and Wall, L.A., in Emmet, P.H. (Editor), Catalysis, Vol. 6, Reinhold Publishing Co. New York, (1958).
736. Brauer, G.M., J. Polym. Sci., C 8, 3 (1965).
737. Voigt, J., Kunststoffe, 54, 2 (1964).
738. Van Schooten, J. and Evenhuis, J.K., Polymer (London), 6, 561 (Nov. 1965).
739. Van Schooten, J. and Evenhuis, J.K., Polymer (London), 6, 343 (1965).
740. Dimbat, M., Gas Chromatogr., Proc. Int. Symp. (Europe), 8, 237 (1970 Pub. 1971).
741. Toader, M., Chivulescu, E., Bader, P. and Bodorodea, M., Mater. Plast. (Bucharest), 10(3), 151 (1973).
742. Moiseev, U.D. and Nieman, M.H., Vys. Soed., 3, 1383 (1961).
743. Pastusha, G. and Jost, U., Angew. Makromol. Chem., 81, 11 (1979).
744. Simha, R., personal communication (Feb. 1962).
745. Wall, L.A., Soc. Petrol. Engrs. J., 16, 1 (1960).
746. Schwenken, R.F. and Zuccarello, R.K., J. Polymer Sci., C 6, 1 (1964).
747. Donald, H.J., Humes, E.S. and White, L.W., J. Poly. Sci., C 6, 93 (1964).
748. Beachall, C. and Beck, D.L., J. Polym. Sci., A 3, 457 (1965).
749. Seeger, M. and Gritter, R.J., J. Polym. Sci., A-1, 15, 1393 (1977).
750. Tsuchiya, Y. and Sumi, K., J. Polym. Sci., A-1, 7, 1599 (1969).
751. Kamida, K. and Yamaguchi, K., Makromol. Chem., 162, 205 (1972).
752. Duswalt, A.A. and Cox, W.W., Polym. Charact. Interdisciplinary Approaches, Craver C.D., Ed., Plenum, New York, P. 147 (1971).
753. Marchetti, A. and Martuscelli, E., J. Polym. Sci., Polym. Phys. Ed.,

12, 1649 (1974).

754.   Bosch, K., Beitr. Gerichtl. Med, 33, 280 (1975).
755.   Neiman, M.B., Ageing and Stabilization of Polymers, (translated
        from Russian), Consultants Bureau, New York, Chap. 4 (1965).
756.   Rugg, F.M., Smith, J.J. and Bacon, R.C., J. Polym. Sci., 13, 535
        (1954).
757.   Meyer, C.S., Ind. Eng. Chem., 44, 1095 (1952).
758.   Carlsson, D.J., Kato, Y. and Wiles, D.M., Macromolecules, 1, 459
        (1968).
759.   Sheehan, W.C. and Cole, T.B., J. Appl. Polym. Sci., 8, 2359 (1964).
760.   Ross, S.E., J. Appl. Polym. Sci., 9, 2729 (1965).
761.   Adams, J.H. and Goodrich, J.E., J. Polym. Sci., A-1, 8, 1269 (1970).
762.   Vogt, H., Siemens-Z, 46(11), 858 (1972).
763.   Iwate, N., Tanaka, H. and Okajima, S., J. Appl. Polym. Sci., 17,
        2533 (1973).
764.   Ballard, D.G.H., Cheshire, P., Longman, G.W. and Schelten, J.,
        Polymer, 19(4), 379 (1978).
765.   Wijga, P.W.O., Van Schooten, J. and Boerma, J., Makromolekulare.
        Chem. 36, 115 (1960).
766..   Van Schooten, J. and Wijga, P.W.O., Makromolekulare Chem., 43, 23
        (1961).
767.   Davis, T.E. and Tobias, R.L., J. Polym. Sci., 50, 227 (1961).
768.   Pegoraro, M., Chim. e Ind. (Milano) 44, 18 (1961).
769.   Shyluk, S., J. Polymer Sci., 62, 317 (1962).
770.   Redlick, O., Jacobson, A.L., and McFadden, W.H., J. Polym. Sci.,
        A1, 393 (1962).
771.   Westerman, L., J. Polym. Sci., A 1, 411 (1963).
772.   Hirota, M., Kanbe, H. and Nakaguchi, K., J. Polym. Sci., B 1, 701
        (1963).
773.   Jacobi' E., Schultenberg, H. and Schulz, R.C., Makromol. Chem.
        Rapid. Comm., 1, 397 (1980).
774.   Haken, J.K. and Obita, J.A., J..Chromato., 213, 55 (1981).
775.   Crouzet, P., Fine and Mangin, P., paper presented at Fifth Int.
        Seminar, London, 1968, preprint 10; J. Appl. Polym. Sci., 13, 205
        (1969).
776.   Ogawa, T., Suzuki, Y. and Inaba, T., J. Polym. Sci., A-1 10, 737
        (1972).
777.   Vaughan, M.F., Ind. Polym; Charact. Mol. Weight. Proc. Meet. 111 (1973).
778.   Wims, A.M. and Swarin, S.J., J. Appl. Polym. Sci., 19(5), 1243 (1975).
779.   Lovric, L., Nafta (Zagreb), 23, (11), 550 (1972).
780.   Prchocova, M. and Pechoc, V., Makromol. Chem., 163, 235 (1973).
781.   Ogawa, T. and Hoshino, S., J. Appl. Polym. Sci., 17, 2235 (1973).
782.   Ogawa, T., Tanaka, S. and Inaba, T., J. Appl. Sci., 18, 1351 (1974).
783.   Ogawa, T., Tanaka, S. and Inaba, T., J. Appl. Polym. Sci., 17,
        779 (1973).
784.   Oth, A. and Desreux , V., Bull. Soc. Chim. Belgique, 63, 261 (1954).
785.   Harris, J. and Miller, R.G., J. Polym. Sci., 7, 377 (1951).
786.   Scholtan, W., Makromolekulare Chem., 27, 104 (1957).
787.   Bischoff, J. and Desreux, V., Bull. Soc. Chim. Belgique, 60, 137
        (1951).
788.   Kyogoku, Y. and Kimura, T., J. Chem. Soc. Japan, Ind. Chem. Sect.
        (Kogyo Kagaku Zassi), 61, 132 (1958).
789.   Gooberman, G., J. Polym. Sci., 40, 469 (1959).
790.   Tanaka, S., Nakamura, A. and Ozawa, J., 10th Annual Meeting of the
        Society of Polymer Science of Japan, May 28th 1961.
791.   Tanaka, S., Nakamura, A. and Morikawa, H., Die Makromoleculare Chemie,
        85, 164 (1965).
792.   Morey, D.R., and Tamblyn, J.W., J. Appl. Phys., 16, 419 (1945).

793. Westerman, L., J. Polym. Sci., A 1, 4511 (1963).
794. Combes, R.L., Slonaker, D.F., Joyner, F.B. and Cooner, H.W., J. Polym. Sci., A-1, 5, 215 (1967).
795. Kinoshito, T., Morinaga, A. and Keir, T., J. Polym. Sci., A-1, 13, 2491 (1975).
796. Wozniak, T., Przegl. Wlok, 30 (1), 18 (1976).
797. Atkinson, C.M.L. and Dietz, R., Nat. Phys. Lab. (U.K.) Div. Chem. Stand., Rep. 32, 18 pp (1974).
798. Crompton, T.R., unpublished work.
799. Samuels, R.J., J. Polym. Sci., Polym. Phys. Ed., 12, 1417 (1974).
800. Radtsig, V.A., Vyo. Soe., Ser. A 17(1), 154 (1975).
801. Van Sickle, D.E., J. Polym. Sci., A-1, 10, 2751 (1972).
802. Russell,C.A., J. Appl. Polym. Sci., 4, No. 11, 219 (1960).
803. Ranalli, F. and Crespi, G., Modern Plastics Encyclopedia, Modern Plastics, New York, 142-145 (1960).
804. Natta, G., Mazzanti, G., Crespi, G. and Moraglio, G., Chim. e Ind. (Milan), 39, 275 (1957).
805. Natta, G., Corradini, P. and Cesari, M., Atti accad. nazl..Lincei, Rend Classe Sci. Fis. Mat. e Nat., 22, 11 (1957).
806. Quynn, R.G., Riley, J.L., Young, D.A. and Noether, H.D., J. Appl. Polym. Sci., 2, 166 (1959).
807. Heinen, W., J. Polym. Sci., 38, 545 (1959).
808. Luongo, J.P., J. Appl. Polym. Sci., 3, 302 (1960).
809. Natta, G., Pino, P. and Mazzanti, G., Gazz. Chim. Ital., 87, 528 (1957).
810. Braun, J.M. and Guillet, J.E., J. Polym. Sci., Polym. Chem. Ed., 13, 1119 (1975).
811. Manjunath, B.R., Venkataraman, A. and Stephen, T., J. Appl. Polym. Sci., 17, 1091 (1973).
812. Maltese, P., Clementini, L. and Panizza, S., Mater. Plaste. Elastomerie, 35, 1669 (1969).
813. Knight, H.S. and Siegel, H., Anal. Chem., 38, 1221 (1966).
814. Styskin, E.L., Gurvic, Ya. A. and Kumak, S.I., Khim. Prom. 5, 359 (1973).
815. Denning, J.A. and Marshall, J.A., Analyst, 97, 710 (1972).
816. Lappin, G.R. and Zannucci, J., Anal. Chem., 41, 2076 (1969).
817. Sedlar, J., Feniokova, E. and Pac, J., Analyst, 99, 50 (1974).
818. Dobies, R.S., J. Chromat., 40, 110 (1969).
819. Yashikawa, T., Ushimi, K., Kimura, K. and Tamuro, N.T., J. Appl. Polym. Sci., 15, 2065 (1971).
820. Kellum, G.E., Anal..Chem., 43, 1843 (1971).
821. Sedlar, J., Pac, J., Foniokova, E., Valtrova, A. and Prudilova, M., Chem. Prum. 23(11), 566 (1973).
822. Peraldo, M., Gazz. Chem. Ital., 89, 798 (1959).
823. McDonald, M.P. and Ward, I.M., Polymer, 2, 341 (1961).
824. Brader, J.J., J. Polymer Science, 3, 370 (1960).
825. Luongo, J.P., J. App. Polymer Sci., 3, 302 (1960).
826. Liang, C.Y. and Pearson, F.G., J. Mol. Spect., 5, 290 (1960).
827. Sibilia, J.P. and Wincklhofer, R.C., J. Appl. Polymer Sci., 6, 557 (1962).
828. Natta, G., Corradini, P. and Bassi, I.W., Makromol. Chem., 21, 240 (1956); Nuovo Cimento Supply., 15, 52 (1960).
829. Danusso, F. and Gianotti, G., Makromol. Chem., 61, 139 (1963).
830. Holland, V.F. and Miller, R.L., J. Appl. Phys., 35, 3241 (1964).
831. Natta, G., Makromol. Chem., 35, 94 (1960).
832. Boor, J. and Mitchell, J.C., J. Polymer Sci., A1, 59 (1963).
833. Geacintov, C., Schotland, R.S. and Miles, R.B., J. Poly. Sci., B1, 587 (1963).

834.   Boor, J. and Youngman, E.A., J. Poly. Sci., B2, 903 (1964).
835.   Luongo, J.P. and Salovey, R., Polymer Letters, 3, 513 (1965).
836.   Higgins, G.M.C. and Turner, D.T., J. Polymer Science, A2, 1713
        (1964).
837.   Toman, L., Marek, M. and Tak, J., J. Polymer Science, A1, 12, 1897
        (1974).
838.   Barrall, E.M., Porter, R.S. and Johnson, J.F., J. Chromatogr. 11,
        177 (1963).
839.   Barrall, E.M., Porter, R.S..and Johnson, J.F., Anal. Chem., 35, 73
        (1963).
840.   Porter, R.S., Hoffman, A.S. and Johnson, J.F., Anal. Chem., 34, 1179
        (1962).
841.   Warren, R.W., Gates, D.S. and Driscoll, G.L., J. Polymer Science,
        A1, 9, 717 (1971).
842.   Mauzac, M., Vairon, J.P. and Sigwalt, P., Polymer, 18(11), 1193
        (1977).
843.   Suzuki, T., Koshiro, S. and Takegami, Y., J. Polym. Sci., Part B,
        10, 829 (1972).
844.   Manatt, S.L., Ingham, J.D. and Miller, J.A., Org. Magn. Reson. 10,
        198 (1977).
845.   Loy, B.R., J. Polymer Science Part A, 1, 2251 (1963).
846.   Voigt, J., Kunststoffe, 2, 54 (1964).
847.   Wall, L.A. and Strauss, S., J. Polymer Sci., 44, 313 (1960).
848.   Barrall, E.M., Porter, R.S. and Johnson, J.F., J..Chromatog., 11,
        177 (1963).
849.   Geacintov, C., Schotland, R.S. and Miles, R.B., J. Polymer Science,
        C 6, 197 (1964).
850.   Holden, H.W., J. Polymer Science C 6, 209, 19 (1964).
851.   Clampitt, B.H. and Hughes, R.H., J. Polymer Science C 6, 43 (1964).
852.   McNeill, J.C., J..Polymer Science A1, 4, 2479 (1966).
853.   Pepper, D.C. and Reilly, P.H., Proc. Chem. Soc., 1, 460 (1961).
854.   McNeill, I.E., Polymer, 4, 15 (1963).
855.   McGuchan, R. and McNeill, I.C., J. Polymer Science A1, 4, 2051
        (1966).
856.   Gallo, S.G., Weise, H.K. and Nelson, J.F., Ind. Eng. Chem., 40, 1277
        (1948).
857.   Lee, T.S., Kolthoff, I.M. and Johnson, E., Anal. Chem., 22, 995
        (1950).
858.   Pastuska, G., Just, U. and August, H., Angew. Makromol. Chem., 107,
        173 (1982).
859.   Dainton, F.S. and Sutherland, G.B.B.M., J. Polymer Sci., 4, 37
        (1949).
860.   Flett, M. St. C. and Plesch, P.H., J. Chem. Soc., 3355 (1952).
861.   Norrish, R.G.W. and Russel, K.E., Trans. Faraday Soc., 48, 91 (1952).
862.   Biddulph, R.H., Plesch, P.H. and Rutherford, P.P., J. Chem. Soc.,
        275 (1965).
863.   Geymer, D.O., Shell Development Company, Emeryville, California,
        private communication.
864.   Lederer, K., Klapp, H., Zipper, P., Wrentschur, E. and Schurz, J.,
        J. Polymer Science, A1, 17, 639 (1979).
865.   Tanaka, T., Chatani, Y. and Tadokoro, H., J. Polym. Sci., Polym.
        Phys. Ed., 12, 515 (1974).
866.   Kenyon, A.C., Salyer, I.O., Kurz, J.E. and Brown, D.R., J. Polymer
        Science, C 8, 205 (1965).
867.   Cantow, M.J.R., Porter, R.S. and Johnson, J.F., J. Polymer Science,
        A 2, 2547 (1964).
868.   Weill, G., Rev. Gen. Caout. Plast., 50, (12), 1003 (1973).
869.   Elliot, J.J. and Kennedy, J.P., J. Polymer Science, A 1, 11, 2993 (1973).

870. Elliott, J.J. and Kennedy, J.P., J. Polym. Sci., Poly. Chem. Ed., 11, 299 (1973).
871. Kissin, Yu. B., Gol'dfarb, Yu. Ya., Novoderzhkin, Yu. V. and Krentsel, B.A., Vysokomol Soedin., Ser. B., 18(3), 167 (1976).
872. McGuchan, R. and McNeill, I.C., J. Polymer Science, A 1, 6, 205 (1968).
873. Gabbay, S.M. and Stivala, S.S., Polymer, 17(2), 121 (1976).
874. Manlus, G.G., J. Polymer Science, 62, 263 (1962).
875. Woodward, A.E., J. Polymer Science, C 8, 137 (1965).
876. Voigt, J., Kunststoffe, 2, 52 (1964).
877. Kelm, J., Kretzschlmar, H.J. and Zimmer, H., Kunststoffe, 71, 514 (1981).
878. Kennedy, J.P. and Johnson, J.E., J. Polym. Sci., Polym. Lett. Ed. 13(8), 465 (1975).
879. Holland-Moritz, K., Modric, I., Heinen, K.U. and Hummel, D.O., Kolloid-Z., Z. Polym., 251 (11), 913 (1973).
880. Holland-Moritz, K., Sausen, E. and Hummel, D.O., Colloid Polym. Sci., 254 (11), 976 (1976).
881. Modric, I., Holland-Moritz, K. and Hummel, D.O., Colloid Polym. Sci., 254(3), 342 (1976).
882. Beebe, D.H., Gordon, C.E., Thudum, R.N., Throckmorton, M.C. and Hanlon, T.L., J. Polymer Science, A 1, 16, 2285 (1978).
883. Elgert, K.F. and Ritter, W., Makromol. Chem., 177(7), 2021 (1976).
884. Hirai, H., Kiraki, K., Hoguchi, I., Inone, T. and Makishima, S., J..Polymer Sci., A 1, 8, 2393 (1970).
885. Takeda, K., Yoshida, H., Hayashi, K. and Okamura, S., J. Polymer Science A 1, 4, 2710 (1966).
886. Dannals, L.E., J. Polymer Science A 1, 8, 2989 (1970).
887. Albert, R. and Malone, W.M., J. Polym. Sci., Symp. No. 42, 1199 (1973).
888. Ayano, S. and Murakawa, T., Kobunshi Kagaku, 29 (10), 723 (1972).
889. Fisch, M.H. and Dannenberg, J.J., Anal. Chem. 49(9), 1405 (1977).
890. Randall, J.C., Macromolecules, 11(3), 592 (1978).
891. Van Schooten, J. and Evenhuis, J.K., Polymer (London), 6, 561 (1965).

892. Van Schooten, J. and Evenhuis, J.K., Polymer (London), 6, 343 (1965).
893. Myers, L.W. and Lord, E.B., Shell Chemical Co., Carrington U.K., private communication.
894. Eggertson, F.T., Dimba, M. and Stross, F.H., Shell Chemical Co., Emeryville, California, private communication.
895. Ho, W., Kissin, Yu. V., Gol'dfarb, Yu. Ya. and Krentsel, B.A., Vysokomol. Soedin, Ser. V. A., 14 (10), 2229 (1972).
896. Bakuyutov, N.G., Kissin, Yu. V., Vavilova, I. and Arkhipova, Z.V., Vysokomol. Soedin. Ser. A. 17, 2163 (1975).
897. Kornet, S., Plasty. A. Kamcuk, 13, 176 (1976).
898. Urbanski, J., Handb. Anal. Synth. Polym. Plast., 372-87 (1977); Ellis Horwood Ltd. Chichester, England.
899. Kalinina, L.S. and Doroshina, L.I., Metody Ispyt. Kontr. Issled. Mashinostroit, Mater. 3, 5 (1973); Chem. Abstr., 80, 54100 (1974).
900. Narasaki, H. and Umezawa, K., Kobunshi Kagaku, 29 (6), 438 (1972).
901. Kanjilal, C., Mitra, B.C. and Palit, S.R., Makromol. Chem. 178(6), 1707 (1977).
902. Kato, K., J. Appl. Polym. Sci., 17, 105 (1973).
903. Pepper, D.C. and Reilly, D.J., Proc. Chem. Soc., 460, (1961).
904. Ghosh, P., Chadha, S.C., Mukherjee, A.R. and Palit, R., J. Polymer Sience, A 2, 4433 (1964).
905. Spetnagel, W.J. and Palit, S.R., J. Polymer Science A1, 15, 945 (1977).
906. Banthia, A.K., Mandal, B.M. and Palit, S.R., J. Polym. Sci. Polym. Chem. Ed., 15(4), 945 (1977).

907.   Stânescu,G., Revista de Chemie, Bucharest, 64, 42 (1963).
908.   Lindley, G., Research and Development, 30, 21 (1964).
909.   Sloane, H.J., Johns, T., Ulrich, W.F. and Cadman, W.J., Applied
        Spectroscopy, 19, 130 (1965).
910.   Yoshina, T. and Shinomuya, M., J. Polymer Science A 3, 2811 (1965).
911.   Jasse, B. and Monnerie, L., J. Phys. D. 8(7), 863 (1975).
912.   Spells, S.J., Shepherd, I.W. and Wright, C.J., Polymer, 18(9), 905
        (1977).
913.   Woodward, K.E., J. Polymer Science C 8, 137 (1965).
914.   Randall, J.C., J. Polymer Sci., 13, 889 (1975).
915.   Inoue, Y.,ᵢ Nishioka, A. and Chujo, R., Makromol. Chem., 156, 207
        (1972).
916.   Matsuzaki, K., Urya, T., Osada, K. and Kawamura, T., Macromolecules,
        5, 816 (1972).
917.   Inoue, Y., Nishioka, A. and Chujo, R., Makromol. Chem., 156, 207
        (1972).
918.   Randall, J.C., J..Polymer Sci. Polymer Phys. Ed., 14, 283 (1976).
919.   Borsa, F. and Lanzi, G., J. Polymer Science, A 2, 2623 (1964).
920.   Florin, R.E. and Wall, L.A., J. Chem. Phys. 57 (4), 1791 (1972).
921.   Lehmann, F.A. and Brauer, G.M., Anal. Chem. 33, 673 (1961).
922.   Strassburger, J., Brauer, G.M., Tyron, M. and Forziati, A.F., Anal.
        Chem. 32, 454 (1960).
923.   Madorsky, S.I. and Straus, S., J. Research Nat'l Bureau of Standards,
        40, 417 (1948).
924.   Straudinger, H. and Steinhoffer, A., Annalen., 35, 517 (1935).
925.   Madorsky, S.L..and Straus, S., J..Research Nat'l Bureau Standards,
        63A, 261 (1959).
926.   Sindelfingen, J.V., Kunststoffe, 51, 18 (1961).
927.   Guillet, J.E., Wooton, W.C. and Combs, R.L., J. Appl. Polymer Sci.,
        3, 61 (1960).
928.   Jones, C.E.R. and Moyles, A.F., Nature, 189, 222 (1961).
929.   Lehrle, R.S. and Robb, J.C., Nature, 183, 1671 (1959).
930.   Esposito, G.G., Anal. Chem., 36, (11), 2183 (1964).
931.   Feuerberg, H. and Weigel, H., Z. Anal. Chem., 199, 121 (1964).
932.   Brauer, O., J. Polymer Science, C 8, 3, (1965).
933.   Klein, J. and Widdecke, H., Ange. Makromol. Chim., 53(1), 145 (1976).
934.   Cabasso, I., Jagur-Grodzinski, J. and Vofsl, D., J. Apply. Polym.
        Sci., 18, 1969 (1974).
935.   Grassie, N. and Weir, N.A., J. Appl. Polymer Science, 9, 963 (1965).
936.   Grassie, N. and Weir, N.A., J. Appl. Polymer Science, 9, 975 (1965).
937.   Schole, R.G., Bednarczyk, J. and Tamanci, T., Anal. Chem., 38, 331
        (1966).
938.   Spitzbergen, J.C. and Beachell, H.C., J. Polymer Sci., A 2, 1205
        (1964).
939.   Miller, A.A. and Mayo, F.R., J. Am. Chem. Soc., 78, 1017 (1956).
940.   Grassie, N. and Weir, N.A., J. Appl. Polymer Sci., 9, 963 (1965).
941.   Shaw, J.N. and Marshall, M.C., J. Polymer Science A 1, 6, 449 (1968).
942.   Ottewill, R.H. and Shaw, J.N., Kolloid-Z., in press.
943.   Ottewill, R.H. and Shaw, J.N., Kolloid-Z., in press.
944.   Shaw, J.N., paper presented before the Division of Polymer Chemistry
        at the Symposium on New Concepts in Emulsion Polymers, American
        Chemical Society Meeting, New York, September 12-16, 1966; J.
        Polym. Sci.
945.   Lemstra, P.J., Schouten, A.J. and Challa, G., J. Polym. Sci., Polym.
        Phys. Ed., 10, 2301 (1972).
946.   Lemstra, P.J., Schouten, A.J. and Challa, G., J. Polym. Sci., Polym.
        Phys. Ed., 12, 1565 (1974).
947.   Lety, A. and Noel, C., J. Chim. Phys. Physicochim. Biol., 69(5),

875 (1972).
948. Taylor, D.L., J. Polymer Science A 2, 611 (1964).
949. Wunderlich, B. and Bodily, D.M., J. Applied Polymer Science C 6, 137 (1964).
950. Shibazaki, Y. and Kamebe, H., Kabish Kagaku, 21, 65 (1964).
951. McNeill, I.C., J. Polymer Science A 1, 4, 2479 (1966).
952. Coloff, S.G. and Vanderborgh, N.E., Anal..Chem., 45, 1507 (1973).
953. Beckewitz, F. and Housinger, H., Angew. Makromol. Chem., 46(1), 143 (1975).
954. Inagaki, N., Kobashi, M. and Katsuura, K., Kobunshi Kagaku, 29(6), 416 (1972).
955. Wall, L.A., J. Elastoplast., 5, 36 (1973).
956. Mitera, J., Kubelka, V., Novak, J. and Mostecky, J., Plasty Kauc, 14(1), 18 (1977); Chem. Abstr. 86, 172232u (1977).
957. Schmitt, C.R., J. Fire Flammability, 3, 303 (1972).
958. Neilsen, L.E., Dahm, D.J., Berger, P.A., Murty, V.S. and Kardos, J.L., J. Polym. Sci., Polym. Phys. Ed., 12, 1239 (1974).
959. Lovell, R. and Windle, A.H., Polymer, 17(6), 488 (1976).
960. Jasse, B..and Monnerie, L., J. Mol. Struct. 39(2), 165 (1977).
961. Hamada, F., Hayashi, H. and Makajima, A., Mem. Fac. Eng., Kyoto Univ., 37 Pt 4, 289 (1975); Chem. Abstr. 85, 22008r. (1975).
962. Moore, J.C., J. Polymer Science, A 2, 835 (1964).
963. Gruber, U. and Elias, H.G., De Makromol. Chemie. 78, 58 (1964).
964. Moore, J.C. and Hendrickson, J.G., J. Polymer Science C 8, 233 (1965).
965. Ambler, M.R., Mate, R.D. and Purdon, J.R., J. Polymer Science, 12, 1759 (1974).
966. Drott, E.E. and Mendelson, R.A., J. Polym. Sci., A 2, 8, 1361 (1970).
967. Drott, E.E. and Mendelson, R.A., J. Polym. Sci., A 2, 8, 1373 (1970).
968. Kraus, G. and Stacey, C.J., J. Polym. Sci., A 2, 10, 657 (1972).
969. Kurata, M., Okamoto, H., Iwama, M., Abe, M. and Homma, T., Polym. J., 3, 739 (1972).
970. Shultz, A.R., Eur. Polym. J., 6, 69 (1970).
971. Peaker, F.W. and Robb, J.C., Nature, No. 4649, 1591 (1958).
972. Loconti, J.D. and Cahill, J.W., J. Polym. Sci., 49, 152 (1961), A 1, 3163 (1963).
973. Ruskin, A.M. and Parravano, G., J. Apply. Polym. Sci., 8, 565 (1964).
974. Bryson, A.P., Hawke, J.G. and Parts, A.G., J. Polymer Science, 12, 1323 (1974).
975. Kranz, D., Pohl, H.U. and Baumann, H., Angew. Makromol. Chem., 26, 67 (1972).
976. Tung, L.H. and Runyon, J.R., J. Appl. Polym. Sci., 17, 1589 (1973).
977. Zhdanov, S.P., Belen'Kii, B.G., Nefedov, P.P..and Koromal'di, E.V., J. Chromatogr., 77(1), 149 (1973).
978. Belen'Kii, B.G., Gankina, E.S., Nefedov, P.P., Kuznetsova, M.A. and Valchikhina,.J. Chromatogr., 77, 209 (1973).
979. Otocka, E.P., J. Chromatogr., 76(1), 149 (1973).
980. Uglea, C.V., Makromol. Chem., 166, 275 (1973).
981. Matsuda, H., Suzaka, Y., Takeshima, M. and Kuroiwa, S., Kobunshi Kagaku, 29(6), 362 (1972).
982. Guenet, J.M., Gallot, Z., Picot, C. and Bennett, J. Appl. Polym. Sci., 21(8), 2181 (1977).
983. Gooberman, G., J. Polymer Sci., 40, 469 (1959).
984. Urwin, J.R., McKenzie, O., Stearne, J., Jordan, D.O. and Mills, R.A., Die Makromol Chemie, 72, 53 (1964).
985. Stearne, J..and Urwin, J.R., Makromol Chemie, 56, 76 (1962).
986. Yamada, S., Prins, W. and Hermans, J.J., J. Polymer Sci., Part A 1, 2335 (1963).
987. Hall, O., "Techniques of Polymer Characterization", P.W. Allen

Edition, Academic Press, New York (1959), Chapter II.

988. Beattie, W.H., J. Polymer Science, Part A, 3, 527 (1965).

989. Beattie, W.H., J. Polymer Science, Private communication.

990. Beattie, W.H., J. Polymer Science, Private communication.

991. Klenin, V.I. and Shchegolev, S. Yu., J. Polym. Sci. Polym. Symp., 42, Part 2, 965 (1973).

992. Rolfson, F.B. and Coll, H., Anal. Chem. 36, 888 (1964).

993. Kallistov, O.V., Zavod Lab., 38(6), 711 (1972).

994. Kamata, T. and Nakahara, T., J. Colloid Interface Sci., 43(1), 89 (1973).

995. Takashima, K., Nakae, K., Shibata, M. and Tamakawa, Macromolecules, 7(5), 641 (1974).

996. Blair, J.E., J. Polymer Science, C 8, 287 (1965).

997. Goodrich, F.C. and Cantow, M.J.T., J. Polymer Science, C 8, 269 (1965).

998. Harrington, R.H. and Pecorano, P.G., J. Polymer Science, A 1, 4, 475 (1966).

999. Morton, M., Milkovich, R., McIntyre, D.B. and Bradley, L.J., J. Polymer Sci., A 1, 443 (1963).

1000. Brewer, F.M. and McCormick, H.W., J. Polymer Science, A 1, 1749 (1963).

1001. Wagner, H.L., Advan. Chem. Ser., 125, 17 (1971) (Pub. 1973).

1002. Spitbergen, J.C. and Beachell, H.C., J. Polymer Science, A 2, 1205 (1964).

1003. King, T.A., Knox, A. and McAdam, J.D.G., Chem. Phys. Lett., 19(3), 351 (1973).

1004. King, T.A., Knox, A. and McAdam, J.D.G., J. Polym. Sci., Symp. No. 44, 195 (1974).

1005. Fourche, G. and Jacq, M.T., Polym. J., 4(5), 465 (1973).

1006. Fourche, G. and Lemaire, B., Polym. J., 4(5), 476 (1973).

1007. Yoshimura, T., Kikkawa, A. and Suzuki, N., Jap. J. Apply. Phys., 11(12), 1797 (1972).

1008. Lange, H., Kolloid-Z. Z. Polym., 250(8), 775 (1972).

1009. Hansen, J. and Hvidt, A., J. Chem. Soc., Faraday Trans. II, 69, 881 (1973).

1010. Gundiah, S., Indian J. Chem., 11(11), 1162 (1973).

1011. Cotton, J.P., Farnoux, B. and Jannink, G., J. Chem. Phys., 57, 290 (1972).

1012. Cotton, J.P., Farnoux, B., Jannink, G. and Strazielle, C., J. Polym. Sci., Symp. No. 42, 981 (1973).

1013. Billmeyer Jr., F.W. and Siebert, L.R., Advan. Chem..Ser., 125, 9 (1971) (Pub. 1973).

1014. Clark, D.T. and Dilks, A., J. Polymer Science, A 1, 15, 15 (1977).

1015. George, G.A., J. Appl. Polym. Sci., 18, 419 (1974).

1016. Raciti, R. and Cuniberti, C., Nuova Chim., 49(11), 35 (1973).

1017. Venediktov, M.V., Dushchenko, V.P., Kolupaev, B.S. and Tarasenko, Yu. G., Mash. Prib. Ispyt. Mater., 99 (1971).

1018. Krause, S. and Stroud, D.E., J. Polym. Sci., Polym. Phys. Ed., 11, 2253 (1973).

1019. Kuwahara, N., Makata, M. and Kaneko, M., Polymer, 14, 415 (1973).

1020. Svishchev, G.M., Zh. Prikl. Spektrosk., 19(1), 155 (1973).

1021. Schelten, J., Wignall, G.D., Ballard, D.G. and Schmatz, W., Colloid Polym. Sci., 252(9), 749 (1974).

1022. Bui, Z.A., Carchano, H., Guastavino, J., Chatain, D., Gautier, P. and Lacabanne, C., Thin Solid Films, 21(2), 313 (1974).

1023. Ranicar, J.H. and Fleming, R.J., J. Polym. Sci., Part A 2, 10, 1979 (1972).

1024. Beevers, R.B., J. Polym. Sci., Polym. Phys. Ed., 12, 1407 (1974).

1025. Looyyenga, H., J. Polym. Sci., Polym. Phys. Ed. 11, 1331 (1973).
1026. Elias, H.G. and Gruber, U., Die Makromolekulare Chemie, 78, 72 (1964).
1027. Kämmerer, H., Sextro, G. and Mozaflari, A.S., J. Polymer Science Polymer
       Chemistry Edition 14, 609 (1976).
1028. Adcock, L.H. Patra J.,3, 5 (1962).
1029. Pfabs, W. and Noffz, D. Z. Anal. Chem., 195, 37 (1963).
1030. Ragelis, E.P. and Gajan, R.J., J.Assos. Offic. Agr. Chemists, 45, 918
       (1962).
1031. Shapras, P. and Claver, G.C., Anal. Chem., 36, (12), 2282 (1964).
1032. Barbul, M., Pop, A. and Bescher, C., Revista de Chimie, 5, (5), 280
       (1964).
1033. Novak, V., Chem. Prum., 22, (6), 298 (1972).
1034. Maeda, S., Kobayashi, H. and Ueno, K., Bunsekl Kagaku, 22, (10), 1365
       (1973).
1035. Vasilev, T., Steva, T., Vranchev, D., Zhilov, N. and Demirev, A., Nauch
       Tr., Plovdivski Univ., Mat. Fiz., Khim., Biol., 10, (3), 37 (1972).
1036. Monita, S. and Nawata, M., Daigaku Rikogakubu Kenkyu Hokoku, 16, 1 (1976).
       Chem. Abstr., 87, 68995e (1977).
1037. Markelov, M.A. and Semenenko, E.I., Plast. Massy., (8), 65 (1973).
1038. De Forero, I.B., De Rascovsky, E.G., De Ruiz, M. and del Carmen, S.,
       Plasticos, 19, (112), 122 (1971).
1039. Corson, B.B., Haintzelman, W.J., Moe, H. and Rousseau, C.R., J. Org.
       Chem., 27, 1636 (1962).
1040. Klesper, E. and Hartmann, W., Eur. Polym. J., 14, (2), 77-88 (1978).
1041. Nowak, P. and Klemm, H., Kunststoffe, 52, 604 (1962).
1042. Shanks, R.A., Pye Unicam Newsletter (1975).
1043. Rohrschneider, L.Z. Analyt. Chem., 255, 345 (1971).
1044. Shapras, P. and Claver, G.C., Anal. Chem., 34, 433 (1962).
1045. Tweet, O. Miller, W.K., Anal. Chem., 35, 852 (1963).
1046. Wilkinson, L.B., Norman, C.W. and Brettner, N.P., Anal. Chem., 36, 1759
       (1964).
1047. Nowak, P. and Klemmtt, O., Kunststoffe, 52, 604 (1962).
1048. Adcock, L.H., Patra, 3, 5 (1962).
1049. Ragelis, E.P. and Gajan, R.J.F., Assoc. Offic. Agric. Chem., 45, 918
       (1962).
1050. Pfab, W. and Noffz, D., Z. Anal. Chem., 37, 195 (1963).
1051. Shapras, P. and Claver, G.C., Anal. Chem., 36, 2282 (1964).
1052. Crompton, T.R., Myers, L.W. and Blair, D., British Plastics, December
       (1965).
1053. Crompton, T.R. and Myers, L.W., European Polymer Journal, 4, 355 (1968).
1054. Schwoetzer, G., Z. Anal. Chem., 260, 10 (1972).
1055. Pozdeeva, R.M., Lukhovitskii, U.I. and Korpov, V.L., Zavod Lab., 37,
       160 (1971).
1056. Crompton, T.R. and Myers, L.W., Plastics and Polymers, 205,(June 1968).
1057. Roberts, C.B. and Swank, J.D., Anal. Chem., 36, (2), 271 (1964).
1058. Simpson, D. and Currell, B.R., Analyst, 96, 515 (1971).
1059. Roberts, C.B. and Swank, J.D., Anal. Chem., 36, 271 (1964).
1060. Bukata, S.W., Zabrucki, L.L. and McLaughlin, M.F., Anal. Chem., 35, 886
       (1963).
1061. Hyden, S., Anal. Chem., 35, 133 (1963).
1062. Brammer, J.A., Frost, S. and Reid, V.W., Analyst, 92, 91 (1967).
1063. Kuta, E.J. and Quackenbush, F.W., Anal. Chem., 32, 1069 (1960).
1064. Armeau, V. and Costian, D., Mater. Plast. (Bucharest), 9, (11), 606
       (1972).
1065. Simonov, E.G. and Semushin, A.M., Zh. Prikl. Khim. (Leningrad), 45, (8),
       1883 (1972).
1066. Kurenkov, V.F., Kukushkina,I.A., Gershman, S.D., Kuznetsov, E.V. and
       Myagchenkov, V.A., Zh., Anal. Khim., 30, (1), 160 (1975).

1067.   Hlavay, J., Inczedy, J. and Fuzes, J., Hung. Sci. Instrum., 42, 13 (1978
1068.   Gordon, J.E., J. Phys. Chem., 66, 1150 (1962).
1069.   Gordon, J.E., Chem. Ind. (London), 267 (1962).
1070.   de Villiérs, J.P. and Parrish, J.R., J.Polymer Science, A 2, 1331 (1964)
1071.   Blasius, E., Haeusler, H. and Lander, H., Talanta, 23, (4), 301 (1976).
1072.   Sutherland, J.E., Research Laboratories Polymer Preparation, 17, 2 (1976)
1073.   Regnier, F.E. and Noel, R., J.Chromatogr. Sci., 14, 316 (1976).
1074.   Okumoto, T. and Takeuchi, T., Makromol. Chem., 167, 305 (1973).
1075.   Durchschlag, H., Kratky, O., Olaj, O.F. and Breitenbach, J.W., J.Polymer
        Science, A 1, 11, 1327 (1973).
1076.   Ayrey, G. and Mazza, R.J., J. Appl. Polym. Sci., 19, (9), 2621 (1975).
1077.   Matsuzaki, K., Uryu, T., Osada, K. and Kawamura, T., Macromolecules,
        5, 816 (1972).
1078.   Evans, D.C., Phillips, L., Barrie, J.A. and George, M.H., J. Polym. Sci.,
        Polym. Lett., 13, 199 (1974).
1080.   Ebdon, J.R. and Huckerby, T.N., Polymer, 17, (2), 170 (1976).
1081.   Morita, S., Shen, M. and Ieda, G.S.M., J. Polym. Sci., Polym. Phys. Ed.,
        14, (10), 1917 (1976).
1082.   Morita, S. and Shen, S., Polym. Prepr., Am. Chem. Soc., Div. Polym.
        Chem., 17, (2), 545 (1976).
1083.   Jasse, B., Laupretre, F. and Monnerie, L., Makromol. Chem., 178, (7),
        1987 (1977).
1084.   Florin, R.E., Wall, L.A. and Brown, D.W., J. Polymer Science, A 1, 1521
        (1963).
1085.   Guk, A.F., Koziova, Z.G. and V.F., Vysokomol. Soedin., Ser. B, 15, (1),
        41 (1973).
1086.   Bullock, A.T., Cameron, G.G. and Smith, P.M., Polymer, 14, 525 (1973).
1087.   Crompton, T.R. and Reid, V.W., J. Polymer Science, Part A, 1, 347 (1963).
1088.   La Cheyney, V.E. and Kelley, E.J., Ind. Eng. Chem. Ind. Ed., 34 (1942).
1089.   Hobbs, A., Hulme, J., Van Dyke, J.D. and Wooden, D.C., Soc. Plast.
        Eng., Tech. Pap., 21, 491 (1975).
1090.   Littke, W.H., Fleber, W., Schmolke, R. and Kimmer, W., Faserforsch.
        Textiltech., 26, (10), 503 (1975).
1091.   Sefton, M.V. and Merrill, E.W., J.Appl. Polym. Sci., 20, (1), 157
        (1976).
1092.   Schmolke, R., Kimmer, W. and Fieber, W., Faserforsch. Textiltech.,29,
        (6), 386 (1978); Chem. Abstr., 89, 90509n (1978).
1093.   Collins, G.C.S., Lowe, A.C. and Nicholas, D., Eur. Polym. J., 9, (11),
        1173 (1973).
1094.   Albert, N.K., Woodbury Research Laboratory, Shell Chemical Co. Ltd.,
        Woodbury, private communication (Oct. 1966).
1095.   Sloane, H.J. and Bramston-Cooke, R., Appl. Spectrosc., 27, (3), 217
        (1973).
1096.   Conti, F., Delfini, M. and Segre, A.L., Polymer, 15, 539 (1974).
1097.   Gronski, W., Murayama, N., Cantow, H.J. and Miyamoto, T., Polymer,
        17, (4), 358 (1976).
1098.   Nowlin, T.E., Ungermah, A.J. and Wallace, S.L., J. Appl. Polym. Sci.,
        20, (8), 2095 (1976).
1099.   Randall, J.C., J. Polym. Sci., Polym. Phys. Ed., 15, (8), 1451 (1977).
1100.   Conti, F., Delfini, M. and Segre, A.L., Polymer, 18, (3), 310 (1977).
1102.   Sindelfingen, J.V., Kunststoffe, 51, 18 (1961).
1103.   Von Virus, W., Chemiken Ztg. Chem. Apparatur, 87, 740 (1963).
1104.   Brauer, G.M., J. Polymer Science, C 8, 3 (1965).
1105.   Dimbat, M. and Shaw, A.W., Emerville Research Laboratories, Shell
        Chemical Co. Ltd., private communication.
1106.   Groten, B., Anal. Chem., 36, (7), 1206 (1964).
1107.   Stanley, C.W. and Peterson, W.R., S.P.E. Trans., 2, 298 (1962).
1108.   Tutorskii, I.A., Boikacheva, E.G., Bukanova, E.F., Guseva, T.V and
        Bukanov, I.G., Izv. Vyssh. Uchebn. Zaved., Khim. Tekhnol., 18,(3),
        460 (1975); Chem. Abstr., 83, 28957e.

1109.   Sokolowska, J. and Maciejowski, F., Chem. Anal., (Warsaw), 22, (2),
        337 (1977).
1110.   Haeusler, K.G., Schroeder, E. and Muehling, P., Plaste Kautsch, 24, (8),
        55 (1977)., Chem. Abstr., 87, 118318 (1977).
1111.   Shimono, T., Tanaka, M. and Shoro, T., Anal. Chem. Acta., 96, 359 (1978).
1112.   McNeill, I.C., J.Polymer Science, Al, 4, 2479 (1966).
1113.   Akutin, M.S., Shabadash, A.N., Salina, Z.I., Golubev, V.A. and Besson-
        ova, N.P., Vysokomol. Soedin., Ser. B, 14, (10), 769 (1972).
1114.   Galliot, B. Colloq. Int. Methodes Anal. Rayonnements X, 2nd, 53,
        (Pub. 1971), (1971).
1115.   Plestil, J. and Baldrian, J. Makromol. Chem., 176, (4), 1009 (1975).
1116.   Donkai, N., Miyamoto, T. and Inagaki, H., Polym. J., 7, (5), 577 (1975).
1117.   Kotaka, T. and White, J.L., Macromolecules, 7,(1), 106 (1974).
1118.   Morton, M., Ann. N.Y. Acad. Sci., 57, 432 (1953).
1119.   Cragg, L.H. and Fern, G.R.H., J. Polymer Sci., 10, 185 (1953).
1120.   Manson, J.H. and Cragg, L.H., Can. J. Chem., 30, 482 (1952).
1121.   Huggins, M.L., J. Am. Chem. Soc., 64, 2716 (1942).
1122.   Blanchford,J. and Robertson, R.F., J. Polymer Science, A 3, 1289 (1965).
1123.   Blanchford,J. and Robertson, R.F., J. Polymer Sci., A 3, 1311 (1965).
1124.   Chang, F.S.C., Advan. Chem. Ser., 125, 154 (1971) (Pub. 1973).
1125.   Welygan, D.G. and Burns, C.M., J. Polym. Sci., Polym. Lett. Ed., 11,
        339 (1973).
1126.   Hoffmann, M. and Urban, H., Makromol. Chem., 178, (9), 268 (1977).
1127.   Stojanov, C., Shirazi, Z.H. and Audu, T.O.K., Chromatographia, 11, (5),
        274 (1978).
1128.   Ambler, M.R., Mate, R.D. and Durdon, J.R., J. Polym. Science, 12, 1771
        (1974).
1129.   Benningfield, Jr., L.V., "A dielectric constant detector for liquid
        chromatography and its application", paper presented at the 30th
        Pittsburgh Conference on Analytical Chemistry and Applied Spectro-
        scopy, Cleveland, March 5-9 p 123 (1979).
1130.   Benningfield, L.V. and Mowery, Jr., R.A., J. Chromatogr. Sci., 19,
        115 (1981).
1131.   Bode, R.K., Benningfield, L.V., Mowrey, R.A. and Fuller, E.N., Int-
        ernational Laboratory, 40, Nov/Dec., (1981).
1132.   Fuller, E.N., Porter, G.T. and Roof, L.B., J. Chromatogr. Sci., 17,
        661 (1979).
1133.   Roof, L.B., Porter, G.T., Fuller, E.N. and Mowrey, Jr., R.A., "On-
        line polymer analysis by liquid chromatography", Instrumentation and
        Automation in the Paper, Rubber, Plastics and Polymerization Indust-
        ries, 4th TFAC Conference, Ghent, Belgium, June 3-5, (1980). A.
        Van Cauwenberghe, Ed. (Pergamon Press, New York, 1980) p 47-53.
1134.   Kratochvil, P., J. Polym. Sci., Polym. Symp. 50 (1975) (Int. Symp.
        Makromol., Invited Lect.,) 487-96 (1974).
1135.   Kratochvil, P., Sedlacek, B., Strakova, D, and Tuzar, Z., Polim. Simp.,
        2, 112 (1971).
1136.   Schneider, I.A., Onu, A. and Aelenei, N., Eur. Polym. J., 10, (3),
        315 (1974).
1137.   Blanchford, J. and Robertson, R.F., J. Polymer Science, A 3, 1303
        (1965).
1138.   Bradford, E.B. and Vanzo, E., J. Polymer Science, A 1, 6, 1661 (1968).
1139.   Bradford, E.B. and Vanzo, E., J. Polymer Science, A 3, 1313 (1965).
1140.   Roth, W., Plaste Kautsch., 22, (12), 954 (1975).
1141.   Stacy, C.J., J. Appl. Polym. Sci., 21, (8), 2231 (1977).
1142.   Novak, V. and Siedl, J., J. Chem. Prum., 28, (4), 186 (1978); Chem.
        Abstr., 89, 111061c (1978).
1143.   Shapras, P. and Claver, G.C., Anal. Chem., 36, 2282 (1964).
1144.   Hilton, C.L., Anal. Chem., 32, 383 (1960).

1145.   Nawakowski, A.C., Anal. Chem., 30, 1868 (1958).
1146.   Brandt, H.J., Anal. Chem., 33, 1390 (1961).
1147.   Parks, C.A., Anal. Chem., 33, 140 (1961).
1148.   Simpson, D. and Currell, B.R., Analyst, 96, 515 (1971).
1149.   Hayes, M.M. and Altenau, A.G., Rubber Age, 102, 59 (1970).
1150.   Sedlar, J., Feniokova, E. and Pac, J., Analyst, 99, 50 (1974).
1151.   Gaeta, L.J., Schleuter, E.W. and Altenau, A.G., Rubber Age, 101, 47
        (1969).
1152.   Fallick, G.J., Talarico, P.C. and McGough, R.R., Soc. Plast. Eng.,
        Tech. Pap., 22, 574 (1976).
1153.   Turner, R.R., Carlson, D.W. and Altenau, A.G., J. Elastomers Plast.,
        9 (April), 94 (1974).
1154.   Scheddel, R.T., Anal. Chem., 30, 1303 (1958).
1155.   Takuchi, T., Tauge, S. and Sugimura, Y., J. Polymer Science, A 1, 6,
        3415 (1968).
1156.   Ol, N., Miyazaki, K., Moriguchi, K. and Shimada, H., Kobunshi Kagaku,
        29, (6), 388 (1972).
1157.   Biyer, P. and Padhye, M.R., Silk Rayon Ind. India, 16, (5), 174 (1973).
1158.   Kimmer, W. and Schmolke, R., Plaste Kaut., 20, (4), 274 (1973).
1159.   Tsefanov, Kh.B., Panalotov, I.M. and Erusalimskii, B.L., Eur. Polym.
        J., 10, (7), 557 (1974).
1160.   Oi, N. and Moriguchi, K., Bunseki Kagaku, 23, (7), 798 (1974); Chem.
        Abstr., 82: 17308x.
1161.   Mikhailov, M., Dirlikov, S.K., Georgieva, Z., Peeva, N. and Panalotova,
        M., Dokl. Bolg. Akad. Nauk., 27, (11), 1525 (1974); Chem. Abstr., 82;
        125737v.
1162.   Simak, P. and Ropte, E., Makromol. Chem., Suppl. 1, 507 (1975).
1163.   Sandner, B., Keller, F. and Roth, H., Faserforsch. Textiltech., 26,
        (6), 278 (1975).
1164.   Elgert, K.F. and Stuetzel, B., Polymer, 16, (10), 758 (1975).
1165.   Voight, J., Kunststoffe, 51, 18 (1961).
1166.   Lebel, P., Rubber Plastics Age., 46, 677 (1965).
1167.   Shibasaki, J., Kobunshi Kagaku, 21, 125 (1964).
1168.   Shibasaki, J., J. Polym. Sci., A 1, 5, 21 (1967).
1169.   Lehman, F.A. and Brauer, G.M., Anal. Chem., 33, 873 (1961).
1170.   Sindefingen, J.V., Kunststoffe, 51, 18 (1961).
1171.   Shibazaki, Y. and Kambe, H., Chem. High Polymers (Tokyo), 21, 71 (1964).
1172.   Shibazaki, Y., Chem. High Polymers (Tokyo), 21, 125 (1964).
1173.   Shibazaki, Y. and Kambe, H., Kabishi Kagaku, 21, 71 (1964).
1174.   Vukovic, R. and Gnjatovic, V., J. Polymer Science, A 1, 8, 139 (1970).
1175.   Deur-Siftar, D., Bistricki, T. and Tandi, T., J.Chromatogr., 24, 404
        (1966).
1176.   Chaigneau, M., Analusis, 5, (5), 223 (1977).
1177.   Grassie, N. and Bain, D.R., J. Polymer Science, A 1, 8, 2683 (1970).
1178.   Grassie, N. and Bain, D.R., J. Polymer Science, A 1, 8, 2669 (1970).
1179.   Gloeckner, G. and Kahle, D., Plaste Kautsch, 23, (8), 577 (1976).
1180.   Gloeckner, G. and Kahle, D., Plaste Kautsch, 23, (5), 338 (1976).
1181.   Teramachi, S. and Fukao, Ti., Polym. J., 6, (6), 532 (1974).
1182.   Belyaev, V.M., Kazanskaya, V.F., Smirnova, S.V., Ivanova, E.E. and
        Frenkel, S. Ya., Vysikomol. Soedin., Ser. B, 15, (3), 154 (1973).
1183.   Hatate, Y. and Naskashio, R., Kagaku Kogaku, 37, (2), 171 (1973).
1184.   Szewczyk, P. and Rajklewicz, M., Polimery, 17, (9), 460 (1972).
1185.   Kambe, Y. and Honda, C., Angew. Makromol. Chem., 25, 163 (1972).
1186.   Roy, S.S., Analyst, 102, 302 (1977).
1187.   Claver, G.C. and Murphey, M.E., Anal. Chem., 31, 1682 (1959).
1188.   Shapras, P. and Claver, G.C., Anal. Chem., 36, 2282 (1964).
1189.   Turner, R.R., Carlson, D.W. and Altenau, A.G., unpublished work.
1190.   Lindsay, G.A., Santee, Jr., E.R. and Harwood, H.J., Polym. Prepr.,Am.

Chem. Soc., Div. Polym. Chem., 14, (2), 646 (1973).

1191.   Araki, K., Fukui-ken Kogyo Shikenjo Nempo, 50, 63 (1975); Chem. Abstr.
        87, 118514v (1977).

1192.   Sindefingen, J.V., Kunststoffe, 51, 18 (1961).

1193.   Voigt, J., Kunststoffe, 51, 18 (1961).

1194.   Voigt, J., Kunststoffe, 51, 314 (1961).

1196.   Kranz, D., Pohl, H.U. and Baumann, H., Angew. Makromol. Chem., 26,
        67 (1972).

1197.   Crompton, T.R., unpublished work.

1199.   Shanks, R.A., Scan, 6, 20 (1975).

1200.   Claver, G.C. and Murphey, M.E., Anal. Chem., 31, 1682 (1959).

1201.   Sedlar, J., Fencokova, E. and Pac, J., Analyst, 99, 50 (1974).

1202.   Yanagisawa, K., Chem. High Polymers (Tokyo), 21, 312 (1964).

1203.   Nishioka, A., Kato, Y. and Ashikari, N., J. Polymer Sci., 62, S 10
        (1962).

1204.   Nishioka, A., Kato, Y. and Mitsuoka, H., J. Polymer Sci., 62, S 9
        (1962).

1205.   Bovey, F.A., J. Polymer Sci., 62, 197 (1962).

1206.   Harwood, H.J. and Pustinger, J.V., unpublished results.

1207.   Ito, K. and Yamashita, Y., J. Polymer Sci., B 3, 625 (1965).

1208.   Overberger, C.G. and Yamamoto, N., J. Polymer Sci., B 3, 569 (1965).

1209.   Harwood, H.J. and Ritchey, W.M., J. Polymer Sci., B 3, 419 (1965).

1210.   Yabumoto, S., Ishi, K. and Arila, K., J. Polymer Science, A 8, 295
        (1970).

1211.   Wang, A., Suzuki, T. and Harwood, H.J., J. Polym. Prepr., Am. Chem.
        Soc., Div. Polym. Chem., 16, (1), 644 (1975).

1212.   Mori, Y., Ueda, A., Tanzawa, H., Matsuzaki, K. and Kobayashi, H.,
        Makromol. Chem., 176, (3), 699 (1975).

1213.   Bockrath, R.E. and Harwood, H.J., Polym. Prepr., Am. Chem. Soc., Div.
        Polym. Chem., 14, (2) 1163 (1973).

1214.   Stroganov, L.B., Plate, N.A., Zubov, V. and Fedorova, S. Yu., Vysok-
        omol. Soedin., Ser. A, 16, (11) 2616 (1974).

1215.   Blouin, F.A., Chang, R.C., Quinn, M.H. and Harwood, H.J., Polym.
        Prepr., Am. Chem. Soc., Div. Polym. Chem., 14, (1), 25 (1973).

1216.   Kato, Y., Ando, I. and Nishioka, A., Kagaku Kaishi, 3, 501 (1975);
        Chem. Abstr., 83: 28076x.

1217.   Katritzky, A.R., Smith, A. and Weiss, D.E., J. Chem. Soc., Perkin
        Trans. 2, (13), 1547 (1974).

1218.   Yakata, K. and Hirabayashi, T., J. Polymer Science, A 1, 14, 57 (1976).

1219.   Podesva, J. and Doskocilova, D., Makromol. Chem., 178, (8), 2382 (1977).

1220.   MacLoed, N., Chromatographia, 5, (9), 516 (1972).

1221.   Hummel, D.O. and Düssel, H.J., Makromol. Chem., 175, 655 (1974).

1222.   Tsuge, S. and Takeuchi, T., Anal. Pyrolysis Proc. Int. Symp., 3rd
        1976 (Pub. 1977), 393-404; Jones, C.E.R. and Cramers, C.A., Ed.
        Elsevier, Amsterdam.

1223.   Milina, R. and Pankova, M., Textilia, 53, (1), 54 (1977); Chem. Abstr.,
        86, 156178a (1977).

1224.   Maciejowski, F. and Sokolowska, J., Polimery (Warsaw), 22, (7), 239
        (1977); Chem. Abstr., 88, 23583v (1978).

1225.   Evans, D.L., Weaver, J.L. and Muklerji, A.K., Anal. Chem., 50, 857
        (1978).

1226.   Evans, D.L., Weaver, J.L., Muklerji, A.K. and Beatty, C.L., Anal.
        Chem., 50, (7), 857 (1978).

1227.   Jaacks, V., Makromol. Chem., 161, 161 (1972).

1228.   Kotaka, T., Uda, T., Tanaka, T. and Inagaka, H., Makromol. Chem., 176,
        (5), 1273 (1975).

1229.   Belen'kii, B.G., Gankina, E.S., Nefedov, P.P., Lazareva, M.A., Savit-
        skaya, T.S. and Voichikhina, M.D., J. Chromatogr., 108, (1), 61 (1975).

406   References

1230.   Gal'perin, V.M., Zak, A.G., Kuznetsov, N.A., Roganova, Z.A. and Smol-
         yanskii, A.L., Vysokomol. Soedin, Ser. A, 17, (3), 575 (1975).
1231.   Lovric, L.J., Grubisic-Gallot, Z. and Kunst, B., Eur. Polym. J., 12,
         (3), 189 (1976).
1232.   Dautzenberg, H., J. Polym. Sci., Part C,, 39, 123 (1972).
1233.   David, C., Baeyens-Volant, D. and Geuskens, G., Eur. Polym. J., 12,
         (2), 71 (1976).
1234.   David, C., Lempereur, M. and Geuskens, G., Eur. Polym. J., 9, 1315
         (1973).
1235.   Rafikov, S.R., Monakov, Yu. B., Duvakina, N.V., Marina, N.G., Budtov,
         V.P., Minchenkova, N. Kh. and Shakirova, A.M., Vysokomol. Soedin.,
         Ser. B, 15, (11), 807 (1973).
1236.   Baranovskaya, I.A. and Eskin, V.E., Vysokomol. Soedin., Ser. A, 17,
         (8), 1875 (1975).
1237.   Lange, H. and Baumann, H., Angew. Makromol. Chem., 43, (1), 167 (1975).
1238.   Srinivasan, K.S.V. and Santappa, M., J. Polym. Sci., Polym. Phys. Ed.,
         11, 331 (1975).
1239.   Han, C.C. and Mozer, B., Macromolecules, 10, (1), 44 (1977).
1240.   Gallo, B.M. and Russo, S., J. Macromol. Sci., Chem., 8, (3), 521 (1974).
1241.   Shima, M., J. Polymer Sci., 56, 213 (1962).
1242.   Kaniewska, D. and Lewandowska, T., Polimery, 17, (2), 104 (1972).
1243.   Holtmann, R. and Souren,J.R., Kunststoffe, 67, (12), 776 (1977).
1244.   Yamaoka, A. and Matsui, T., Himeji Kogyo Daigaku Kenkyu Hokoku, 30A,
         101 (1977); Chem. Abstr., 89, 7000w (1978).
1245.   Acosta, J.L. and Sastre, R., Rev., 29, (224), 212 (1975).
1246.   Tweet, O. and Miller, W.K., Anal. Chem., 35, (7), 852 (1963).
1247.   Kreshkov, A.P., Balyatinskaya, L.N. and Chesnokova, S.M., Tr. Mosk.
         Khim-Tekhnol. Inst., 70, 146 (1972); Chem. Abstr., 80, 19144 (1974).
1248.   Kreshkov, A.P., Balyatinskaya, L.N. and Chesnokova, S.M, Zh. Anal.
         Khim., 28, (8), 1571 (9173).
1249.   Sakurada, I., Ikada, Y. and Kawahara, T., J. Polymer Sciemce, 11,
         2329 (1973).
1250.   Sakurada, I., Ikada, Y. and Kawahara, T., J. Polymer Science, Polym.
         Chem. Ed., 11, 2329 (1973).
1251.   Hori, F., Ikada, Y. and Sakurada, I.J., Polymer Science, Polymer Chem-
         istry, Ed., 13, 755 (1975).
1252.   Sakurada, I., Ikada, Y. and Kawahara, T., J. Polymer Science, 11, 2329
         (1973).
1253.   Ang, T.L. and Harwood, H.J., Amer. Chem. Soc. Polymer Preprints, 5,
         (1), 306 (1964).
1254.   Buchak, B.E. and Ramey, K.C.J., Polym. Sci., Polym. Lett. Ed., 14,
         (7), 401 (1976).
1255.   Chow, C.D., J. Appl. Polym. Sci., 20, (6),619 (1976).
1256.   Goel, S.R. and Kumar, K., Chem. Era, 9, (10), 25 (1974).
1257.   Douglas, J., Timnick, A. and Guile. R.L., J. Polymer Science, A 1,
         1609 (1963).
1259.   Hen, J., J. Colloid Interface Sci., 49, (3), 425 (1974).
1260.   Garrett, S.R. and Guile, R.L., J. Am. Chem. Soc., 73, 4533 (1951).
1261.   Bamford, C.H. and Barb, W.G., Discussions Faraday Soc., 14, 208 (1953).
1262.   Saha, M.K., Ghosh, P. and Palit, S.R., J. Polymer Science, A 2, 1365
         (1964).
1263.   Palit, S.R. and Ghosh, P., J. Polymer Sci., 58, 1225 (1962).
1264.   Palit, S.R., Makromol. Chem., 36, 89 (1959); ibid., 38, 96 (1960).
1265.   Fuoss, R.M. and Strauss, V.P., J. Polymer Sci., 3, 246 (1948).
1266.   Fuoss, R.M., Watanabe, M. and Coleman, B.D., J. Polymer Sci, 48, 5
         (1960).
1267.   Alfrey, T. Jr., and Harrison, J.G. Jr., J. Am. Chem. Soc., 68, 299
         (1946).

1268 Doak, K.W., J. Am. Chem. Soc., 70, 1525 (1948).
1269 Breitenbach, J.W., Schindler, A. and Pflug, C., Monatsh. Chem., 81, 21 (1950).
1270 Mayo, F.R., Lewis, F.M. and Walling, C., J. Am. Chem. Soc., 70, 1529 (1948).
1271 Agron, P., Alfrey, T. Jr., Bohrer, J., Haas, H. and Wechsler, H., J. Polymer Sci., 3, 157 (1948).
1272 Fuoss, R.M. and Cathers, G.I., J. Polymer Sci., 4, 97, 121 (1949).
1273 Tweet, O. and Miller, W.K., Anal. Chem., 35, 852 (1963).
1274 Cobler, J.G. and Samsel, E.P., SPE Transactions, 145 April (1962).
1275 Parrish, J.R., Anal. Chem., 47, (12), 1999 (1975).
1276 Koval'chuk, V.M., Ivanov, A.I. and Samenenko, E.I., Khim.-Farm. Zh., 10, (8), 114 (1976); Chem. Abstr., 86, 44179v (1977).
1277 Blasius, E., Lohde, H. and Haeusler, H., Fresenius Z Anal. Chem., 264, (4), 290 (1973).
1278 Parrish, J.R., Anal. Chem., 45, 1659 (1973).
1279 Blasius, E. and Haeusler, H., Fresenius' Z Anal. Chem., 277, (1), 9 (1975); Chem. Abstr., 83:206850z.
1280 Oehme, G., Baudisch, H. and Mix, H., Makromol. Chem., 177, (9), 2657 (1976).
1281 Benoit, H., Decker, D., Duplessix, R., Picot, C., Rempp, P., Cotton, J.P. Farnoux, B., Jannink, G. and Ober, R., J. Polym. Sci., Polym. Phys. Ed., 14, (12), 2119 (1976).
1282 Park, W.S. and Graessley, W.W., J. Polym. Sci., Polym. Phys. Ed., 15, (1), 85 (1977).
1283 Yuasa, T. and Kamiya, K., Japan Analyst, 13, 966 (1964).
1284 Kobayashi, S., Kato, Y., Watanabe, H. and Nishioka, A., J. Polymer Science, A 1 4, 245 (1966).
1285 Eda, B., Nunome, K. and Iwasaki, M., J. Polymer Science, A 1, 8, 1831 (1970).
1286 Forette, J.E. and Rozek, A.L., J. Appl. Polym. Sci., 18, 2973 (1974).
1287 Yamashita, Y., Yoshida, M., Kawase, J. and Ito, K., Asahi Garasu Kogyo Gijutsu Shoreikal Kenkyo Hokoku, 25, 87 (1974); Chem. Abstr., 84: 60070s.
1288 Kharas, G.B., Kissin, Yu. V., Kieiner, V.I., Krentsel, B.A., Stotskaya, L.L. and Zakharyan, R.Z., Eur. Polym. J., 9, (4), 315 (1973).
1289 Cabasso, I., Jagur-Grodzinski, J. and Vofsi, D., J. Appl. Polym. Sci., 18, 1969 (1974).
1290 Okumoto, T., Takeuchi, T. and Tsuge, S., Macromoleculres, 6, 922 (1974).
1291 Cabasso, I., Jagur-Grodzinski, J. and Vofsi, D., J. Appl. Polym. Sci., 18, 1969 (1974).
1292 Tashmukhamedov, S.A., Khasankhanova, M.N. and Tillaev, R.S., Uzb. Khim. Zh., 17, (2), 35 (1973).
1293 Regel, W., Westfelt, L. and Cantow, H.J., Angew. Chem., Int. Ed. Engl., 12, (5), 434 (1973).
1294 Work, R.N. and Trehu, Y.M., J. Appl. Physics, 27, 1003 (1956).
1295 Obukhova, E.N. and Voloshin, G.A., Visn. L'viv. Politekh. Inst., No. 58, 55 (1971).
1296 Aliev, S.M., Bairamov, M.R., Azizov, A.G., Aliev, S.A. and Akhmedov, S.T., Azerb. Khim. Zh., 3, 70 (1976); Chem. Abstr., 86, 90366b (1977).
1297 Patel, C.M. and Patel, V.M., Staerke, 25, (2), 47 (1973).
1298 Kalal, J., Houska, M., Seycek, O. and Adamek, P., Makromol. Chem., 164, 249 (1973).
1299 Merrett, F.M., Trans. Faraday Soc., 50, 759 (1954).
1300 Yasuda, H., Wray, J.A. and Stannett, V., in Fourth Cellulose Conference (J. Polym. Sci., C 2) Marchessault, R.H., Ed., Interscience, New York, 1963, p. 387.
1301 Magat, E.E., Miller, I.K., Tanner, D. and Zimmerman, J., in Makromol.

Chem. Paris 1963 (J. Polym Sci., C 4) Magat, M., Ed., Interscience, New York, 1963, p. 615.

1302. Sumitomo, H., Takakura, S. and Hachihama, Y., Kōgyō Kagaku Zasshi, 66, 269 (1963).

1303. Sakurada, I., Matsuzawa, S. and Kubota, Y., Makromol. Chem., 69, 115 (1963).

1304. Sakurada, I., Ikada, Y. and Uesaki, Y., Bull. Inst. Chem. Res., Kyoto Univ., 47, 49 (1969).

1305. Sakurada, I., Ikada, Y. and Horii, F., Bull. Inst. Chem. Res., Kyoto Univ., 47, 58 (1969).

1306. Hamburger, C.J., J. Polym. Sci., A 1, 7, 1023 (1969).

1307. Ende, H.A. and Stannett, V., J. Polym. Sci., A 2, 4047 (1964).

1308. Inagaki, H., Matsuda, H. and Kamiyama, F., Macromolecules, 1, 520 (1968)

1309. Belenkii, B.G. and Gankina, E.S., Dokl. Akad. Nauk SSSR, 186, 857 (1969)

1310. Belenkii, B.G. and Gankina, E.S., J. Chromatogr., 53, 3 (1970).

1311. Kamide, K., Manabe, S. and Osafune, E., Makromol. Chem., 168, 173 (1973)

1312. White, J.L., Salladay, D.G., Quinsenberry, D.O. and MacLean, D.L., J. Appl. Polym. Sci., 16, 2811 (1972).

1313. Kotaka, T. and White, J.L., Macromolecules, 7, 106 (1974).

1314. Kamiyama, F., Matsuda, H. and Inagaki, H., Makromol. Chem., 125, 286 (1969).

1315. Inagaki, H., Miyamoto, T. and Kamiyama, F., J. Polym. Sci., B 7, 329 (1969).

1316. Miyamoto, T. and Inagaki, H., Macromolecules, 2, 554 (1969).

1317. Buter, R., Tan, Y.Y. and Challa, G., Polymer, 14, 171 (1973).

1318. Inagaki, H. and Kamiyama, F., Macromolecules, 6, 107 (1973).

1319. Donkai, N., Murakami, N., Miyamoto, T. and Inagaki, H., Makromol. Chem., 175, 187 (1974).

1320. Inagaki, H., Bull. Inst. Chem. Res., Kyoto Unov., 47, 196 (1969).

1321. Otocka, E.P. and Hellman, M.Y., Macromolecules, 3, 362 (1970).

1322. Otocka, E.P., Macromolecules, 3, 691 (1970).

1323. Belenkii, B.G. and Gankina, E.S., J. Chromatogr., 53, 3 (1970).

1324. Kamiyama, F., Matsuda, H. and Inagaki,H., Polym. J., 1, 518 (1970).

1325. Horii, F. and Ikada, Y., J. Polym. Sci. Polym. Letters Ed., 12, 27 (1974).

1326. Ebdon, J.J., Kandil, S.H. and Morgan, K.J., J. Polymer Science, A 1, 17, 2783 (1979).

1327. Svec, F., Houska, M. and Kalal, J., Sb. Prednasek, "Makrotest 1973", 1, 15 (1973).

1328. Kratochvil, P., Sedlacek, B., Strakova, D. and Tuzar, Z., Makromol. Chem., 166, 265 (1973).

1329. Mirabella, F.M., Jr., Barrall, E.M., II. and Johnson, J.F., Polymer, 17, 17 (1976).

1330. Dawkins, J.V.and Hemming, M., J. Appl. Polym. Sci., 19, (11), 3107 (1975).

1331. Bogomolova, T.B., Gantmakher, A.R. and Lyudvig, E.B., Vysokomol. Soedin., Ser. A, 14, (10), 2210 (1972).

1332. Woodward, A.E., J. Polymer Science, C 8, 137 (1965).

1333. Elgert, K.I., Wicke, R., Stüszel, B. and Ritter, W., Polymer, 16, 465 (1975).

1334. Richards, D.H. and Williams, R.L., J. Polym. Sci., Polym. Chem. Ed., 11, 89 (1973).

1335. Leonard, J. and Malhotra, S.L., J. Polym. Sci., Polym. Chem. Ed., 12, (10), 2391 (1974).

1337. Elgert, K.F., Cantow, H.J. and Stutzel, B., Frenzel, Angew. Chem., Int. Ed., 12, 427 (1973).

1338. Cobler, J.G., J. Polymer Science, A 1, 4, 2479 (1966).

1339. McNeill, I.C., J. Polymer Science, A 1, 4, 2479 (1966).

1340.   Geymer, D.O., Shell Development Company, Emeryville, California, private communication.
1341.   Kato, Y., Kido, S. and Hashimoto, T., J. Polym. Sci., Polym. Phys. Ed., 11, 2329 (1973).
1342.   Bradley, J.H., J. Polymer Science, C 8, 305 (1965).
1343.   Klenin, V.I., Podol'skii, A.F., Shchegolev, S. Yu., Shvartsburd, B.I. and Petrova, N.E., Vysokomol. Soedin., Ser. A, 16, (5), 974 (1974).
1344.   Chaula, A.S. and Huang, R.Y.M., J. Polymer Science, A 1, 13, 1271 (1975).
1345.   Allen, G., Wright, C.J. and Higgins, J.S., Polymer, 15, 319 (1974).
1346.   Yamamoto, M., Hayashi, K., Sasamoto, K. and Takemura, F., Polym. J., 8, (3), 307 (1976).
1347.   Gankina, E.S., Val'chikhina, M.D. and Belen'kii, B.G., Vysokomol. Soedin., Ser. A 18, (5), 1170 (1976).
1348.   Zizin, V.G. and Grigor'eva, L.A., Plast. Massy, 8, 55, (1976); Chem. Abstr., 85, 143765u (1976).
1349.   Stojanov, K., Shirazi, Z.H. and Audu, T.O.K., Ber. Bunsenges. Phys. Chem., 81, (8), 767 (1976).
1350.   Elgert, K.F., Seiler, E., Puschendorf, G. and Cantow, H.J., Makromole. Chem., 165, 245 (1973).
1351.   Elgert, K.F., Seiler, E., Puschendorf, G. and Cantow, H.J., Makromole. Chem., 165, 261 (1973).
1352.   Zwierzak, A. and Pines, H., J. Org. Chem., 28, 3392 (1963).
1353.   Fieber, W., Schmolke, R. and Kimmer, W., Fasserforsch. Textiltech., 27, (6), 311 (1976).
1354.   Baras, E.M. and Juveland, O.O., J. Polymer Science, A 1, 5, 397 (1967).
1355.   Shirakawa, H., Yamagaki, N. and Kambara, S., personal communication from Dr. N. Yamazaki, Tokyo Institute of Technology.
1356.   Gritter, A.J., Gipstein, E. and Adams, G.E., J. Polymer Science, A 1, 17, 3959 (1979).
1358.   Robila, G., Buruiana, E.C. and Caraculacu, A.A., Eur. Polym. J., 13, (1), 22 (1977).
1359.   Mitterberger, W.D. and Gross, H., Kunststofftecknik, 12, (7), 176; (8),219; (9), 281; (10), 277; (11), 303 (1973).
1360.   Tanaka, Y. and Morikawa, T., Kagaku To Kogyo (Osaka), 48, (10), 387 (1974); Chem. Abstr. 82, 98702w.
1361.   Narasaki, H. and Umezawa, K., Kobunshi Kagaku, 29, (6), 438 (1972).
1362.   Getmanenko, E.N. and Perepletchikova, E.M., Fiz-Khim. Metody Anal., 1, 75 (1976); Chem. Abstr., 87, 53794q (1977).
1363.   Petiaud, R., Makromol. Chem., 178, (3), 741 (1977).
1364.   Girgis Takla, P., Analyst, 103, 122 (1978).
1365.   Purdon, J.R. and Mate, R.D., J. Polym. Sci., B, 1, 451 (1963).
1366.   Purdon, J.R. and Mate, R.D., J. Polym. Sci., A, 1, 8, 1306 (1970).
1367.   Rogozinski, M. and Kramer, M., J. Polym. Sci., A, 1, 10, 3110 (1972).
1368.   Braun, D., Pure Appl. Chem., 26, 173 (1971).
1369.   Marks, G.C., Benton, J.L. and Thomas, C.M., Soc. Chem. Ind. Monograph., 26, 204 (1967).
1370.   Daniels, V.D. and Rees, N.H., Polym. Sci., 12, 2115 (1974).
1371.   Krimm, S., Folt, V.L., Shipman, J.J. and Berens, A.R., J. Polym. Sci., A 1, 2621 (1963).
1372.   Krimm, S. and Enomoto, S., J. Polym. Sci., A 2, 669 (1964).
1373.   Enomoto, S., Koguro, M. and Asakina, M., J. Polym. Sci., A 2 5355 (1964).
1374.   Germar, M., Makromolekulare Chem., 86, 89 (1965).
1375.   Grisenthwaite, R.J. and Hunter, R.F., Chem. Ind. (London), 719 (1958).
1376.   Fordham, J.W.L., Burleigh, P.H. and Sturm, C.L., J. Polym. Sci., 41, 73 (1959).
1377.   Shimanouchi, T., Tsuchiya, S. and Mizushima, S., J. Chem. Phys., 30, 1365 (1959).
1378.   Imura, K. and Takeda, M., J. Polym. Sci., 51, 551 (1961).

410   References

1379.   Takeda, M. and Imura, K., J. Polym. Sci., 57, 383 (1962).
1380.   Germar, H., Kolloid-Z, 193, 25 (1963).
1381.   Schneider, B., Stokr, J., Doskocilova, D., Kolinsky, M., Sykora, S.
        and Lim, D., paper presented at International Symposium Macromolecular
        Chemistry Prague, (1965).
1382.   Schneider, J.S.B., Kolinsky, M., Ryska, M., and Lim, D., J. Polym. Sci.,
        A, 1, 5, 2013 (1967).
1383.   Park, G., Macromol. Sci., Phys., B 14, (1), 151 (1977).
1384.   Enomoto, S., Asahina, M. and Satoh, S., J. Polym. Sci., A 4, 1373 (1966)
1385.   Baker, C., Maddams, W.F., Park, G.S. and Robertson, B., Makromol. Chem.,
        165, 321 (1973).
1386.   Millan, J.L. and De la Pena, J.L., Rev. Plast. Mod., 26, (206), 232
        (1973).
1387.   Stanciu, I.E., Ind. Usoara, 22, (8-9), 359 (1975); Chem. Abstr., 84,
        91435u.
1388.   Bezadea, E., Braun, D., Buruiana, E., Caraculacu, A. and Istrate-Robila,
        G., Angew. Makromol. Chem., 37, 35 (1974).
1389.   Brandmueller, J. and Schroetter, H.W., Fortschr. Chem. Forsch., 38,
        85 (1972).
1390.   Gerrard, D.L. and Maddams, W.F., Macromolecules, 8, (1), 54 (1975).
1391.   Maddams, W.F., J. Macromol. Sci., Phys., B 14, (1), 87 (1977).
1392.   Bovey, F.A., Anderson, E.W., Douglas, D.C. and Mason, J.A., J. Chem.
        Phys., 39, 1199 (1963).
1393.   Ramey, K.C., J. Phys. Chem., 70, 2525 (1966).
1394.   Bovey, F.A., Hood, F.P., Anderson, E.W. and Kornegay, R.L., J. Polym.
        Chem., 71, 312 (1967).
1395.   Heatley, F. and Bovey, F.A., Macromolecules, 2, 241 (1969).
1396.   Cavalli, L., Borsini, G.C., Carraro, G. and Confalonieri, G., J. Polym.
        Sci., A, 1, 8, 801 (1970).
1397.   Okuda, K., J. Polym. Sci., A,2, 1749 (1964).
1398.   Chujo, R., Satoh, S. and Nagai, E., J. Polym. Sci., A, 2 895 (1964).
1399.   McClanahan, J.L. and Previtera, S.A., J. Polym. Sci., A, 3, 3919 (1965).
1400.   Schaefer, J., J. Phys. Chem., 70, 1975 (1966).
1401.   Fischer, F., Kiisinger, J.B. and Wilson, C.W., III., J. Polym. Sci.,
        B, 4, 379 (1966).
1402.   Yamashita, Y., Ito, K., Ikuma, S. and Koda, H., J. Polym. Sci., B,6,
        219 (1968).
1403.   Ito, K. and Yamashita, Y., J. Polym. Sci., B, 6, 227 (1968).
1404.   Wilkes, C.E., Westfahl, J.C. and Backderf, R.H., J. Polym. Sci., A,
        1, 7, 23 (1969).
1405.   Carman, C.J., Tarpley, A.R., Jr., and Goldstein, J.H., Macromolecules,
        4, 445 (1971).
1406.   Carman, C.J., Macromolecules, 6, 725 (1973).
1407.   Abe, Y., Tasumi, M., Shimanouchi, T., Satoh, S. and Chujo, R., J.
        Polym. Sci., A, 1, 4, 1413 (1966).
1408.   Satoh, S., J. Polym. Sci., A 2, 5221 (1964).
1409.   Tho, P.Q. and Taieb, M., J. Polym. Sci., A, 1, 10, 2925 (1972).
1410.   Ando, I., Nishioka, A. and Watanabe, S., Polym. J., 3, (3), 403 (1972).
1411.   Abe, A. and Nishioka, A., Kobunshi Kagaku, 29, (6), 402, 448 (1972).
1413.   Bezdadea, E.C., Buriana, E.C., Istrate-Robila, G. and Caraculacu, A.A.,
        private communication.
1414.   Abdel-Alim, A.H., J. Appl. Polym. Sci., 19, (5), 1227 (1975).
1415.   Cavalli, L., Relaz. Corso Teor-Prat. Risonanza Magna Nucl., 351 (1973).
1416.   Repko, E., Tureckova, M. and Halamek, J., Petrochemia, 14, (1), 25
        (1974).
1417.   Bovey, F.A., Abbas, K.B., Schilling, F.C. and Starnes, W.H., Macromole-
        cules, 8, (4), 437 (1975).
1418.   Bezdadea, E., J. Polym. Sci., Polym. Chem. Ed., 15, (3), 611 (1977).

1419.   Caraculacu, A. and Bezdadea, E., J. Polym. Sci., A, 1, 15, 611 (1977).
1420.   Liebman, S.A., Renwer, J.F., Gollatz, K.A. and Nauman, C.D., J. Polym.
        Sci., A, 1, 9, 1823 (1971).
1421.   Ohnishi, S.I., Sugimoto, S.I. and Nitta, I., J. Polym. Sci., A, 1, 625
        (1963).
1422.   Ouchi, I., J. Polym. Sci., A, 3, 2685 (1965).
1424.   Tokikai, A., Takahashi, Y. and Kuri, Z.I., J. Polym. Sci., A, 1, 15,
        1519 (1977).
1425.   Cobler, J.G., and Samsel, E.P., SPE Transactions 145 April (1962).
1426.   Feuerberg, H. and Weigel, H., Z. Anal. Chem., 199, 121 (1964).
1427.   Levi, D.W., Literature Survey on Thermal Degradation, Thermal Oxidation
        and Thermal Analysis of High Polymers, June 1963. Clearinghouse for
        Fed. Sci. and Tech. Info. U.S. Dept. of Commerce, AD423546.
1428.   Teetsel, D.A. and Levi, D.W., Literature Survey on Thermal Degradation,
        Thermal Oxidation and Thermal Analysis of High Polymers II, January
        1966, Clearinghouse for Fed. Sci. and Tech. Info. U.S. Dept. of Comm-
        erce, AD631655.
1429.   Stromberg, R.R., J. Polym. Sci., 35, 355 (1959).
1430.   Ohtani, S. and Ishikawa, T., Kogyo Kagaku Zasshi, 65, 1617 (1962).
1431.   Noffz, D., Benz, W. and Pfab, W., Z. Anal. Chem., 235, 121 (1968).
1432.   Coleman, E.H. and Thomas, C.H., J. Appl. Chem., 4, 379 (1954).
1433.   Coleman, E.H., Plastics, 24, 416)1959).
1434.   Boettner, E.A. and Weiss, B., Amer. Ind. Hyg. Assoc. J., 28, 535 (1967).
1435.   Boettner, E.A., J. Appl. Polym. Sci., 13, 377 (1969).
1436.   O'Mara, M.M., J. Polym. Sci., A, 1, 8, 1887 (1970).
1437.   Watson, J.T. and Biemann, K., Anal. Chem., 36, 1135 (1964).
1438.   O'Mara, M.M., J. Polym. Sci., A, 1, 7, 1887 (1970).
1439.   O'Mara, M.M., J. Polym. Sci., A,1, 9, 1387 (1971).
1440.   Frye, A.H., Horst, R.W. and Paliobagix, M.A., J. Polym. Sci., A, 2,
        1801 (1964).
1441.   Frye, A.H. and Horst, R.W., J. Polym. Sci., 45, 1 (1960).
1442.   Stromberg, R.R., Straus, S. and Achhammer, B.G., J. Polym. Sci., 35,
        355 (1959).
1443.   Tsuchiya, F. and Sumi, K., J. Appl. Chem., 17, 364 (1967).
1444.   Ohta, M., Kogyo Kagaku Zasshi, 55, 31 (1952).
1445.   O'Mara, M.M., J. Polym. Sci., A, 1, 8, 1887 (1970).
1446.   Boettner, E.A., Ball, G. and Weiss, B., J. Appl. Polym. Sci., 13, 377
        (1969).
1447.   Noffz, D., Benz, W. and Pfab, W., Z. Anal. Chem., 235, 121 (1968).
1448.   Liebman, S.A., Ahlstrom, D.H., Quinn, E.J., Geigley, A.G. and Meluskey,
        J.T., J. Polym. Sci., A- 1, 9, 1921 (1971).
1449.   Iida, T., Nakanishi, M. and Goto, K., J. Polym. Sci., 12, 737 (1974).
1450.   Cotman, J.D., Jr., J. Amer. Chem. Soc., 77, 2790 (1955).
1451.   Carrega, M., Bonnebat, C. and Zednik, G., Anal. Chem., 42, 1807 (1970).
1452.   de Vries, A.J., Bonnebat, C. and Carrega, M., Pure Appl. Chem., 26, 209
        (1971).
1453.   Bovey, F.A., Abbas, K.B., Schilling, F.C. and Starnes, W.H., Macro-
        molecules, 8, 437 (1975).
1454.   Ahlstrom, D.H., Liebman, S.A. and Abbas, K.B., J. Polym. Sci., Polym.
        Chem. Ed., 14, 2479 (1976).
1455.   Chang, E.P. and Salovey, R., J. Polym. Sci.Polym. Chem. Ed., 12, 2927
        (1974).
1456.   Suzuki, M., Tsuge, S. and Takeuchi, T., J. Polym. Sci., A-1, 10, 1051
        (1972).
1457.   O'Mara, M.M., J. Polym. Sci., A-1, 8, 1887 (1970).
1458.   Noffz, D., Henz, W. and Pfab, W., Z. Anal. Chem., 235, 121 (1968).
1459.   Chang, E.P. and Salovey, R., J. Polym. Sci. Polym. Chem. Ed., 12, 2927
        (1974).

412   References

1460. McNeill, I.C., J. Polym. Sci., A, 1, 4, 2479 (1966).
1461. Danforth, J.D. and Takeuchi, T., J. Polym. Sci., A,1, 11, 2091 (1973).
1462. Bataille, P. and Van, B.T., J. Polym. Sci., A-1, 10, 1007 (1972).
1463. Gupta, V.P. and St. Pierre, L.E. J. Polym. Sci., A-1, 11, 1841 (1973).
1464. Gupta, V.P. and St. Pierre, L.E., J. Polym. Sci., A-1, 8, 37 (1970).
1465. Guyot, A., Bert, A. and Spitz, R., J. Polym. Sci., A,1, 8, 1596 (1970).
1466. Chany, E.P. and Salovey, R., J. Polym. Sci., A-1, 12, 2927 (1974).
1467. Gedemer, T.J., J. Macromol. Sci., Chem., 8, (1), 95 (1974).
1468. Bosche, K., Reitr. Gerichtl. Med., 33, 280 (1975).
1469. Iida, T., Nakanishi, M. and Goto, K., J. Polym. Sci. Polym. Chem. Ed.,
       12, 737 (1974).
1470. Geddes, W.C., Rubber Chem. Technol., 40, 177 (1967).
1471. Braun, D., Pure Appl. Chem., 26, 173 (1971).
1472. Guyot, A., Benevise, J.P. and Trambouze, Y., J. Appl. Polym. Sci., 6,
       103 (1962).
1473. Braun, D. and Thallmaier, M., Makromol. Chem., 99, 59 (1966).
1474. Geddes, W.C., Eur. Polym. J., 3, 747 (1967).
1475. Kelen, T., Balint, G., Galambos, G. and Tüdös, F., Eur. Polym. J., 5,
       597 (1969).
1476. Tüdös, F., Lecture presented at the IUPAC International Symposium on
       Macromolecules, Helsinki (1972).
1477. Bengough, W.I. and Varma, J.K., Eur. Polym. J., 2, 61 (1966).
1478. Thallmaier, M. and Braun, D., Makromol. Chem., 108, 241 (1967).
1479. Shindo, Y. and Hirai, T., Makromol. Chem., 155, 1 (1972).
1480. Bohlmann, F. and Mannhard, H.J., Chem. Ber., 89, 1309 (1956).
1481. Sadron, C., Parrod, J. and Roth, P., C.R. Hebd. Seances Acad. Sci.,
       250, 2206 (1960).
1482. Sondheimer, F., Ben-Efraim, D.A. and Wolovsky, R ., J. Amer. Chem. Soc.,
       83, 1675 (1961).
1483. Popov, K.R. and Smirnov, L.V., Optika Spektroskopiya, 14, 787 (1963).
1484. Minsker, K.S., Krats, E.O. and Pakhomova, I.K., Polym. Sci. USSR, 12,
       545 (1970).
1485. Smirnov, L.V. and Popov, K.R., Polym. Sci. USSR, 13, 1356 (1971).
1486. Minsker, K.S. and Krats, E.O., Polym. Sci. USSR, 13, 1358 (1971).
1487. Kelen, T., Galambos, G., Tudos, F. and Balint, G., Eur. Polym. J., 5,
       617 (1969).
1488. Kelen, T., Galambos, G., Tudos, F. and Balint, G., Eur. Polym. J., 5,
       629 (1969).
1489. Kelen, T., Galambos, G., Tudos, F. and Balint, G., Eur. Polym. J., 6,
       127 (1970).
1490. Abbas, K.B. and Lawrence, R.L., J. Polym. Sci. Polym. Sci. Ed., 13, 1889
       (1975).
1491. Abney, G., Head, B.C. and Pollen, R.C., J. Polym. Sci. Macromolecular
       Review, 8, 1, (1974).
1492. Ahlstrom, D.H. and Foltz, C.R., J. Polym. Sci., A-1, 16, 2703 (1978).
1493. Shtarkman, B.P., Lebedev, V.P., Kosmynin, T.L., Gerasimov, V.I., Genin,
       Ya.V. and Tsvankin, D.Ya., Vysokomol. Soedin. Ser. A, 14, (7), 1629
       (1972).
1494. Neilson, G.F. and Jabarin, S.A., J. Appl. Phys., 46, (3), 1175 (1975).
1495. Gouinlock, E.V., J. Polym. Sci. Polym. Phys. Ed., 13, (5), 961 (1975).
1496. Baker, C., Maddams, W.F. and Preedy, J.E., J. Polym. Sci. Polym. Phys.
       Ed., 15, (6), 1041 (1977).
1497. Gaidarova, L.L. and Polyakova, K.A., Izv. Vyssh. Uchebu. Zaved. Tekhnol.
       Legk. Prom-sti, 2, 43 (1978); Chem. Abstr. 89, 44495g (1978).
1498. Buriana, E.C., Barbinta, V.T. and Caraculacu, A.A., Eur. Polym. J., 13,
       (4), 311 (1977).
1499. Schroeder, E. and Byrdy, M., Plaste Kautsch, 24, (11), 757 (1977).
1500. Jisova, V., Janca, J. and Kolinsky, M., J. Polym. Sci., A-1, 15,553 (1977).

1501. Lin, T.F., Tung, L.Z., Liu, F.Y. and Hsu, W.W., J. Chin. Inst. Chem.
       Eng., 4, (1), 43 (1973).
1502. Ambler, M.R. and Mate, R.D., J. Polym. Sci., Part A-1, 10, (9),2677
       (1972).
1503. Epton, R., Holding, S.R. and McLaren, J.V., Polymer, 15, 466 (1974).
1504. Belen'kii, B.G., Gankina, E.S., Nefedov, P.P., Kuznetsova, M.A. and
       Valchikhine, M.D., J. Chromatography, 77, (1), 209 (1973).
1505. Daley, L.E., J. Polymer Sci., C, 8, 253 (1965).
1506. Freeman, M. and Manning, P.P., J. Polym. Sci., A,2, 2017 (1964).
1507. Feldman, D., Uglea, C.V., Nuta, V. and Popa, M., Analusis, 1, (4),305
       (1972).
1508. Feldman, D., Uglea, C.V., Nuta, V. and Popa, M., Rev. Roum. Chim., 17,
       1033 (1972).
1509. Cantow, H.H., Kowalski, M. and Krozer, S., Angew. Chem., Int. Ed. Engl.,
       11, (4), 336 (1972).
1510. Janca, J. and Kolinsky, M., (Ustav Makromol. Chem. Cesk. Akad. Ved.,
       Prague, Czech.) Plasty Kauc., 13, (5), 138 (1976).
1511. Nakao, K. and Kuramoto, K., Nippon Secchaku Kyokal Shi, 8, (4), 186
       (1972).
1512. Kahn, P., Marechal, C. and Verdu, J., Anal. Chem., 51, 1000 (1979).
1513. Muinov, T.M., Dokl. Akad. Nauk. Tadzh. SSR, 15, (7), 25 (1972).
1514. Eliassaf, J., J. Macromol. Sci., Chem., 8, (2),459 (1974).
1515. Roszkowski, Z., Legocki, M., Plochocka, K., Iskra, K., Karwat, I.,
       Bankowska, L. and Obioj, M., Przem. Chem., 52, (3), 200 (1973).
1516. Szczepanska, M., Ochr. Przeciwpozarowa. Przem. Chem., 1,1 (1973).
1517. Venediktov, M.V., Dushchenko, V.P., Kolupaev, B.S. and Tarasenko, Yu.
       G., Mash. Prib. Ispyt. Mater., 99 (1971).
1518. Liso, J.C., Anal. Chem., 50, 1683 (1978).
1519. Berens, A.R., Crider, L.B., Tomanek, C.M. and Whitney, J.M., BP Good-
       rich Tyre Centre, Brecksville, Ohio. M/S circulated to members of
       the Vinyl Chloride Safety Association. Nov. 14th 1974 entitled "Anal-
       ysis of Vinyl Chloride in PVC powders by head-space chromatography".
1520. Berens, A.R., Paper delivered to the 168th ACS Meeting, Atlantic City,
       New Jersey, Sept. 1974 entitled "The Solubility of Vinyl Chloride
       in Polyvinylchloride".
1521. Steichen, R.J., Anal. Chem., 48, (9), 1398 (1976).
1522. Shanks, R.A., Scan, 6, 20 (1975).
1523. Baba, T., Shokuhin Eiseigaku Zasshi, 18, (6), 500 (1977); Chem. Abstr.,
       88, 153342e (1978).
1524. Gilbert, S.G., Giacin, J.R., Morano, J.R. and Rosen, J.D., Package
       Dev. Syst., 5, 20 (1975).
1525. Simpson, D. and Currell, B.R., Analyst, 96, 515 (1971).
1526. Vasil'eva, A.A., Vodzinkii, Yu.V. and Korshunov, I.A., Zavod Lab., 34,
       1304 (1968).
1527. Protivova, J., Pospisil, J. and Zikmund, L., J. Polym. Sci. Symp. No.
       40, 233 (1973).
1528. Simpson, D. and Currell, B.R., Analyst, 96, 515 (1971).
1529. Stafford, C., Anal. Chem., 34, 794 (1962).
1530. Thinus, K., Stabilizierung und Alterung von Plastwerkstoffen. Akademie
       Venlag Berlin (1969).
1531. Schmidt, W., Beckman Report, 4, 6 (1961); 3, 13 (1962) and 3, 13 (1962).
1532. Neubert, G., Beckman Report, 15, 203, 265 (1964).
1533. Burger, K., Beckman Report, 15, 192, 280 (1963).
1534. Franzen, V. and Neubert, G., Chemiber. Zeitung, Chem. Apparatur., 89,
       801 (1965).
1535. Berger, K.G., Rudt, U. and Mack, D., Deutsch. Lebensmittell Rdsch.,
       6, 180 (1967).
1536. Mazur, H., Rocnz. panst. Zakl. Hig., 24, 551 (1973).

1537.   Udris, J., Analyst, 96, 130 (1971).
1538.   Groagova, A. and Pribyl, M.Z., Anal. Chem., 234, 423 (1968).
1539.   Stapfer, C.H. and Dvorkin, R.D., Anal. Chem., 40, 1891 (1968).
1540.   Schroeder, E., Pure Appl. Chem., 36, 233 (1973).
1541.   Mazur, H., Rocz. panst. Zakl. Hig., 23, 263 (1972).
1542.   Hagen, E., Deutsch. Lebensmittel. Rdsch., 14, 158 (1967).
1543.   Schroeder, E., Hagen, E. and Frimel, S., Deutsch Lebensmittel Rdsch.,
        31, 814 (1967).
1544.   Korn, O. and Waggon, H., Nahrung, 8, 351 (1964).
1545.   Waggon, H., Kohler, U. and Korn, O., Ernahrungforschung., 11, 548
        (1966).
1546.   Simpson, D. and Currell, B.R., Analyst, 96, 515 (1971).
1547.   Veres, L., Kunststoffe, 59, 241 (1969).
1548.   Weatherhead, R.G., Analyst, 91, 445 (1966).
1549.   Kreiner, J.G., J. Chromatography, 75, 271 (1973).
1550.   Schroeder, E. and Hagen, E., Plaste Kautsch, 15, 625 (1968).
1551.   Kucera, J., Collect. Czechoslovakia Chem. Commun., 28, 1344 (1963).
1552.   Wandel, M., Tengler, H. and Ostromov, H., Die Analyse von Weichmach-
        ern, Springer Verlag, Berlin, Heidelberg, New York (1967).
1553.   Zilio-Grandi, F., Libralesso, G., Sassu, G. and Svegidao, G., Mater.
        Plast. Elast., 30, 643 (1964).
1554.   Korn, O. and Waggon, H., Plaste Kautschuk, 11, 278 (1964).
1555.   Schroeder, E. and Hagen, E., Plaste Kautsch., 15, 625 (1968).
1556.   Verfungen and Mitteilungen des ministeriums für Gesundleitwesau, No.
        2 (1965).
1557.   Schroeder, E. and Thinius, K., Deutsch. Farben Z., 14, 146 (1960).
1558.   Schroeder, E. and Malz, S., Deutsch. Farben, 5, 417 (1958).
1559.   Nalik, W.U., Hague, R. and PalVerma, S., Bull. Chem. Soc. Japan, 36,
        746 (1963).
1560.   Schmidt, W., Beckman Report, 4, 6 (1961); 3, 13 (1962).
1561.   Mal'kove, L.N., Kalanin, A.I. and Perepletchikova, E.M., Zhur. Analit.
        Khim, 27, 1924 (1972).
1562.   Schroeder, E., Hagen, E. and Zysik, M., Beckman Report, 13, 720 (1966).
1563.   Mitterberger, D. and Gross, R., Kunststoffechnik, 12, (7), 176 (1973):
        (8), 219; (9), 261; (10), 277; (11),
1564.   Vasil'yanova, L.S., Gutsalyuk, V.G. and Yatsenko, E.A., Plast. Massy.,
        2, 65 (1977); Chem. Abstr., 86, 122203q.(1977).
1565.   Kohman, Z. and Ciecierski, G., Polimery (Warsaw), 22, (5), 155 (1977);
        Chem. Abstr., 88, 23757h (1978).
1566.   Hasles, J. and Soppet, W.W., J. Chem. Soc., 67, 33 (1948).
1567.   Dahring, H., Kunststoffe, 28, 230 (1938).
1568.   Robertson, M.W. and Rowley, R.M., British Plastics, January 26 (1960).
1569.   Haslam, J. and Squirrell, D.C.M., Analyst, 80, 871 (1955).
1570.   Wake, W.C., The Analysis of Rubber and Rubber-like Polymers, McLaren,
        London (1958).
1571.   Hasse, H., Kaut. Gummi Kunststoffe, 20, 501 (1967).
1572.   Schroeder, E. and Hagen, E., Plaste Kautsch., 15, 625 (1968).
1573.   Zulake, J. and Guichon, G., Revista de Plastices Modernos, 13, 13
        (1963).
1574.   Zulake, J., Landault, C. and Guichon, G., Bull. Soc. Chim. France,
        1294 (1962).
1575.   Cook, C.D., Elgood, E.J. and Solomon, D.H., Anal. Chem., 34, 1177 (1962).
1576.   Veres, L., Kunststoffe, 59, 13 (1969).
1577.   Veres, L., Kunststoffe, 59, 241 (1969).
1578.   Turnstall, F.I.H., Anal. Chem., 42, 542 (1970).
1579.   Zulaica, J. and Guichon, G., Anal. Chem., 35, 11 1724 (1963).
1580.   Yoshita, T., Toyoda Gosei Giho, 17, (2), 64 (1975); Chem. Abstr.,
        88, 170952w (1978).

1581. Barla, F., Muanyag Gumi, 14, (11), 330 (1977); Chem. Abstr., 88,
       1705955z (1978).
1582. Takada, T., Fukui-ken Kogyo Shikenjo Nempo, 49, 66 (1974); Chem. Abstr.,
       87, 118636m, (1977).
1583. Takada, T., Fukui-ken Kogyo Shikenjo Nempo, 50, 77 (1975); Chem. Abstr.,
       87, 118638p (1977).
1584. Krishen, A., Anal. Chem., 43, 1130 (1971).
1585. Guichon, G. and Henniker, J., British Plastics, 37, 74 (1964).
1586. Steuerle, H. and Pfab, W., Deutsch Lebensmittel Rdsch., 65, 113 (1964).
1587. Peereboom, J.W.C., J. Chromatography, 3, 323 (1960).
1588. Von Nagy, F.Z., Lebensmittel Unterschung und Forschung, Part 4, 126
       (1965).
1589. Ligotti, I., Piazantini, R. and Bonomi, G., Hass Chim., No. 1, 16
       (1964).
1590. Von Braun, D., Chimia, 19 February (1965).
1591. Jaminel, F., I.C. Formaco Edizons Practica. Anno XVIII No. 12 December
       (1963).
1592. Burns, W., J. Appl. Chem., 5, 599 (1955).
1593. Schevschenko, Z.A. and Favorskaya, I.A., Vestu. Leningr. Univ., 19,
       2 (1964).
1594. Ruddle, L.H., Swift, S.D., Udris, J. and Arnold, P.E. in P.W. Shallis
       (Editor) "Proceedings of the SAC Conference, Nottingham, 1955", W.
       Heffer and Sons Ltd. Cambridge Press.
1595. Walker, W.H. and Gaenshirt, K.H., Z.Analyt Chem., 267, 127 (1973).
1596. Veres, L., Kunststoffe, 59, 13 (1969).
1597. Veres, L., Kunststoffe, 59, 241 (1969).
1598. Cambell, G.E., Foxton, A.A. and Wordsall, R.L., Lab. Pract., 19, 369
       (1970).
1599. Struele, W. and Pfab, W., Deutsch Lebensmittel Rundschau. sch., 65,
       113 (1969).
1600. Burns, W., Polymer J. Singapore, 3, 58 (1971).
1601. Von Braun, D., Kunststoffe, 52, 2 (1962).
1602. Gomaryova, A., Elekrviz Kahl. Tech., 23, 207 (1970).
1603. Sokolowska, R., Roczn. panst. Zakl. Hig, 595, 20 (1969).
1604. Kreiner, J.G., J. Chromatography, 75, 271 (1973).
1605. Mazur, H., Rocz. panst. Zakl. Hig., 23, 263 (1972).
1606. Kula, H., Rocz. panst. Zakl. Hig., 20, 307 (1969).
1607. Guichon, G. and Henniker, J., British Plastics, February 74 (1974).
1608. Udris, J., Analyst, 96, 130 (1971).
1609. Robertson, M.W. and Rowley, R.M., British Plastics, January 26 (1960).
1610. Mansfield, P.B., Chemy. Ind. 28, 792 (1971).
1611. Hurtibese, R.J. and Latz, H.W., J. Agric. Food Chem., 18, 377 (1970).
1612. Zilio-Grandi, F., Libraesso, J., Sassu, G. and Svegidao, G., Mater
       Plast. Elast., 30, 643 (1964).
1613. Wexler, T., Jaks, I. and Jucan., Mater. Plast. (Bucharest), 9, (6),
       268 (1972).
1614. Mol'kova, L.N., Kalinin, A.I. and Perepletchikova, E.M., Zh. Anal.
       Khim., 27, (10), 1924 (1972).
1615. Komeleva, V.N. and Kalinin, A.I., Tr. Khim. Khim Tekhnol., 1, 170
       (1973).
1616. Laverty, J.J. and Gardlund, Z.G., J. Polymer Sci., A-1, 15, 2001 (1977).
1617. Private Communication, British Standards Institution, London.
1618. Gemanenko, E.N. and Perepletchikova, E.M., Zh. Anal. Khim., 29, (4),
       830 (1974).
1619. Fleischer, G. and Hellebrand, J., Wiss. Z. Karl Marx-Univ. Leipzig,
       Math-Naturwiss, Reihe, 21, (6), 653 (1972).
1620. Helmuth, J., Polim. Vehomarium Plast., 3, (2), 7 (1973).
1621. Chia, K.S. and Chen, G.M.S., J. Chin. Chem. Soc. (Taipei), 20,4,241(1973).

1622.   Aslanova, R., Yul'chivaev, A.A. and Usmanov, Kh.U., Uzb. Khim. Zh.,
        17, (3), 31 (1973).
1623.   Campbell, D.R., Anal. Chem., 47, (8), 1477 (1975).
1624.   Kalal, J., Marousek, V. and Svec, F., Sb. Vys. Sk. Chem-Technol. Praze.
        Org. Chem. Technol. C 22, 57 (1975); Chem. Abstr., 83, 132443v.
1625.   Janca, J. and Kolinsky, M., J. Appl. Polym. Sci., 21, (1) 83 (1977).
1626.   Hirai, T. and Tanaka, M., Angew. Makromol. Chem., 48, (1), 45 (1975).
1627.   Wilkes, C.E., J. Polym. Sci., Polym. Symp., 60, 161 (1977); Chem. Abstr.,
        89, 24952a (1978).
1628.   Keller, F., Faserforsch. Textiltech., 28, (10), 515 (1977); Chem. Abstr.,
        88, 105888r (1978).
1629.   Kleinpeter, E. and Keller, F., Z. Chem., 18, (6), 222 (1978); Chem.
        Abstr., 89, 130030h (1978).
1630.   Chujo, S., Satoh, S., Ozeki, T. and Nagai, E., J. Polymer Sci., 61,
        512 (1962).
1631.   Chujo, R., Satoh, S. and Nagai, E., J. Polymer Sci., A,2, 895 (1964).
1632.   Germar, M., Makromolekulare Chem., 84, 36 (1965).
1633.   Enomoto, S., J. Polym. Sci., 55, 95 (1961).
1634.   McClanahan, O. and Previtera, S.A., J. Polym. Sci., A,3, 3919 (1965).
1635.   Okuda, K., J. Polym. Sci., A,2, 171 (1964).
1636.   Revillon, A., Dumont, B. and Guyot, A., J. Polym. Sci., Polym. Chem.
        Ed., 14, 2263 (1976).
1637.   Cais, R.E. and O'Donnell, J.H., Makromol. Chem., 176, (12), 3517 (1975).
1638.   Cais, R.E. and O'Donnell, J.H., J. Makromol. Sci. Chem., A, 10, (5),
        763 (1976).
1639.   Brück, D. and Hummel, D.O., Makromol. Chem., 163, 245 (1973).
1640.   Brück, D. and Hummel, D.O., Makromol. Chem., 163, 259 (1973).
1641.   Brück, D. and Hummel, D.O., Makromol. Chem., 163, 271 (1973).
1642.   Albarino, V., Otacka, E.P. and Luongo, J.P., J. Polym. Sci., A-1, 15,
        1517 (1977).
1643.   Kubik, I., Singliar, M. and Navratil, M., Petrochemia, 17, (1-2), 10
        (1977); Chem. Abstr., 88, 51308f (1978).
1644.   Kubik, I., Singliar, M. and Navratil, M., Petrochemia, 17, (1-2), 15
        (1977).
1645.   Wasilewska, W., Grzywa, E. and Rajkiewicz, M., Chem. Anal.(Warsaw),
        19, (1), 89 (1974).
1646.   Cascaval, C.N., Schneider, I.A. and Poinescu, I., J. Polym. Sci., Polym.
        Chem. Ed., 13, 2259 (1975).
1647.   Hay, J.N., J. Polym. Sci., A-1, 8, 1201 (1970).
1648.   Bovey, F.A., Schilling, F.C., Kwei, T.K. and Frisch, H.L., Macromol-
        ecules, 10, (3), 559 (1977).
1649.   Roberts, D.R. and Gatzke, A.L., J. Polym. Sci., A-1, 16, 1211 (1978).
1650.   Agostina, D.E. and Gatzke, A.L., J. Polym. Sci., A-1, 11, 649 (1973).
1651.   Brinkman, K., Kolloid-Zeitung Z. Polym., 251, 962 (1973).
1652.   Doskocilova, D., Schneider, B., Draoradova, E. and Stokr, J., J. Polym.
        Sci., A-1, 9, 2753 (1971).
1653.   Keller, F., Opitz, H., Hoesselbarth, B., Beckert, D. and Reicherdt, W.,
        Faserforsch. Textiltech., 26, (7), 329 (1975).
1654.   Lucas, R., Kolinsky, M. and Doskocilova, D., J. Polym. Sci., Polym.
        Chem. Ed., 16, (5), 889 (1978).
1655.   Keller, F., Faserforsch. Textiltech., 28, (10), 515 (1977); Chem. Abstr.,
        88, 105888r (1978).
1656.   Keller, F. and Hoesselbarth, B., Faserforsch. Textiltech., 29, (2),
        152 (1978); Chem. Abstr., 88, 170672e (1978).
1657.   Keller, F., Zepnik, S. and Hoesselbarth, B., Faserforsch. Textiltech.,
        28, (6), 287 (1977).
1658.   Froix, M.F., Goedde, A.O. and Pochau, M.J., Macromolecules, 10,778(1977).
1659.   Gasan-Zade, V.G., Sutovskii, S.M. and Kaplan, M.Y., Lakokras. Mater.

Ikh. Primen, 5, 11 (1976); Chem. Abstr., 86, 44253q (1977).

1660. Grinevich, T.V., Shupik, A.N., Korovina, G.V. and Entelis, S.G., Vysokomol. Soedin., Ser. B, 17, (6), 459 (1975).

1661. Szap, P., Kesse, I. and Klapp, J., J. Liq. Chromatogra., 1, (1), 89 (1978).

1662. Kuzaev, A.I., Susiova, E.N. and Entelis, S.G., Dokl. Akad. Nauk SSR, 208, (1), 142 (1973).

1663. Myers, G.E. and Dagon, J.R., J. Polym. Sci., A-1, 2, 2631 (1964).

1664. Tsuji, K., Iwamoto, T., Hayashi, K. and Yoshida, H., J. Polym. Sci., A-1, 5, 265 (1967).

1665. Tsuji, K., Takeshita, T. and Seiki, T., J. Polym. Sci., A-1, 10, 123 (1972).

1666. Ghosh, P., Chadha, S.C., Mukoyee, A.R. and Palit, R., J. Polym. Sci., A, 2, 4433 (1964).

1667. Bartlett, P.D. and Nozaki, K., J. Polym. Sci., 3, 216 (1948).

1668. Palit, S.R. and Ghosh, P., Microchem. J. Symp. Ser., 2, 663 (1961).

1669. Palit, S.R. and Ghosh, P., J. Polym. Sci., 58, 1225 (1962).

1670. Palit, S.R. and Mukherjee, A.R., J. Polym. Sci., 58, 1243 (1962).

1671. St.Pierre, L.E. and Price, C.C., J. Am. Chem. Soc., 78, 3432 (1956).

1672. Ghosh, P., Mukherjee, A.R. and Palit, S.R., J. Polym. Sci., A,2, 2807 (1964).

1673. Ghosh, P., Chadha, S.C. and Palit, S.R., J. Polym. Sci., A,2, 4441 (1964).

1674. Ghosh, P., Sengupta, P.K. and Pramanik, A., J. Polym. Sci., A, 3, 1725 (1965).

1675. Ghosh, P., Chadha, S.C., Mukherjee, A.R. and Palit, S.R., J. Polym. Sci., A,2, 4433 (1964).

1676. Maiti, S. and Saha, M.K., J. Polym. Sci., A-1, 5, 151 (1967).

1677. Palit, S.R., Makromol. Chem., 36, 89 (1959); ibid, 38, 96 (1960).

1678. Maiti, S., Ghosh, A. and Saha, M.K., Nature, 210, 513 (1966).

1679. Arai, K., Kominc, S. and Negishi, M., J. Polym. Sci., A-1, 8, 917 (1970).

1683. Negishi, M., Arai, K. and Okada, S., J. Appl. Polym. Sci., 11, 2427 (1967).

1684. Arai, K., Negishi, M. and Okabe, T., J. Appl. Polym. Sci., 12, 2585 (1968).

1685. Kabayashi, O., J. Polym. Sci., A-1, 17, 293 (1979).

1686. Yoshina, T. and Shinomya, M., J. Polym. Sci., A,3, 2811 (1965).

1687. Spells, S.J. and Shepherd, I.W., J. Chem. Phys., 66, (4), 1427 (1977).

1688. Woodward, A.E., J. Polym. Sci., C, 8, 137 (1965).

1689. Peat, I.R. and Reynolds, W.F., Tetrahedron Letters, No. 14, 1359 (1972).

1690. Girad, H. and Monjol, P., C.R. Hebd. Seances Acad. Sci. Ser. C., 279, (13), 553 (1974).

1691. Matsuzaki, H., Kansi, T., Kawamura, T., Matsumoto, S. and Uryu, T., J. Polym. Sci., Polym. Chem. Ed., 11, 961 (1973).

1692. Nagai, M. and Nishioka, A., J. Polym. Sci., A-1, 6, 1655 (1968).

1693. Brosio, E., Delfini, M. and Conti, F., Nuova Chim., 48, 35 (1972).

1694. Kusakov, M.M., Koshevnik, A.Yu., Mekenitskaya, L.I., Shul'pine, L.M., Amerik, Yu.B. and Golova, L.K., Vysokomol. Soedin., Ser.B., 15, (3), 150 (1973).

1695. Amiya, S., Ando, I. and Chujo, R., Polym. J., 4, (4), 385 (1973).

1696. Amiya, S., Ando, I., Watanabe, S. and Chujo, R., Polym. J., 6, (2), 194 (1974).

1697. Miyamoto, T. and Cantow, H.J., Makromol. Chem., 162, 43 (1972).

1698. Cavalli, L., Relaz. Corso Teor-Prat. Risonanza Magn. Nucl., 351 (1973).

1699. Spevacek, J. and Schneider, B., Makromol. Chem., 175, (10), 2939 (1974).

1700. Spevacek, J. and Schneider, B., J. Polym. Sci., Polym. Lett. Ed., 12, (6), 349 (1974).

1701.   Spevacek, J. and Schneider, B., Makromol. Chem., 176, (3), 729 (1975).
1702.   Suzuki, T. and Harwood, H.J., Polym. Prepr., Am. Chem. Soc., Div.
        Polym. Chem., 16, (1), 638 (1975).
1703.   Schaefer, J., Stejskal, E.O. and Buchdahl, R., Macromolecules, 8, (3),
        291 (1975).
1704.   Strasilla, D. and Klesper, E., J. Polym. Sci., Polym. Lett. Ed., 15,
        (4), 199 (1977).
1705.   Hatada, K., Ohta, K., Okamoto, Y., Kitayama, T., Umemura, Y. and Yuki,
        H., J. Polym. Sci., Polym. Lett. Ed., 14, (9), 531 (1976).
1706.   Hatada, K., Ishikawa, H., Kitayama, T. and Yuki, H., Makromol. Chem.,
        178, (9), 2753 (1977).
1707.   Hatada, K., Kitayama, T., Okamoto, Y. and Ohta, K., Makromol. Chem.,
        179, (2), 485 (1978).
1708.   Hajimoto, Y., Tamura, N. and Okamoto, S., J. Polym. Sci., A,3, 255
        (1965).
1709.   Michel, R.E., Chapman, F.W. and Mao, T.J., J. Polym. Sci., A-1, 5,
        1077 (1967).
1710.   Iwasaki, M. and Sakai, Y., J. Polym. Sci., A, 1, 7, 1537 (1969).
1711.   Harris, J.A., Hinojosa, O. and Arthur, J.C.,Jr., J. Polym. Sci., Polym.
        Chem. Ed., 11, 3215 (1973).
1712.   Sakaguchi, M., Kodama, S., Ediund, O. and Sohma, J., J. Polym. Sci.,
        Polym. Lett. Ed., 12, 609 (1974).
1713.   Bullock, A.T., Cameron, G.G. and Elsom, J.M., Polymer, 15, 74 (1974).
1714.   Geuskens, G. and David, C., Makromol. Chem., 165, 273 (1973).
1715.   Yashioka, H., Matsumotoh, H., Uno, S. and Higashide, F., J. Polym.
        Sci., A-1, 14, 1331 (1976).
1716.   Torikai, A. and Okamoto, S.I., J. Polym. Sci., A-1, 16, 2689 (1978).
1717.   de Angelis, G., Ippolita, P. and Spina, N., La Richenha Scienfica,
        28, 144 (1958).
1718.   Haslam, J., Hamilton, J.B. and Jeffs, A.R., Analyst, 83, 983, 66 (1958).
1719.   Haslam, J. and Jeffs., J. Anal. Chem., 7, 24 (1957).
1720.   Radell, E.A. and Strutz, H.C., Anal. Chem., 31, 1890 (1959).
1721.   Guillet, J.E., Wooten, W.C. and Combs, R.L., J. Appl. Polym. Sci.,
        7, 61 (1960).
1722.   Lehrle, R.S. and Robb, J.C., Nature, 183, 1671 (1959).
1723.   Strassburger, J., Brauer, G.M., Trijon, M. and Forziati, A.F., Anal.
        Chem., 32, 454 (1961).
1724.   Lehmann, F.A. and Brauer, G.M., Anal. Chem., 33, 673 (1961).
1725.   Brauer, G.M., J. Polym. Sci., C, 8, 3 (1965).
1726.   Stanley, E.W. and Peterson, W.R., SPE Transactions, 288 (October 1962).
1727.   Cobler, J.G. and Samsel, E.P., SPE Transactions, 145 (April 1962).
1728.   McNeill, I.C., J. Polym. Sci., A, 1, 4, 2479 (1966).
1729.   MacCallum, J.R., Makromol. Chem., 83, 137 (1965).
1730.   Clark, J.E. and Jellinek, H.H.G., J. Polym. Sci., A, 3, 1171 (1965).
1731.   Kiran, E., Gillham, J.K. and Gipstein, E., J. Makromol. Sci., Phys.,
        B 9, (2), 341 (1974).
1732.   Bosch, K., Beitr. Gerichtl. Med., 33, 280 (1975).
1733.   Meyerhoff, G., J. Chromatogr. Sci., 9, (10), 596 (1971).
1734.   Miyamoto, Y., Tomoshige, S. and Inagaka, H., Polym. J., 6, (6), 564
        (1974).
1735.   Grubisie, Z., Rempp, P. and Benoit, H., J. Polym. Sci., B,5, 753 (1967).
1736.   Weiss, A.R. and Cohn-Ginsberg, E., J. Polym. Sci., A-2, 8, 148 (1970).
1737.   Belinkii, B.G. and Nefedov, P.P., Vysokomol. Soedin., A-14, 1568 (1972).
1738.   Kolinsky, M. and Janca, J., J. Polym. Sci., A-1, 12, 1181 (1974).
1739.   Janca, J., Vlcek, P., Trekoval, J. and Kolinsky, M., J. Polym. Sci.,
        Polym. Chem. Ed., 13, 1471 (1975).
1741.   Inagaki, H. and Kamiyama, F., Macromolecules, 6, (1), 107 (1973).
1742.   Buter, R., Tan, Y.Y. and Challa, G., Polymer, 14, 171 (1973).

1743. Byl'skii, B.Ya., Pozdynyakov, O.F., Regel, V.R. and Redov, B.P., Mekh.
       Polim., 5, 835 (1973).
1744. Yasina, L.L. and Pudov, V.S., Vysokomol. Soedin., Ser. A, 15, (6), 1291
       (1973).
1745. Gardella, J.A. and Hercules, D.M., Anal. Chem., 52, 226 (1980).
1746. Kuo, J.F. and McIntyre, D., J. Chin. Inst. Chem. Eng., 4, (1), 33 (1973).
1747. Fischer, E.W., Wendorff, J.H., Dettenmaier, M., Lieser, G. and Voigt-
       Martin, I., Polym. Prepr., Am. Chem. Soc., Div. Polym. Chem., 15, (2),
       8 (1974).
1748. Patterson, G.D., J. Polym. Sci., Polyn. Lett. Ed., 13, (7), 415 (1975).
1749. Kalinin, A.I., Komleva, V.N. and Mol'kova, L.N., Metody Anal. Kontrol-
       ya Proizvod. Khim. Prom-sti., 11, 62 (1977); Chem. Abstr., 88, 153329f
       (1978).
1750. Waters, D.N., Proc. Conf. Raman Spectrosc. 5th, 500 (1976); Chem. Abstr.,
       87, 185223n (1977).
1751. Cobler, J.G. and Samsel, E.D., SPE Transactions, (April 1962).
1752. Haslam, J., Hamilton, J.B. and Jeffs, A.R., Analyst, 83, 66 (1958).
1753. Harris, I. and Miller, R.G.J., J. Polym. Sci., 7, 377 (1951).
1754. Gruber, U. and Elias, H.G., Die Makromol. Chemie, 78, 58 (1964).
1755. Scholten, W. and Lange, H., Kolloid-Z.Z. Polym., 250, (8), 782 (1972).
1756. Tomono, T., Yao, T. and Tsuichida, E., Kobunshi Kagaku, 29, (6), 353
       (1972).
1757. Tsitorovskii, K.B., Ivanov, A.P., Kazimirov, L.M. and Ivanov, R.A.,
       Prom. Sin. Kauch., Nauch-Tokh. Sb., No.9, 10 (1972).
1758. Schelten, J., Kruse, W.A. and Kirste, R.G., Kolloid-Z.Z.Polym., 251,
       (11), 919 (1973).
1759. Allen, G., Wright, C.J. and Higgins, J.S., Polymer, 15, 319 (1974).
1760. Solunov, H. and Vassilev, T., J. Polym. Sci., Polym. Phys. Ed., 12,
       1273 (1974).
1761. Weill, G., Rev. Gen. Caout. Plast., 50, (12), 1003 (1973).
1762. Gedemer, T.J., Plast. Eng., 31, (2), 28 (1975).
1763. Venediktov, M.V., Dushchenko, V.P., Kolupaev, B.S. and Tarasenko, Yu.G.,
       Mash. Prib. Ispyt. Mater., 99 (1971).
1764. Ross, G. and Addleman, R.L., J. Phys. D, 6, (13), 1537 (1973).
1765. Ettre, K. and Varadi, P.F., Anal. Chem., 35, 69 (1963).
1766. Bevington, J.C., Eaves, D.E. and Vale, R.L., J. Polym. Sci., 32, 317
       (1958).
1767. Bevington, J.C., Trans. Faraday Soc., 56, 1762 (1960).
1768. Baines, F.C. and Bevington, J.C., J. Polymer Sci., A-1, 6, 2433 (1968).
1769. Takeuchi, T. and Kakugo, M., Kogyo Kagaku Zasshi, 67, 308 (1964).
1770. Johnson, D.E., Lyerla, J.R., Horikawa, T.T. and Pederson, L.A., Anal.
       Chem., 49, 77 (1977).
1771. Klesper, E., Johnson, A., Gronski, W. and Wehrilr, F.W., Makromol. Chem.,
       176, (4), 1071 (1975).
1772. Johnson, A., Klesper, E. and Wirthlin, T., Makromol. Chem., 177, (8),
       2398 (1976).
1773. Johnson, D.E., Lyeria, J.R., Horikawa, T.T. and Pederson, L.A., Anal.
       Chem., 49, (1), 77 (1977).
1774. Grassie, N. and Fortune, J.D., Makromol. Chem., 168, 1 (1973).
1775. Grassie, N. and Torrance, B.D.J., J. Polym. Sci., A,1, 6, 3303 (1968).
1776. Grassie, N. and Torrance, B.D.J., J. Polym. Sci., A, 1, 6, 3315 (1968).
1777. Hirai, H., Koinuma, H., Tanabe, T. and Takeuchi, K., J. Polym. Sci.,
       A-1, 17, 1339 (1979).
1778. Hill, R., Lewis, J.R. and Simenson, J.L., Trans. Faraday Soc., 35,
       1073 (1939).
1780. Ebdon, J.R., J. Macromol. Sci., Chem., 8, (2), 417 (1974).
1781. Suzuki, T., Mitani, K., Takegami, Y., Furukawa, J., Kobayashi, E.
       and Arai, Y., Polym. J., 6, (6), 496 (1974).

1782.   Sokolowska, J. and Maciejowski, F., Polimery (Warsaw), 22, (5), 160
        (1977).
1783.   Elias, H.G. and Gruber, U., Die Makromolekulare Chemie, 78, 72 (1964).
1784.   Fujimoto, T., Kababata, N. and Furukawa, J., J. Polym. Sci., A-1, 6,
        1200 (1968).
1785.   Udipi, K., Harwood, H.J., Friebolin, H. and Cantow, H.J., Makromol.
        Chem., 164, 283 (1973).
1786.   Furukawa, J., Kobayashi, E., Arai, Y., Suzuki, T. and Takegami, Y.,
        J. Polym. Sci., Polym. Chem. Ed., 14, (10), 2553 (1976).
1787.   Mihra, R. and Pankova, M., Faserforsch. Textiltech., 26, (5), 217 (1975).
1788.   Fritzsche, P., Klug, P. and Groebe, V., Nuova, Chim., 49, (6), 39 (1973).
1789.   Kambe, H., Kambe, Y. and Honda, C., Polymer, 14, (9), 460 (1973).
1790.   Roussel, R. and Galin, J.C., J. Macromol. Sci., Chem., A,11, (2), 347
        (1977).
1791.   Alekseeva, K.V., Zh. Anal. Khim., 33, (2), 348 (1978); Chem. Abstr.,
        88, 170702q (1978).
1792.   Okada, T. and Otsuru, M., J. Polym. Sci., Polym. Lett. Ed., 14, (10),
        595 (1976).
1793.   Williams, E.A., Corgalli, J.D. and Hobbs, S.Y., Makromolecules, 10,
        782 (1977).
1794.   Ebdon, J.R., Polymer, 15, (12), 782 (1974).
1795.   Paul, S. and Ranby, B., Anal. Chem., 47, (8), 1428 (1975).
1796.   Amerik, Yu.B., Bairamov, Yu.Yu. and Krentsel, B.A., Zhur. Pribl. Spec-
        trosc., 18, (8), 1023 (1973).
1797.   Chiang, T.C., Pham, Q.T. and Guyot, A., J. Polym. Sci., A,1, 15, 2173
        (1977).
1798.   Revillon, A., Dumont, B. and Guyot, A., J. Polym. Sci., Polym. Chem.
        Ed., 14, 2263 (1976).
1799.   Zaitseva, N.K., Klimova, O.M. and Kostenko, V.V., Zh. Prikl. Khim.
        (Leningrad), 46, (5), 1160 (1973).
1800.   Heublein, G., Freitag, W. and Schuetz, H., Faserforsch. Textiltech.,
        26, (10), 498 (1975).
1801.   Hori, F., Ikada, Y. and Sakurada, I., J. Polym. Sci., Polym. Chem.
        Ed., 13, 775 (1975).
1802.   Heublein, G., Boerner, R. and Schuetz, H., Faserforsch. Textiltech.,
        29, (5), 317 (1978); Chem. Abstr., 88, 191695c (1978).
1803.   Frunze, T.M., Petrovskii, P.V., Ermakova, I.P., Evdokomov, A.M. and
        Sakharova, A.A., Vysokomol. Soedin, B, 15, (2), 110 (1973).
1805.   Wesslen, B., Gunneby, G., Heilstrom, G. and Svedling, P., J. Polym.
        Sci., Symp. No. 42, 457 (1973).
1806.   Bowden, M.J. and O'Donnell, J.H., J. Polym. Sci., A-1, 7, 1665 (1969).
1807.   Sakai, Y. and Iwasaki, M., J. Polym. Sci., A-1, 7, 1749 (1969).
1808.   Harris, J.A., Honojosa, O. and Arthur, J.C., J. Polym. Sci., A-1, 11,
        3215 (1973).
1809.   Esposito, G.G., Anal. Chem., 36, (11), 2183 (1964).
1810.   Guillet, J.E., Wooten, W.C. and Coombs, R.L., J. Appl. Polym. Sci.,
        3, (7), 61 (1960).
1811.   Ettre, K. and Varadi, P.F., Anal. Chem., 34, (7), 752 (1962).
1812.   Ettre, K. and Varadi, P.F., Anal. Chem., 35, (1), 69 (1963).
1813.   Feuerberg, H. and Weigel, H., Z. Anal. Chem., 199, 121 (1964).
1814.   Grassie, N., Scotney, A. and MacKinnon, L., J. Polym. Sci., A-1, 15,
        251 (1977).
1815.   Brinkman, U.A.T., Van Schaik, T.A.M., DeVries, G. and DeVisser, A.C.,
        ACS Symp. Ser. 31 (Hydrogels Med. Relat. Appl. Symp. 1975), 105-18
        (1976).
1816.   Pasteur, G.A., Anal. Chem., 49, (3), 363 (1977).
1817.   Clark, D.T. and Thomas, H.R., J. Polym. Sci., A-1, 14, 1701 (1976).
1818.   Wendorff, J.H., Finkelmann, H. and Ringdorf, H., ACS Symp. Ser. 74

(Mesomorphic Order Polym. Liq. Cryst. Media), 12-21 (1978); Chem. Abstr., 89, 75563g (1978).

1819. Kuryatnikov, E.I., Vizgert, R.V. and Berlin, A.A., Visn. L'viv. Politekh. Inst. No. 57, 36 (1971).

1820. Suzuki, T., Santee, E.R., Jr., Harwood, H.J., Vogl, O. and Tanaka, T., J. Polym. Sci., Polym. Lett. Ed., 12, 635 (1974).

1821. Matsuzaki, K., Kanai, T., Kawamura, T., Matsumoto, S. and Uryu, T., J. Polym. Sci., Polym. Chem. Ed., 11, 961 (1973).

1822. Kulkarni, S.Y. and Pansare, V.S., J. Appl. Chem. Biotechnol., 23, (6), 479 (1973).

1823. Radell, E.A. and Strutz, H.C., Anal. Chem., 31, 1890 (1959).

1824. Haken, J.K. and Houghton, E., J. Polym. Sci., A-1, 12, 1163 (1976).

1825. Grant, D.H. and McPhee, V.A., Anal. Chem., 48, (12), 1820 (1976).

1826. Belopol'skaya, T.V., Vestn. Leningr. Univ., Fiz. Khim., 22, (4), 44 (1976); Chem. Abstr., 86, 90477p (1977).

1827. Bovey, F.A., J. Polym. Sci., A, 1, 843 (1963).

1828. Chachaty, C. and Latimer, A., J. Polym. Sci., A,1-13, 189 (1975).

1829. Furukawa, J., Kobayashi, E. and Nagata, J., J. Polym. Sci., A-1, 14, 237 (1976).

1830. Koma, Y., Iimura, K., Kondo, S. and Takeda, M., J. Polym. Sci., A,1, 15, 1697 (1977).

1831. Kalinina, L.S. and Doroshina, L.I., Metody Ispyf. Kontr. Issled. Mashinostroit. Mater., 3,5 (1973); Chem. Abstr., 80, 54100 (1974).

1832. Coe, G.R., Soc. Plast. Eng. Tech. Pap., 23, 496 (1977).

1833. Coe, G.R., Aust. Plast. Rubber, 28,(8), 36 (1977); Chem. Abstr., 88, 74757r (1978).

1834. Kennett, A.C. and Stanton, D.W., Rep-N.Z.Dep. Sci.Ind.Res., Chem.Div. C.D. 2217 (1976); Chem. Abstr., 86, 17350n (1977).

1835. Guillet, J.E., Wooten, W.C. and Combs, R.L., J. Appl. Polym. Sci., 3, 61 (1960).

1836. Haslam, J., Hamilton, J.B. and Jeffs, A.R., Analyst, 83, 66 (1958).

1837. Miller, D.L., Samsel, E.P. and Cobler, J.G., Anal. Chem., 33, 677 (1961).

1838. Samsel, E.P. and McHard, J.A., Ind. Eng. Chem., Anal. Ed., 14, 750 (1942).

1839. Miller, D.L., Samsel, E.P. and Cobler, J.G., Anal. Chem., 33, (6),677 (1961).

1840. Nelson, D.F., Yee, J.L. and Kirk, P.L., Microchem. J., 6, 225 (1962).

1841. Wada, A., Mol. Phys., 3, 400 (1960).

1842. Nagasawa, M. and Holtzer, A., J. Am. Chem. Soc., 86, 538 (1964).

1843. Leyte, J.C., and Mandel, M., J. Polym. Sci., A,2, 1879 (1964).

1844. Combet, S., Compt. Rend., 254, 2961 (1962).

1845. Jacobson, A.L., J. Polymer Sci., 57, 321 (1962).

1846. Mathieson, A.R. and McLaren, J.V., J. Polym. Sci., A,3, 2555 (1965).

1847. Katehalsky, A, Shavit, N. and Eisenberg,H., J. Polym. Sci., 13, 69 (1954).

1848. Mathieson, A.R. and McLaren, J.V., J. Chem. Soc., 3581 (1960).

1849. Mandel, M. and Leyte, J.C., J. Polym. Sci., 56, 823 (1962).

1850. Brako, F.D. and Wexler, A.S., Anal. Chem., 35, 1944 (1963).

1851. Clay, M.R. and Charlesby, A., Eur. Polym. J., 11, (2), 187 (1975).

1852. Forobioni, A. and Chachaty, C., J. Polym. Sci., A,1, 10, 1923 (1972).

1853. Nagata, S. and Moritani, T., J. Polym. Sci., A-1, 12, 1799 (1974).

1854. Shioji, Y., Ohnishi, S. and Nitta, I., J. Polym. Sci., A,1, 3373 (1963).

1855. Guillet, J.E., Wooten, W.C. and Combs, R.L., J. Appl. Polym. Sci., 3, 61 (1960).

1856. Lehrle, R.S. and Robb, J.C., Nature, 183, 1671 (1959).

1857. Jones, C.E.R. and MOyles, A.F., Nature, 189, 222 (1961).

1858. Strassburger, J., Brauer, G.M., Tryon, M. and Forziati, A.F., Anal. Chem., 32, 454 (1961).

1859. Gedemer, T.J., J. Macromol. Sci., Chem., 8,(1), 95 (1974).
1860. Pogorel'skii, K.V., Asanov, A. and Akhmedov, K.S., Dokl. Akad. Nauk
       Uzh. SSR, 27,(3), 28 (1970).
1861. Hobden, F.W., J. Oil and Colour Chemists Assoc., 41, (1), 24 (1958).
1862. Brunn, J., Doerffel, K., Much, H. and Zimmermann, G., Plaste Kautsch.,
       22,(6), 485 (1975).
1863. Brown, L., Analyst, 104, 1165 (1979).
1864. Mel'nikova, S.L., Tishchenko, V.T. and Sazonenko, V.V., Lakokras.
       Mater. Ikh. Primen 4, 56 (1977); Chem. Abstr., 87, 118118u (1977).
1865. Cobler, J.G. and Samsel, E.P., SPE Transactions (April 1962).
1866. Schwoetzer, G., Z. Anal. Chem., 10, 260 (1972).
1867. Shiryaev, B.V. and Kozhukhova, E.B., Zavod Lab., 38, 1303 (1972).
1868. Clark, D.T. and Thomas, H.R., J. Polym. Sci., A-1, 14, 1671 (1976).
1869. Klenin, V.I., Shchegolev, S.Yu., Severinov, A.V. and Bulkin, Yu.I.,
       Lakokrasoch. Mater. Ikh Primen., 4, 56 (1972).
1870. Miller, D.F., Samsel, E.P. and Cobler, J.G., Anal. Chem., 33, 677
       (1961).
1871. Budlyina, V.V., Marinin, V.G., Vodzinskii, Yu.V., Kalinina, A.I. and
       Korschunov, I.A., Zavod Lab., 36, 1051 (1970).
1872. Stanley, C.W. and Peterson, O., Midwest Research Institute, Kansas
       City, Missouri and United Carbon Co., Texas, private communication.
1873. Streichen, R.J., Anal. Chem., 48, 1398 (1976).
1874. Dever, G.R., Karasz, F.E., Macknight, W.J. and Lenz, R.W., J. Polym.
       Sci., Polym. Chem. Ed., 13, (8), 1803 (1975).
1875. Merrill, L.J., Sauer, J.A. and Woodward, A.E., J. Polym. Sci., A,3,
       4243 (1965).
1876. Rosenthal, R.W., Schwartzman, L.H., Greco, N.P. and Proper, P., J.
       Org. Chem., 28, 2835 (1963).
1877. Tal'roze, R.V., Shibaev, V.P. and Plate, N.A., J. Polym. Sci., Symp.
       No. 44, 35 (1974).
1878. Newman, B.A., Frosini, V. and Magagnini, P.L., J. Polym. Sci., Polym.
       Phys. Ed., 13,(1), 87 (1975).
1879. Czaja, K., Nowakowska, M. and Zubek, J., Polimery (Warsaw), 21,(4),
       158 (1976); Chem. Abstr., 85, 124396h.
1880. Keller, F. and Roth,H., Plaste Kautsch., 22, (12), 956 (1975).
1881. Roth, H., Roth, H.K. and Keller, F., Plaste Kautsch., 22,(3),256(1975).
1882. Keller, F., Plaste Kautsch., 22, 8 (1975).
1883. Keller, F., Findeisen, G., Raetzsch, M. and Roth, H., Plaste Kautsch.,
       22,(9), 722 (1975).
1884. Furukawa, J., Kobayashi, E., Arai, Y., Suzuki, T. and Takegami, Y.,
       J. Polym. Sci., A,1, 14, 2553 (1976).
1885. McGrath, J.E. and Robeson, L.M., Polym. Prepr., Am. Chem. Soc., Div.
       Polym. Chem., 17,(2), 706 (1976).
1886. Barrall, E.M., Porter, R.S. and Johnson, J.F., Anal. Chem., 35,(1),
       73 (1963).
1887. Bombaugh, K.J., Cook, C.E. and Clampitt, B.H., Anal. Chem., 35,(12),
       1834 (1963).
1888. Sugimura, Y., Tsuge, S. and Takeuchi, T., Anal. Chem., 50, 1173 (1978).
1889. Mcgaugh, M.C. and Kottle, S., J. Polymer Sci., A,1, 6, 1243 (1968).
1890. Macknight, W.J., Taggart, W.P. and Stein, R.S., J. Polym. Sci., Symp.
       No. 45, 113 (1974).
1891. Hull, D. and Gilmore, J., Division of Fuel Chemistry, 141st Meeting
       ACS, Washington, D.C. (March 1962).
1892. Mukhopadhyay, S., Mitra, B.C. and Palit, S.R., J. Polym. Sci., A-1,
       7, 2442 (1969).
1893. Palit, S.R. and Ghosh, P., J. Polym. Sci., 58, 1225 (1962).
1894. Saha, M.K., Ghosh, P. and Palit, S.R., J. Polym. Sci., A,2, 1365 (1964).
1895. Aydin, O. and Schulz, R.C., Makromol. Chem., 177,(12), 3537 (1976).

1896.   Garratt, T.S. and Pauk, G.S., J. Polym. Sci., A,4, 2714 (1966).
1897.   Sharp, J.L. and Paterson, G., Analyst, 105, 517 (1980).
1898.   Tan, J.S. and Gasper, S.P., Macromolecules, 6,(5), 741 (1973).
1899.   Akhmedov, K.S., Aref'eva, M.M. and Apazidl, A.I., Dokl. Akad. Nauk
        Uzb. SSR, 29, (6), 33 (1972).
1900.   Muresan, I. and Zador, L., Stud. Univ. Babes-Bolval. Ser. Chem., 18,
        (1), 89 (1973).
1901.   Crisp, S., Lewis, B.G. and Wilson, A.D., J. Dent. Res., 54,(6), 1238
        (1975).
1902.   Keller, F., Fuhrmann, C., Roth, H., Findetsen, G. and Raezsch, M.,
        Plaste Kautsch., 24, (9), 626 (1977); Chem. Abstr., 88, 38232g (1978).
1903.   Kvasnikov, Yu.P., Filippychev, G.F. and Maksimov, E.G., Sb. Nauch. Tr.,
        Yaroslav. Tekhnol. Inst., 22,(2), 136 (1972); Chem. Abstr., 81, 106169
        (1974).
1904.   Smith, O., Chem. Eng. News, 38,(28), 37 (1960).
1905.   Mrkvickova-Vaculova, L. and Kratochvil, P., Collect., Czech. Chem.
        Commun., 37,(6), 2015 (1972).
1906.   Mrkvickova-Vaculova, L. and Kratochvil, P., Collect. Czech. Chem.
        Commun., 37,(6) 2029 (1972).
1907.   Greene, B.W., J. Colloid Interface Sci., 43,(2), 449 (1973).
1908.   Greene, B.W., J. Colloid Interface Sci., 43,(2), 462 (1973).
1909.   Haslam, J., Hamilton, J.B. and Jeffs, A.R., Analyst, 83, 66 (1958).
1910.   Haslam, J. and Jeffs, A.R., J.Appl. Chem., 7, 24 (1957).
1911.   Radell, E.A. and Strutz, H.C., Anal. Chem., 31, 1890 (1959).
1912.   Strassburger, J., Brauer, G.M., Tryon, M. and Forziati, A.F., Anal.
        Chem., 32, 454 (1960).
1913.   Cook, W.S., Jones, C,O. and Altenan, A.G., Canadian Spectroscopy, 13,
        64 (1968).
1914.   Bergmann, J.S., Ehrhart, C.H., Grantelli, L. and Janik, L.J., 153rd,
        National ACS Meeting, Miami Beach, Florida, (April 1967).
1915.   Rowe, W.A. and Yates, K.P., Anal. Chem., 35, 368 (1963).
1916.   Shott, J.E.,Jr., Garland, T.J. and Clark, R.O., Anal. Chem., 33, 506
        (1961).
1917.   Gorsuch, T.T., Analyst, 84, 135 (1959).
1918.   Gamble, L.W. and Jones, W.H., Anal. Chem., 27, 1456 (1955).
1919.   Purdon, J.R. and Mate, R.D., J. Polym. Sci., A-1, 8, 1306 (1970).
1920.   Purdon, J.R., Jr. and Mate, R.D., in First Biannual American Chemical
        Society Polymer Symposium (J. Polym. Sci., C, 1), H.W. Starkweather,
        Jr., Ed., Interscience, New York, p. 451 (1963).
1921.   Kranz, D., Dinges, K. and Wendling, P., Angew Makromol. Chem., 51,(1),
        25 (1976).
1922.   Hill, R., Lewis, J.R. and Simonsen, J.L., Trans. Faraday Soc., 35,
        1067 (1937).
1923.   Aleekseeva, E.N., J. Gen. Chem., (U.S.S.R.), 11, 353 (1941).
1924.   Rabjohn, N., Bryan, C.E., Inskip, G.E., Johnston, H.W. and Lawson, J.K.,
        Jr., J. Amer. Chem. Soc., 69, 314 (1947).
1925.   Yakubehik, A.I., Spasskova, A.I., Zak, A.G. and Shotatskaya, I.D.,
        Zh. Obshsh. Khim., 28, 3090 (1958).
1926.   Lorenz, O. and Parks, C.R., J. Org. Chem., 39, 1131 (1967).
1927.   Bailey, P.J., Chem. Rev., 58, 925 (1958).
1928.   Beroza, M. and Bierl, B.A., Anal. Chem., 39, 1131 (1967).
1929.   Hackathorn, M.J. and Brock, M.J., J. Polym. Sci., Polym. Chem. Ed.,
        13, 945 (1975).
1930.   Hackathorn, M.J. and Brock, M.K., Rubber Chem. Technol., 45, 1295
        (1972).
1931.   Furukawa, J., Haga, K., Kobayashi, E., Iseda, Y., Yoshimoto, T. and
        Sakamoto, K., Polym. J., 2, 371 (1971).
1934.   Binder, J.L., J. Polym. Sci., Part A,1, 47 (1963).

1935.   Natta, G., 15th Annual Lechnical Conference of the Society of Plastic
        Engineers, New York (Jan. 1959).
1936.   Bellamy, L.J., Infrared Spectra of Complex Molecules, 2nd Ed., Wiley,
        New York, (1958).
1937.   Jones, L.N. and Sanderfly, C., Techniques of Organic Chemistry, Vol.
        19, Interscience, New York-London, ch. IV (1956).
1939.   Hampton, R.R., Anal. Chem., 21, 923 (1949).
1940.   Binder, J.L., Anal. Chem., 26, 1877 (1954).
1941.   Natta, G., paper presented at 15th Annual Technical Conference of
        Society of Plastics Engineers, Stereospecific Polymers Session, New
        York, Jan 27-30 (1959).
1942.   Silas, R.S., Yates, J. and Thornton, V., Anal. Chem., 31, 529 (1959).
1943.   Morero, D., Santambrogio, A., Porri, L. and Ciampelli, F., Chim.Ind.
        (Milan), 41, 758 (1959).
1944.   Binder, J.L., J. Polym. Sci., A,1, 47 (1963).
1945.   Binder, J.L., J. Polym. Sci., A,3, 1587 (1965).
1946.   Brako, F.D. and Wexler, A.S., Anal. Chem., 35, 1944 (1963).
1947.   Fraga, D.W., Emeryville Research Laboratories, Shell Chemical Co. Ltd,
        Emeryville, U.S.A., private communication.
1948.   Clark, J.K. and Chen, H.Y., J. Polym. Sci., 12, 925 (1974).
1949.   Harwood, H.J. and Ritchey, W.M., J. Polym. Sci., B,2, 601 (1964).
1950.   Silas, R.D., Yates, J. and Thornton, V., Anal. Chem., 31, 529 (1959).
1951.   Cornell, S.W. and Koenig, J.L., Macromolecules, 2, 540 (1969).
1952.   Neto, N. and diLauro, C., Euro. Polym. J., 3, 645 (1967).
1953.   Binder, J.L., Anal. Chem., 26, 1877 (1954).
1954.   Hummel, K., Steizer, F., Kathan, W., Demel, H., Wedam, O.A. and Kar-
        aoulis, C., Recl. Trav. Chim. Pays-Bas, 96,(11), 75 (1977); Chem.
        Abstr., 88, 6269 1b.
1955.   Ast, W., Zott, C., Bosch, H. and Kerber, R., Recl. Trav. Chim. Pays-
        Bas., 96,(11), 81 (1977); Chem. Abstr., 88, 62692c (1978).
1956.   Braun, D. and Canji, E., Angew, Makromol. Chem., 33, 143 (1973).
1957.   Braun, D. and Canji, E., Angew, Makromol. Chem., 35, 27 (1974).
1958.   Hast, M. and Deur-Siftar, D., Chromatographia, 5,(9), 502 (1972).
1959.   Canji, E. and Perner, H., Recl. Trav. Chim. Pays-Bas., 96,(11), 70
        (1977); Chem. Abstr., 88, 51286x (1978).
1960.   Allars, D.L., Pap. Meet.-Am. Chem. Soc., Div. Org. Coat. Plast. Chem.,
        36,(1), 399 (1976).
1961.   Quack, G. and Fetters, L., J. Macromolecules, 11,(2), 369 (1978).
1962.   Carlson, D.W. and Altenan, A.G., Anal. Chem., 41, 969 (1969).
1963.   Dinsmore, H.L. and Smith, D.C., Rubber Chem. Technol., 22, 572 (1949).
1964.   MacKillop, D.A., Anal. Chem., 40, 607 (1968).
1965.   Harms, D.L., Anal. Chem., 25, 1140 (1953).
1966.   Hummel, D., Rubber Chem. Technol., 32, 854 (1959).
1967.   Lerner, M. and Gilbert, R.C., Anal. Chem., 36, 1382 (1964).
1968.   Tryon, M., Horowitz, E. and Mandel, J.J., J. Res. Nat. Bur. Std., 55,
        219 (1955).
1969.   Mochel, V.D., Rubber Chem. Technol., 40, 1200 (1967).
1970.   Carlson, D.W., Ransaw, H.C. and Altenan, A.G., Anal. Chem., 42, 1278
        (1970).
1971.   Binder, J.L., Anal. Chem., 26, 1877 (1954).
1972.   Binder, J.L., J. Polym. Sci., Part A,1, 47 (1963).
1973.   Binder, J.L., J. Polym. Sci., Part A,3, 1587 (1965).
1974.   Elgert, K.F., Stützel, B., Frenzel,P., Cantow, H.J. and Streck, R.,
        Makromol. Chem., 170, 257 (1973).
1975.   Conti, F., Segre, A., Pini, P. and Porri, L., Polymer, 15, 5 (1974).
1976.   Tanaka, Y., Sato, H., Ogawa, M., Hatada, K. and Terawaki, Y., J. Polym.
        Sci., Polym. Lett.Ed., 12, 369 (1974).
1977.   Elgert, K.F., Quack, G. and Stützel, B., Makromol. Chem., 175, 1955

(1974).

1978. Grant, D.M. and Paul, E.G., J. Amer. Chem. Soc., 86, 2984 (1964).

1980. Hatada, K., Terawaki, Y., Okuda, H., Tanaka, Y. and Sato, H., J. Polym. Sci., Polym. Lett.Ed., 12,(6), 305 (1974).

1981. Eng, S.B., Diss. Abstr. Int. B., 35,(7), 3263 (1975); Chem. Abstr., 82, 112393e.

1982. Eng, S.B. and Woodward, A.E., J.Macromol. Sci., Phys., B,10,(4), 627 (1974).

1983. Chen, H.Y., J. Polym. Sci., Polym. Lett.Ed., 12, 85 (1974).

1984. Elgert, K.F., Quack, G. and Stützel, B., Makromol. Chem., 176,(3), 759 (1975).

1985. Elgert, K.F., Quack, G. and Stützel, B., Polymer, 16,(3), 154 (1975).

1986. Toy, M.S. and Stringham, R.S., Polym. Prepr.Am. Chem. Soc., Div. Polym. Chem., 17,(2), 672 (1976).

1987. Ritter, W., Elgert, K.F. and Cantow, H.J., Makromol. Chem., 178,(2), 557 (1977).

1988. Elgert, K.F., Quack, G. and Stützel, B., Makromol. Chem., 175,(6), 1955 (1974).

1989. Shibata, C., Yamazaki, M. and Takeuchi, T., Bull. Chem. Soc. Jpn., 50, (1), 311 (1977); Chem. Abstr., 86, 90540d (1977).

1990. Shahab, Y.A. and Basheer, R.A., J. Polym. Sci., A,1, 16, 2667 (1978).

1991. Shahab, Y.A. and Basheer, R.A., J. Polym. Sci., A-1, 17, 919 (1979).

1992. Suman, P.T. and Werstler, D.D., J. Polym. Sci., A-1, 13, 1963 (1975).

1993. Mochel, V.D., J. Polym. Sci., A-1, 10, 1009 (1972).

1994. Gemmer, R.V. and Golub, M.A., J. Polym. Sci., A-1, 16, 2985 (1978).

1995. Tompa, A.S., Barefoot, R.D. and Price, E., J. Polym. Sci., A-1, 6, 2785 (1968).

1996. Brame, E.G. and Khan, A.A., Rubber Chem. Technol., 50,(2), 272 (1977).

1997. Tanaka, Y., Sato, H. and Gonzalez, I.G., J. Polym. Sci., A-1, 17, 3027 (1979).

1998. Hiraki, K., Inoue, T. and Hirai, H., J. Polym. Sci., A-1, 8, 2545 (1970).

1999. Cobler, J.G. and Samsel, E.P., SPE TRansactions, 145 (April 1962).

2000. Tamura, S. and Gillham, J.K., J. Appl. Polym. Sci., 22,(7), 1867 (1978).

2001. Von Raven, A. and Heusinger, H., Angew. Makromol. Chem., 42,(1), 183 (1975).

2002. Donkai, N., Murayama, N., Miyamoto,T. and Inagaki, H., Makromol. Chem., 175, 187 (1974).

2003. Harman, D.J., J. Polym. Sci., C,8, 243 (1965).

2004. Kössler, J.P. and Vodchnal, J., J. Polym. Sci., A,3, 2511 (1965).

2005. Baker, C.A. and Williams, R.J.P., J. Chem. Soc., 2852 (1956).

2006. Francis, P.S., Cooke, R.C. and Elliott, J.H., J. Polym. Sci., 31, 453 (1958).

2007. Henry, P.M., J. Polym. Sci., 36, 3 (1959).

2008. Davis, T.E. and Tobias, R.L., J. Polym. Sci., 50, 227 (1964).

2009. Schneider, N.S., Loconti, J.D. and Holmes, L.G., J.Appl. Polymer Sci., 3, 251 (1960); ibid, 5, 354 (1961).

2010. Weakley, T.J., Williams, R.J.P. and Wilson, J.D., J. Chem. Soc., 3963 (1960).

2011. Jungnickel, J.L. and Weiss, F.T., J. Polym. Sci., 49, 437 (1961).

2012. Hall, R.W., in Techniques of Polymer Characterization, P.W.Allen, Ed, Butterworths, London (1959).

2013. Cooper, W., Vaughan, G., Eaves, D.E. and Madden, R.W., J. Polym. Sci., 50, 159 (1961).

2014. Cooper, W., Vaughan, G. and Yardley, J., J. Polym. Sci., 59, 52 (1962).

2015. Hulme, B. and McLeod, M. and L.A., Polymer, 3, 153 (1962).

2016. Chang, M.S., French, D.M. and Rogers, P.L., J. Macromol. Sci., Chem., 7, (8), 1727 (1973).

2017. Law, R.D., J. Polym. Sci., Polym. Chem. Ed., 11, 175 (1973).

2018.   Kuzaev, A.I., Susiova, E.N. and Entelis, S.G., Dokl. Akad. Nauk SSSR,
        208,(1), 142 (1973).
2019.   Maley, I.E., J. Polym. Sci., C,8, 253 (1965).
2020.   Gaeta, L.J., Schleuter, E.W. and Altenan, A.G., Rubber Age, 47, 101
        (1969).
2021.   Beevers, R.B., Lab. Pract., 22,(4), 272 (1973).
2023.   Park, W.S. and Graessley, W.W., J. Polym. Sci., Polym. Phys. Ed., 15,
        (1), 85 (1977).
2024.   Law, R.D., J. Polym. Sci., A-1, 9, 589 (1971).
2025.   Muenker, A.H. and Hudson, B.E., "Functionality Determination of Binder
        Prepolymers", AFRPL-TR-68-327, Oct.1, 1966-Sept.30, 1968, ESSO Research
        and Engineering Report GR-8-FbP-68.
2026.   Muenker, A.H., "Determination of Prepolymer Functionality and its
        Relationship to Binder Properties", AFRPL-TR-69-214, Feb.3, 1969-
        Aug.31, 1969.
2027.   Adiscoff, A. and Martin, E.C., J. Polym. Sci., A-1, 10, 1877 (1972).
2028.   Law, R.D., J. Polym. Sci., A-1, 7, 2097 (1969).
2029.   Law, R.D., J. Polym. Sci., A-1, 9, 589 (1971).
2030.   Law, R.D., J. Polym. Sci., A-1, 7, 2097 (1969).
2031.   Law, R.D., J. Polym. Sci., A-1, 11, 175 (1973).
2032.   Natta, G., Farina, M., Donati, M. and Peraldo, M., Chim. Ind.(Milan),
        42, 1363 (1960).
2033.   Natta, G., Farina, M. and Donati, M., Makromol. Chem., 43, 251 (1961).
2034.   Natta, G., Corradini, P. and Ganis, P., J. Polym. Sci., A,3, 11 (1965).
2035.   Hackathorn, M.J. and Brock, M.J., J. Polym. Sci., Polym. Chem. Ed., 13,
        (4), 945 (1975).
2036.   Valuev, V.I., Shiyakhter, R.A., Dmitrieva, T.S. and Tsvetovskii, I.B.,
        Zh. Anal. Khim., 30,(6), 1235 (1975); Chem. Abstr., 83, 132156d.
2037.   Shipmen, J.J. and Golub, M.A., J. Polym. Sci., A,1, 832 (1963).
2038.   Carmen, C., J. Macromolecules, 7, 789 (1974).
2039.   Pilar, J., Toman, L. and Marek, M., J. Polym. Sci., A,1, 14, 2399 (1976).
2040.   Hackathorne, M.J. and Brock, M.J., J. Polym. Sci., Polym. Chem. Ed.,
        13, 945 (1975).
2041.   Kawasaki, A., paper presented at 27th Autumn Meeting, Japan Chemical
        Society, 1972, Preprints, 2, 20 (1972).
2042.   Iida, T., Nakanishi, M. and Goto, K., J. Polym. Sci., Polym. Chem. Ed.,
        13, 1381 (1975).
2043.   Narasaki, H. and Umezawa, K., Kobunshi Kagaku, 29,(6), 438 (1972).
2044.   Berger, W. and Ewert, K., Faserforsch. Textiltech., 24(4), 169 (1973).
2045.   Pohorelsky, L. and Heran, J., Chem. Prum., 27,(12), 630 (1977).
2046.   Oprea, N. and Pogorevici, A., Rev. Chim. (Bucharest), 25,(3), 244
        (1974); Chem. Abstr., 82, 58641u.
2047.   Majewska, J., Polimery, 18,(3), 142 (1973).
2048.   Nissen, D., Rossbach, V. and Zahn, H., J.Appl. Polym. Sci., 18, 1953
        (1974).
2050.   Alter, U. and Bonart, R., Colloid Polym. Sci., 254,(3), 348 (1976).
2051.   D'Esposito, L. and Koenig, J.L., J. Polym. Sci., Polym. Phys. Ed., 14,
        (10), 1731 (1976).
2052.   Stas'kov, N.I., Golovachev, V. and Gusev, S.S., Vestsi Akad. Navuk SSSR,
        Ser Fiz-Mat. Navuk 5, 76 (1977); Chem. Abstr., 88, 7769h (1978).
2053.   Stas'kov, N.I., Golovachev, V.I. and Gusev, S.S., Vysokomol Soedin,
        Ser.A, 19,(10), 2291 (1977); Chem. Abstr., 87, 202266y (1977).
2054.   Gusev, S.S. and Golovachev, V.I., Zh. Prikl. Spektrosk., 20,(2), 314
        (1974).
2055.   Afanas'eva, N.I., Vitsnudel, M.B. and Zhizhin, G.N., Dokl. Akad. Nauk
        SSSR, 213,(3), 611 (1973).
2056.   Mocheria, K.K. and Bell, J.P., J. Polym. Sci., Polym. Phys. Ed., 11,
        1779 (1973).

2057. Mehta, R.E. and Bell, J.P., J. Polym. Sci., Polym. Ohys. Ed., 11, 1793
       (1973).
2058. Purvis, J. and Bower, D.I., J. Polym. Sci., Polym. Phys. Ed., 14,(8),
       1461 (1976).
2059. Boerio, F.J., Bahl, S.K. and McGraw, G.E., J. Polym. Sci., Polym. Phys.
       Ed., 14,(6), 1029 (1976).
2060. Bahl, S.K., Cornell, D.D. and Boerio, F.J., J. Polym. Sci., Polym. Lett.
       Ed., 12, 13 (1974).
2061. Boerio, F.J. and Bailey, R.A., J. Polym. Sci., Polym. Lett.Ed., 12,
       433 (1974).
2062. Derouault, J., Gall, J., Hendra, P.J., Ellis, V., Cudby, M.E.A. and
       Willis, H.A., Quad. Ric. Sci., 84, 85 (1973).
2063. Derouault, J., Hendra, P.J., Cudby, M.E.A., Fraser, G., Walker, J. and
       Willis, H.A.W., Advan. Raman Spectrosc., 1, 277 (1972).
2064. Derouault, J., Hendra, P.J., Cudby, M.E.A. and Willis, H.A., J. Chem.
       Soc., Chem. Commun., 21, 1187 (1972).
2065. Purvis, J., Bower, D.I. and Ward, I.M., Polymer, 14, 398 (1973).
2066. Tirpak, G.A. and Sibilia, J.P., J.Appl. Polym. Sci., 17, 643 (1973).
2067. Bryce-Smith, D., Chem. and Ind., 244 (1953).
2068. Michael, F. and Schweppe, H., Angew, Chem., 66, 137 (1954).
2069. Franc, J., Collection Czech. Chem. Commun., 23, 655 (1958).
2070. Golosova, L.V., Zhur. Anal. Khim., 14, 748 (1959).
2071. Komers, R. and Bazant, V., Dok. Akad. Nauk SSSR, 126, 1268 (1959).
2072. Grizenthwaite, R.J., Brit. Plastics, 32, 428, 439 (1959).
2073. Benisek, L., Textile Res.J., 32, 539 (1962).
2074. Nishioka, A., Kato, Y. and Mitsuoka, H., J. Polymer Sci., 62, S10 (1962).
2075. Murano, M., Kaneishi, Y. and Yamadera, R., J. Polym. Sci., A,3, 2693
       (1965).
2076. Zachmann, H.G., Kolloid-Z.Z.Polym., 251,(11), 951 (1973).
2077. Sauer, W., Kuzay, P., Kimmer, W. and Jahn, H., Plaste Kautsch., 23,(5),
       331 (1976).
2078. Cambell, D., Araki, K. and Turner, D.T., J. Polym. Sci., A-1, 4, 2597
       (1966).
2079. Cambell, D. and Turner, D.T., J. Polym. Sci., A-1, 5, 2199 (1967).
2080. Janssen, R., Ruysschaert, H. and Vroom, R., Die Makromolekulare Chemie,
       21, 153 (1963).
2081. Schulkin, R.M., Boy, R.E. and Cox, R.H., J. Polym. Sci., 6, 17 (1964).
2083. Miller, G.W., Thermochimica Acta, 8,(1-2), 129 (1974).
2084. Troitskii, B.B., Varyukhin, V.A. and Khokhlova, L.V., Tr. Khim. Khim.
       Tekhnol., 2, 115 (1974).
2085. Gilland, J.C. and Lewis, J.S., Angew. Makromol. Chem., 54, (1), 49
       (1976).
2086. Peebles, L.H., Huffman, M.W. and Ablett, C.T., J. Polym. Sci., A-1, 7,
       479 (1969).
2087. Repina, L.P. and Khalatur, P.G., Zavod. Lab., 41,(3), 287 (1975); Chem.
       Abstr., 83, 60034n.
2088. Zaborsky, L.M., Anal. Chem., 49,(8), 1166 (1977).
2089. Shiono, S., Anal. Chem., 51, 2398 (1979).
2090. Lindner, W.L., Polymer, 14, 9 (1973).
2091. Desai, A.B. and Wiles, G.L., J. Polym. Sci., Polym. Lett. Ed., 12, 113
       (1974).
2092. Wlochowicz, A. and Przygocki, W., J. Appl. Polym. Sci., 17, 1197 (1973).
2093. Groeninckx, G., Reynaers, H. and Berghmans, H., Polymer, 15, 61 (1974).
2094. Kashmagä, M., Cunningham, A., Manuel, A.J. and Ward, I.M., Polymer,
       14, 111 (1973).
2095. Northolt, M.G. and Stuut, H.A., J. Polym. Sci., Polym. Phys. Ed., 16,
       (5), 939 (1978).
2096. Paschke, E.E., Bidingmeyer, B.A. and Bergmann, J.G., J. Polym. Sci.,

Polym. Chem. Ed., 15,(4) 983 (1977).

2097.   Minarik, M., Sir, Z. and Coupek, J., Angew. Makromol. Chem., 64,(1),
         147 (1977).

2098.   Van der Maeden, F.P.B., Biemond, M.E.F. and Janssen, P.C.G.M., J.
         Chromatogr., 149, 539 (1978).

2099.   Benicka, E. and Ciganekova, V., Sb. Prednasek. "MAKROTEST1973", 2, 69
         (1973) (Pub. 1973).

2100.   Vasile, C., Onu, A., Popa, O. and Matel, T., Mater. Plast. (Bucharest),
         10, (12), 631 (1973).

2101.   Cooper, D.R. and Semlyen, J.A., Polymer, 14, 185 (1973).

2102.   Shioni, S., J. Polym. Sci., A-1, 17, 4120 (1979).

2103.   Sutherland, J.E., Research Laboratories, Polymer Prepr., 17,(2) (1976).

2104.   Simonova, M.I. and Alzenshtein, E.M., Zavod. Lab., 40,(4), 435 (1974).

2105.   Steinke, J. and Vogel, I., Fresenius' Z. Anal. Chem., 285,(3) 268 (1977).

2106.   Walent, Z., Majewska, J., Grams, W. and Krzystek, H., Hem. Vlakna,
         12,(3-4), 5 (1972); Chem. Abstr., 80, 133942 (1974).

2107.   Dinse, H.D. and Tucek, E., Faserforsch. Textiltech., 21, 205 (1970).

2108.   Schmalz, E.O., Fas.forsch. Tex.tech., 21, 209 (1970).

2109.   Schmalz, E.O., Anal. Abstr., 17, 2249 (1969).

2110.   McGraw, G.E., Polym. Sci., Technol., 1, 97 (1973).

2111.   Burow, S.D., Turner, D.T., Pezdirtz, G.F. and Sands, G.D., J. Polym.
         Sci., A,4, 613 (1966).

2112.   Mueller, E.H., Colloid Polym. Sci., 252,(9), 696 (1974).

2113.   Ruscher, C., Seganow, I. and Teichgraeber, M., Faserforsch. Textiltech.,
         25,(12), 544 (1974).

2114.   Razumova, L.L., Rudalova, T.E., Moiseev, Yu.V., Mel'nikov, L.A. and
         Zaikov, G.E., Vysokomol. Soedin., Ser.A, 17,(4), 861 (1975).

2115.   Berghmans, H., Groeninckx, G. and Hautecler, S., J. Polym. Sci., Polym.
         Phys. Ed., 13,(1), 151 (1975).

2116.   Urbanski, J., Chem. Anal. (Warsaw), 22,(4), 749 (1977); Chem. Abstr.,
         88, 23666c (1978).

2117.   Radovici, A., Radovici, R. and Ciornei, T., Mater. Plast. (Bucharest),
         11, (7), 360 (1974); Chem. Abstr., 83, 11093x.

2118.   Shvyrkova, L.A., Rogatinskaya, S.L. and Burakova, T.P., Plast. Massy,
         9, 68 (1976); Chem. Abstr., 86, 17005d (1977).

2119.   Kalinina, L.S. and Doroshina, L.I., Metody Ispyf. Kontr. Issled. Mash-
         inostroit. Mater., 3, 5 (1973); Chem. Abstr., 80, 54100 (1974).

2120.   Moiseeva, L.D., Saifullin, A.S., Emelin, E.A. and Shvetsova, L.N.,
         Zavod. Lab., 39,(2), 154 (1973).

2121.   King, D., Stanonis, D.J. and Reid, J.D., Text. Chem. Color, 7,(12),
         218 (1975).

2122.   Vigneron, J.M. and Deschreider, A.R., Lebensm-Wiss. Technol., 5,(6),
         198 (1972).

2123.   Smolyanskii, A.L. and Gusakova, G.V., Tr. Vologod. Moloch. Inst., No.
         60, 105 (1970).

2124.   Holland-Moritz, K., Kolloid-Z.Z.Polym., 251,(11), 906 (1973).

2125.   Holland-Moritz, K, and Hummel, D.O., J.Mol. Struct., 19,(1), 289 (1973).

2126.   Holland-Moritz, K. and Hummel, D.O., Quad.Ric.Sci., 84, 158 (1973)
         (Pub. 1971).

2127.   Kadyrmatova, F.N. and Maklakov, L.I., Zh. Prikl. Spektrosk., 18,(5),
         928 (1973).

2128.   Funke, W. and Schuh,H., J. Polym. Sci., Symp., No. 42, 379 (1973).

2129.   Iyer, P.B., Iyer, K.R.K. and Patil, N.B., J.Appl. Polym. Sci., 20,(3),
         591 (1976).

2130.   Vance, J.A. and Brakke, N.B. and Quinney, P.R., Anal.Chem., 51, 499
         (1979).

2131.   Yamadira, R. and Murano, M., J. Polym. Sci., A-1, 5, 2259 (1967).

2132.   Murano, M., Polym. Sci., A-1, 9, 567 (1971).

2133. Urman, Ya. G., Khramova, T.S., Avdeeva, G.M., Sedov, L.N. and Sionim, .
      I.Ya., Vysokomol. Soedin., Ser.A, 14,(12), 2597 (1972).
2134. Sionim, I.Ya., Urman, Ya.G. and Klyuchnikov, V.N., Plast. Massy, 4,
      68 (1973).
2135. Barshtein, R.S., Urman, Ya. G., Gorbunova, V.G., Khramova, T.S., Bulai,
      A.Kh. and Sionim, I.Ya., Dokl. Akad.Nauk SSSR, 206,(5), 1140 (1972).
2136. Wiechec, L., Chem. Anal. (Warsaw), 18,(4), 853 (1973); Chem. Abstr.
      80, 134136 (1974).
2137. Urman, Y.G., Alekseeva, S.G. and Sionim, I.Y., Vysokomol. Soedin.,
      Ser.A, 19,(2), 299 (1977); Chem. Abstr., 86, 140576t (1977).
2138. Yeager, F.W. and Becker, J.W., Anal. Chem., 49, 722 (1977).
2139. de Angelis,G., Ippalita, P. and Spina, N., La Richerha Sciefica, 28,
      144 (1958).
2140. Nelson, D.F., Yee, J.L. and Kirk, P.L., Microchem.J., 6, 225 (1962).
2141. Esposito, C.G., Anal. Chem., 34,(9), 1173 (1962).
2142. Esposito, C.G. and Swann, M.H., Anal. Chem., 34,(9), 1048 (1962).
2143. Percival, D.F., Anal. Chem., 35,(2), 236 (1963).
2144. Allen, B.J., Elsea, G.N., Keller, K.P. and Kinder, H.D., Anal. Chem.,
      49,(6), 741 (1977).
2145. Miejnek, O. and Cveckova, L., J. Chromatogr., 94, 135 (1974).
2146. Bloom, P.J., J. Chromatogr., 115,(2), 571 (1975).
2147. Tsarfin, Ya.A. and Kharchenkova, V.D., Zh. Anal. Khim., 30,(2), 391
      (1975); Chem. Abstr., 82, 157078x.
2148. Fukuda, A., Nakahara, R. and Horluchi, H., Kagaku To Kogyo (Osaka),
      50,(2), 43 (1976); Chem. Abstr., 85, 6329r.
2149. Bosch, K., Beitr. Gerichtl.Med., 33, 280 (1975).
2150. Mol, G.J., Thermochim.Acta, 10,(3), 259 (1974).
2151. Hughes, J.C., Wheals, B.B. and Whitehouse, M.J., Analyst (London),
      102,(1211), 143 (1977).
2152. Zaborsky, L.M., Anal. Chem., 49, 1166 (1977).
2153. Wagner, E.R. and Greef, R.J., J. Polym. Sci., A-1, 9, 2193 (1971).
2154. Gilbert, M. and Hybart, F.J., J. Polym. Sci., A-1, 9, 227 (1971).
2155. Eremeeva, T.V., Evrenov, V.V., Davydova, E.V., Karyakina, M.I.,
      Sarynina, L.I. and Entelis, S.G., Plaste Kautsch., 25,(3), 192 (1978).
2156. Bradley, A. and Heagnay, T.R., Anal. Chem., 42, 894 (1970).
2157. Manjunath, B.R., Venkataraman, A. and Stephen, T., J.Appl. Polym.
      Sci., 17, 1091 (1973).
2158. Tazuke, S. and Matsuyama, Y., Macromolecules, 8,(3), 280 (1975).
2159. Skorokhodov, S.S., Nuova Chim., 49,(3), 57 (1973).
2160. Silbey, R. and Deutch, J.M., J. Chem. Phys., 57,(11), 5010 (1972).
2161. Barzykina, R.A., Komratov, G.N., Korovina, G.V. and Entelis, S.G.,
      Vysokomol. Soedin., Ser.A, 16,(4), 906 (1974).
2162. Roczniak, K., Biernacka, T. and Wertz,Z., Polimery (Warsaw), 21, 306
      (1976).
2163. Matsui, Y., Kubota, T., Tadokoro, H. and Yoshihara, T., J. Polym.Sci.,
      A,3, 2775 (1965).
2164. Yuki, H., Hatada, K. and Takeshita, M., J. Polym. Sci., A-1, 7, 667
      (1969).
2165. Pruckmayr, G. and Wu, T.K., Macromolecules, 6, 33 (1973).
2166. Kawamura, I.H. and Uryu, T., J. Polym. Sci., A-1, 11, 971 (1973).
2167. Matsuzaki, K., Ito, H., Kawamura, T. and Uryu, T., J. Polym. Sci.,
      Polym. Chem. Ed., 11, 971 (1973).
2168. Kruglova, V.A., Ratovskii, G.V., Borisenko, A.A. and Kalabina, A.V.,
      Vysokomol. Soedin., Ser.A., 15,(9), 1967 (1973).
2169. Hiyashimura, T., Hoshino, M., Hirokawa, Y., Matsuzaki, K. and Uryu,
      T., J. Polym. Sci., A-1, 15, 2691 (1977).
2170. Matsuzaki, K., Okazono, S. and Kanai, T., J. Polym. Sci., A-1, 17,
      3447 (1979).

2171. Wiley, R.H., J. Polym. Sci., A-1, 9, 129 (1977).
2172. Hutton, J.F. and Phillips, M.C., Nature Phys. Sci., 245, 15 (Set.3.
       (1973).
2173. Lerner, L.R., J. Polym. Sci., A,1, 12, 2477 (1974).
2174. Montaudo, G., Przybski, M. and Ringsdorf, H., Makromol. Chem., 176,
       (6), 1763 (1975).
2175. Ehlers, G.F.C., Fisch, K.R. and Powell, W.R., J. Polym. Sci., A-1, 7,
       2969 (1969).
2176. Ehlers, G.F.C., Fisch, K.R. and Powell, W.R., J. Polym. Sci., A-1, 7,
       2955 (1969).
2177. Ehlers, G.F.C., Fisch, K.R. and Powell, W.R., J. Polym. Sci., A-1, 7,
       2931 (1969).
2178. Lanikova, J. and Hlousek, M., Chem. Prum., 23,(6), 10 (1973).
2179. Dinsoin, V.I., Goncharov, L.V., Filatov, I.S. and Yudkin, B.I., Plast.
       Massy, 1, 66 (1973).
2180. Fewell, L.L., J. Chromatogr. Sci., 14,(12), 564 (1976).
2248. Motorina, M.A., Kalinina, L.S. and Metalkina, E.I., Plast. Massy., 6,
       74 (1973).
2249. Senetskaya, L.P., Smirnova, O.V. and Domozhakova, L.M., Plast. Massy.,
       1, 68 (1974).
2250. Gedemer, T.J., Appl. Spectroscopy, 19, 141 (1965).
2251. Wielgosz, Z., Boranowska, Z. and Janicka, K., Plast Kaut., 19,(2),902
       (1972).
2252. Jansson, J.F. and Yannas, I.V., Int. Congr. Rhelo., 7th 1976, 274;
       C. Klason, and J. Kubat, Ed. Chalmers Uni. Technol. Goeteborg, Sweden;
       Chem. Abstr., 86, 55843h (1977).
2253. Andronikashvili, G.G., Samsoniya, S.A., Zhamierashvili, M.G. and Soob-
       shch, I.M., Akad, Nauk Gruz. SSSR, 85,(1), 73 (1977); Chem. Abstr.,
       86, 190545c (1977).
2254. Larsen, D.W. and Strange, J.J., J. Polym. Sci., Polym. Phys.Ed., 11,
       65 (1973).
2255. Larsen, D.W. and Strange, J.J., J. Polym. Sci., Polym. Phys.Ed., 11,
       449 (1973).
2256. Larsen, D.W. and Strange, J.J., J. Polym. Sci., Polym. Phys. Ed., 11,
       1453 (1973).
2257. Chandler, J.F. and Stark, B.P., Angew. Makromol. Chem., 27, 159 (1972).
2258. Stefan, D. and Williams, H.L., J. Appl. Polym. Sci., 18, 1279 (1974).
2259. Stefan, D. and Williams, H.L., J. Appl. Polym. Sci., 18, 1451 (1974).
2260. Nagumanova, E.I.. Sagitov, R.Ya. and Voskresenskii, V.A., Kolloid.
       Zh., 37,(3), 571 (1975).
2261. Smirnova, O.V., Kirshak, V.V., Slonim, I.Ya, Urman, Ya.G., Alekseeva,
       S.G. and Bairamov, V.A., Vysokomol. Soedin., Ser.A, 17,(11), 2415
       (1975).
2262. Placek, J., Szocs, F. and Borsig, E.J., J. Polymer Science, A,1, 14,
       1549 (1976).
2263. Hama, Y. and Shinohara, K., J. Polymer Science, A,1, 8, 651 (1970).
2264. Bosch, K., Beitr. Gerichtl. Med., 33, 280 (1975).
2265. Gedemer, T.J., Plast. Eng., 31,(2), 28 (1975).
2266. Korshak, V.V., Baochinitser, T.M., Tsvankin, D.Ya., Gribova, I.A.,
       Chumaevskaya, A.N. and Gventsadze, D.I., Vysokomol. Soedin., Ser.A,
       14,(11), 2306 (1972).
2267. Di Pasquale, G., Di Iorio, G. and Capaccioli, T., J. Chromatogr., 152,
       (2), 538 (1978).
2268. Kalinina, L.S. and Doroshina, L.I., Metody Ispyt. Kontr. Issled. Mash-
       inostroit, Mater., 3, 5 (1973); Chem. Abstr., 80, 54100 (1974).
2367. Grassie, N. and MacCallum, J.R., J. Polym. Sci., 1, 551 (1963).
2560. Spirin, Yu.L. and Yatsimirskaya, T.S., Vysokomol. Soedin., Ser.A,15,
       (11), 2595 (1973).

2561. Groom, T., Babiec, J.S., Jr. and Van Leuwen, B.G., J. Cell Plast., 10,(1), 43 (1974).
2562. Fritz, D.F., Sahil, A., Keller, H.P. and Kovat, E., Anal. Chem., 51, 7 (1979).
2563. Yokoyama, M., Ochi, H., Tadokoro, H. and Price, C.C., Macromolecules, 5,690(1972).
2564. Killmann, E. and Strasser, Hans J., Angew. Makromol. Chem., 31, 169 (1973).
2565. Hartley, A., Leung, Y.K., Booth, C. and Shepherd, I.W., Polymer, 17, (4), 354 (1976).
2566. Matsura, H., Miyazawa, T. and Machida, K., Spectrochim. Acta., Part A, 29,(5), 771 (1973).
2567. Hartley, A.J., Leung, Y.K., McMahon, J., Booth, C. and Shepherd, I.W., Polymer, 18,(4), 336 (1977).
2568. Cross, C.K. and Mackay, A.C., J. Am. Oil Chem. Soc., 50,(7), 249 (1973).
2569. Mank, V.V., Solomentseva, I.M., Baran, A.A. and Kurilenko, O.D., Ukr. Khim. Zh. (Russ. Ed.), 40,(1), 28 (1974).
2570. Slonim, I.Ya., Alekseeva, S.G., Aksel'rod, B.Ya. and Urman, Ya.G., Vysokomol. Sodein., Ser. B., 17,(12), 919 (1975).
2571. Matsuzaki, K. and Ito, H., J. Polymer Science, A,1, 15, 647 (1977).
2572. Okada, T., J. Polymer Science, A,1, 17, 155 (1979).
2573. Thyron, F.C. and Baijal, M.D., J. Polymer Science, A,1, 6, 505 (1968).
2574. Sakaguchi, M., Yamakawa, H. and Sohma, J., J. Polymer Sci., Polym. Lett.Ed., 12, 193 (1974).
2575. Esposito, G.G. and Swann, M.H., Anal. Chem., 33,(13), 1854 (1961).
2576. Schole, R.G., Bednarczyk, J. and Yamanchi, T., Anal. Chem., 38, 331 (1966).
2577. Kuzaev, A.I., Susiova, E.N. and Entelis, S.G., Dokl. Akad. Nauk. SSSR, 208,(1), 142 (1973).
2578. Kato, Y., Sasaki, H., Aiura, M. and Hashimoto,T., J. Chromatogr., 153, (2), 546 (1978).
2579. Case, L.C., Makromol. Chem., 51, 61 (1960).
2580. Nakagawa, T. and Nakata, I., Kogyo Kagaku Zasshi, 59, 710 (1956); Chem. Abstr., 52, 4418d (1958).
2581. Burger, K., Z. Anal. Chem., 196, 259 (1963).
2582. Mikkelsen, L., Characterization of High Molecular Weight Substances, Pittsburgh Conf. Anal. Chem., March 5-9 (1962).
2583. Puschmann, Fette, Seife, Anstrichmittel, 65 (1963).
2584. Celades, R. and Pacquot, C., Rev. Franc. Corps Gras., 9, 145 (1962).
2585. Sephadex,G-10 and G-15 for Fractionation of Low Molecular Weight Solutes, brochure from Pharmacia Fine Chemicals, Inc.
2586. Sweeley, C.C., Bentley, R., Makita, M. and Wells, W.W., J. Am. Chem. Soc., 85, 2497 (1963).
2587. Crabb, N.T., unpublished work (1964).
2588. Fletcher, J.P. and Persinger, H.E., J. Polymer Science, A,1, 6, 1025 (1968).
2589. Gruber, U. and Elias, H.G., Die Makromol. Chemie, 78, 58 (1964).
2590. Grechanovskii, V.A., Vysokomol. Soedin., 17, 2721 (1975).
2591. Moore, J.C. and Handrickson, J.G., J. Polymer Science, C,8, 233 (1965).
2592. Kalinina, L.S. and Motorina, M.A., USSR Patent 340, 962 (6/5/72).
2593. Ingham, J.D. and Lawson, D.D., J. Polym. Sci., A,3, 2707 (1965).
2594. Kuresevic, V. and Samsa, M., Hem. Ind., 26, 99 (1972).
2595. Suzuki, H. and Leonis, C.G., Brit. Polym.J., 5,(6), 485 (1973).
2596. Gervais, M. and Gallot, B., Makromol. Chem., 171, 157 (1973).
2597. Bergmann, K., Kolloid-Z.Z. Polym., 251,(11), 962 (1973).
2598. Steiger, F.H., Amer. Lab., 5,(1), 19 (1973).
2599. Takahashi, Y., Sumita, I. and Tadokoro, H., J. Polym. Sci., Polym. Phys. Ed., 11, 2113 (1973).

2600. Klein, J. and Widdecke, H., Angew. Makromol. Chim., 53,(1), 145 (1976).
2601. Lee, A.K. and Sedgwick, R.D., J. Polym. Sci., Polym. Chem. Ed., 16, (3), 685 (1978).
2602. Bowden, M.J. and Thompson, L.F., J. Appl. Polym. Sci., 19,(3), 905 (1975).
2603. Stetzler, R.S. and Smullin, C.F., Anal. Chem., 34, 194 (1962).
2604. Fritz, J.S. and Scherk, G.H., Anal. Chem., 31, 1808 (1959).
2605. Price, C.E. and Osgan, M., private communication.
2606. Vandenberg, E.J., J. Polymer Sci., 47, 486 (1960); J. Am. Chem. Soc., 83, 3538 (1961); J. Polymer Sci., B,2, 1085 (1965); Abstracts, Tokyo IUPAC Polymer Symposium (1966).
2607. Belen'kii, B.G., Vakhtina, I.A. and Tarakanov, O.G., Vysokomol. Soedin., Ser.A, 17,(9), 2116 (1975).
2608. Kumpanenko, I.V. and Kazanskii, K.S., J. Polym. Sci., Polym. Symp., No. 42, Pt. 2 973-80 (1973).
2609. Lapeyre, W., Cheradame, H., Spassky, N. and Sigwait, P., J. Chim. Phys. Physio Chim. Biol., 70,(5), 838 (1973).
2610. Uryu, T., Shimazu, H. and Matsuzaki, K., J. Polym. Sci., Polym. Lett. Ed., 11, 275 (1973).
2611. Schaefer, J., Macromolecules, 5,(5), 590 (1972).
2612. Gnauck, R., Plaste Kautsch., 23,(9), 649 (1976); Chem. Abstr., 85, 178237a (1976).
2613. Vakhtina, I.A., Khrenova, R.I. and Tarakanov, O.G., Zh. Anal. Khim., 28,(8), 1625 (1973).
2614. Bode, R.K., Bennington, L.V., Mowrey, R.A. and Fuller, E.N., International Laboratory 40 Dec/nov. (1981).
2615. Vakhtina, I.A. and Tarakanov, O.G., Plaste Kaut., 21,(1), 28 (1974).
2616. Berek, D. and Nova, L., Chem. Prum., 23,(2), 91 (1973).
2617. Leopold, H. and Trathnigg, B., Angew, Makromol. Chem., 68,(1), 185 (1978).
2618. Huang, Y.Y. and Wang, C.H., J. Chem. Phys., 62,(1), 120 (1975).
2619. Cervenka, A. and Merrall, G.T., J. Polym. Sci., Polym. Chem. Ed., 14, 2135 (1976).
2620. Gibson, U.H. and Quick,Q., J. Appl. Polym. Sci., 15, 2667 (1971).
2621. Toyzo, U., Koichiro, M. and Yoshie, M., Chem. Pharm. Bull., 15, 77 (1967).
2622. Stead, J.B. and Hindley, A.H., J. Chromatogr., 42, 470 (1969).
2623. Mathias, A. and Mellor, N., Anal. Chem., 38, 472 (1966).
2624. Hewitt, G.C. and Whitham, B.T., Analyst, 86, 643 (1961).
2625. Neumann, E.W. and Nadeau, H.G., Anal. Chem., 35, 1454 (1963).
2626. Whipple, E.B. and Green, P.J., Macromolecules, 6, 38 (1973).
2627. Neumann, E.W. and Nadeau, H.G., Anal. Chem., 10, 1454 (1963).
2628. Swann, W.B. and Dux, J.P., Anal. Chem., 33, 654 (1961).
2629. Zeman, I., Novak, L., Mitter, L., Stekla, J. and Holendova, O., J. Chromatogr., 119, 581 (1976).
2630. Crompton, T.R., unpublished work.
2631. Myers, L.W., Shell Research Ltd, Carrington, Cheshire, U.K., private communication.
2632. Price, C.C., Spector, R. and Tumulo, A.L., J. Polym. Sci., A,1, 5, 407 (1967).
2633. Pruitt, M.E. and Baggett, J.B., U.S. Pat., 2,706, 181 (April 12 1955).
2644. Hackathorn, M.J. and Brock, M.J., J. Polym. Sci., Polym. Chem.Ed., 13, 945 (1975).
2645. Richardson, W.S. and Sacher, A., J. Polym. Sci., 10, 353 (1953).
2646. Plkrovskiu, E.I. and Volkenstejn, M., Dokl. Akad. SSSR, 95, 301 (1954).
2647. Binder, J.L. and Ransaw, H.C., Anal. Chem., 29, 503 (1957).
2648. Corish, P.J., Spectrochim. Acta., 15, 598 (1959).
2649. Sutherland, G.B.B.M. and Jones, A.V., Discussions Faraday Soc., 9, 281

(1950).
2650. Saunders, R.A. and Smith, D.C., J. Appl. Phys., 20, 953 (1949).
2651. Binder, J.L., J. Polym. Sci., Part A, 37 (1963).
2652. Binder, J.L. and Ransaw, H.C., Anal. Chem., 29, 503 (1957).
2653. Golub, M.A., J. Polym. Sci., 36, 10 (1959).
2654. Maynard, J.T. and Moobel, W.E., J. Polym. Sci., 13, 251 (1954).
2655. Ferguson, R.C., Anal. Chem., 36, 2204 (1964).
2656. Kössler, I, and Vodchnal, J., Polymer Letters, 4, 415 (1963).
2657. Binder, J.L. and Ransaw, H.C., Anal. Chem., 29, 503 (1957).
2658. Binder, J.L., Rubber Chem. and Technol. 35, 57 (1962).
2659. Binder, J.L., J. Polym. Sci., A,1, 37 (1963).
2660. Golub, M.A., J. Polym. Sci., 36, 523 (1959).
2661. Kössler, I. and Vodchanl, J., Collection Czech. Chem. Communs., unpublished.
2662. Vodchnal, J. and Kössler, I., Coll. of Czech. Communs., 29, 2428(1964).
2663. Vodchnal, J. and Kössler, I., Coll. of Czech. Communs., 29, 2859(1964).
2664. Fraga, D.W. and Beason, L.H., Shell Chemical Co., Torrance, California, private communication.
2665. Van Stratum, P.G.M. and Dvorak, J., J. Chromatogr., 71,(1), 9 (1972).
2666. Motorina, M.A., Kalinina, L.S. and Metalkina, E.I., Plast. Massy, 6, 74 (1973).
2667. Yuki, H. and Okamoto, Y., J. Polym. Sci., A,1,9, 1247 (1971).
2668. Golub, M.A., Hsu, M.S. and Wilson, L.A., Rubber Chem. Technol., 48, (5), 953 (1975).
2669. Gronski, W., Murayamo, N., Carlow, H.J. and Miyamoto, T., Polymer, 17, 358 (1976).
2670. Morese-Seguela,B., St.Jacques, M., Renaud, J.M. and Prud'homme, J., Macromolecules, 10,(2), 431 (1977).
2671. Duck, M.W. and Grant, D.M., Macromolecules, 3, 165 (1970).
2672. Tanaka, Y., Sato, H. and Sumiya, T., Polymer Journal, 7, 264 (1975).
2673. Tanaka, Y., Sata, H., Ogura, A. and Nagoya, I., J. Polym. Sci., Polym. Chem. Ed., 14, 73 (1976).
2674. Okada, T. and Ikushige, T., J. Polym. Sci., Polym. Chem.Ed., 14, 2059 (1976).
2675. Gemmer, R.V. and Golub, M.A., J. Polym. Sci., A,1, 16, 2985 (1978).
2676. Beebe, D.H., Polymer, 19,(2), 231 (1978).
2677. Dolinskaya, E.R., Khachaturov, A.S., Poletaeva, I.A. and Kormer, V.A., Makromol. Chem., 179,(2), 409 (1978).
2678. Galin, M., J. Macromol. Sci.-Chem., A,7, 783 (1973)
2679. Hackathorn, M.J. and Brock, M.J., J. Polym. Sci., Polym. Lett. Ed., 8, 617 (1970).
2680. Tanaka, Y., Sato, H., Ogura, A. and Nagoya, I., J. Polym. Sci., Polym. Chem. Ed., 14, 73 (1976).
2681. Kössler, J.P. and Vodchnal, J., J. Polym. Sci., A,3, 2511 (1965).
2682. Baker, C.A. and Williams, R.J.P., J. Chem. Soc., 2852 (1956).
2683. Francis, P.S., Cooke, R.C. and Elliot, J.H., J. Polym. Sci., 31, 453 (1958).
2684. Henry, P.M., J. Polym. Sci., 36, 3 (1959).
2685. David, T.E. and Tobias, T.L., J. Polym. Sci., 50, 227 (1961).
2686. Schneider, N.S., Loconti, J.D. and Holmes, L.G., J. Appl. Polymer Sci., 3, 251 (1960); ibid, 5, 354 (1962).
2687. Weakley, T.J., Williams, R.J.P. and Wilson, J.D., J. Chem. Soc., 3963 (1960).
2688. Jungnickel, J.L. and Weiss, F.T., J. Polymer Sci., 49, 437 (1961).
2689. Hall, R.W., in Techniques of Polymer Characterization, P.W. Allen Ed., Butterworths, London (1959).
2690. Cooper, W., Vaughan, G., Eaves, D.E. and Madden, R.W., J. Polym. Sci., 50, 159 (1961).

2691. Cooper, W., Vaughan, G. and Yardley, J., J. Polym. Sci.,59, S2 (1962).
2692. Hulme, J.M. and McLeod, L.A., Polymer, 3, 153 (1962).
2693. Caldeson, N. and Scott, K.W., J. Polym. Sci., A,3, 551 (1965).
2694. Mayer, R., Polymer, 15, 137 (1974).
2695. Gruber, U. and Elias, H.G., Die Makromol. Chemie, 78, 58 (1964).
2696. Elias, H.G. and Gruber, U., Die Makromol. Chemie, 78, 72 (1964).
2697. Ferguson, R.C., J. Polym. Sci., A,2, 4735 (1964).
2698. Coleman, M.M., Painter, P.C., Tabb, D.L. and Koenig, J.L., J. Polym.
       Sci., Polym. Lett.Ed., 12,(10), 577 (1974).
2699. Painter, P.C. and Koenig, J.L., Crit. Rev. Tech. Charact. Polym. Mater.,
       ADA 036082, 11 (1976); Chem. Abstr., 88, 105922x (1978).
2700. Dreyfuss, D., Fraser, G.V., Keller, A. and Pope, D.P., Proc. Int. Conf.
       Raman Spectrosc., 5th, 492-3 (1976).
2701. Coleman, M.M., Painter, P.C. and Koenig, J.L., J. Raman Spectrosc.,
       5,(4), 417 (1976).
2702. Okada, T. and Ikushige, T., J. Polym. Chem., Polym. Chem. Ed., 14,(8),
       2059 (1976).
2703. Kössler, J.D. and Vodchnal, J., J. Polym. Sci., A,3, 2511 (1965).
2704. Baker, C.A. and Williams, R.J.P., J. Chem. Soc., 2852 (1956).
2705. Francis, P.S., Cooke, R.C. and Elliott, J.H., J. Polym. Sci., 31, 453
       (1958).
2706. Henry, P.M., J. Polym. Sci., 36, 3 (1959).
2707. Davis, T.E. and Tobias, R.L., J. Polym. Sci., 50, 227 (1961).
2708. Schneider, N.S., Loconti, J.D. and Holmes, L.G., J. Appl. Polym. Sci.,
       3, 251 (1960); ibid, 5, 354 (1961).
2709. Weakley, T.J., Williams, R.J.P. and Wilson, J.D., J. Chem. Soc., 3963
       (1960).
2710. Jungnickel, J.L. and Weiss, F.T., J. Polym. Sci., 49, 437 (1961).
2711. Hall, R.W., in Techniques of Polymer Characterization, P.W. Allen Ed.,
       Butterworths, London (1959).
2712. Cooper, W., Vaughan, G., Eaves, D.E. and Madden, R.W., J. Polym. Sci.,
       50, 159 (1961).
2713. Cooper, W., Vaughan, G. and Yardley, J., J. Polymer Sci., 59, S 2(1962).
2714. Hulme, J.M. and McLeod, L.A., Polymer, 3, 153 (1962).
2715. Coleman, M.M., Painter, P.C., Tabb, D.L. and Koenig, J.L., J. Polym.
       Sci., Polym. Lett. Ed., 12, 577 (1974).
2716. Kawasaki, N. and Hashimoto, T., J. Polym. Sci., A,1, 11, 671 (1973).
2717. Lange, H., Kolloid-Z.Z. Polym., 250,(8), 775 (1972).
2718. Gallo, S.G., Weiss, H.K. and Nelson, J.F., Ind. Eng. Chem., 40, 1277
       (1948).
2719. Lee, T.S., Kolthoff, I.M. and Johnson, E., Anal. Chem., 22, 995 (1950).
2720. McNeill, I.C., Polymer, 4, 15 (1963).
2721. Falcon, J.Z., Love, J.L., Gaeta, L.T. and Altenau, A.G., Anal. Chem.,
       47, 171 (1975).
2726. Chen, H.Y. and Field, J.E., J. Polym. Sci., B,5, 501 (1967).
2727. Clark, D.T., Kilcast, D., Feast, W.J. and Musgrave, W.K.R., J. Polym.
       Sci., A,1, 10, 1637 (1972).
2728. Zitch, P. and Zelinger, J., J. Polym. Sci., A,1, 6, 467 (1968).
2729. McNeill, I.C. and Straiton, T., J. Polym. Sci., A,1, 12, 2369 (1974).
2730. Katritzky, A.R. and Smith, A., Rubber J., 154, 30 (1972).
2731. Campos Lopez, E. and Palacios, J., J. Polym. Sci., A,1,14, 1561 (1976).
2732. Shahab, Y.A. and Basheer, R.A., J. Polym. Sci., A,1, 16, 2667 (1978).
2733. Shahab, Y.A. and Basheer, R.A., J. Polym. Sci., A,1, 17, 919 (1979).
2734. Still, H.H. and Whitehead, A., J. Appl. Polym. Sci., 16, 3207 (1972).
2735. Kusumoto, H. and Gutowsky, H.S., J. Polym. Sci., A,1, 2905 (1963).
2736. Jamroz, M., Kozyowski, K., Sieniakowski, M. and Jackym, B., J. Polym.
       Sci., A,1, 15, 1359 (1977).
2737. Feuerberg, H. and Weigel, H., Kautschuk Gummi, 15, WT276-WT282 (1962).

2738.   Kubinova, M. and Mikl, O., Przemysl. Chem., 39, 552 (1960).
2739.   Baba, T. and Tokumaru, S., Nippon Gomu Kyokaishi, 35, 162 (1962).
2740.   Krishen, A., in "ASTM STP 554", American Society for Testing and Materials, p. 74 (1974).
2741.   Coupe, N.B., Jones, C.E.R. and Stockwell, P.B., Chromatographia, 6, (11), 483 (1973).
2742.   Jones, C.E.R., Perry, S.G. and Coupe, N.B., Gas Chromatogr., Proc. Int. Symp. (Europe), 8, 399 (Pub. 1971) (1970).
2743.   Chih-An, Hu.J., Anal. Chem., 49, 537 (1977).
2744.   Van Schooten, J. and Evenhuis, J.K., Polymer (London) 561, Nov. 1965 and Polymer (London), 6, 343 (1965).
2745.   Bristow, G.M., Moore, C.G. and Russell, R.M., J. Polym. Sci., A,3, 3893 (1965).
2746.   Burfield, D.R. and Gan, S.N., J. Polym. Sci., A,1, 15, 2721 (1977).
2747.   Takashima, S. and Okada, F., Himeji Kogyo Daigaku Kenkyu Hokoku, 12, 34 (1960).
2748.   Brodsky, J., Kunststoffe, 51, 20 (1961).
2749.   Nelson, F.M., Eggertsen, F.T. and Holst, J.J., Anal. Chem., 33,(9), 1150 (1961).
2750.   Brock, M.J. and Louth, G.D., Anal. Chem., 35, 1575 (1955).
2751.   Wandel, M. and Tengler, H., Fette, Seifen, Anstrichmittel, 66, 815 (1964).
2752.   Varmier, J.P., Suryanaraya, N.P. and Sircar, A.K., J. Sci. Ind. Res., India, 20, 79 (1961).
2753.   Yuasa, T. and Kamiya, K., Japan Analyst, 13, 966 (1964).
2754.   Varma, J.P., Suryanaraya, N.P. and Sircar, A.K., J. Sci. Ind. Res., India, 21, 49 (1962).
2755.   Parker, C.A. and Barnes, W.J., Analyst, 82, 606 (1957).
2756.   Mocker, F., Kautchuk Gummi, 11, 281 (1958).
2757.   Mocker, F., Kautchuk Gummi, 12, 155 (1959).
2758.   Mocker, F., Kautchuk Gummi, 13, 91 (1960).
2759.   Adams, R.N., Rev. Polarog., 11, 71 (1963).
2760.   Zweig, A., Lancaster, E., Neglis, M.T. and Jura, W.H., J. Am. Chem. Soc., 86, 413 (1964).
2761.   Vodzinskii, Yu. and Semchikova, G.S., T. Po Khim i Khim Teknol., 272 (1963).
2762.   Vermillion, F.J. and Pearl, T.A., J. Electrochem. Soc., 111, 1392(1964).
2763.   Nawakoswki, A.C., Anal. Chem., 30, 1868 (1958).
2764.   Protivova, J. and Pospisil, J., J. Chromatography, 88, 99 (1974).
2765.   Coupek, J., Pekorny, S. and Pospisil, J., IUPAC Microsymposium on Makromolecules, 11th Prague, September (1972).
2766.   Coupek, J., Pokorny, S., Jirachova, L. and Pospisil, J., J. Chromat., 75, 87 (1973).
2767.   Coupek, J., Kalovec, M., Krivakova, M. and Pospisil, J., Angew. Makromol. Chem., 15, 137 (1971).
2768.   Coupek, J., Pokorny, S., Protivova, J., Holcik, J., Karvas, M. and Pospisil,J., J. Chromatogr., 65, 279 (1972).
2769.   Fiorenza, A., Bonomi, G. and Seradi, A., Materie. Plastiche ed Elastomerie, 31, 1045 (1965).
2770.   Majers, R.E., J. Chromatogr. Sci., 8, 338 (1970).
2771.   Wize, R.W. and Sullivan, A.B., Rubber Age, 91, 773 (1962).
2772.   Tswrugi, J., Murakaus, S. and Goda, K., Rubber Chem. Technol., 44, 857 (1971).
2773.   Wize, R.W. and Sullivan, A.B., Rubber Chem. Technol., No 3, 35 July-September (1962).
2774.   LiGotti, L., Materie Plast. Elastomerie, 39, 889 (1973).
2775.   Bellamy, L.J., Laurie, J.H. and Pren, E.W.S., Trans. Inst. Rubber Industry, 15 (1947).

2776.   Mann, J., Trans. Inst. Rubber Industry, 27, 232 (1951).
2777.   Parker, C.A. and Berriman, J.M., Trans. Rubber Industry, 28, 279 (1952).
2778.   Kriner, J.G. and Warner, W.C., J. Chromatography, 44, 315 (1969).
2779.   Kueda, I. and Fernandez, G., Revta. Plast., 24, 82 (1973).
2780.   Zijp, J.W.H., Rec. Trav. Chim., 75, 1155 (1956).
2781.   Burger, V.L., Rubber Chem. Technol., 32, 1452 (1959).
2782.   Miksch, R. and Prolsi, O., Gummi Asbest., 13, 250 (1960).
2783.   Carlson, D.W., Hayes, M.W., Ransaw, H.C., McFadden, R.S. and Altenau,
        A.G., Anal. Chem., 43, 1874 (1971).
2784.   Zijp, J.W.H., Rec. Trav. Chim., 75, 1155 (1956).
2785.   Gaczynski, R. and Stephen, M., Przemylek Chem., 38, 9 (1959).
2786.   Burger, V.L., Rubber Chem. Technol., 32, 1452 (1959).
2787.   Freiner, J.G., Rubber Chem. Technol., 44, 381 (1971).
2788.   Yuasa, T. and Kamiya, K., Benseki Kagaku, 13, 966 (1964).
2789.   Kreiner, J.G. and Warner, W.C., J. Chromatogr., 44, 315 (1969).
2790.   Hilton, A.S. and Altenau, A.G., Rubber Chemistry and Technology, 46,
        1035 (1973).
2791.   Parks, C.R. and Brown, R.J., J. Appl. Polym. Sci., 18, 1079 (1974).
2792.   Hulot, H. and Lebel, P., Rubber Chem. and Tech., 37,(1), 297 (1964).
2793.   Hackathorn, M.J. and Brock, M.J., J. Polym. Sci., B,8, 617 (1971).
2794.   Crompton, T.R., unpublished work.
2795.   Jeffs, A.R., Analyst (London), 94, 249 (1969).

INDEX